RENEWALS 458-4574

IET RADAR, SONAR, NAVIGATION AND AVIONICS SERIES 21

Series Editors: Dr N. Stewart
Professor H. Griffiths

Principles of Space-Time Adaptive Processing 3rd Edition

Other volumes in this series:

Volume 1	**Optimised radar processors**	A. Farina (Editor)
Volume 2	**Radar development to 1945**	R. W. Burns (Editor)
Volume 3	**Weibull radar clutter**	M. Sekine and Y. H. Mao
Volume 4	**Advanced radar techniques and systems**	G. Galati (Editor)
Volume 7	**Ultrawideband radar measurements**	L. Yu. Astanin and A. A. Kostylev
Volume 8	**Aviation weather surveillance systems: advanced radar and surface sensors for flight safety and air traffic management**	P. Mahapatra
Volume 10	**Radar techniques using array antennas**	W. D. Wirth
Volume 11	**Air and spaceborne radar systems**	P. Lacomme, J. C. Marchais, J. P. Hardange and E. Normant
Volume 12	**Principles of space-time adaptive processing, 2nd edition**	R. Klemm
Volume 13	**Introduction to RF Stealth**	D. Lynch
Volume 14	**Applications of space-time adaptive processing**	R. Klemm (Editor)
Volume 15	**Ground penetrating radar, 2nd edition**	D. J. Daniels (Editor)
Volume 16	**Target detection by marine radar**	J. Briggs
Volume 17	**Strapdown inertial navigation technology, 2nd edition**	D. H. Titterton and J. L. Weston
Volume 18	**Introduction to radar target recognition**	P. Tait
Volume 19	**Radar imaging and holography**	A. Pasmurov and S. Zinoviev

Principles of Space-Time Adaptive Processing
3rd Edition

By Richard Klemm

The Institution of Engineering and Technology

Published by: The Institution of Engineering and Technology, London, United Kingdom

© 2006: The Institution of Engineering and Technology

This publication is copyright under the Berne Convention and the Universal Copyright Convention. All rights reserved. Apart from any fair dealing for the purposes of research or private study, or criticism or review, as permitted under the Copyright, Designs and Patents Act, 1988, this publication may be reproduced, stored or transmitted, in any forms or by any means, only with the prior permission in writing of the publishers, or in the case of reprographic reproduction in accordance with the terms of licences issued by the Copyright Licensing Agency. Inquiries concerning reproduction outside those terms should be sent to the publishers at the undermentioned address:

The Institution of Engineering and Technology,
Michael Faraday House,
Six Hills Way, Stevenage,
Herts. SG1 2AY, United Kingdom

www.theiet.org

While the authors and the publishers believe that the information and guidance given in this work are correct, all parties must rely upon their own skill and judgment when making use of them. Neither the authors nor the publishers assume any liability to anyone for any loss or damage caused by any error or omission in the work, whether such error or omission is the result of negligence or any other cause. Any and all such liability is disclaimed.

The moral right of the authors to be identified as authors of this work have been asserted by them in accordance with the Copyright, Designs and Patents Act 1988.

British Library Cataloguing in Publication Data

Klemm, Richard
  Principles of space-time adaptive processing. – 3rd ed. –
  (Radar, sonar, navigations & avionics; no. 21)
  1. Moving target indicator radar 2. Adaptive signal processing
  I. Title II. Institution of Electrical Engineers
  621.3'848

ISBN (10 digit) 0 86341 566 0
ISBN (13 digit) 978-086341-566-1

Typeset in India by Newgen Imaging Systems (P) Ltd, Chennai, India
Printed in the UK by MPG Books Ltd, Bodmin, Cornwall

To my dear Eka

Contents

Biography   xv

Preface to the first edition   xvii

Preface to the second edition   xxi

Preface to the third edition   xxiii

1 Introduction   **1**
    1.1 Preliminary remarks 1
       1.1.1 Basics of MTI radar 3
       1.1.2 One-dimensional clutter cancellation 4
       1.1.3 Aspects of air- and spaceborne radar 5
       1.1.4 Impact of platform motion 6
       1.1.5 Some notes on phased array radar 11
       1.1.6 Systems and experiments 12
       1.1.7 Validity of models 14
       1.1.8 Historical overview 15
    1.2 Radar signal processing tools 17
       1.2.1 The optimum processor 17
       1.2.2 Orthogonal projection 25
       1.2.3 Linear subspace transforms 27
       1.2.4 Clutter suppression with digital filters 34
       1.2.5 Examples 39
       1.2.6 Angle or frequency domain processing 41
    1.3 Spectral estimation 44
       1.3.1 Signal match (SM) 45
       1.3.2 Minimum variance estimator, MVE 46
       1.3.3 Maximum entropy method, MEM 47
       1.3.4 Orthogonal projection, MUSIC 48
       1.3.5 Comparison of spectral estimators 49
    1.4 Summary 50

2 Signal and interference models — 51
- 2.1 Transmit and receive process — 51
- 2.2 The Doppler effect — 52
- 2.3 Space-time signals — 53
  - 2.3.1 The spatial dimension: array geometry — 53
  - 2.3.2 The temporal dimension: pulse trains — 55
- 2.4 Interference — 57
  - 2.4.1 Ground clutter — 57
  - 2.4.2 Moving clutter — 59
  - 2.4.3 Jamming — 59
  - 2.4.4 Noise — 60
- 2.5 Decorrelation effects — 60
  - 2.5.1 Temporal decorrelation — 61
  - 2.5.2 Spatial decorrelation: effect of system bandwidth — 63
  - 2.5.3 Doppler spread within range gate — 64
  - 2.5.4 System Doppler spread — 67
  - 2.5.5 Total correlation model — 67
- 2.6 The standard parameter set — 68
  - 2.6.1 Multiple-time around clutter — 68
  - 2.6.2 Remark on image quality — 69
- 2.7 Summary — 69

3 Properties of airborne clutter — 71
- 3.1 Space-Doppler characteristics — 71
  - 3.1.1 Isodops — 71
  - 3.1.2 Doppler-azimuth clutter trajectories — 73
- 3.2 The space-time covariance matrix — 79
  - 3.2.1 The components — 82
  - 3.2.2 The displaced phase centre antenna (DPCA) principle — 86
  - 3.2.3 Eigenspectra — 98
- 3.3 Power spectra — 102
  - 3.3.1 Fourier spectra — 103
  - 3.3.2 High-resolution spectra — 105
- 3.4 Effect of radar parameters on interference spectra — 106
  - 3.4.1 Array orientation — 106
  - 3.4.2 Temporal and spatial sampling — 107
  - 3.4.3 Decorrelation effects — 109
  - 3.4.4 Clutter and jammer spectra — 114
- 3.5 Aspects of adaptive space-time clutter rejection — 114
  - 3.5.1 Illustration of the principle — 114
  - 3.5.2 Some conclusions — 115
- 3.6 Summary — 118

4 Fully adaptive space-time processors — 121
- 4.1 Introduction — 121
- 4.2 General description — 122
  - 4.2.1 The optimum adaptive processor (OAP) — 122
  - 4.2.2 The orthogonal projection processor (OPP) — 127
- 4.3 Optimum processing and motion compensation — 129
  - 4.3.1 Principle of RF motion compensation — 129
  - 4.3.2 Correction patterns — 131
  - 4.3.3 Interrelation with the optimum processor — 133
- 4.4 Influence of radar parameters — 139
  - 4.4.1 Transmit beamwidth — 139
  - 4.4.2 Array and sample size — 141
  - 4.4.3 Sampling effects — 142
  - 4.4.4 Influence of the CNR — 145
  - 4.4.5 Bandwidth effects — 146
  - 4.4.6 Moving clutter — 153
- 4.5 Range-Doppler IF matrix — 153
- 4.6 Summary — 155

5 Space-time subspace techniques — 159
- 5.1 Principle of space-time subspace transforms — 160
- 5.2 The auxiliary eigenvector processor (AEP) — 163
  - 5.2.1 Comparison with the optimum adaptive processor (OAP) — 165
  - 5.2.2 Reduction of the number of channels — 166
  - 5.2.3 Bandwidth effects — 167
- 5.3 Auxiliary channel processor (ACP) — 170
  - 5.3.1 Comparison with optimum processor — 172
  - 5.3.2 Reduction of the number of channels — 172
  - 5.3.3 Bandwidth effects — 172
- 5.4 Other space-time transforms — 173
  - 5.4.1 Single auxiliary elements and echo samples transform — 173
  - 5.4.2 Space-time sample subgroups — 173
  - 5.4.3 Space-time blocking matrices — 174
  - 5.4.4 The JDL-GLR — 174
- 5.5 Aspects of implementation — 175
  - 5.5.1 General properties — 175
  - 5.5.2 Auxiliary eigenvector processor — 176
  - 5.5.3 Auxiliary channel processor — 177
- 5.6 Summary — 177

6 Spatial transforms for linear arrays — 179
- 6.1 Subarrays — 180
  - 6.1.1 Overlapping uniform subarrays (OUS) — 181
  - 6.1.2 Effect of subarray displacement — 183
  - 6.1.3 Non-uniform subarrays — 187
- 6.2 Auxiliary sensor techniques — 191

		6.2.1	Symmetric auxiliary sensor configuration (SAS)	191
		6.2.2	Bandwidth effects	199
		6.2.3	Asymmetric auxiliary sensor configuration	200
		6.2.4	Optimum planar antennas	202
	6.3	Other techniques		204
		6.3.1	Spatial blocking matrix transform	204
		6.3.2	Σ-Δ-processing	206
		6.3.3	CPCT Processing	209
	6.4	Summary		211
7	**Adaptive space-time digital filters**			**213**
	7.1	Least squares FIR filters		215
		7.1.1	Principle of space-time least squares FIR filters	215
		7.1.2	Full antenna array	224
		7.1.3	Spatial transforms and FIR filtering	226
	7.2	Impact of radar parameters		230
		7.2.1	Sample size	230
		7.2.2	Decorrelation effects	233
		7.2.3	Depth of the clutter notch	236
		7.2.4	Computation of the filter coefficients	236
	7.3	Other filter techniques		236
		7.3.1	FIR filters for spatial and temporal dimension	236
		7.3.2	The projection technique	237
		7.3.3	Space-time IIR filters	238
		7.3.4	Adaptive DPCA (ADPCA)	238
	7.4	Summary		239
8	**Antenna related aspects**			**241**
	8.1	Introduction		241
	8.2	Non-linear array configurations		242
		8.2.1	Circular planar arrays	242
		8.2.2	Randomly spaced arrays	257
		8.2.3	Conformal arrays	262
	8.3	Array concepts with omnidirectional coverage		264
		8.3.1	Four linear arrays	265
		8.3.2	Circular ring arrays	265
		8.3.3	Horizontal planar arrays	267
	8.4	STAP and conventional MTI processing		272
		8.4.1	Introduction	272
		8.4.2	Linear arrays	273
		8.4.3	Circular planar array	275
		8.4.4	Volume array	280
	8.5	Other antenna related aspects		280
		8.5.1	Sparse arrays for spacebased radar	280
		8.5.2	Polarisation-space-time processing	284
		8.5.3	Radome effects	286

	8.5.4	Alternating transmit approach	288
8.6	Summary		288

9 Space-frequency processing — 297
- 9.1 Introduction 297
- 9.2 The auxiliary space-time channel processor (ACP) 300
- 9.3 The symmetric auxiliary sensor/echo processor 301
  - 9.3.1 Computing the inverses of the spectral covariance matrices .. 303
- 9.4 Frequency domain FIR filter (FDFF) 304
- 9.5 Frequency-dependent spatial processing (FDSP) 306
  - 9.5.1 Spatial blocking matrices 310
- 9.6 Comparison of processors 310
- 9.7 Angle-Doppler subgroups 311
  - 9.7.1 General description 311
  - 9.7.2 Comparison angle-Doppler subgroup architectures with other techniques 314
  - 9.7.3 Other post-Doppler techniques 315
- 9.8 Summary 315

10 Radar ambiguities — 317
- 10.1 Range ambiguities 318
  - 10.1.1 Multiple-time-around clutter, linear arrays 318
  - 10.1.2 Multiple-time-around clutter, circular planar arrays .. 326
- 10.2 Doppler ambiguities 329
  - 10.2.1 Preliminaries 329
  - 10.2.2 Clutter and target models 331
  - 10.2.3 Pseudorandom staggering 332
  - 10.2.4 Quadratic staggering 334
  - 10.2.5 Impact of platform acceleration 335
  - 10.2.6 Space-time FIR filter processing 336
- 10.3 Summary 338

11 STAP under jamming conditions — 343
- 11.1 Introduction 343
- 11.2 Simultaneous jammer and clutter cancellation 344
  - 11.2.1 Optimum adaptive processing (OAP) 345
  - 11.2.2 Coherent repeater jammers 347
  - 11.2.3 Space-time auxiliary channel processors 347
  - 11.2.4 Spatial auxiliary channel processors 349
- 11.3 Circular arrays with subarray processing 352
  - 11.3.1 Adaptive space-time processing versus temporal clutter filtering 352
  - 11.3.2 Two-dimensional arrays in multi-jammer scenarios 353
- 11.4 Separate jammer and clutter cancellation 356
  - 11.4.1 Optimum jammer cancellation and auxiliary channel clutter filter 357
  - 11.4.2 Jammer and clutter auxiliary channel filters cascaded 361
- 11.5 Jamming in the range-Doppler IF matrix 369

	11.6	Terrain scattered jamming	370
		11.6.1 Transmit waveform	370
		11.6.2 Adaptive multipath cancellation	371
	11.7	Summary	373

12 Space-time processing for bistatic radar — 377

	12.1	Effect of bistatic radar on STAP processing	378
		12.1.1 Discussion of the bistatic clutter Doppler	378
		12.1.2 Numerical examples	380
	12.2	Realistic bistatic geometries (tandem configuration)	385
		12.2.1 Two aircraft with aligned flight paths	385
		12.2.2 Two aircraft with parallel flight paths (horizontal across-track)	386
		12.2.3 Two aircraft, transmitter above receiver (vertical across-track)	387
		12.2.4 A note on range dependence	388
	12.3	Ambiguities in bistatic STAP radar	389
		12.3.1 Range ambiguities	389
		12.3.2 Range and Doppler ambiguities	391
	12.4	Use of sparse arrays in bistatic spaceborne GMTI radar	393
		12.4.1 Introduction	393
		12.4.2 DPCA in bistatic configurations	394
		12.4.3 Some numerical examples	395
		12.4.4 Comparison with fully filled array	396
	12.5	Summary	396

13 Interrelated problems in SAR and ISAR — 403

	13.1	Clutter rejection for multichannel ISAR	405
		13.1.1 Models	406
		13.1.2 Space-time FIR filtering	410
		13.1.3 Effect of clutter cancellation on ISAR resolution	411
	13.2	Jammer nulling for multichannel radar/SAR	413
		13.2.1 Models	413
		13.2.2 Comparison of models	416
		13.2.3 MV spectra of jammers and noise	418
		13.2.4 Space-TIME FIR filter approach	419
		13.2.5 Effect of broadband jammer cancellation on SAR resolution	420
	13.3	Summary	422

14 Target parameter estimation — 425

	14.1	CRB for space-time ML estimation	425
		14.1.1 The principle	426
		14.1.2 The Cramér–Rao bound	428
		14.1.3 Some properties of the Cramér–Rao bound	430
	14.2	Impact of radar parameters on the CRB	436
		14.2.1 Environmental effects	437
		14.2.2 Impact of system parameters	439
	14.3	Order reducing transform processors	442

			Contents	xiii

	14.3.1	Conventional processing	443
	14.3.2	Space-time transforms	444
	14.3.3	Spatial transforms	446
	14.3.4	Auxiliary sensor/echo processing (ASEP)	447
14.4	Space-time monopulse processing		450
	14.4.1	Nickel's approach	450
	14.4.2	Numerical examples	455
	14.4.3	Ground target tracking with monopulse radar	459
14.5	Summary		465

15 Influence of the radar equation — 477
- 15.1 Fundamentals ... 477
  - 15.1.1 From notional radar concepts to realistic operation ... 477
  - 15.1.2 The radar equation ... 478
  - 15.1.3 SNIR and probability of detection ... 479
  - 15.1.4 Mapping SNIR onto probability of detection ... 480
- 15.2 Numerical examples ... 481
  - 15.2.1 Optimum space-time processing at subarray level ... 481
  - 15.2.2 Suboptimum space-time processors ... 485
  - 15.2.3 One-dimensional processing ... 486
- 15.3 Summary ... 486

16 Special aspects of airborne MTI radar — 491
- 16.1 Antenna array errors ... 491
  - 16.1.1 Tolerances of sensor positions ... 492
  - 16.1.2 Array channel errors ... 495
  - 16.1.3 Channel equalisation ... 500
- 16.2 Range dependence of clutter Doppler ... 505
  - 16.2.1 Impact on adaptation and filtering ... 505
  - 16.2.2 Doppler compensation ... 506
- 16.3 Aspects of implementation ... 512
  - 16.3.1 Comparison of techniques in terms of computational complexity 512
  - 16.3.2 Comparison of pre- and post-Doppler architectures ... 516
  - 16.3.3 Effect of short-time data processing ... 518
  - 16.3.4 Inclusion of signal in adaptation ... 518
  - 16.3.5 Homogeneity of clutter background ... 521
  - 16.3.6 Non-adaptive space-time filtering ... 524
  - 16.3.7 Further limitations ... 525
- 16.4 Adaptive algorithms ... 529
  - 16.4.1 Approximations of the optimum processor ... 529
  - 16.4.2 QR-decomposition ... 532
  - 16.4.3 Orthogonal projection algorithms ... 534
- 16.5 Alternative processor concepts ... 540
  - 16.5.1 Least squares predictive transform ... 540
  - 16.5.2 Direct data domain (D^3) approaches ... 540
  - 16.5.3 Frequency hopping ... 542

xiv *Contents*

    16.6 Summary 542

A Sonar applications   **545**
    A.1 Introduction 545
    A.2 Signal processing in the modal environment 546
        A.2.1 Signal models 547
        A.2.2 Extension to space-time matched field processing 552
    A.3 Active sonar application: suppression of reverberation 554
    A.4 Estimation of target position and velocity 556
        A.4.1 Influence of surface fluctuations 558
        A.4.2 Application: a multistatic CW surveillance system 558
    A.5 Summary 558

Bibliography   **559**

Glossary   **615**

Index   **620**

Biography

Richard Klemm was born in 1940 in Berlin, Germany. He received his Dipl.-Ing. and Dr.-Ing. degrees in communications from the Technical University of Berlin in 1968 and 1974, respectively.

From 1968 to 1976 he was with FGAN-FFM (Research Institute for Radio and Mathematics, a branch of the German Defence Research Establishment) at Wachtberg, Germany. His main field of activity has been adaptive clutter and jammer suppression for radar.

From 1977 to 1980 he was with SACLANT ASW Research Centre, La Spezia, Italy, where he conducted studies on spatial signal processing for active and passive sonar, with emphasis on matched field processing for shallow water applications.

Since 1980 he has again been with FGAN-FFM (nowadays FGAN-FHR) where he has been working on aspects of anti-jamming and detection of moving targets for moving sensor platforms. He has published numerous articles on various aspects of radar and sonar signal processing, with particular emphasis on space-time adaptive processing. He is a permanent reviewer of renowned journals such as *IEEE Transactions on Aerospace Systems* and *IEE Proc. Radar, Sonar and Navigation*. Richard Klemm has given numerous invited seminars on STAP in different countries.

Dr. Klemm has been a member of the Sensor and Propagation Panel of AGARD (nowadays the RTO-SET panel) and is a member of Commission C (Signals and Systems) of URSI. He chaired several AGARD and RTO symposia and was chairman of the AGARD Avionics Panel. He initialised and chaired the first European Conference on Synthetic Aperture Radar (EUSAR'96), March 1996, in Königswinter, Germany. He won the RTO-SET Excellence Award (2001), the IEE Clarke-Griffiths Premium (2003), and the RTO von Kármán Medal (2004), and was nominated Honorary Professor of UESTC (University of Electronics Science and Technology of China, Chengdu, China).

In his spare time Richard Klemm is a passionate classical pianist. Under the headline *Science and Music* he likes to combine his technical and musical sides by giving piano recitals at technical conferences. Dr. Klemm is married and has three children and four grandchildren.

Preface to the first edition

Space-time processing has become very popular in the past few years. In particular space-time adaptive processing (STAP) with application to airborne MTI radar has become a key topic of international radar conferences. Although mainly used for airborne MTI (moving target indication) radar other applications are possible.

Air- and spaceborne radar plays an important role for civilian and military use. There are numerous applications such as earth observation, surveillance, reconnaissance and others. Among a large number of facets of air- and spaceborne radar the capability of moving target detection plays an important role. This book focuses on a specific part of air- and spaceborne MTI radar: the suppression of clutter returns.

Look at the following simple formula

$$f_D = \frac{2v_P}{\lambda} \cos \alpha$$

It shows the dependence of the clutter Doppler frequency on the angle between an individual clutter scatterer and the flight direction. It seems hard to believe that one can write a thick book on such a simple matter. But it's true – this book deals with the consequences arising from this little formula.

The directional dependence of the clutter Doppler frequency as given by the above formula leads to the concept of *adaptive space-time clutter filtering* which is expanded in some detail. The design of suboptimum processors promising near optimum performance at low cost, reduced computing time, weight, energy consumption, etc., in other words, enabling real-time operation on small radar platforms, is the main goal of this book. The basis for space-time signal processing is a *multichannel* phased array antenna which provides spatial sampling of the backscattered echo field.

Since I started this work the radar world has changed considerably. While 15 years ago phased array antennas were still considered to be futuristic because the technology was still in an early stage they are nowadays state of the art. In the same timeframe dramatic progress in digital technology took place which led to powerful programmable signal processors, including parallel processing architectures. Moreover, a lot of theoretical work has been devoted to the development of new array processing techniques and algorithms. For a large number of applications the next generation of radar will include active multichannel phased array antennas. Such radar offers high flexibility, reliability, and, in conjunction with appropriate array processors, the potential of multichannel signal processing. In other words, the technology for real-time airborne MTI is there.

When I started working on airborne MTI there was not much material available in the open literature. The first publications date back to the 1970s (papers on DPCA even earlier). The first paper on adaptive clutter cancellation for airborne radar is the one by BRENNAN et al. [62]. In the meantime research on space-time adaptive processing, nowadays widely known by its acronym 'STAP', has grown tremendously. There are activities all over the world, especially in the USA and China.

A lot of the existing literature is concerned with the problem of suboptimum processor design. 'Suboptimum' means that the clutter rejection performance should be as close as possible to the optimum processor, however at much less cost and complexity. It was not my intention to summarise and compare all the various technical solutions proposed in the literature – this would exceed the limits of such book, and at the end the reader is still left with the problem of selecting his favourite solution. I tried instead to present as much insight as possible into the problems in order to help the reader understand the phenomena associated with adaptive space-time processing. I hope the reader will appreciate this.

Besides taking a lot of literature by other authors into consideration I tried to summarise mainly the insights and results of about 15 years of my own research in the field of air- and spaceborne MTI radar. In that sense this book certainly does not cover the whole STAP area. However, there are nowadays so many proposals for STAP processor architectures in the literature that it is hardly possible to describe all of them or even to compare them. By following the line of my own research and communicating the insights I obtained I hope to contribute to a better understanding of the associated problems. Maybe this makes this book even more valuable than presenting an overview of the existing literature.

The text includes in essence a selection of numerical results obatined for linear antenna arrays. These results are based on a specific parameter constellation which is given in Chapter 2. Consequently, the results are not exhaustive and are not meant to be. It is certainly not possible to address all potential combinations of radar parameters. Moreover, linear arrays are normally not used in radar applications. However, they are relatively easy to handle numerically and exhibit some fundamental properties which form a good basis for understanding the various effects. In that sense the examples presented are meant to be a guideline for the radar designer on how to proceed finding the solution for his actual problem.

I should mention that all beforementioned results have been obtained by calculations based on simple signal and clutter models. I agree that beyond the information presented to the reader many open questions will remain, which are in part related to the 'real world', i.e., the limited validity of computer models. Such limitations are mostly caused by differences between an ideal and a real radar sensor and also by the statistical behaviour of the target and clutter environment. Some of the results obtained may turn out to be too optimistic and need experimental verification. Nevertheless, I am convinced that in general such model studies are very useful to understand the mechanisms behind the observed phenomena, and can be used to pre-select promising practical solutions. Following this line some of the results presented here have contributed to international projects such as EUCLID, SWORD and AMSAR.

It is assumed that the reader has some basic knowledge of pulse Doppler radar, including the radar range equation, conventional MTI techniques, antenna arrays, signal and array processing, and is familiar with some fundamentals of statistical detection theory and complex matrix algebra.

My book is intended to be a systematic introduction to airborne MTI system design. It is organised in the following way: First a brief overview of potential array and signal processing techniques and the signal and clutter models used is presented (Chapters 1–3). Some basic properties of airborne clutter are summarised. Secondly the discussion of space-time adaptive clutter filters starts with optimum processing as given by array processing theory (Chapter 4). In this part the basic behaviour of space-time clutter suppression techniques is discussed, keeping in mind that the implementation of such techniques in practical systems is not realistic. In the sequel (Chapters 5–7) suboptimum processing techniques based on linear arrays will be presented which are more and more adapted to practical requirements, including aspects of on-board and real-time processing. The techniques discussed in these chapters are based on signal processing in a vector subspace. Such techniques have been frequently referred to as *partially adaptive*.

Chapter 8 deals with non-linear antenna configurations. Circular planar arrays with realistic numbers of sensors are discussed. Chapter 9 deals with space-frequency techniques. In Chapter 10 aspects of clutter suppression under jamming conditions are considered. In particular the question is raised whether clutter and jamming should be filtered simultaneously or by cascaded anti-jamming/anti-clutter filters. In Chapter 11 two applications for space-time processing in SAR or ISAR are presented. Although the applications are quite different in nature it is shown that they are based on the same mathematical background. In Chapter 12 finally miscellaneous aspects (array errors, aspects of implementation, computational complexity of processing techniques, etc.) and issues of space-time filtering are presented. Each of the twelve chapters concludes with a short summary highlighting the major findings of the individual chapter. The book concludes with a listing of more than 350 references which I collected up to the last moment before delivering the manuscript. By doing so I intended to include recent literature as much as possible.

Although this book contains a lot of numerical results it is not meant to be a handbook from which the user can extract technical data. It is merely intended to present the principles of space-time processing and to show the radar designer how to find his personal solution. In that sense I also anticipate that the book may serve as a background document for courses at universities or other technical institutions. I hope sincerely that the reader will find this book useful and will be stimulated to do his own research in this magic area.

The motivation to summarise all the work done on air- and spaceborne MTI radar in the past 15 years originates from several seminars which I gave on invitation by Ericsson, Sweden, and by three universities in China. The interest in this field is currently very strong so that I felt I should communicate as many details as I know to the public. There is at least one item which I have contributed to the world of space-time adaptive processing: to my knowledge all authors working in this field denote the number of antenna elements by N and the number of echo pulses by M as I did in my paper of 1983 [300]. I will keep with that tradition.

Acknowledgements. I would like to thank Dr. habil. J. Grosche, director of FGAN-FFM, for his generous and enthousiastic support of this work.

Dr. Wirth, head of the Electronics Department of FFM, suggested that I start research into adaptive airborne MTI as early as 1980. His continuous encouragement and numerous fruitful discussions and suggestions improved my understanding of the problems considerably. Dr. Wirth's co-operation is gratefully acknowledged.

I would like to thank my colleagues Dr. U. Nickel and Dr. J. Ender for contributing a number of valuable ideas. Furthermore, I owe thanks to O. Kreyenkamp for numerous discussions and helpful suggestions. I thank Dr. W.-D. Wirth and O. Kreyenkamp for their help in proofreading the manuscript.

In particular I would like to thank Mrs. G. Gniss. She carried out all the numerical evaluation and listened to my explanations of various physical matters with patience. Thus she helped me to understand at lot of the phenomena just by listening. It would have been impossible to write such a book without her continuous and ambitious assistance. More than 25 years of excellent co-operation with Mrs. Gniss are gratefully acknowledged.

I am indebted to H. Wolff of the German Ministry of Defence for his continuous support of my research activities over many years which have now been summarised in this book.

I feel obliged to thank Östen Erikmats of Ericsson, Mölndal, Sweden, who invited me to give a seminar at the radar division of his company. This invitation encouraged me to compile all my material on space-time processing which in turn stimulated the idea of summarising my work in a book.

I owe many thanks to the Chinese Professors Peng Yingning (Beijing), Wu Shunjun and Bao Zheng (Xi'an), and Huang Shun-Ji (Chengdu), for offering me the opportunity to give lectures on space-time adaptive processing at their universities. The discussions with them as well as with their colleagues and students were of high value for this undertaking.

Writing a book means first of all *learning* a great deal. I wish to thank all those authors who contributed to the field of space-time adaptive processing with publications and contributions to international conferences. I apologize for any publication I may have overlooked.

The excellent co-operation with the publisher, in particular the commissioning editor, J. Simpson, should be emphasised. I feel indebted to Prof. H. Griffiths, editor of Proc. IEE on Radar, Sonar and Navigation, for his encouraging support. I owe many thanks to the editors of the IEE radar series, Prof. E. D. R. Shearman and P. Bradsell, and also C. Wigmore and Dr. N. Steward of Siemens-Plessey for reviewing the manuscript and giving many helpful suggestions. In particular I am grateful to Dr. John Mather of DERA for carefully reviewing the manuscript. His numerous comments and suggestions improved considerably the quality of this book.

R. Klemm, Wachtberg, 1997

Preface to the second edition

The worldwide interest in space-time adaptive processing continues to be strong. The little Doppler formula on which all STAP activities (and, moreover, SAR) are based (see the preface to the first edition) has not lost its magic fascination. No future military air- or spaceborne observation radars will be designed without the capability of slow moving target detection. We felt that a second edition of this book would be useful as an introduction to this still fascinating field. In the meantime many new results and additional insight into the various problems have been gained.

The second edition has been enlarged by almost 30%. In detail the following major changes have been made:

- A few errata have been corrected. I am grateful for all the advice I received from several readers.

- A new Chapter 10 has been added which deals with the effect of radar ambiguities in range and Doppler.

- The new Chapter 12 gives an introduction to space-time adaptive clutter suppression for bistatic radar configurations

- The new Chapter 14 addresses the problem of estimating the target azimuth and velocity in the presence of Doppler coloured clutter. The Cramér–Rao bounds for various processor architectures are treated as well as the impact of radar parameters on the estimation accuracy. The performance of adaptive monopulse is studied.

- The degradation of the performance of the space-time FIR filter by temporal decorrelation due to internal clutter motion has been added (Chapter 7).

- The effect of range walk has been incorporated in Chapters 2, 3, 4 and 7.

- The effect of Doppler spread in large range bins has been analysed (Chapters 2 and 4).

- Some discussion of the range dependence of the clutter Doppler and related compensation techniques has been added to Chapter 10.

- The use of the space-time FIR filter with staggered PRI has been added to Chapter 10.

- The bibliography has been updated by including over 210 recent publications.

I hope that the reader will appreciate this effort and that the second edition will be received by the international radar community in the same way as the first one.

Acknowledgements. I would like to thank Prof. Dr. K. Krücker, Director of FGAN-FHR, for his support of this work. Dr. J. Ender, head of FGAN-FHR/EL, contributed numerous knowledgeable suggestions which improved my insight into certain problems. I am indebted to my colleague Dr. U. Nickel who contributed a number of valuable ideas. As in the first edition Mrs. G. Gniss performed the numerical evaluation. I am grateful for her enthusiastic assistance in completing this work.

I feel obliged to thank Prof. G. Galati (Italy), R. T. Hill, Dr. E. J. Ferraro (USA), Dr. P. Bhartia (Canada), G. P. Quek (Singapore), Dr. E. Velten, Prof. K. Krücker (Germany), Dr. W. Klembowski (Poland), Prof. B. Kutuza (Russia), and Dr. S. Kent (Turkey) for offering me the opportunity to give seminars on space-time adaptive processing in their organisations and to discuss details with their colleagues and students.

I am particularly grateful to to P. Richardson who thoroughly reviewed the entire manuscript. He made a number of valuable suggestions which have all been incorporated in the text. Finally, the excellent cooperation with the publisher, especially the commissioning editor, Dr. R. Harwood, and the production editor, Diana Levy, is greatly appreciated.

R. Klemm, Wachtberg, 2001

Preface to the third edition

Although space-time adaptive processing (STAP) has been invented about 30 years ago the worldwide interest seems still to be growing. This is reflected by the large number of publications in journals and conference proceedings and also by the popularity of the second edition of this book. The success of the second edition motivated me to provide an update which hopefully will find a similar enthusiastic audience as the second edition.

The third edition has been extended by including a number of new topics:

- Several errata have been corrected for;

- A new Chapter 15 on the impact of the radar range equation on the STAP performance has been added. In this chapter the role of the target strength is specifically addressed;

- In Chapter 14 the section on the use of adaptive monopulse for azimuth and Doppler estimation has been extended, including sensor specific aspects of ground target tracking;

- Two sections have been added to Chapter 8 which discuss issues arising with special antennas (arrays with 360° coverage, and arrays with multi-polarised elements);

- In Chapter 11 a section dealing with circular arrays in a multi-jammer scenario has been added. The use of the space-time FIR filter with staggered PRI has been added to Chapter 10;

- The bibliography has been updated by including more than 150 recent publications.

- The total number of pages has increased by about 20 %.

I hope that the reader will appreciate this effort, and I hope that the third edition will be received by the international radar community in the same way as the predecessors.

Acknowledgements. I would like to thank Prof. Dr. J. Ender, Director of FGAN-FHR, for supporting this work and a number of useful discussions on certain issues. I am grateful to Dr. U. Nickel and Dr. W. Bürger for numerous helpful suggestions during our co-operation. I like to thank Mrs. I. Gröger who performed the numerical evaluation of the new parts. I am also grateful to Dr. W. Bürger and F. Schulz who carefully read the manuscript and gave me valuable hints.

I am particularly grateful to the anonymous reviewer for carefully reading the manuscript and for all the valuable suggestions which have all been taken into account. Finally, the excellent cooperation with the publisher, especially the commissioning editor, Sarah Kramer, and the production editors, Wendy Hiles and Phil Sergeant, is greatly appreciated.

R. Klemm, Wachtberg, 2005

Chapter 1

Introduction

1.1 Preliminary remarks

Sensor systems such as radar or sonar receivers are tools for the interpretation of waves. A wave is by definition a function of space and time. Therefore, wave processing techniques basically include the spatial or temporal dimension or both. Synthesis of an antenna directivity pattern can be interpreted as *spatial* processing. Operations such as demodulation, filtering or Fourier analysis of the antenna output signal are *temporal* processing techniques.

Space-time signal processing is required whenever there is a functional dependency between the spatial and temporal variable. This is fulfilled in several applications, for example:

- moving pulse Doppler radar: dependency of the clutter Doppler frequency on the direction of arrival;

- frequency dependency of the directional response of an antenna array with narrowband beamforming;

- ambient noise in sonar (different frequencies arriving from different directions).

Space-time processing techniques can typically be applied in areas such as

- airborne MTI radar (BRENNAN *et al.* [63]);

- synthetic aperture radar (SAR), see ENDER [131], BARBAROSSA and FARINA [34], FARINA and LOMBARDO [158], DONG *et al.* [120], PETTERSON [536];

- shipborne MTI radar (LESTURGIE [389]);

- space-based MTI radar (SEDWICK *et al.* [595], MAHER and LYNCH [427], SCHINDLER *et al.* [588]), ZHANG *et al.* [749];

- Clutter and jammer rejection in ground-based radar (FARINA and TIMMONERI [159]);

- clutter cancellation for SAR/ISAR (ENDER [131], MERIGEAULT et al. [459], GENELLO et al. [179]), SOUMEKH and HIMED [615];
- interference suppression in narrowband radar through artificial array motion (LEWIS and EVINS [391]);
- interference suppression for broadband radar (COMPTON [97, 96]);
- wideband interference rejection in GPS receive arrays (FANTE and VACCARO [148], REED et al. [556], MCDONALD et al. [444]);
- terrain scattered jamming (GABEL et al. [171], KOGON et al. [359, 360], KOGON [364], ABRAMOVICH et al. [1, 2], NELANDER et al. [477]);
- cancellation of mainbeam interference in the presence of multipath (KOGON et al. [361]);
- combinations: suppression of terrain scattered jamming and clutter[1] (RABIDEAU [548], TECHAU et al. [637]), clutter rejection with wideband radar (FANTE et al. [143], RABIDEAU [551]);
- ground target tracking with STAP (KLEMM, KOCH, in KLEMM, ed., [350]);
- separation of closely spaced sources (SYCHEV [631, 632]);
- space-time coding of an array for simplification of beamforming (GUYVARCH [233]);
- clutter mitigation in over-the-horizon (OTH) radar (ABRAMOVICH et al. [1, 2], KRAUT et al. [365]), ABRAMOVICH et al., in KLEMM, ed., [350];
- terrain height estimation with GMTI radar in support of knowledge aided space-time adaptive processing (MORGAN et al. [469]);
- superresolution and anti-jamming with broadband arrays (NICKEL [495], MOORE and GUPTA [468]);
- adaptive space-time monopulse (NICKEL [496]);
- suppression of reverberation for active sonar;
- simultaneous localisation and Doppler estimation for passive sonar, see Appendix A (matched field processing);
- seismics (MANN et al., in KLEMM, ed., [350]);
- communications (PAULRAY and LINDSKOG [532], SEE et al. [597], HOCHWALD et al. [271], SANDHU et al., in KLEMM, ed., [350], BÖHME and WEBER, in KLEMM, ed., [350], WALKE et al. in KLEMM, ed., [350]), WALKE and REMBOLD [665], KUZMINSKIY and HATZINAKOS [372]);

---

[1] Frequently referred to as 'hot and cold clutter'

- specific use of STAP for GPS receiver synchronisation in presence of jamming;

- simultaneous frequency and DOA estimation in a multiple source environment (ROBINSON [573], LUKIN and UDALOV [420]);

- navigation (MYRICK and ZOLTOWSKI, in KLEMM, ed., [350]);

An overview of some of the techniques listed above is given by WARD et al. [695], MELVIN [457], RANGASWAMY et al. [554], and FARINA and LOMBARDO [160]. Further detailed information can be gathered in the special issues of *ECEJ* (KLEMM ed., [325]) and *IEEE Trans. AES* (MELVIN [454]). The book by GUERCI [228] is an excellent introduction. It is complementary to this book in that a number of aspects of space-time adaptive processing which not included here are covered. The book 'Applications of space-time adaptive processing' (KLEMM ed. [350]) summarises the practical work on STAP conducted by 45 authors, including chapters on radar, sonar, seismics, and communications.

Some authors propose to extend the signal vector space from space-time to space-time-waveform (SEED et al. [598]) or space-time-polarisation (PARK and WANG [521], SHOWMAN et al. [608]).

Applications in other areas may be possible. As the author's expertise is mainly in the field of airborne MTI radar the major part of this book is focused on this subject. Some of the other aspects, however, will be touched on in later chapters.

1.1.1 Basics of MTI radar

MTI radar (Moving Target Indication) has by definition the capability of detecting moving targets before an interfering background (usually called *clutter*). More specifically, by MTI radar, pulse Doppler radar (PDR) is understood which uses the Doppler effect to detect moving targets before a clutter background (Doppler discrimination). The difference between target and clutter velocities is exploited for target detection. The pulse Doppler radar transmits phase-coherent pulses and measures the phases of the backscattered echoes. The radial velocity of any moving object results in phase advances of successive echo pulses. By spectral analysis of the phase history of an echo sequence the Doppler frequency and, hence, the radial velocity of the reflecting object can be found. For spectral analysis either DFT, FFT or a bank of Doppler filters may be used. In synthetic aperture radar (SAR) velocity matching techniques involving the Wigner-Ville distribution and other techniques are used to focus the echoes reflected by a moving targets (e.g., LIPPS et al. [404], BARBAROSSA and FARINA [34]).

MTI techniques are based on the temporal coherence and, hence, the phases of echo signals. This is in contrast to change detection methods as known from imaging radar (e.g., KIRSCHT [289]) which compare successive amplitude images.

KOCH [355] and more recently KOCH and VAN KEUK [356] demonstrate that tracking of air targets in a densely cluttered environment can be successfully accomplished by smoothing via retrodiction without using any anti-clutter device before tracking. It can be expected, however, that MTI pre-filtering will alleviate the tracking workload which may result in reduced time delays due to multiple hypothesis testing.

There are several types of clutter which differ in their spectral parameters. The most important kind of clutter is *ground clutter* caused by echoes scattered from the ground. Ground clutter received by a stationary radar exhibits a symmetric Doppler spectrum centred at zero Doppler. The clutter power and the Doppler bandwidth depend on the type of background. Hard objects (buildings, urban areas) will produce high-power narrowband Doppler spectra while areas with a high degree of roughness or internal motion (agriculture, vegetation, forests) cause less clutter power at larger bandwidth. The bandwidth increases with wind speed and radar frequency, see NATHANSON [475, p. 274]. Moreover, antenna rotation causes additional broadening through spatial decorrelation of clutter echoes.

Weather clutter may show a shift of the Doppler spectra towards non-zero frequencies. This frequency shift reflects the average radial motion of weather (clouds, rain, snow, hale, chaff) due to the wind speed. Positive Doppler frequencies are an indication for approaching, negative for receding weather. Usually the Doppler bandwidth of weather clutter is larger than that of ground clutter. There are four mechanisms that are responsible for the shape of the weather clutter spectrum: wind shear, Doppler spread in the cross-wind direction, fluctuating currents, and a fall velocity distribution of reflecting particles, see NATHANSON [475, p. 205]. For sea clutter a shift of the Doppler spectrum occurs due to the velocity of the ocean waves relative to the radar. Moreover, there are effects such as individual wavelets, foam and spray which broaden the clutter spectrum, see NATHANSON [475, p. 241].

1.1.2 One-dimensional clutter cancellation

Ground clutter echoes can be cancelled by use of a discrete filter (FIR or IIR) operating on a pulse-to-pulse basis, with a notch at zero Doppler frequency. The clutter notch may be formed adaptively so as to match the individual shape of the clutter Doppler spectrum. Normally, the variations of the clutter bandwidth are taken care of by a fixed filter centred at zero frequency whose clutter notch can be matched to the actual clutter bandwidth. Of course, low Doppler targets, i.e., targets whose radial velocity component relative to the radar is small, may be buried in the clutter bandwidth and, hence, are hard to detect.

For sea and weather clutter the Doppler shift of the spectrum due to the radial velocity components of ocean waves or clouds may result in an offset between the maximum of the Doppler spectrum and the clutter notch of a fixed zero Doppler clutter filter. There are techniques for aligning the centre frequency of the clutter spectrum with the clutter notch of the pre-filter. This may be done by adjusting the Coho frequency in such a way that the clutter energy maximum falls into the filter notch (TACCAR, see SKOLNIK [609, p. 17-32]). Alternatively, the motion induced phase progression of clutter returns may be used to compensate for the clutter motion in the baseband (clutter locking, see VOLES [663]).

Since the bandwidth and the centre frequency of sea and weather clutter returns varies within a wide range, adaptive clutter filtering is the appropriate way of solving this problem. A lot of solutions have been reported in the literature. Basically optimum clutter cancellation can be formulated as a binary hypothesis test ('target present' or 'no target') which is given by the Neyman–Pearson test. It requires knowledge of the covariance matrix of temporal clutter samples which has a Toeplitz form if the pulse repetition frequency (PRF) is constant. In [75] (BÜHRING and KLEMM) an adaptive filter operating with a surveillance radar with rotating antenna has been described (for details see Sections 1.2.4.4 and 1.2.5.2). In this system the Levinson algorithm as used by Burg in his papers on maximum entropy spectral analysis (BURG [79, 80]) has been implemented to adaptively estimate the clutter correlation function and to calculate the associated least squares FIR filter which minimises the clutter power.

A lot of literature on clutter and clutter suppression for stationary radar has been published. For the sake of brevity, only a few are quoted here. In SKOLNIK'S *Radar Handbook* [609, Chapter 17] and [610, Chapter 15] an overview of the problems and techniques for clutter suppression is given. An even more detailed and comprehensive description is given by SPAFFORD [616] and in the book by SCHLEHER [589]. In the book by NATHANSON [475] a collection of measured clutter backscatter data can be found, with conclusions on detection of signals in clutter.

1.1.3 Aspects of air- and spaceborne radar

Observation of the earth's surface by air- and spaceborne radar has several advantages compared with ground-based radar which are caused by the elevated position of the radar and the radar platform motion:

- The horizon, i.e., the visible range, is extended through the elevated position.

- The elevated position leads to a reduction of terrain masking effects. The effect of shadowing due to hilly terrain is mitigated for a radar looking from above.

- For air- and spaceborne radar the wave propagation conditions are more favourable than for ground-based radar. The beam of a high-flying radar has to *cross* the lower atmosphere layer while the beam of a ground-based radar is entirely buried in the lower atmosphere. Moreover, the interaction of the transmitted wave with the ground is reduced. This interaction causes additional sidelobes in the directivity pattern.

- The platform motion offers the potential of high-resolution imaging through synthetic aperture processing, including interferometric imaging and change detection.

There are some specific problems associated with a flying radar platform as far as clutter is concerned. The extended horizon means an extended visible range but also an extended clutter area. For *ground-based* radar ground clutter ends at about 50 km, but for *spaceborne* radar clutter will be present in the whole visible range. *Airborne* radar clutter will cover some distance in between, depending on the platform altitude. Furthermore, since the depression angle of the radar beam to the ground is larger for air- and spaceborne than for ground-based radar higher clutter power can be expected. In particular, from underneath the platform high-power clutter returns with zero Doppler frequency are received which are commonly called the spectral *altitude line*. Since the altitude line is generated through *specular* reflection the associated clutter power is particularly high.

1.1.4 Impact of platform motion

1.1.4.1 *From DPCA to STAP*

Another, even more important property of clutter echoes received by a moving radar is the motion-induced Doppler spread. A radar mounted on a moving platform receives clutter echoes that are Doppler shifted. The Doppler shift depends on the radial velocity of the individual scatterer relative to the radar which in turn is a function of the direction, i.e., azimuth and depression angle, see the formula at the beginning of the preface. The maximum clutter Doppler frequency occurs in flight direction while in the cross-flight direction the Doppler frequency is zero. The total of clutter arrivals from all possible directions sum up in a Doppler broadband clutter signal whose bandwidth is determined by the platform speed and the wavelength. Any kind of Doppler spread of the clutter spectrum leads to a degradation of the detectability of low Doppler targets. Conventional *temporal* clutter filters operating on echo data sequences can be designed in such a way that the clutter is suppressed optimally. Low Doppler targets, however, will be suppressed as well.

These problems can in principle be circumvented by appropriate radar antenna and signal processing design. This is illustrated by the following consideration. As mentioned before for clutter echoes there is an equivalence between the direction of arrival and the Doppler frequency, whereas the Doppler frequency due to a moving target depends also on the target velocity. This fact can be exploited for target detection in a motion-induced Doppler coloured clutter environment.

Suppose, for example, a beam is steered in a certain direction. Then the clutter bandwidth at the radar output is determined by the antenna beamwidth. If this clutter part of the spectrum is suppressed any moving target whose Doppler frequency falls outside the received clutter spectrum can be detected. Assuming an infinitely narrow beam then the received clutter spectrum is just a single frequency line. In this ideal case any target Doppler frequency different from the clutter line can be detected. There is obviously no limitation due to the platform motion. Target detection is limited only by the *spatial* resolution of the beam and the *spectral* resolution of the Doppler analysis.

It should be emphasised that the processing described in the previous paragraph consists of two parts, a spatial and a temporal part. Likewise, in the Fourier domain one may speak of directional and spectral processing. In the example given above (beamformer + Doppler analysis) the temporal part depends on the spatial part because the beamformer determines the clutter spectrum.

Since beamforming and Doppler filtering are linear operations the succession of spatial and temporal processing can basically be interchanged. This might not be relevant for realistic radar operation but helps for further clarification of the two-dimensional processing principle. Consider an omnidirectional transmitter and an array antenna with individual sensor outputs. First the echo signals of *all* antenna channels pass narrowband filters in order to select one specific Doppler frequency (temporal filtering). The second stage is a multibeamformer with one beam pointing in the clutter direction. The clutter beam will be discarded (spatial filtering) while all the others will find those moving targets which have the same Doppler frequency as the narrowband filters.

It should be noted that in the above discussion on the relations between Doppler frequency and angular direction no assumptions on the orientation of the antenna were made. It follows that basically any kind of array configuration may be used[2]. This will play a major role in the discussion of sideways and forward-looking antenna arrays, that means, antenna arrays aligned with the flight or cross-flight direction.

From these considerations one can see that basically the platform motion does not limit the detection of slow targets, provided that appropriate space-time processing is applied. The first approach to space-time processing of clutter echoes was the DPCA technique (displaced phase centre antenna), see ANDREWS [17, 18, 19], SKOLNIK [609, pp. 18-1], [610, pp. 16-1] and TSANDOULAS [653]. The DPCA technique is based on a sidelooking antenna arrangement with two phase centres displaced along the flight axis. This can be realised by use of two identical antennas or a monopulse antenna (ANDERSON [15]). The PRF, the displacement of the phase centres and the platform speed are adjusted in such a way that the second phase centre assumes the position of the first one after one PRI (pulse repetition interval). Therefore, any two successive pulses received by the two different antenna parts appear to come from one phase centre fixed in space so as to compensate for the influence of the platform motion[3]. A subsequent two-pulse canceller subtracts the second echo received by the first antenna from the first echo of the second antenna. In essence this is a kind of space-time processing.

As pointed out above conventional clutter filters as used in stationary radar are by nature *temporal*, that is, they operate on a echo-to-echo basis. By talking about space-time processing we enter the area of signal processing for array antennas. Fortunately here we can refer to a large choice of literature on adaptive arrays. Since the early publications by BRYN [73], MERMOZ [460], SHOR [606] and WIDROW et al. [709] a lot of authors have contributed to this subject. There are several textbooks, for example those by MONZINGO and MILLER [466], SCHARF [587], HUDSON [275], NICOLAU and ZAHARIA [499], HAYKIN [253], and FARINA [149], LACOMME [374]. Fundamentals of statistical signal processing can be found in SCHARF [587].

There are two principal applications of adaptive array processing which are closely related to each other: 1. cancellation of directional noise (adaptive nulling), 2. superresolution of targets (adaptive beamforming). First applications of adaptive arrays were in the field of underwater acoustics, mainly with application to submarine localisation, and in geophysics, to locate earthquakes and subterraneous nuclear explosions. With the advent of phased array technology the theory of adaptive arrays has been adopted by the radar community, see BRENNAN and REED [62]. Predominant tasks are cancellation of unwanted radiation, such as jamming or directional clutter, and resolution of close targets.

---

[2]However, as will be shown below, the antenna geometry has a dramatic influence on the clutter spectra and the kind of signal processing.

[3]Actually, it is not the positions of the two sensors that have to coincide but the phases of two subsequent clutter returns. Due to the two-way propagation the distance the antenna has to move forward is only half the sensor spacing.

8 *Introduction*

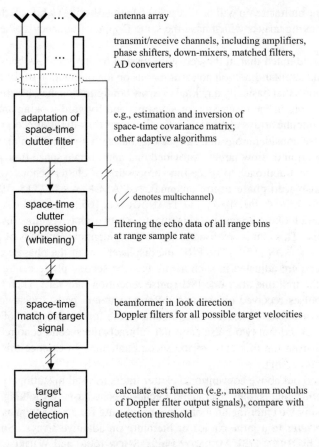

Figure 1.1: Functional block diagram of STAP radar

After a few years of initial enthusiasm over adaptive systems, it was recognised that adaptive array processing may be seriously limited by several effects. In sonar applications, various stochastic irregularities of the propagation medium lead to a lack in spatial and temporal correlation of received waves. This lack of correlation limits the capability of noise cancellation. Moreover, the environmental conditions of the propagation path may cause a mismatch between the expected and the received signal which leads to degradation of the performance of superresolution techniques. Since ambient noise is not normally highly directive, spatial noise cancellation requires a high number of degrees of freedom which in turn means high computational complexity and cost.

In the radar world the propagation medium is much more predictable than in underwater acoustics. There are, however, 'home-made' problems in the radar receive process. Receiver noise, instabilities of the amplification and down-converting chain as well as tolerances in the transmission characteristics of individual array channels may limit the performance of adaptive radar array antennas. The impact of channel errors on the performance of adaptive radar arrays has been analysed in some detail by NICKEL [486].

DPCA techniques are basically non-adaptive. As pointed out earlier adaptive processing is necessary whenever parameters of the received clutter spectrum are unknown, like for instance the centre Doppler frequency and bandwidth of weather or sea clutter. For such application the TACCAR loop (SKOLNIK [610, pp. 17-32]) provides an adaptive shift of the centre frequency of the clutter spectrum towards the MTI clutter notch.

Even for ground clutter adaptive processing may offer some advantages. On the one hand the clutter characteristics may change while being passed by the radar beam. On the other hand airborne platforms are subject to irregular motion due to atmospheric turbulences, wind drift, and vibrations. An adaptive filter will be able to compensate for such perturbations provided that the adaptive algorithm converges fast enough.

Space-time processing has already been recognised as a technique for broadband jammer cancellation. In 1976 BRENNAN et al. [63] proposed the principle of space-time adaptive processing (referred to by several authors as STAP) for suppression of airborne radar clutter. They analysed the performance of the *optimum processor*, see Section 4. More details on the clutter suppression performance of the optimum processor can be found in a paper by KLEMM [300].

Since then a lot of papers on suboptimum techniques for clutter suppression have been published by ENDER [122], ENDER and KLEMM[133], KLEMM [302, 303, 304, 306, 308, 315, 341], LIAO et al. [396, 397, 398], NOHARA [500], SU et al. [620], WANG H. et al. [667, 668, 672], WANG Y. and PENG [681], WICKS et al. [706] and ZEGER and BURGESS [743], and others. The report by WARD [690] and the papers by KLEMM [318, 319] can be used as an introduction to the area of space-time adaptive processing for airborne radar. A recent overview paper on space-time adaptive processing has been published by WARD et al. [695]. The digest of the IEE STAP colloquium (1998) (KLEMM [343]) gives an overview of the state of the art in space-time adaptive processing. It contains contributions by KLEMM, WARD, FARINA and LOMBARDO, ENDER, RICHARDSON, GABEL, and WRIGHT and WELLS.

The optimum detector (maximum likelihood) concept leads to a fully adaptive space-time processor which in practice can be difficult to implement for realistic antenna dimensions (possibly several thousands of sensor elements, hundreds of echo pulses per CPI). Therefore suboptimum techniques which promise near-optimum clutter suppression at reduced computational expense are of great interest. The discussion of suboptimum techniques for air- and spaceborne clutter suppression will form a significant part of this book.

An alternative use of space-time processing has been proposed by LEWIS and EVINS [391]. The authors propose to generate an artificial Doppler by shifting the illumination of a subarray of a linear (or rectangular) transmit array during the transmitted pulse length. Therefore, sidelobe clutter appears at frequencies outside the radar bandwidth and will be suppressed. A similar technique involving sequential transmission of individual transmit elements leads to improved beamforming and target search, and simplified processing (MAHAFZA and HEIFNER [425]).

In most of the published papers on space-time clutter suppression the authors assume a sidelooking antenna in which case the function of the motion compensated MTI can be explained by the DPCA principle (see above and Chapter 3).

Besides sideways looking radar, forward-looking radar plays an important role. A review of the references shows that so far very little information is available on forward-looking MTI. In SKOLNIK'S *Radar Handbook* [610, Chapter 16] one finds a plot which indicates that forward-looking MTI can function for an antenna arranged in the cross-flight direction. Two or more phase centres of the antenna have to be generated along the flight axis by appropriate illumination so as to create the DPCA effect. In [315, 321] (KLEMM) and [563] (RICHARDSON and HAYWARD) some properties of forward-looking MTI radar as compared with sidelooking radar are discussed. In the subsequent discussions a major part of our attention will be devoted to forward-looking radar. Forward-looking radar plays a significant role in aircraft nose radar as well as spaceborne radar. Even synthetic aperture imaging may be accomplished with a forward-looking antenna (MAHAFZA et al. [424]). MTI for forward-looking radar will play a major role in this book.[4]

The content of this book consists mainly of numerical examples which illustrate the function of the various processors discussed. Although a large number of numerical results have been presented it should be emphasised that the material presented is not exhaustive. It is merely thought that the examples show the radar system designer a way to analyse his specific problem under his actual system requirements.

1.1.4.2 The principle of space-time adaptive processing

In this subsection the principle of space-time processing and the role of this book in this context are briefly explained. Figure 1.1 shows a coarse block diagram of a MTI radar including space-time adaptive processing for cancellation of clutter with motion-induced Doppler colouring. A sequence of coherent pulses is transmitted via the array antenna. The associated echo sequence enters the antenna, eventually through a network of pre-beamformers.[5] Echo signals are amplified, down-mixed and digitised. The next step includes adaptation of the clutter filter to the actual clutter echoes received. Adaptation is done by estimating the space-time covariance matrix from secondary training data or other covariance matrix-free algorithms (TUFTS *et al.* [655, 656, 657], HUNG and TURNER [276]).

The received echo signals are filtered so as to cancel the clutter component in the radar returns. The remaining signal + noise mixture is then matched with a space-time matched filter which in essence is a beamformer cascaded with a Doppler filter bank since the target Doppler frequency is unknown. The output signals of the Doppler filter bank are used to design a test function which has to be compared with a detection threshold. Usually the maximum of the squared magnitudes of the output signals of the Doppler filter bank is chosen for threshold comparison.

Notice that the STAP techniques are contained in the two blocks 'adaptation' and 'clutter suppression'. The design of suitable architectures of adaptive space-time filters

---

[4]Many papers on space-time processing do not indicate whether they deal with sideways or forward-looking array geometry. Presumably, since most of the ongoing experiments are based on sidelooking arrays, these papers address the sidelooking geometry.

[5]This depends on the actual STAP algorithm used.

is the main subject of this book.

1.1.5 Some notes on phased array radar

Space-time processing requires a phased array radar with an array antenna which has several output channels so as to provide a *spatial* processing dimension. In future years phased array radar will play an increasing role, especially for military applications. A phased array radar has a number of unique capabilities and properties:

- Inertialess beamsteering;

- Multifunction operation (e.g., search, track, terrain following, guidance, mapping, SAR, ISAR);

- potential of energy management strategies;

- advanced search techniques such as the sequential test (WIRTH [716, 713, 719, 721]);

- multiple target tracking (VAN KEUK and BLACKMAN [661]);

- potential of multiple beamforming (WIRTH [720, 718]);

- potential of *spotlight* SAR;

- correction for motion-induced azimuth errors of moving targets in SAR images (ENDER [124, 125, 126]);

- high reliability of active arrays;

- high efficiency of active arrays;

- design of antennas with 360° coverage (WILDEN and ENDER [711]);

- potential of spatial signal processing on receive (anti-jamming, superresolution), see for instance WIRTH and NICKEL [724], WIRTH [714];

- potential of spatial filtering on transmit;

- space-time, space-TIME, space-time-TIME, space-frequency processing for suppression of various kinds of interference.[6]

As one can see from the above list the reasons for using phased arrays in future radar systems are numerous. The development is not necessarily driven by the requirement for space-time processing.

---

[6]By 'time' we mean the slow (pulse-to-pulse) time, while 'TIME' denotes the fast (range) time.

1.1.6 Systems and experiments

Several experimental or operational radar systems with multichannel antenna for space-time processing currently exist or are in the planning phase:

1. The operation of a phased array based DPCA antenna has been tested experimentally by TSANDOULAS [653]. The PRF was automatically adjusted to variations of the aircraft attitude.

2. The AN/APG-76 (see TOBIN [648], NORDWALL [506]) by Westinghouse Norden (now Northrop Grumman Aerospace) is a fielded and operational multimode radar system with SAR and GMTI (ground moving target indication) capability.

3. The JOINT-STARS radar has space-time MTI capability based on an array antenna with three subarrays, see HAYSTEAD [255], SHNITKIN [604, 605]. The clutter suppression achieved is of the order of 20 dB. The MTI function has proven to be useful during the NATO operations in Kuwait and Bosnia, see COVAULT [103], ENTZMINGER et al. [137].

4. The Multichannel Airborne Radar Measurements program MCARM by Rome Laboratories (see SURESH BABU et al. [626, 627], FENNER and HOOVER [162]) and SOUMEKH and HIMED [614]) uses an L-band airborne phased array radar testbed. Studies on real-time processing architectures are part of this program (see BROWN and LINDERMAN [70], LINDERMAN and LINDERMAN [403], LITTLE and BERRY [405]).

5. NRL used an eight-element UHF linear array under an EP-3A aircraft (BRENNAN et al. [66] and LEE and STAUDAHER [382]).

6. The experimental C-band SAR by DRA (UK), see COE and WHITE [94, 95], uses three antennas in an along-track arrangement on an aircraft.

7. The MOUNTAINTOP program is an ARPA/NAVY sponsored initiative started in 1990 to study advanced processing techniques and technologies required to support the mission requirements of next generation airborne early warning (AEW) platforms (TITI [646], TITI and MARSHALL [647]).

8. The motion of an airborne radar can be emulated by use of a linear array of transmit antennas which transmit radar pulses successively. Then the clutter returns received by a stationary radar are equivalent to clutter echoes received by a moving radar. This technique has been described by AUMANN et al. [23], and has been used in the framework of the MOUNTAINTOP program.

    To emulate a sidelooking array the transmit array has to be aligned with the receive array. For forward-looking operation the transmit array has to be rotated by 90°.

9. The NFMRAD by GOGGINS and SLETTEN [195] (Air Force Cambridge Research Labs., USA) has been designed to analyse experimentally the principle of space-frequency clutter nulling. First experiments used a truck as carrier.

Preliminary remarks 13

10. The ADS-18S antenna (Loral-Randtron), operating at UHF (DAY [112]).

11. The DOSAR by Dornier (Germany). HIPPLER and FRITSCH [270] use two antennas in the along-track configuration for space-time MTI operation.

12. The AER II developed at FGAN-FFM (Germany) by ENDER [123, 127, 129]; ENDER and SAALMANN [135], RÖSSING and ENDER [575], is a four-channel experimental airborne X-band SAR. It has the capability of sideways and forward-looking operation. For testing the aircraft was replaced by a truck. By driving over highway bridges and looking down into the valley a flight geometry was simulated. Flight experiments were conducted later.

13. The PAMIR experimental high resolution SAR (subdecimeter resolution) by BRENNER and ENDER [67, 68], ENDER and BRENNER [132, 136]. Flight experiments confirmed geometrical resolution below 10×10 cm.

14. The APY-6 by Northrop-Grumman (GROSS and HOLT [217]), an airborne X-band radar with SAR and GMTI (ground moving target indication) capability. The antenna is subdivided into a large part for transmit and SAR receive, and three smaller panels for MTI receive. The radar can be operated in sidelooking and forward-looking configuration. APY-6 is a low-cost approach using COTS components.

15. There are deliberations going on to install surveillance functions such as AWACS or JointSTARS in space. The Ground MTI is a key function in these concepts (COVAULT [104]).

16. The Airborne Stand-Off Radar (ASTOR) developed in the UK will have a GMTI mode in conjunction with high-resolution SAR imagery in such a way that moving target scenes can be overlaid on SAR images (NORDWALL [507]).

17. AMSAR (Airborne Multirole Solid-state Active array Radar) is a current European project (Thomson-CSF, France; GEC Marconi, UK; DASA, Germany) concerning a nose radar for future fighter aircraft (ALBAREL *et al.* [13], GRUENER *et al.* [218]). The antenna is a circular planar array subdivided into two-dimensional subarrays. The concept includes anti-jamming and STAP capabilities.

18. SOSTAR (Stand Off Surveillance and Target Aquisition Radar) is a planned European project (HOOGEBOOM *et al.* [273]). A sidelooking linear X-band array will be mounted under the fuselage of an aircraft. The system will include both SAR imaging and moving target indication. Participating companies: Dornier (Germany), Thomson-CSF Detexis (France), FIAR (Italy), TNO-FEL (The Netherlands).

19. The UESA program (UHF Electronically Scanned Array) is an experimental radar system including a circular ring array being fabricated by Raytheon, USA. Application might be Airborne Early Warning (AEW), possibly as a successor to systems such as AWACS (ZATMAN [741]).

20. RADARSAT-2 (Canada, to be launched in 2003) will be the first earth observation satellite with MTI capability (MEISL et al. [447], THOMPSON and LIVINGSTONE [643], EVANS and LEE [138], GIERULL and LIVINGSTONE [192]).

1.1.7 Validity of models

This textbook is meant to be a primary tutorial and reports on the status quo in research in space-time processing. In order to make the various effects associated with the suppression of moving clutter as clear as possible I have tried to simplify our models as much as possible. This may have the problem of losing contact with the 'real world'. I believe, however, that the underlying principles should be isolated and considered separately as a superposition of various effects causes problems in understanding. It is certainly up to the system designer to refine such models up to a level which is sufficient for his requirements.

Most of the considerations presented in this book have been carried out for one single range increment. This is justified by the fact that space-time clutter filtering is basically carried out along each individual range ring. In practice, however, clutter statistics of an individual range ring depend on other range increments. Such effects are listed below.

1. **Adaptation** of the space-time covariance matrix has not been considered in this book. It is assumed that the clutter covariance matrix is known for any individual range increment. In practice the clutter covariance matrix has to be estimated, e.g., by averaging clutter dyadics over various range rings. This may cause problems induced by range dependency of the clutter Doppler and non-homogeneity of clutter returns. These aspects exceed the scope of this book. In Chapter 16 related literature has been quoted.

2. **Mutual coupling** effects (GUPTA and KSIENSKI [234]) between array elements on the performance of the array have been neglected.

3. It is assumed that the clutter returns are gaussian. Whenever we talk about 'optimality' this refers to gaussian statistics. The space-time technique and its application to non-gaussian clutter has been investigated by RANGASWAMY and MICHELS [552]).

4. **Multiple-time around clutter** occurs whenever the PRF is chosen such that the radar is range ambiguous within the visible radar range. In most of the examples presented in this book the multiple clutter returns have been neglected. This is in accordance with almost all available references on space-time processing. There are currently only a very few papers in the open literature that include the effect of ambiguous clutter returns. The reason is that most papers on space-time processing focus on sidelooking radar. As will be shown in Chapter 3 sidelooking radar is insensitive to ambiguous clutter. In Chapter 10 the effect of ambiguous clutter for forward-looking radar is specifically addressed. It is shown that an array antenna which is adaptive in two dimensions can suppress the multiple returns.

5. I have assumed that the **reflectivity** of the ground is independent of the depression angle. In practice there is a strong dependence which is in turn associated with the kind of clutter background (roughness). This assumption implies also that the effect of altitude return is not specifically emphasised.

6. Most examples given in this book are based on **linear arrays**. As will be seen in Chapter 6 linear arrays have favourable properties concerning space-time processing. They are also most useful for illustrating the fundamentals. Therefore, most of the existing literature is focused on linear arrays. Moreover, linear arrays include all kinds of cylindrical array configurations whose axes are horizontal, and rectangular planar arrays. In Chapter 8 some considerations on circular arrays have been included.

1.1.8 Historical overview

A brief historical overview of the evolution of STAP is presented in Table 1.1. This table includes a number of major publications, programs or events which contributed to the development of STAP. This listing is certainly not complete but includes to our understanding the major steps in the development of STAP.

Table 1.1 Milestones in the evolution of STAP

Year Milestones	Author(s)	Country	Ref.
1958 first paper on motion compensated MTI	Anderson	USA	[14]
1970 DPCA techniques	Skolnik	USA	[609]
1973 DPCA experiment	Tsandoulas	USA	[653]
1973 Theory of adaptive radar	Brennan and Reed	USA	[62]
1976 first paper on STAP	Brennan *et al.*	USA	[63]
JointStars development			[137]
1983 clutter power spectra eigenspectra	Klemm	Germany	[300]
1987 subspace techniques	Klemm	Germany	[302, 303]
	Klemm and Ender		[306, 340]
1990 worldwide interest in STAP		USA,UK Italy,China	
1990 MCARM program	Suresh Babu	USA	[626]
Mountaintop program	Titi	USA	[646]
DOSAR	Hippler and Fritsch	Germany	[270]
1992 Brennans rule	Brennan *et al.*	USA	
1993 STAP with SAR	Ender	Germany	[122]
1993 Clutter and jamming	Klemm	Germany	[308]
1994 first ASAP workshop	MIT Lincoln Lab.	USA	
1994 Report on STAP	Ward	USA	[690]
1995 STAP processing with systolic arrays	Farina *et al.*	Italy	[151]
1995 forward-looking	Richardson, Hayward,	UK	[563]
STAP	Klemm	Germany	[315]
1995 Parameter estimation	Ward	USA	[693]
1996 Hot clutter	Kogon *et al.*	USA	[359]
1996 AER II experiment	Ender	Germany	[127]
1996 Subspace techniques	Goldstein, Reed	USA	[197, 201]
1997 STAP at DERA	Murray *et al.*	UK	[471]
1997 Array orientation	Wang *et al.*	China	[682]
1998 APY-6	Gross and Holt	USA	[217]
1998 Textbook on STAP	Klemm	Germany	[323, 324]
1999 Bistatic STAP	Brown et al.	USA	[71]
2000 Large scale research at –Universities –Research est. –Industry		USA	

Table 1.1 Milestones in the evolution of STAP cont'd

Year Milestones	Author(s)	Country	Ref.
2000 UESA program	Zatman	USA	[741]
2001 KASSPER project	Guerci	USA	[229]
2001 Tracking with STAP	Koch and Klemm	Germany	[357]
2003 RADARSAT 2 (first spaceborne GMTI radar)	Meisl et al. Thompson and Livingstone	Canada Canada	[447] [643]
2003 Textbook on STAP	Guerci	USA	[612]
2004 Textbook Applications of STAP	Klemm ed.	Germany	[350]

1.2 Radar signal processing tools

In the following Sections of this chapter a few well-known signal processing tools which will be used in this book are summarised.

1.2.1 The optimum processor

The principle of detecting a signal vector s before a noisy background given by q is briefly summarised. Let us define the following complex vector quantities:

$$\mathbf{q} = \begin{pmatrix} q_1 \\ q_2 \\ \vdots \\ q_N \end{pmatrix} ; \quad \mathbf{s} = \begin{pmatrix} s_1 \\ s_2 \\ \vdots \\ s_N \end{pmatrix} ; \quad \mathbf{x} = \begin{pmatrix} x_1 \\ x_2 \\ \vdots \\ x_N \end{pmatrix} \quad (1.1)$$

We assume that the noise interference vector q is a zero mean random vector with a multivariate complex gaussian distribution.[7] In general the noise vector consists of a correlated part c (e.g., jammer, clutter) and an uncorrelated part n (e.g., receiver noise):

$$\mathbf{q} = \mathbf{c} + \mathbf{n} \quad (1.2)$$

The signal vector s is assumed to be deterministic. x is the actual data vector which may be noise only ($\mathbf{x} = \mathbf{q}$) or signal-plus-noise ($\mathbf{x} = \mathbf{q} + \mathbf{s}$). The problem of extracting s optimally out of the background noise q is solved by applying the well-known linear weighting

$$\mathbf{w}_{\mathrm{opt}} = \gamma \mathbf{Q}^{-1} \mathbf{s} \quad (1.3)$$

---

[7]This assumption is verified by the following consideration. Let the clutter echo contributions of individual scatterers on the ground be statistically independent, with arbitrary distributions. Since the radar beam integrates over a large number of scatterers the beam output echo becomes approximately gaussian according to the central limit theorem of statistics.

to the received signal vector **x**. $\mathbf{Q} = E\{\mathbf{qq}^*\}$ is the noise covariance matrix

$$\mathbf{Q} = \begin{pmatrix} q_{11} & q_{12} & \cdots & q_{1N} \\ q_{21} & q_{22} & \cdots & q_{2N} \\ \vdots & \vdots & \ddots & \vdots \\ q_{N1} & q_{12} & \cdots & q_{NN} \end{pmatrix} \quad (1.4)$$

The asterisk means complex conjugate transpose. The matrix $\mathbf{Q}$ is hermitian, i.e., $q_{ik} = q_{ki}^*$ or $\mathbf{Q} = \mathbf{Q}^*$. In addition we may define a covariance matrix of signal + noise:

$$\mathbf{R} = E\{(\mathbf{s}+\mathbf{q})(\mathbf{s}+\mathbf{q})^*\} \quad (1.5)$$

The linear processor given by (1.3) is optimal under several criteria:

- **Maximum likelihood performance measure.** Given a signal vector of the form $\mathbf{s}(t) = s(t)\mathbf{v}$ where $s(t)$ is the scalar temporal part and $\mathbf{v}$ the vector valued spatial part, then $\mathbf{w}_{\mathrm{ML}}$ maximises the likelihood function of $\mathbf{x} = \mathbf{s}(t) + \mathbf{q}$ with respect to $s(t)$. The constant in (1.3) becomes $\gamma = (\mathbf{v}^*\mathbf{R}^{-1}\mathbf{v})^{-1}$ (MONZINGO and MILLER [466, p. 100]). $\mathbf{w}_{\mathrm{ML}}$ also maximises the likelihood ratio of the *signal + noise* versus *noise only* hypotheses, see HUDSON [275, p. 75] for array processing, and SCHLEHER [589, pp. 587] for MTI applications.

- **Maximum signal-to-noise power ratio** (SNR). Generally the optimum weighting is given by the eigenvector associated with the maximum eigenvalue of the matrix $\mathbf{B} = \mathbf{Q}^{-\frac{1}{2}}\mathbf{ss}^*\mathbf{Q}^{-\frac{1}{2}}$. Since **s** is deterministic the optimum weighting becomes $\mathbf{w}_{\mathrm{SNR}} = \mathbf{Q}^{-1}\mathbf{s}$ (HUDSON [275, p. 60]). The constant γ has no influence on the SNR and can be chosen arbitrarily. Among other optimisation problems this result has been presented by LO Y. T. *et al.* [411].

- **Linearly constrained minimum noise variance.** The output power is minimised subject to the constraint $\mathbf{w}^*\mathbf{s} = g$. The constraint is necessary to avoid the trivial solution $\mathbf{w} = \mathbf{0}$. The constant γ in (1.3) becomes $g/(\mathbf{s}^*\mathbf{Q}^{-1}\mathbf{s})$ (HUDSON [275, p. 70], MONZINGO and MILLER [466, p. 101]).

- **Least mean square error criterion.** The data vector **x** is weighted so that the squared difference between the output and a known signal $s(t)$ is minimised. The result is

$$\mathbf{w}_{\mathrm{LMSE}} = \mathbf{R}^{-1}\mathbf{v} \quad (1.6)$$

where

$$\mathbf{v} = E\{\mathbf{x}s^*(t)\} \quad (1.7)$$

is the cross-correlation between the data vector and the desired signal. Equations (1.6) and (1.7) have the same form as (1.3), except that here: 1. the signal is included in the covariance matrix, 2. for the signal reference the cross-correlation has to be used (HUDSON [275, p. 64], MONZINGO and MILLER [466, p. 90]). In many practical applications the cross-correlation will be identical with the desired signal vector so that (1.6) becomes identical with (1.3). Equations (1.6) and (1.7) are known as the *Wiener* filter.

A more general form of the optimum processor is given by

$$\mathbf{w}_{\text{opt}} = \mathbf{Q}^{-1}\mathbf{C}(\mathbf{C}^*\mathbf{Q}^{-1}\mathbf{C})^{-1}\mathbf{f} \tag{1.8}$$

where $\mathbf{C}$ is a constraint matrix which imposes linear constraints on the processor

$$\mathbf{C}^*\mathbf{w}_{\text{opt}} = \mathbf{f} \tag{1.9}$$

In the simplest case the constraint matrix reduces to a signal vector so that the constraint equation becomes

$$\mathbf{s}^*\mathbf{w}_{\text{opt}} = 1 \tag{1.10}$$

This constraint takes care that the maximum signal energy is preserved while the remaining part of the covariance matrix is minimised. Notice that by applying this constraint the processor (1.8, 1.9) reduces to (1.3) except for a normalisation factor.

The output of the linear processor (1.3) is a scalar given by

$$y = \mathbf{x}^*\mathbf{w} = \mathbf{x}^*\mathbf{Q}^{-1}\mathbf{s} \tag{1.11}$$

The output power is

$$E\{yy^*\} = E\{\mathbf{w}^*\mathbf{x}\mathbf{x}^*\mathbf{w}\} \tag{1.12}$$

Specifically for signal and noise one gets

$$P_s = \mathbf{w}^*\mathbf{s}\mathbf{s}^*\mathbf{w}; \qquad P_n = \mathbf{w}^*\mathbf{Q}\mathbf{w} \tag{1.13}$$

The signal-to-noise power ratio becomes

$$\text{SNR} = \frac{P_s}{P_n} = \frac{\mathbf{w}^*\mathbf{s}\mathbf{s}^*\mathbf{w}}{\mathbf{w}^*\mathbf{Q}\mathbf{w}} \tag{1.14}$$

The efficiency of any linear processor $\mathbf{w}$ can be characterised by the improvement factor[8] which is defined as the ratio of signal-to-noise power ratios at output and input, respectively

$$\text{IF} = \frac{\frac{P_s^{\text{out}}}{P_n^{\text{out}}}}{\frac{P_s^{\text{in}}}{P_n^{\text{in}}}} = \frac{\frac{\mathbf{w}^*\mathbf{s}\mathbf{s}^*\mathbf{w}}{\mathbf{w}^*\mathbf{Q}\mathbf{w}}}{\frac{\mathbf{s}^*\mathbf{s}}{\text{tr}(\mathbf{Q})}} = \frac{\mathbf{w}^*\mathbf{s}\mathbf{s}^*\mathbf{w} \cdot \text{tr}(\mathbf{Q})}{\mathbf{w}^*\mathbf{Q}\mathbf{w} \cdot \mathbf{s}^*\mathbf{s}} \tag{1.15}$$

Specifically for the optimum processor $\mathbf{w}_{\text{opt}} = \mathbf{Q}^{-1}\mathbf{s}$, (1.3),[9] one gets

$$\text{IF}_{\text{opt}} = \frac{\mathbf{s}^*\mathbf{Q}^{-1}\mathbf{s}\mathbf{s}^*\mathbf{Q}^{-1}\mathbf{s} \cdot \text{tr}(\mathbf{Q})}{\mathbf{s}^*\mathbf{Q}^{-1}\mathbf{Q}\mathbf{Q}^{-1}\mathbf{s} \cdot \mathbf{s}^*\mathbf{s}} = \mathbf{s}^*\mathbf{Q}^{-1}\mathbf{s} \cdot \frac{\text{tr}(\mathbf{Q})}{\mathbf{s}^*\mathbf{s}} \tag{1.16}$$

---

[8] The expression *improvement factor* is commonly used for characterising *temporal* (i.e., pulse-to-pulse) filters for clutter rejection. The same formula may be used for *spatial* applications (array processing) in the context of interference or jammer suppression. Then instead of *improvement factor* the term *gain* is used. Since our main objective is clutter rejection we prefer the term *improvement factor* or its abbreviation IF.

[9] Here we assume that the processor is perfectly matched to the expected target signal. In practice some losses occur due to the finite number of scan positions of the search beam (spatial application) or a finite set of Doppler filters (temporal).

20 Introduction

In (1.3) we assumed that the signal vector **s** is entirely known. In practical applications, however, several parameters (phase, angle of arrival and Doppler frequency) of the target are unknown. Therefore, the linear operation (1.11) is replaced by

$$y = \max_{\Theta} \left| \mathbf{x}^* \mathbf{Q}^{-1} \mathbf{s}(\Theta) \right|^2 \tag{1.17}$$

where Θ is a vector which includes the unknown parameters, and $\mathbf{s}(\Theta)$ is the *steering vector*. Accordingly, the improvement factor (1.15) becomes a function of Θ

$$\mathrm{IF}(\Theta) = \frac{\mathbf{w}^*(\Theta)\mathbf{s}(\Theta)\mathbf{s}^*(\Theta)\mathbf{w}(\Theta) \cdot \mathrm{tr}(\mathbf{Q})}{\mathbf{w}^*(\Theta)\mathbf{Q}\mathbf{w}(\Theta) \cdot \mathbf{s}^*(\Theta)\mathbf{s}(\Theta)} \tag{1.18}$$

where we assumed that the processor **w** is matched to the signal vector **s**. For the optimum processor (1.3) one obtains

$$\mathrm{IF}_{\mathrm{opt}}(\Theta) = \mathbf{s}^*(\Theta)\mathbf{Q}^{-1}\mathbf{s}(\Theta) \cdot \frac{\mathrm{tr}(\mathbf{Q})}{\mathbf{s}^*(\Theta)\mathbf{s}(\Theta)} \tag{1.19}$$

This form of the improvement factor takes individual target parameters into account. Plots showing the improvement factor versus the Doppler frequency will play a major role in the following chapters.

In many of the following plots the improvement factor as described by (1.18, 1.19) is replaced by the improvement factor normalised to the theoretical maximum $IF_{max} = IF/(CNR \cdot N \cdot M)$ (see, for instance, Figure 1.3). This quantity is very useful to compare the clutter suppression performance of different STAP processors[10].

Notice that the optimum improvement factor after (1.19) is just the Rayleigh quotient of $\mathbf{Q}^{-1}$, multiplied by $\mathrm{tr}(\mathbf{Q})$. Therefore, the bounds for the optimum improvement factor are

$$\frac{1}{\lambda_{\max}} \mathrm{tr}(\mathbf{Q}) \leq \mathrm{IF}_{\mathrm{opt}} \leq \frac{1}{\lambda_{\min}} \mathrm{tr}(\mathbf{Q}) \tag{1.20}$$

where $\lambda_{\min}$ and $\lambda_{\max}$ are the minimum and maximum eigenvalues of $\mathbf{Q}$. For Toeplitz matrices as occur in the context of stationary processes a lower bound of $\lambda_{\min}$ has been found by MA and ZAROWSKI [422].

It is also possible to optimise the improvement factor by averaging over all unknown target parameter values, i.e., all possible values of Θ. In the book by SCHLEHER [589, p. 295] the signal is represented by a covariance matrix which includes all possible Doppler frequencies. The covariance matrix turns out to become the identity matrix, and the improvement factor is simply $1/\lambda_{\min}$ where $\lambda_{\min}$ is the smallest eigenvalue of the noise covariance matrix $\mathbf{Q}$. In this way the performance of a processor is characterised by just a single number. Since we are very much interested in the dependence of the IF on the target parameters (to analyse for instance the response to slowly moving targets) we use the form given in (1.18).

A similar optimisation as given in [589, p. 295] has been applied to space-time data received by a moving radar by PARK et al. [528] and SUNG et al. [624].

---

[10]The normalised IF is in the literature frequently referred to as 'loss factor'. Some other authors use the quotient IF/IF_{opt} as loss factor. This version, however, does not show the clutter notch of the processor under consideration and, therefore, is not used here

1.2.1.1 Some properties of the optimum processor

Let us discuss briefly what the physical function of the optimum processor is. The role of **s** in the processor $\mathbf{Q}^{-1}\mathbf{s}$ is a filter matched to the desired signal. In case of white noise we have $\mathbf{Q} = \mathbf{I}$ so that the SNR after (1.14), becomes

$$\text{SNR} = \frac{P_s}{P_n} = \frac{\mathbf{s}^*\mathbf{s}\mathbf{s}^*\mathbf{s}}{\mathbf{s}^*\mathbf{s}} = \mathbf{s}^*\mathbf{s} = N \tag{1.21}$$

if the signal amplitude is unity. The role of $\mathbf{Q}^{-1}$ in (1.11) can easily be explained. Since **Q** is positive semi-definite it can be factorised into two matrices

$$\mathbf{Q} = \mathbf{D}^*\mathbf{D} \tag{1.22}$$

This factorisation can be carried out for example by the Cholesky algorithm in which case **D** is an upper triangular matrix. Any uniform transform **UD** satisfies (1.22) as well. Inserting (1.22) into (1.11) gives

$$y = \mathbf{x}^*\mathbf{w} = \mathbf{x}^*\mathbf{D}^{-1}\mathbf{D}^{*-1}\mathbf{s} \tag{1.23}$$

The product $\mathbf{x}^*\mathbf{D}^{-1}$ whitens the noise component **q** in **x**:

$$E\{\mathbf{D}^{*-1}\mathbf{q}\mathbf{q}^*\mathbf{D}^{-1}\} = \mathbf{D}^{*-1}\mathbf{Q}\mathbf{D}^{-1} = \mathbf{D}^{*-1}\mathbf{D}^*\mathbf{D}\mathbf{D}^{-1} = \mathbf{I} \tag{1.24}$$

where **I** is the identity matrix.

The product $\mathbf{D}^{*-1}\mathbf{s}$ means a pre-distortion of the matched filter vector **s** by a linear transform $\mathbf{D}^{*-1}$. This transform ensures that the matched filter vector **s** is in fact matched to a potential signal component in the transformed data vector $\mathbf{x}^*\mathbf{D}^{-1}$.

The optimum processor (1.11) is based on the noise covariance matrix **Q**. In practice the processor will be based on an estimate of **Q** or some iterative algorithm that approximates $\mathbf{Q}^{-1}$. Since such a processor is adapted to the noise we call it an *adaptive processor*, even if no adaptive algorithm is under discussion.

1.2.1.2 Simulation of interference data

The factorised covariance matrix $\mathbf{Q} = \mathbf{D}^*\mathbf{D}$ offers a way of simulation of interference data whose statistical properties are given by **Q**

$$\mathbf{z} = \mathbf{D}^*\mathbf{n} \tag{1.25}$$

where **n** is a vector of complex independent gaussian numbers with covariance matrix **I**. The covariance matrix of the transformed gaussian vector is

$$E\{\mathbf{z}\mathbf{z}^*\} = E\{\mathbf{D}^*\mathbf{n}\mathbf{n}^*\mathbf{D}\} = \mathbf{D}^*\mathbf{D} = \mathbf{Q} \tag{1.26}$$

It has been pointed out by BRENNAN and MALLETT [64] that for simulating an *infinite* number of noise sources distributed in space (continuous noise) radiating on an array only N complex gaussian numbers have to generated.

1.2.1.3 Examples

Consider a simple noise model of the form $\mathbf{q} = \mathbf{c} + \mathbf{n}$ according to (1.2), where $\mathbf{c}$ consists of a deterministic part and a complex random amplitude A_c. The vector $\mathbf{c}$ may stand for interference while $\mathbf{n}$ denotes uncorrelated receiver noise.

$$\mathbf{c} = A_c \begin{pmatrix} 1 \\ 1 \\ \vdots \\ 1 \end{pmatrix}; \quad \mathbf{C} = P_c \begin{pmatrix} 1 & 1 & \cdots & 1 \\ 1 & 1 & \cdots & 1 \\ \vdots & \vdots & \ddots & \vdots \\ 1 & 1 & \cdots & 1 \end{pmatrix}; \quad \mathbf{N} = E\{\mathbf{nn}^*\} = P_w \mathbf{I} \quad (1.27)$$

$P_c = E\{A_c A_c^*\}$ is the power of the interference and P_w the white noise power. After divison by P_c the normalised covariance matrix becomes

$$\mathbf{Q} = (\mathbf{C} + P_w \mathbf{I})/P_c = \begin{pmatrix} 1+a & 1 & \cdots & 1 \\ 1 & 1+a & \cdots & 1 \\ \vdots & \vdots & \ddots & \vdots \\ 1 & 1 & \cdots & 1+a \end{pmatrix} \quad (1.28)$$

where $a = P_w/P_c$ is the noise-to-interference power ratio (NIR). Equivalently we have

$$a = \frac{1}{\text{INR}} \quad (1.29)$$

or, in the case of clutter

$$a = \frac{1}{\text{CNR}} \quad (1.30)$$

where CNR is the clutter-to-noise ratio and INR the interference-to-noise ratio. The eigenvalues of $\mathbf{Q}$ are

$$\lambda_1 = N + a, \quad \lambda_2, \ldots, \lambda_N = a \quad (1.31)$$

The inverse of $\mathbf{Q}$ is

$$\mathbf{Q}^{-1} = \frac{1}{a(N+a)} \begin{pmatrix} N-1+a & -1 & \cdots & -1 \\ -1 & N-1+a & \cdots & -1 \\ \vdots & \vdots & \ddots & \vdots \\ -1 & -1 & \cdots & N-1+a \end{pmatrix} \quad (1.32)$$

with eigenvalues

$$\lambda_1 = \frac{1}{N+a}, \quad \lambda_2, \ldots, \lambda_N = \frac{1}{a} \quad (1.33)$$

Let us assume a signal vector

$$\mathbf{s} = \begin{pmatrix} \exp(j2\pi F) \\ \exp(j2\pi 2F) \\ \vdots \\ \exp(j2\pi NF) \end{pmatrix} \quad (1.34)$$

where F is a normalised frequency. According to (1.20) the range of the improvement factor is determined by the range of eigenvalues of $\mathbf{Q}$ multiplied by $\text{tr}(\mathbf{Q}) = N(1+a)$:

$$\frac{N(1+a)}{N+a} \le \text{IF}_{\text{opt}} \le \frac{N(1+a)}{a} \tag{1.35}$$

In the noise-free case ($a \to 0$) the minimum IF is 1 while the maximum is infinity. For large noise power ($a \to \infty$) both the minimum and maximum IF assume the value N which is the white noise gain. Notice that in the latter case the IF is independent of the steering parameter F.

Covariance matrices of the form (1.28) play a major role in radar. There are two typical applications:

Spatial domain. The vector space can be used to describe a narrowband sensor array. The interference vector **c** describes an interfering source radiating on the array at $\varphi = 0$ where φ is the angle of arrival ($\varphi = 0$ means *broadside*). The elements of the noise vector denote the uncorrelated noise contributions of the individual receive channels. $\mathbf{Q}^{-1}$ is a *spatial* filter which places a null in the directivity pattern of the array in the broadside direction. The signal vector **s** (1.34) includes the beamformer weights with the normalised *spatial* frequency

$$F_s = \frac{d}{\lambda} \sin \varphi \tag{1.36}$$

where λ is the wavelength and d the sensor spacing.

Temporal domain. The same kind of covariance matrix may occur in adaptive MTI applications. Assume a stationary radar with fixed antenna look direction. In this case the elements of **c** denote successive echo signals due to a stationary non-fluctuating clutter background (zero Doppler frequency). **n** denotes again the receiver noise. $\mathbf{Q}^{-1}$ is a clutter filter that places a null in the Doppler spectrum at zero frequency. For the signal vector (1.34) the normalised Doppler frequency becomes $F = f_d T$ where f_d denotes the Doppler frequency and T the pulse repetition interval.

For example, for $N = 2$ the covariance matrix inverse becomes

$$\mathbf{Q}^{-1} = \frac{1}{a(2+a)} \begin{pmatrix} 1+a & -1 \\ -1 & 1+a \end{pmatrix} \tag{1.37}$$

Normalising with $1/(a(2+a))$ and setting $a = O$ (zero noise) leads to

$$\mathbf{K}_{2\text{pc}} = \begin{pmatrix} 1 & -1 \\ -1 & 1 \end{pmatrix} \tag{1.38}$$

Notice that both columns of $\mathbf{K}_{2\text{pc}}$ contain the familiar weights of a *two-pulse clutter canceller*. For a spatial interpretation these weightings form a dipole directivity pattern with the null steered in the broadside direction. The difference between the two above equations is that the optimum processor including the matrix (1.37) suppresses the interference only down to the noise level (in other words, only as much as necessary)

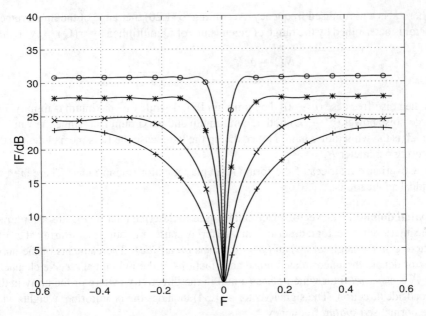

Figure 1.2: Improvement factor (one interfering source at $F = 0$, INR = 20 dB): $+\ N = 2$; $\times\ N = 3$; $*\ N = 6$; $\circ\ N = 12$

while the matrix (1.38) places a perfect null at zero Doppler (or, in the broadside direction):

$$\mathbf{1}^* \begin{pmatrix} 1+a & -1 \\ -1 & 1+a \end{pmatrix} \mathbf{1} = 2a; \quad \mathbf{1}^* \begin{pmatrix} 1 & -1 \\ -1 & 1 \end{pmatrix} \mathbf{1} = 0 \qquad (1.39)$$

The vector $\mathbf{1}^* = (1\ 1)$ on the left side includes interference samples while the vector $\mathbf{1}$ on the right denotes a filter matched to the interference. In this case the improvement factor according to (1.19, 1.34, 1.36) with $\varphi = 0$ becomes

$$\mathrm{IF}_{\mathrm{opt}}(0) = \frac{1}{a(2+a)} \cdot 2a \cdot \frac{2(1+a)}{2} = \frac{2+2a}{2+a} \qquad (1.40)$$

which is close to unity for small values of a. For the two-pulse canceller (right hand side of (1.39)) the improvement factor becomes 0. The optimum processor has a certain capability of target detection in the interference direction while the approximation for $a \to 0$ cancels interference *and* target perfectly. For $N = 3$ echo pulses (or sensors) one gets

$$\mathbf{Q}^{-1} = \frac{1}{a(3+a)} \begin{pmatrix} 2+a & -1 & -1 \\ -1 & 2+a & -1 \\ -1 & -1 & 2+a \end{pmatrix} \qquad (1.41)$$

It can be noticed that for $a \to 0$ the columns of $\mathbf{Q}^{-1}$ approach the well-known three-pulse canceller weightings.

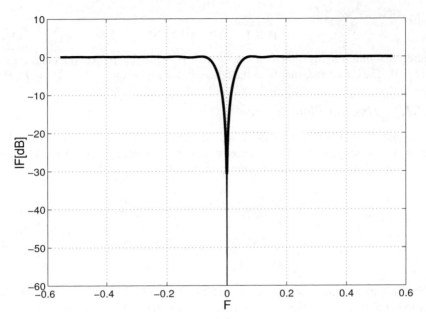

Figure 1.3: Cancellation of an interfering source at $F = 0$ ($N = 12$). Fat: optimum processor; thin: orthogonal projection

In Figure 1.2 the improvement factor for the optimum processor has been plotted versus a normalised (temporal or spatial) frequency F. The interference + noise covariance matrix was chosen according to (1.28) with $a = -20$ dB. The interference is a single narrow line at F = 0. Therefore, at F = 0 all IF curves exhibit a minimum. For $N = 2$ the well-known dipole pattern (or transmission characteristics of a two-pulse canceller) is shown. For higher values of N the interference resolution capability increases so that the notch becomes narrower.

1.2.2 Orthogonal projection

In the following two techniques are described briefly which are based on a projection of the received signal vector orthogonal to the interference vector subspace of $\mathbf{Q}$.

1.2.2.1 Known interference

Consider a covariance matrix of the form

$$\mathbf{Q} = \mathbf{C} + P_\mathrm{w} \mathbf{I} \tag{1.42}$$

with $\mathbf{C} = \mathbf{DD}^*$. $\mathbf{D}$ is a $N \times L$ matrix with $L < N$ whose columns may denote signals due to L independent noise sources radiating on an array (or L frequency components in a discrete signal). If the powers and the positions of the noise sources are known $\mathbf{D}$ is known, and the noise sources can be cancelled by use of the following projection

matrix
$$P = I - D(D^*D)^{-1}D^* \qquad (1.43)$$
Notice that multiplying the interference matrix with the projection matrix gives $PD = 0$. The covariance matrix of the remaining noise becomes $P_w P^* I P = P_w P$.

1.2.2.2 Unknown interference

If the interference is unknown an orthogonal projection technique can be found by eigenvalue decomposition of the interference + noise covariance matrix Q. Q can be expressed by its eigenvalues and eigenvectors as follows

$$\begin{aligned} Q &= \sum_{l=1}^{L} \lambda_l e_l^{(i)} e_l^{(i)*} + \sum_{l=L+1}^{N} \lambda_l e_l^{(n)} e_l^{(n)*} \\ &= E^{(i)} \Lambda^{(i)} E^{(i)*} + E^{(n)} \Lambda^{(n)} E^{(n)*} \qquad (1.44) \\ &= Q^{(i)} + Q^{(n)} \end{aligned}$$

where λ_l are the eigenvalues and e_l the eigenvectors of Q. The first L eigenvalues are due to interference while the eigenvalues $L+1,...,N$ denote the noise component. Λ is a diagonal matrix of eigenvalues while E is a matrix whose columns are eigenvectors of Q. The indices i and n denote 'interference' and 'noise'.

The eigenvectors of Q are mutually orthogonal which means that the operation

$$PD = 0 \qquad (1.45)$$

with

$$P = E^{(n)} E^{(n)*} \qquad (1.46)$$

cancels perfectly the interference while preserving the dimension of the vector space. Equivalently one may use a projection matrix orthogonal to the interference subspace of the form

$$P = I - E^{(i)} (E^{(i)*} E^{(i)})^{-1} E^{(i)*} \qquad (1.47)$$

The use of orthogonal projection techniques based on eigenvalue decomposition is referred to as eigencanceller. For more details on projection techniques for interference cancellation see HAIMOVICH and BAR-NESS [236], GIERULL [188], and LEE and LEE [384], HAIMOVICH and BERIN [240].

1.2.2.3 Examples

Consider again the simple example given in (1.37). The interference component is given by a vector

$$d = \begin{pmatrix} 1 \\ 1 \end{pmatrix} \qquad (1.48)$$

The projection matrix according to (1.47) becomes

$$P = \begin{pmatrix} 1 & 0 \\ 0 & 1 \end{pmatrix} - \frac{1}{2} \begin{pmatrix} 1 & 1 \\ 1 & 1 \end{pmatrix} = \frac{1}{2} \begin{pmatrix} 1 & -1 \\ -1 & 1 \end{pmatrix} \qquad (1.49)$$

As can be seen by comparison with (1.38) this projection matrix is proportional to $\mathbf{Q}^{-1}$ for the zero noise case.

The eigendecomposition of the covariance matrix in (1.37) gives two eigenvectors

$$\mathbf{e}_1 = \begin{pmatrix} 1 \\ 1 \end{pmatrix} \text{ and } \mathbf{e}_2 = \begin{pmatrix} -1 \\ 1 \end{pmatrix} \tag{1.50}$$

and the associated eigenvalues $\lambda_1 = 2 + a$, $\lambda_2 = a$. The index '1' denotes interference + noise while index '2' means noise only. The projection matrix after (1.46) becomes

$$\mathbf{P} = \begin{pmatrix} -1 \\ 1 \end{pmatrix} \begin{pmatrix} -1 & 1 \end{pmatrix} = \begin{pmatrix} 1 & -1 \\ -1 & 1 \end{pmatrix} \tag{1.51}$$

For this simple example both projection methods lead to the same result (except for a factor of 2).

1.2.2.4 Comparison with optimum processing

In Figure 1.3 the optimum and orthogonal projection type processors (OPP) are compared. As can be seen the nulls formed by orthogonal projection are much deeper.[11] In other words, the optimum processor still achieves some improvement in SNIR in the interference direction.

It can easily be verified that the minimum possible IF occurs if the steering vector is matched to the interference (if, for example, a beamformer points in the jammer direction). Consider again an interference + noise covariance matrix of the form (1.28). By setting $\mathbf{s} = \mathbf{e}^{(i)}$ (where $\mathbf{e}^{(i)}$ is the interference eigenvector of $\mathbf{Q}$) the improvement factor is calculated according to (1.19), using the eigendecomposition (1.44). The result is equal to the lower bound on the IF as given in (1.35).

For the orthogonal projection technique the processor becomes, according to (1.46)

$$\mathbf{w}_{\text{OP}} = \mathbf{E}^{(n)} \mathbf{E}^{(n)*} \mathbf{s} \tag{1.52}$$

which, by virtue of the orthogonality of different eigenvectors, becomes a zero vector if the steering vector is matched to the interference, i.e., if $\mathbf{s} = \mathbf{e}^{(i)}$.

If there is more than one interfering source the optimum IF in the interference directions becomes larger as the number of sources increases.

1.2.3 Linear subspace transforms

The following subspace techniques have traditionally been used for cancellation of interfering sources radiating on a sensor array. The optimum processor discussed before involves the inversion of a covariance matrix which may become a cumbersome procedure for large sample size. On-line and real-time processing might be difficult to achieve. The idea of subspace techniques is to reduce the vector space given by

---
[11] Ideally they should be zero, but this requires that the nulls are perfectly met by a point on the abszissa.

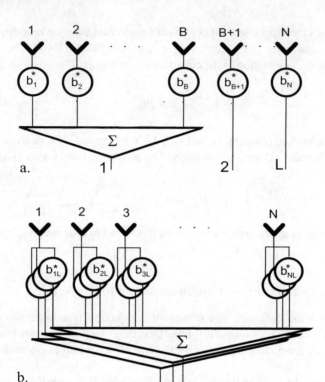

Figure 1.4: Antenna configurations: a. Beamformer and auxiliary elements, b. multiple weightings

the array size down to a suitable subspace and perform interference rejection at the subspace level. This can be done by a suitable linear transform $\mathbf{T}$

$$\mathbf{q}_T = \mathbf{T}^*\mathbf{q}; \quad \mathbf{s}_T = \mathbf{T}^*\mathbf{s}; \quad \mathbf{x}_T = \mathbf{T}^*\mathbf{x}; \quad \mathbf{Q}_T = \mathbf{T}^*\mathbf{Q}\mathbf{T} \qquad (1.53)$$

The optimum processor in the transformed domain becomes, according to (1.3)

$$\mathbf{w}_T = \gamma \mathbf{Q}_T^{-1} \mathbf{s}_T \qquad (1.54)$$

The improvement factor, according to (1.15), becomes

$$\mathrm{IF} = \frac{\frac{\mathbf{w}_T^* \mathbf{s}_T \mathbf{s}_T^* \mathbf{w}_T}{\mathbf{w}_T^* \mathbf{Q}_T \mathbf{w}_T}}{\frac{\mathbf{s}^* \mathbf{s}}{\mathrm{tr}(\mathbf{Q})}} = \frac{\mathbf{w}_T^* \mathbf{s}_T \mathbf{s}_T^* \mathbf{w}_T \cdot \mathrm{tr}(\mathbf{Q})}{\mathbf{w}_T^* \mathbf{Q}_T \mathbf{w}_T \cdot \mathbf{s}^* \mathbf{s}} \qquad (1.55)$$

The achievable improvement factor depends on the kind of transform $\mathbf{T}$. There are a few criteria which are essential to achieve near-optimum interference suppression performance:

1. No signal energy must be lost. This means that $\mathbf{T}$ should contain a beamformer weighting or parts of it that are matched to the expected signal $\mathbf{s}$, usually called

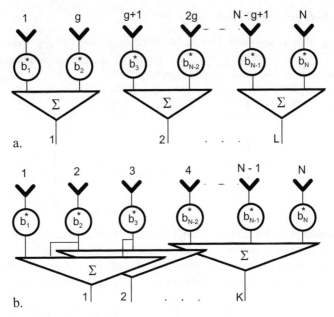

Figure 1.5: Antenna configurations: a. Disjoint subgroups; b. overlapping subgroups

a *search beam*. This means that in a search mode $\mathbf{T}$ depends on the signal direction. Adaptation of $\mathbf{Q}_T^{-1}$ has to be conducted separately for all individual directions.

2. There should be additional *auxiliary* weightings to estimate the power and direction of the interference (reference channels).

3. The interference-to-noise ratio (INR) in the auxiliary channels should be as large as in the search beam.

4. The number of channels L (dimension of the transformed vector space) should be small compared with the number of antenna array elements so that significant saving in processing power is achieved.

5. The number of channels L should be at least as large as the number of *interference* eigenvalues of $\mathbf{Q}$. This means technically that the number of coefficients of the interference suppression filter given by $\mathbf{Q}_T^{-1}$ has to be matched to the number of degrees of freedom of the interfering scenario.

6. The columns of the transform have to be chosen so that $\mathbf{T}$ is regular. Otherwise, $\mathbf{Q}_T^{-1}$ does not exist. However, problems arising with the inversion of ill-conditioned covariance matrices can be mitigated by adding *artificial noise* (diagonal loading).

In Figures 1.4 and 1.5 four typical kinds of pre-transform are depicted.

1.2.3.1 Sidelobe canceller

Figure 1.4a shows the well-known sidelobe canceller which may consist of a beamformer and additional auxiliary elements as reference channels. This configuration can be described by a transform

$$\mathbf{T} = \begin{pmatrix} b_1 & & & & & \\ \vdots & & & & 0 & \\ b_B & & & & & \\ & b_{B+1} & & & & \\ & & b_{B+2} & & & \\ & & & \ddots & & \\ & 0 & & & & b_{B+N} \end{pmatrix} \quad (1.56)$$

where the entries b_i are the weighting coefficients of a beamformer and B is the number of sensors used for beamforming. These weights may be phase only or include some taper function. This structure is followed by the adaptive algorithm (inverse of the clutter+noise covariance matrix). Therefore, the diagonal coefficients $b_{B+1} \ldots b_{B+N}$ can be set to one.

To illustrate the function of the sidelobe canceller consider the following special case (one search beam, one auxiliary element), $L = N + 1 - B = 2$,

$$\mathbf{T} = \begin{pmatrix} b_1 & 0 \\ \vdots & \vdots \\ b_B & 0 \\ 0 & 1 \end{pmatrix} \quad (1.57)$$

For interference and noise let us assume again the simple covariance model according to (1.28). Let us further assume that the beamformer weights are phase only. After transformation of the normalised covariance matrix with (1.57) we get

$$\mathbf{Q}_T = \mathbf{T}^* \mathbf{Q} \mathbf{T} = \begin{pmatrix} \mathbf{b}^* \mathbf{1} \mathbf{1}^* \mathbf{b} + Ba & \mathbf{b}^* \mathbf{1} \\ \mathbf{1}^* \mathbf{b} & 1 + a_{\text{RE}} \end{pmatrix} \quad (1.58)$$

where $\mathbf{1}^* = (1, 1 \ldots 1)$ and a_{RE} is the noise-to-interference ratio in the reference channel. While in (1.28) the INR was equal in all channels the INR's in the beam and the reference element are now different. Obviously the INR in the beam depends on the scalar product $\mathbf{b}^* \mathbf{1}$ which denotes the mismatch between beamformer and interference direction, i.e., on the directivity pattern of the array beam. If the interference meets a null of the directivity pattern ($\mathbf{b}^* \mathbf{1} = 0$) the INR in the reference channel becomes larger than in the beam. If the interference meets a sidelobe maximum or even the main beam the INR may be larger in the beam channel.

In Figure 1.6 the effect of different noise levels in the reference element is illustrated. It can be noticed that losses in SNIR (signal-to-noise + interference ratio) occur wherever the INR in the reference element is lower than at the beamformer output. This results in some sidelobe ripple of the improvement factor curves which is stronger the noisier the reference channel is. The effect shown reflects the requirement for sufficient INR in the reference channels according to criterion 2. (see above).

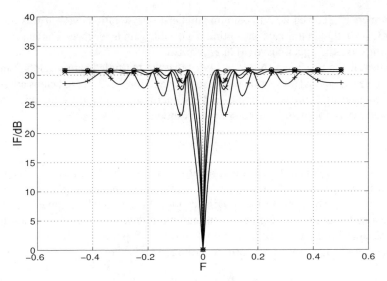

Figure 1.6: Sidelobe canceller with one reference element (one interfering source at $F = 0$, INR = 20 dB ($a = 0.01$)). ○ Optimum processor; ∗ SLC with $a_{\text{RE}} = 0.01$ at the reference element; × SLC with $a_{\text{RE}} = 0.02$; + SLC with $a_{\text{RE}} = 0.1$

1.2.3.2 Multiple beams

Multiple receive beams are formed for example in the OLPI.[12] radar (WIRTH [720, 718]) which uses an omnidirectional antenna on transmit and a multibeam antenna on receive for simultaneous search in 64 azimuth directions.

Figure 1.4b shows a multiple beam weighting transform

$$\mathbf{T} = \begin{pmatrix} b_{11} & b_{12} & \cdots & b_{1L} \\ b_{21} & b_{22} & \cdots & b_{2L} \\ \vdots & \vdots & \ddots & \vdots \\ b_{N1} & b_{N2} & \cdots & b_{NL} \end{pmatrix} = \begin{pmatrix} \mathbf{b}_1 & \mathbf{b}_2 & \cdots & \mathbf{b}_L \end{pmatrix} = \begin{pmatrix} \mathbf{b} & \mathbf{B} \end{pmatrix} \quad (1.59)$$

The related matrix structure is given by (1.59) where the vectors $\mathbf{b}_i$ include the weighting coefficients. According to criterion 1, given in Section 1.2.3, one of the weightings should form a search beam while the others take care of the interference. In the right hand part of (1.59) **b** denotes a search beam while **B** includes all the auxiliary beams. In principle each of the columns of **T** may assume the role of a search beam.

Assume a situation with $L - 1$ jammers radiating on an N-element array. Let $\mathbf{b}_1$ form a search beam and $\mathbf{b}_2, \ldots, \mathbf{b}_L$ be a bunch of *auxiliary* or *reference* beams. If the jammer directions are known the reference beams may be steered into the jammer directions. Obviously, by steering $L - 1$ reference beams in the jammer directions the resulting vector space is just the interference subspace of **Q**. Moreover, each of

---

[12]OLPI = Omnidirectional Low Probability of Intercept. In this experimental radar beamforming is accomplished by a 64 point Butler matrix.

the jammers is received with maximum jammer-to-noise ratio (JNR). It was found by KLEMM [290] that in this case the optimum possible gain is reached. The improvement factor runs like the upper curve (o) in Figure 1.6.

In fact, it can be shown (NICKEL [490]) that the multiple beam sidelobe canceller leads to the orthogonal projection processor (see Section 1.2.2) if the beams are steered on the interfering sources. A multisource scenario as observed by the array can be described by the following covariance matrix model

$$\mathbf{Q} = \mathbf{AWA}^* + \mathbf{I} \quad (1.60)$$

where $\mathbf{W}$ is the intersource covariance matrix, $\mathbf{A}$ contains the source–antenna phase transform, and $\mathbf{I}$ is the covariance matrix of white receiver noise. Applying the sidelobe canceller transform (1.59) to the covariance model (1.60) leads to quite complicated expressions for the matrix elements. Instead we follow here the formulation for the sidelobe canceller as has been given by HUDSON [275, pp. 64].

The output signals of a transform according to (1.59) are

$$y = \mathbf{b}^*\mathbf{x} \qquad \mathbf{z} = \mathbf{B}^*\mathbf{x} \quad (1.61)$$

where y is the search beam response and $\mathbf{z}$ the auxiliary channel response. The sidelobe canceller weight vector becomes according to (1.6, 1.7),

$$\mathbf{w}_{\text{slc}} = E\{\mathbf{zz}^*\}^{-1}E\{\mathbf{z}y^*\} = (\mathbf{B}^*\mathbf{QB})^{-1}\mathbf{B}^*\mathbf{Qb} \quad (1.62)$$

Let us assume that the auxiliary beams are steered on the interfering sources. This means that in (1.60)

$$\mathbf{B} = \mathbf{A} \quad (1.63)$$

Inserting this relation into (1.60) and (1.62) gives

$$\begin{aligned}\mathbf{w}_{\text{slc}} &= (\mathbf{B}^*\mathbf{B} + \mathbf{B}^*\mathbf{BWB}^*\mathbf{B})^{-1}(\mathbf{B}^* + \mathbf{B}^*\mathbf{BWB}^*)\mathbf{b} \\ &= (\mathbf{B}^*\mathbf{B})^{-1}(\mathbf{I} + \mathbf{B}^*\mathbf{BW})^{-1}(\mathbf{I} + \mathbf{B}^*\mathbf{BW})\mathbf{B}^*\mathbf{b}\end{aligned} \quad (1.64)$$

so that

$$\mathbf{w}_{\text{slc}} = (\mathbf{B}^*\mathbf{B})^{-1}\mathbf{B}^*\mathbf{b} \quad (1.65)$$

and

$$y - \mathbf{w}_{\text{slc}}^*\mathbf{x} = \mathbf{b}^*(\mathbf{I} - \mathbf{B}(\mathbf{B}^*\mathbf{B})^{-1}\mathbf{B}^*)\mathbf{x} \quad (1.66)$$

In other words, if the auxiliary channels given by $\mathbf{B}$ are perfectly matched to the interference then the sidelobe canceller (condition (1.63)) becomes identical to the orthogonal projection processor treated in Section 1.2.2.

If the reference beams are steered in arbitrary directions the jammers will be received through the sidelobes of the reference beams. This causes losses in the signal-to-noise plus interference power ratio (SNIR) whenever the INR in the search beam exceeds the INR in the reference sidelobe. This happens in particular if a jammer falls into the search main beam or a major sidelobe while meeting a null in the pattern of a reference beam. Consequently, losses in INR will occur. If the reference beams move together with the search beam (i.e., they are all used as search beams) then the resulting

IF pattern is determined by the sidelobes of all beams which leads to an irregular IF loss pattern.

The interference subspace may be obtained in a more rigorous way by eigenvalue decomposition of the jammer + noise covariance matrix $\mathbf{Q}$

$$\mathbf{Q} = \sum_{l=1}^{L-1} \lambda_l^{(i)} \mathbf{e}_l^{(i)} \mathbf{e}_l^{(i)*} + \sum_{l=L}^{N} \lambda_l^{(n)} \mathbf{e}_l^{(n)} \mathbf{e}_l^{(n)*} = \mathbf{E}^{(i)} \mathbf{\Lambda}^{(i)} \mathbf{E}^{(i)*} + \lambda^{(n)} \mathbf{I} \qquad (1.67)$$

where λ_l are the eigenvalues and $\mathbf{e}_l$ the associated eigenvectors of $\mathbf{Q}$. The superscripts (i) and (n) denote 'interference' and 'noise'. The sum on the left describes the jammer subspace while the second term consists of noise eigenvalues and eigenvectors. The transform matrix is then obtained by replacing $\mathbf{B}$ in (1.59) by the eigenvector matrix

$$\mathbf{T} = \begin{pmatrix} \mathbf{b} & \mathbf{E}^{(i)} \end{pmatrix} \qquad (1.68)$$

Since (1.67) has the same form as (1.60) the conclusion given by (1.66) holds for a sidelobe canceller with eigenvectors as auxiliary channels as well, which means the eigenvector sidelobe canceller is identical to the orthogonal projection processing described in Section 1.2.2.

If the positions of the interfering sources are unknown[13] then the eigenchannels have to be calculated adaptively based on the estimated covariance matrix. For this application the eigenvector processor is only of theoretical value because the eigenvalue decomposition of $\mathbf{Q}$ is more complex than its inversion.

1.2.3.3 Disjoint subgroups

For the disjoint subgroup arrangement in Figure 1.5a the transform becomes

$$\mathbf{T} = \begin{pmatrix} b_1 & & & & & \\ \vdots & & & 0 & & \\ b_g & & & & & \\ & b_{g+1} & & & & \\ & \vdots & & & & \\ & b_{2g} & & \ddots & & \\ & & & & b_{N-g+1} & \\ & & 0 & & \vdots & \\ & & & & b_N & \end{pmatrix} \qquad (1.69)$$

As can be seen from (1.69) all subgroups are weighted with segments of a beamformer vector $\mathbf{b}$ so that their main beams are all steered in the same direction. If the array is linear as suggested by Figure 1.4b and has constant inter–element spacing of $\lambda/2$ (λ = wavelength) then the phase centres of the subgroups are displaced by $g\lambda/2$. The use of

---

[13] It is the task of the adaptive processor to cope with unknown source directions.

subgroup concepts for jammer suppression in planar phased array antennas has been discussed by NICKEL [484, 489]. Random disjoint subgroups have been implemented in the ELRA system,[14] see GRÖGER et al. [216].

1.2.3.4 Overlapping subgroups

Overlapping subgroups as shown in Figure 1.5b can be designed in such a way that the transformed array has again $\lambda/2$-spacing like the original array. The transform becomes

$$\mathbf{T} = \begin{pmatrix} b_1 & & & & & & 0 \\ \cdot & b_2 & & & & & \\ \cdot & \cdot & & & & & \\ b_g & \cdot & & & & & \\ & b_{g+1} & & & & & \\ & & \cdot & & & & \\ & & & \ddots & & & \\ & & & & b_{N-g+1} & & \\ & & & & \cdot & & \\ 0 & & & & \cdot & & \\ & & & & & & b_N \end{pmatrix} \tag{1.70}$$

Such an array configuration is unusual in adaptive arrays but plays an important role in suboptimum space-time filters as described in Chapter 6. Overlapping subarrays have been used by LEE and LEE [385] to design a robust subspace based adaptive beamformer.

1.2.4 Clutter suppression with digital filters

Figure 1.7a shows a block diagram of the optimum processor after equation (1.11). The covariance matrix $\mathbf{Q}$ has been factorised into a product of two (eventually triangular) matrices

$$\mathbf{Q}^{-1} = \mathbf{D}^{-1}\mathbf{D}^{*-1} \tag{1.71}$$

$\mathbf{D}^{-1}$ whitens the clutter component in $\mathbf{x}(t)$ while $\mathbf{D}^{*-1}$ equalises the matched filter s. After taking the squared magnitude of the output signal y target detection is done by threshold comparison. As pointed out above, this scheme can basically be used in the space or time domain. In the first case x_n are the output signals of a sensor array, while in the latter one x_n denote temporal samples of an echo data sequence.

Let us assume a stationary sequence of signal samples as typically received by a pulse Doppler radar transmitting at constant PRF. For this special case the whitening operation $\mathbf{x}^*\mathbf{D}^{-1}$ may be replaced by a digital filter whose impulse response has to be convolved with the input signal $x(t)$ (KLEMM [291]). In order to achieve perfect signal match, the signal replica has to be convolved with the whitening filter as well. Such a processor is shown in Figure 1.7b. The whitened input signal is convolved with

---

[14]ELRA = Electronic Radar (experimental ground-based multifunction phased array radar, operating at S-band, developed by FGAN, Germany).

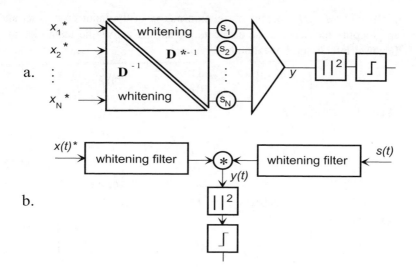

Figure 1.7: Processor schemes: a. optimum processor; b. temporal processor using digital filters

the equalised replica signal to yield the output signal $y(t)$. As in the case of the array processor shown in Figure 1.7a, a potential target signal will be detected by comparing $|y(t)|^2$ with a detection threshold.

By using such pre-whitening filters some simplification of the processor may be achieved. First of all, the order of the digital filter (number of coefficients) can be chosen independently of the length of the data sequence. Of course, by using a short filter the clutter notch will become broader than by using the whitening matrix $\mathbf{D}^{-1}$ which leads to loss in detection of slow targets. The dependence of the clutter notch on the sample size was shown in Figure 1.2. On the other hand, for large sample size N the inversion of $\mathbf{Q}$ becomes very difficult (numerical accuracy, computing time, cost) and unrealistic so that suboptimum approximations have to be considered.

1.2.4.1   FIR filters

Consider an L-element segment $x_t, x_{t-1}, \ldots, x_{t-L+1}$ of a stationary signal sequence x_t. The output of a FIR (finite impulse response) filter with coefficients h_n becomes

$$y_t = \sum_{n=0}^{L-1} h_n{}^* x_{t-n} \qquad (1.72)$$

or, in vector notation,

$$y_t = \mathbf{h}^* \mathbf{x}_t \qquad (1.73)$$

where $\mathbf{h}$ is the vector of filter coefficients and $\mathbf{x}_t$ contains the data segment $x_t, x_{t-1}, \ldots, x_{t-N+1}$. The discrete autocorrelation function of the output process is

$$q_{yy}(\tau) = E\{y_t y_{t+\tau}{}^*\} = \mathbf{h}^* \mathbf{Q}_{xx}(\tau) \mathbf{h} \qquad \tau = -\infty, \ldots, \infty \qquad (1.74)$$

where $\mathbf{Q}_{xx}(0)$ is the $L \times L$ Toeplitz correlation matrix of the input process $\mathbf{x}_t$ and $\mathbf{Q}_{xx}(\tau)$ are Toeplitz matrices whose elements are shifted across the main diagonal towards the lower left corner. For example,

$$\mathbf{Q}_{xx}(0) = \begin{pmatrix} q_{xx}(0) & q_{xx}(1) & \cdots & q_{xx}(L-1) \\ q_{xx}^*(1) & q_{xx}(0) & \cdots & q_{xx}(L-2) \\ \vdots & \vdots & \ddots & \vdots \\ & & & q_{xx}(1) \\ q_{xx}^*(L-1) & \cdots & q_{xx}^*(1) & q_{xx}(0) \end{pmatrix}$$

(1.75)

$$\mathbf{Q}_{xx}(1) = \begin{pmatrix} q_{xx}(1) & q_{xx}(2) & \cdots & q_{xx}(L) \\ q_{xx}(0) & q_{xx}(1) & \cdots & q_{xx}(L-1) \\ \vdots & \vdots & \ddots & \vdots \\ & & & q_{xx}(2) \\ q_{xx}^*(L-2) & \cdots & q_{xx}(0) & q_{xx}(1) \end{pmatrix}$$

The cross-correlation between input and output signals leads to a relation between the cross-correlation and the input autocorrelation

$$\mathbf{q}_{xy}(\tau) = E\{\mathbf{x}_t y_{t+\tau}^*\} = \mathbf{Q}_{xx}(\tau)\mathbf{h} \qquad \tau = -\infty, \ldots, \infty \qquad (1.76)$$

equation (1.76) can be solved for $\mathbf{h}$ if $\mathbf{Q}_{xx}(0)$ is regular:

$$\mathbf{h} = \mathbf{Q}_{xx}^{-1}(0)\mathbf{q}_{xy}(0) \qquad (1.77)$$

Comparing (1.74) with (1.76) one obtains the relation between cross-correlation and output autocorrelation

$$q_{yy}(\tau) = \mathbf{h}^* \mathbf{q}_{xy}(\tau) \qquad \tau = -\infty, \ldots, \infty \qquad (1.78)$$

Notice that after filtering the sample size is reduced by $L-1$.

1.2.4.2 IIR filters

We consider only the recursive part of IIR filters (infinite impulse response). The output of a recursive filter with L coefficients h_m becomes

$$y_1(t) = h_0^* x_t - \sum_{m=1}^{L-1} h_m^* y_{t-m} \qquad (1.79)$$

After normalisation by h_0^* one obtains

$$x(t) = \mathbf{h}^* \mathbf{y}_t \qquad (1.80)$$

which is obviously the same equation as (1.73), but with input x_t and output y_t interchanged. Then the associated correlation equations become

$$\begin{aligned} q_{xx}(\tau) &= \mathbf{h}^* \mathbf{Q}_{yy}(\tau)\mathbf{h} & \tau = -\infty, \ldots, \infty & \quad (1.81) \\ \mathbf{q}_{yx}(\tau) &= \mathbf{Q}_{yy}(\tau)\mathbf{h} & \tau = -\infty, \ldots, \infty & \quad (1.82) \\ q_{xx}(\tau) &= \mathbf{h}^* \mathbf{q}_{yx}(\tau) & \tau = -\infty, \ldots, \infty & \quad (1.83) \end{aligned}$$

1.2.4.3 Whitening filters based on the autocorrelation equations

FIR filters. Equation (1.74) can be used to calculate a filter that produces an uncorrelated output sequence y_t, given an input process x_t with known autocorrelation. One needs L equations to solve for an L-dimensional filter vector. Hence, we postulate

$$\mathbf{h}^*\mathbf{Q}_{xx}(\tau)\mathbf{h} = \begin{cases} 1 & : \quad \tau = 0 \\ 0 & : \quad \tau = 1,\ldots,L-1 \end{cases} \tag{1.84}$$

Perfect whitening of the input process can be achieved only under certain conditions. Let K be the autocorrelation length of the input process so that $q_{xx}(\tau) \neq 0$, $|\tau| \leq K$, $q_{xx}(\tau) = 0$, $|\tau| > K$. In order to achieve perfect decorrelation the number of filter coefficients has to be $L \geq 2K - 1$. This follows from the shifted Toeplitz matrices $\mathbf{Q}_{xx}(\tau)$ in (1.84). If $L < 2K - 1$ then $2K - 1 - L$ non-zero correlation values will remain.

IIR filters. Using (1.81) we can define a system of equations to determine a recursive (all-pole) filter that decorrelates a process with given autocorrelation function

$$\mathbf{h}^*\mathbf{I}_{yy}(\tau)\mathbf{h} = q_{xx}(\tau) \qquad \tau = 0,\ldots,L-1 \tag{1.85}$$

where $\mathbf{I}_{yy}(\tau)$ are shifted identity matrices. For $\tau = 1$ we get for example

$$\mathbf{I}_{yy}(1) = \begin{pmatrix} 0 & & & 0 \\ 1 & \ddots & & \\ & \ddots & & \\ 0 & & 1 & 0 \end{pmatrix} \tag{1.86}$$

It follows from (1.86) that an L-point recursive filter can decorrelate perfectly an input process with finite correlation length L.

Equations (1.84, 1.85) are systems of quadratic equations which can be solved only numerically. It has turned out that this procedure is quite cumbersome and needs a lot of arithmetic operations.

1.2.4.4 Whitening filters based on the cross-correlation equations

FIR Filters. In equations (1.76, 1.77) a FIR filter was defined as the linear relation between the autocorrelation of the input process and the cross-correlation between input and output. This filter can be used to predict the future input sample with least square error (LEVINSON [390], BURG [79, 80], MAKHOUL [431] GALATI [175, p. 62])

$$\hat{x}_{t+1} = \mathbf{h}^*\mathbf{x}_t. \tag{1.87}$$

In this case the cross-correlation vector becomes a shifted version of the autocorrelation

$$\mathbf{q}^*_{xy}(1) = (q^*_{xx}(1),\ldots,q^*_{xx}(L)) \tag{1.88}$$

so that the prediction (or Wiener) filter is determined as

$$\mathbf{h}_P = \mathbf{Q}^{-1}_{xx}(0)\mathbf{q}_{xx}(1) \tag{1.89}$$

The prediction error is $x_{t+1} - \hat{x}_{t+1} = x_{t+1} - \mathbf{h}_P^* \mathbf{x}_t$. A prediction error filter, i.e., a filter whose output sequence is the prediction error, can be written as follows

$$\mathbf{h}_{PE}^* = (1 \quad -\mathbf{h}_P^*) \tag{1.90}$$

Notice that the dimension of the prediction error filter is $L+1$. The $(L+1)$-dimensional cross-correlation equation becomes

$$\begin{pmatrix} P_{t+1} & \mathbf{q}_{xx}^*(1) \\ \mathbf{q}_{xx}(1) & \mathbf{Q}_{xx}(0) \end{pmatrix} \begin{pmatrix} 1 \\ -\mathbf{h}_P \end{pmatrix} = \begin{pmatrix} P_{PE} \\ \mathbf{0} \end{pmatrix} \tag{1.91}$$

The matrix on the left is the $(L+1) \times (L+1)$-input correlation matrix where P_{t+1} is the power of the $(t+1)$th input sample, $\mathbf{0}$ is an $(L \times 1)$-zero vector and P_{PE} is the prediction error power. Some observations can be made:

- As the prediction error is minimised in the least mean square sense this filter can be used for minimising the interference power.

- The prediction error filter is the solution of a *linear* system of equations while a perfect whitening filter according to (1.84) requires the solution of a system of quadratic equations.

- The prediction error filter is equal to the first column of $\mathbf{Q}^{-1}$. It should be noted that all columns of $\mathbf{Q}^{-1}$ have the least squares property and can be used for interference suppression. In this case the unity element in $\mathbf{h}_{PE}$ and the prediction error power P_{PE} in the cross-correlation vector assume different positions (the term *prediction* might then better be replaced by *interpolation*). However, the various columns of $\mathbf{Q}^{-1}$ differ in their noise suppression characteristics. The first column uses the whole data length for prediction of a future value of the input process, whereas the inner columns are based on less correlation values (as can be seen from the Toeplitz structure of $\mathbf{Q}$). This is equivalent to a smaller aperture or observation time. The associated interference notches, as evident in Figure 1.2, become narrower for the outer than for the inner columns. This is in contrast to KRETSCHMER and LIN [367] who found that the inner columns may achieve better clutter suppression performance for certain clutter models.

- The least squares filter does not whiten the input sequence perfectly. Some correlation remainders may be observed. In practice they are negligible.

- The concept of prediction is not required for the problem of clutter suppression. The filter vector $\mathbf{h} = \mathbf{Q}^{-1}\mathbf{e}_n$ ($\mathbf{e}_n$ is the n-th column of the identity matrix) can also be obtained by minimising the output power of a FIR filter under the constraint that one of the filter coefficients is unity to avoid the trivial solution $\mathbf{h} = \mathbf{0}$. For complex optimisation problems see BRANDWOOD [60] and VAN DEN BOS [59].

- The prediction error filter can be calculated very efficiently via a recursive algorithm which computes the filter coefficients in such a way that the filter dimension is increased with every iteration (see BURG [79, 80], BÜHRING and KLEMM [75]).

A hardware implementation of the above-mentioned algorithm for near real-time clutter suppression in a surveillance radar with rotating antenna has been described by BÜHRING and KLEMM [75]. For theoretical details on linear prediction the reader is referred to the book by HAYKIN [253, pp. 122].

1.2.4.5 Improvement factor for FIR whitening filters

Let us recall that the inverse of the covariance matrix included in the optimum processor whitens the received interference vector and equalises the signal reference vector in order to provide a perfect match between the matched filter and the signal after passing the pre-whitening operation (see Figure 1.7 and (1.71)). Accordingly, the FIR filter processor becomes $\mathbf{w} = \mathbf{HH}^*\mathbf{s}$. The filter matrix $\mathbf{H}$ contains shifted vectors of filter coefficients $\mathbf{h}$

$$\mathbf{H} = \begin{pmatrix} h_1 & & & & \\ \cdot & h_1 & & 0 & \\ \cdot & \cdot & & & \\ h_L & \cdot & \ddots & & \\ & h_L & & h_1 & \\ & & & \cdot & \\ & 0 & & \cdot & \\ & & & h_L & \end{pmatrix} \quad (1.92)$$

The product $\mathbf{H}^*\mathbf{x}$ describes the convolution of the filter vector $\mathbf{h}$ with the data vector $\mathbf{x}$ according to (1.73) in such a way that the filter always remains inside the data length. Therefore, the number of columns of $\mathbf{H}$ is $M - L + 1$ where L is the dimension of $\mathbf{h}$, i.e., the number of filter coefficients, and M the length of the data record.

The maximum improvement factor achieved by a processor after Figure 1.7b and equation (1.73) is obtained by inserting $\mathbf{w} = \mathbf{HH}^*\mathbf{s}$ in (1.15),

$$\text{IF} = \frac{\mathbf{s}^*\mathbf{HH}^*\mathbf{ss}^*\mathbf{HH}^*\mathbf{s} \cdot \text{tr}(\mathbf{Q})}{\mathbf{s}^*\mathbf{HH}^*\mathbf{QHH}^*\mathbf{s} \cdot \mathbf{s}^*\mathbf{s}} \quad (1.93)$$

where we assumed that the processor is perfectly matched to the target signal.[15]

1.2.5 Examples

1.2.5.1 Comparison of optimum and FIR filter processors

In Figure 1.8 the IF achieved by the optimum processor (including the inverse of a 12 × 12-covariance matrix and a 12-point steering vector) is compared with the performance of a FIR least squares (or prediction error) filter as depicted in Figure 1.7b. In the example the filter has $L = 3$ coefficients and has been cascaded with a 10-point signal integrating matched filter (or beamformer) which is steered over the frequency F. Notice that in the latter case the matching vector has length $N - L + 1$ so that the FIR filter does not move over the edges of the data sequence.

---
[15] For example, in direction or Doppler or both.

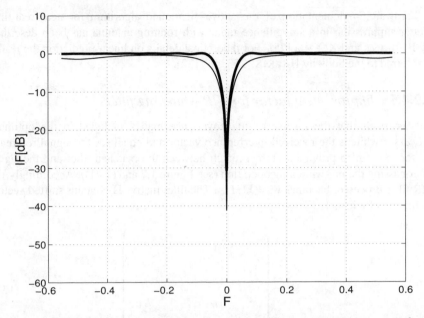

Figure 1.8: Comparison of the optimum and FIR filter processor (one interfering source at $F = 0$, $SNR = 20$ *dB*, $N = 12$). *Fat: optimum processor; thin: FIR least squares clutter filter with* $L = 3$ *coefficients cascaded with a 10-point matched filter*

It can be noticed that the optimum gain is almost reached by the FIR filter processor outside the interference region. Slight differences in IF can be noticed in the interference direction and in its neighbourhood.

1.2.5.2 Practical application: adaptive clutter filter for surveillance radar

The following experiment conducted in the early 1970s is decribed briefly to illustrate the principle of adaptive clutter filtering. This filter was designed to suppress weather clutter with unknown centre Doppler frequency and bandwidth (BÜHRING AND KLEMM [75]). Figure 1.9 shows the block diagram of an adaptive FIR filter based on the prediction error filter principle. The filter coefficients were estimated in real-time and the echo data were clutter filtered during the following revolution of the antenna. In a recent publication by ANDERSON [16] the use of such adaptive techniques for suppression of moving clutter for application to modern air traffic control radars is suggested. Usually the shape of the clutter Doppler spectrum is assumed to be gaussian (see also Chapter 2), and thus has two parameters, the centre frequency and the bandwidth (standard deviation). (LOMBARDO et al. [415]) have shown that windblown terrain may show merely an exponential spectral shape. This model has been implemented for GMTI data cube generation (MOUNTCASTLE [470]).

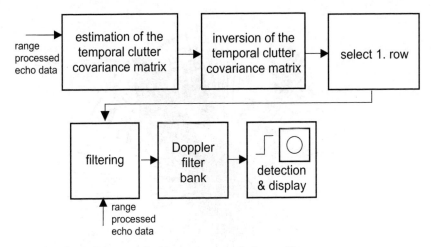

Figure 1.9: Block diagram of adaptive temporal clutter filter

In Figure 1.10 the filtering effect is demonstrated using simulated clutter. The picture shows a photograph of a PPI (plan position indicator) screen. The zero Doppler filter has been switched off so that a lot of ground clutter can be noticed. The spokes are the simulated clutter (for simplicity this clutter was simulated independent of range). The available radar operated in the L-band which is quite insensitive to weather. Therefore, to have reproduceable clutter conditions a hardwired simulator was developed. The clutter filter was adapted based on clutter data in the window on the right. As can be seen the clutter has been removed and a (true) target is visible.

Figure 1.11 and 1.12 show suppression of a real weather cloud before and after filtering. Except for a few false alarms the weather clutter has been removed. The air targets (big spots on the left) did not do us the favour of entering our measurement window, they just bypassed it.[16]

1.2.6 Angle or frequency domain processing

The discrete Fourier transform plays an important role in radar signal processing. In certain applications it may be attractive to apply first the Fourier transform to the received data and continue subsequent processing in the frequency (or angular) domain. The implementation via the fast Fourier transform (FFT) algorithm makes the DFT a key tool for a variety of radar functions, for example

- In real aperture radars target echoes are normally harmonic functions so that the FFT is used as a Doppler filter bank to detect moving targets with unknown Doppler frequency.

- In broadband radar pulse compression is carried out by means of the FFT.

---

[16]No responsible pilot will enter a thunderstorm deliberately.

42 Introduction

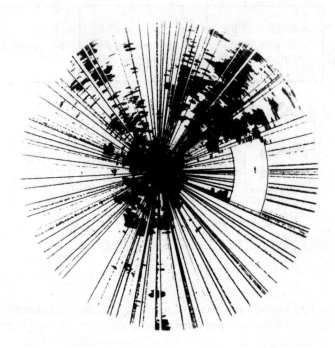

Figure 1.10: Adaptive suppression of simulated weather clutter

- In synthetic aperture radar it is worthwhile to perform the convolution operations associated with range and azimuth compression in the frequency domain to exploit the efficiency of the fast Fourier transform (FFT) algorithm.

- Clutter echoes can be suppressed by means of FIR filters as pointed out in the previous Section. The convolution operation inherent in FIR filtering may be carried out efficiently in the frequency domain by using FFT techniques (overlap-add, overlap-save, see OPPENHEIM and SCHAFER [512, pp. 110]).

- One can make use of the fact that the individual spectral outputs of the Fourier transform become mutually uncorrelated as the observation interval (length of transformed data sequence) approaches infinity (OPPENHEIM and SCHAFER, [512, pp. 544]). Therefore, transforming a covariance matrix into the spectral domain leads to a diagonalisation of the covariance matrix which may result in saving arithmetic operations. For example, if the output signals of different Fourier channels are sufficiently uncorrelated the matrix inversion is reduced to the inversion of a diagonal matrix, i.e., simple divisions by the main diagonal elements.

Figure 1.11: Weather clutter before adaptation

The discrete Fourier transform (DFT) can be written as a $N \times N$-Matrix $\mathbf{F}$ with the entries

$$f_{nm} = \frac{1}{\sqrt{N}} \exp(-j2\pi(n-1)(m-1)/N) \quad n,m = 1,\ldots,N \quad (1.94)$$

where the column index n denotes *time* or *space* and the row index m means *temporal* or *spatial frequency*. The inverse DFT matrix $\mathbf{F}^*$ has the entries

$$f_{nm}^* = \frac{1}{\sqrt{N}} \exp(j2\pi(n-1)(m-1)/N) \quad n,m = 1,\ldots,N \quad (1.95)$$

The DFT matrix $\mathbf{F}$ is unitary, i.e.,

$$\mathbf{F}^* = \mathbf{F}^{-1} \text{ so that } \mathbf{F}\mathbf{F}^* = \mathbf{F}\mathbf{F}^{-1} = \mathbf{I} \quad (1.96)$$

Transforming the various data vectors according to (1.53) yields

$$\mathbf{q}_F = \mathbf{F}\mathbf{q}; \quad \mathbf{s}_F = \mathbf{F}\mathbf{s}; \quad \mathbf{x}_F = \mathbf{F}\mathbf{x}; \quad \mathbf{Q}_F = \mathbf{F}^*\mathbf{Q}\mathbf{F} \quad (1.97)$$

Similar to (1.54) the optimum processor becomes

$$\mathbf{w}_F = \gamma \mathbf{Q}_F^{-1} \mathbf{s}_F \quad (1.98)$$

and the associated improvement factor corresponding to (1.55)

$$\text{IF} = \frac{\mathbf{w}_F^* \mathbf{s}_F \mathbf{s}_F^* \mathbf{w}_F \cdot \text{tr}(\mathbf{Q})}{\mathbf{w}_F^* \mathbf{Q}_F \mathbf{w}_F \cdot \mathbf{s}^*\mathbf{s}} = \frac{\mathbf{w}^* \mathbf{F}^* \mathbf{F}\mathbf{s}\mathbf{s}^* \mathbf{F}^* \mathbf{F}\mathbf{w} \cdot \text{tr}(\mathbf{Q})}{\mathbf{w}^* \mathbf{F}^* \mathbf{F}\mathbf{Q}\mathbf{F}^* \mathbf{F}\mathbf{w} \cdot \mathbf{s}^*\mathbf{s}} \quad (1.99)$$

Notice that the IF in (1.99) is identical with the time (or space) domain formulation (1.16), since $\mathbf{F}$ is unitary, see (1.96).

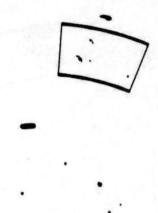

Figure 1.12: Weather clutter after adaptation

1.3 Spectral estimation

In this Section some tools for spectral analysis are summarised. These techniques differ from the above-described signal detection methods in that the covariance matrix includes *signal and noise*. Spectral estimation techniques are based on the covariance matrix of the received signal $\mathbf{R} = E\{\mathbf{xx}^*\}$, i.e., target signal, interference and noise. The covariance matrix represents the measured data which are to be analysed in terms of a variable steering vector $\mathbf{s}(\Theta)$. Θ is a parameter vector to be estimated and may include angular direction, Doppler frequency, and time of arrival. As far as the covariance matrix is concerned there is no distinction between signal and interference while the previously described techniques are based on the signal-free interference + noise covariance matrix $\mathbf{Q}$.

Power spectral estimators can be subdivided into the *signal match* and the *high-resolution* type. The first one is frequently used in radar and SAR applications, for instance as beamformer for target search, matched filter for range estimation, and Doppler filters for velocity estimation. The latter class may serve for special tasks such as angular (adaptive beamforming) or spectral superresolution. They have been applied originally in geophysics (CAPON) [83, 84] to detect subsurface sound sources (earthquakes, nuclear explosions). Later on high-resolution techniques have been discussed for application in sonar (COX [105, 106, 107]) and even later in radar (NICKEL [479, 480, 487, 485, 489]). Application of high-resolution techniques to resolve radar multipath has been presented by WIRTH and NICKEL [724], and MATHER [440].

High-resolution techniques play a special role in the context of acoustic matched field processing (for an overview see BAGGEROER and KUPERMAN [26] and the references given there). It has been shown by KLEMM [296] that range and depth of acoustic sources in shallow water can be estimated with a passive sensor array by exploiting modal interferences of the acoustic wave field. High-resolution spectral estimation techniques have been used for adaptive array processing by GABRIEL [172, 173]. Application of different spectral estimators to three-dimensional feature extraction in interferometric SAR images has been described by LI et al. [392].

There are different underlying principles of signal match and high-resolution type. The signal match type is based on the collinearity of two vectors while the high-resolution methods are based in some way on the principle of orthogonality. In other words, by matching the signal the output is maximised, whereas high-resolution estimators look for a minimum. The maximum of a directivity pattern is used for information transfer, whereas a minimum in the antenna directivity pattern is used in radio location applications.

The reason for using minimum search for source location is that minima of directivity patterns are generally point shaped and, hence, narrower than maxima.

This can be illustrated for the special case of a two-element sensor array. The beamformer is then a two-dimensional weighting vector whose coefficients are varied until a signal match is obtained. This happens when the squared output signal is a maximum. The beamformer can also be used to minimise the output power. The scalar product between the signal and the weighting vectors is proportional to the cosine of the angle between them which becomes 0° for match and 90° for orthogonality. The derivative of the cosine function is minimum at 0° and maximum at 90°. It follows from this consideration that for a dipole a null in the directivity pattern is narrower than the beamwidth. This causes the difference in resolution between minimum and maximum based search techniques. Unfortunately, it is not easy to generalise this example to a N-dimensional vector space. Interrelations between different high-resolution power estimators are pointed out by NICKEL [483].

1.3.1 Signal match (SM)

Let us consider a covariance matrix of the form

$$\mathbf{R} = E\{\mathbf{xx}^*\} = \mathbf{S} + \mathbf{N} \qquad (1.100)$$

where $\mathbf{N}$ is the noise component and $\mathbf{S}$ includes all kinds of signal or interference. Then the output of a signal matched filter is simply

$$y_{\text{SM}}(\Theta) = \mathbf{x}^*\mathbf{s}(\Theta) \qquad (1.101)$$

and the normalised power output is

$$P_{\text{SM}}(\Theta) = \frac{\mathbf{s}^*(\Theta)\mathbf{R}\mathbf{s}(\Theta)}{\mathbf{s}^*(\Theta)\mathbf{s}(\Theta)} \qquad (1.102)$$

$s(\Theta)$ is a steering vector which seeks for signal components $s(\Theta_i)$ in $\mathbf{R}$. $P_{\text{SM}}(\Theta)$ attempts to become a maximum wherever the steering vector $s(\Theta)$ coincides with a signal vector $s(\Theta_i)$ in $\mathbf{R}$. The position of such maximum may, however, be corrupted by noise. In the case of multiple signals, lack of resolution may lead to biased estimates for Θ_i, and even to cancellation of weaker signals.

Typical-known applications are

- **Fourier analysis.** The steering vector consists of columns of the DFT matrix as given by (1.94). $\mathbf{R}$ is a temporal covariance matrix of signal echo samples and noise. One prominent application is the Doppler filter bank.

- **Beamforming.** The steering vector includes beamformer weights for an array antenna, and $\mathbf{R}$ is the spatial signal + interference + noise covariance matrix across the array outputs. For an equidistant linear array the coefficients of a beamformer are identical to those of a Doppler filter bank.

- **Wavefront matching.** Passive range estimation can basically be accomplished by use of (1.102). In this case Θ denotes the curvature of a spherical wavefront.

- **Matched field processing.** In underwater acoustics (1.102) can basically be used for passive localisation of acoustic sources. The steering vector $s(\Theta)$ consists of coefficients corresponding to a wavefront due to an acoustic source at a hypothetic location Θ. $\mathbf{R}$ is again the spatial signal + interference + noise covariance matrix. Moreover, the vector Θ is not constrained to the source coordinates. It may include for example environmental parameters which have inflence on the propagation in the dispersive sound channel.

1.3.2 Minimum variance estimator, MVE

In analogy with (1.3) the minimum variance estimator[17] becomes

$$\mathbf{w}_{\mathrm{MV}} = \gamma \mathbf{R}^{-1} \mathbf{s} \tag{1.103}$$

with $\gamma = (\mathbf{s}^* \mathbf{R}^{-1} \mathbf{s})^{-1}$. The output power is

$$P_{\mathrm{MV}}(\Theta) = (\mathbf{s}^*(\Theta) \mathbf{R}^{-1} \mathbf{s}(\Theta))^{-1} \tag{1.104}$$

Consider a covariance matrix of the simple form $\mathbf{R} = \mathbf{s}(\Theta_0)\mathbf{s}^*(\Theta_0) + \sigma^2 \mathbf{I}$. Assuming $\sigma^2 \ll 1$ one obtains $P_{\mathrm{MV}}(\Theta) \approx N$ for $\Theta = \Theta_0$ while for all $\Theta \neq \Theta_0$ $P_{\mathrm{MV}}(\Theta)$ is minimised. This property makes (1.104) a high-resolution estimator. COX [107] compared both the processors (1.104) and (1.3) in terms of resolution and sensitivity to mismatch. It turns out that the resolution capability of the optimum detector (1.3) is determined by the beamwidth given by the steering vector. Therefore the resolution is the same as that of the signal match spectral estimator (1.102) while the processor (1.104) is capable of resolving signals within the beamwidth. It should be noted that (1.104) is just proportional to the inverse of the improvement factor of the optimum detector (1.16). In fact, looking at the curves in Figure 1.2 this processor provides an optimum (high-resolution) clutter notch.

It should be noticed that the resolution capability of the minimum variance estimator depends on the noise level if additional noise is present, i.e. $\mathbf{R} = \mathbf{s}(\Theta_0)\mathbf{s}^*(\Theta_0) + \mathbf{N}$. The resolving capability of the optimum detector (1.3) and the signal match estimator (1.102) do not depend on the noise power.

Both properties are useful for individual purposes. While low resolution is useful for detection purposes (search) high-resolution estimators are well adapted to

---

[17] Also referred to as the *maximum likelihood* estimator, see CAPON [83, 84].

separation of sources (or frequencies) that are close together. Some properties of the minimum variance estimator have been discussed in the tutorial paper by GABRIEL [172]. The performance of the minimum variance beamformer has been analysed in some detail by WAX and ANU [699].

1.3.3 Maximum entropy method, MEM

The maximum entropy power method (MEM) proposed by BURG [79, 80] is given by

$$P_{\mathrm{ME}}(\Theta) = \frac{1}{\mathbf{s}^*(\Theta)\mathbf{rr}^*\mathbf{s}(\Theta)} \quad (1.105)$$

where $\mathbf{r}$ is the first column of $\mathbf{R}$. The idea behind the ME principle is that for data records of finite length the autocorrelation function is extrapolated to infinity by use of a maximum entropy criterion. In this way the spectral resolution is extended beyond the limit given by the length of the data record.

The maximum entropy power spectral estimator has the following properties:

- It is assumed that the underlying signals are stationary so that $\mathbf{R}$ is a Toeplitz matrix. In practice this condition is fulfilled for echo sequences with constant PRF (application: high-resolution Doppler estimation) or equidistant linear arrays (application: high-resolution beamforming).

- Practical applications of the P_{ME} have demontrated that ME-spectra may achieve higher peak-to-sidelobe levels than the SM or the MVE.

- MEM spectra tend to suffer from high spurious sidelobes, including splitting of the principal peaks.

- SM and MVE produce a *linear* signal output. The MEM is a non-linear function of steering vector and measured data.

- The MEM is strongly related to the prediction error filter (1.90). Therefore, the MEM spectrum can also be seen in the light of least squares optimisation. The prediction error (FIR) filter minimises the output power of the actual signal process. Therefore, the filter coefficients contain the signal information in the least square sense.

- The MEM technique, normally used with stationary temporal signal sequences, can basically be extended to use with two-dimensional signals, see ROUCOS and CHILDERS [578]. The 3-D clutter plots in a paper by the author (KLEMM [300]), however, are quite 'spiky' and do not reflect the true spectrum properly. As will be shown in Chapter 3 the MV estimator is more appropriate.

Some properties of the maximum entropy method estimator have been discussed by GABRIEL [172].

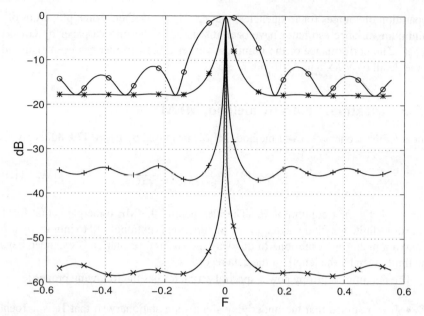

Figure 1.13: Power spectra: one source at $F = 0, N = 6, \text{SNR} = 20$ dB. ○ *Fourier;* ∗ *minimum variance;* + *maximum entropy;* × *MUSIC*

1.3.4 Orthogonal projection, MUSIC

The principle of orthogonal projection which has been discussed earlier for suppression of interference can also be used for high-resolution power estimation (PISARENKO [546], SCHMIDT [590, 591]). It has been used for high-resolution radar imaging by ODENDAAL et al. [508]. The principle is widely known as MUSIC (multiple signal classification). It has a similar form to the MVE

$$P_{\text{MU}}(\Theta) = (s^*(\Theta)E^n E^{n*} s(\Theta))^{-1} \tag{1.106}$$

where the matrix E^n includes the noise eigenvectors of R. Since the noise subspace of R is orthogonal to the signal subspace the expression (1.106) becomes infinity whenever the steering vector $s(\Theta)$ coincides with a signal component in R. It should be noted that a separation of noise and signal subspace is possible only if there is a clear distinction between the eigenvalues of signal and noise. This means that the technique is basically confined to resolving a number of discrete sources by means of an array antenna (or discrete frequencies in a data sequence). This signal model has been used by SCHMIDT [590].

According to HAENDEL [235] some pioneering work on MUSIC was contributed at about the same time as Schmidt's work by ZIEGENBEIN [751] and KLEMM [295].

While MUSIC can be applied with arbitrarily spaced arrays (or staggered signal sequences) the method by PISARENKO [546] and the ESPRIT technique (ROY et al. [579]) are both based on equispaced data sequences. The IFRELAX algorithm (LI et al. [392]) has proven to be superior to the MUSIC in terms of accuracy and computational efficiency when applied to 3-D feature extraction in interferometric SAR images.

1.3.5 Comparison of spectral estimators

Figure 1.13 shows a numerical example in which the above four power estimators are compared. As in the previous examples one source at $F = 0$ was assumed. For signal matching one obtains the well-known $\sin x/x$ directivity pattern of a linear array.

The MVE gives a moderately sharp peak and a smooth sidelobe region. For the MEM one obtains an even higher peak-to-sidelobe ratio. Both MVE and MEM are based on the signal + noise covariance matrix. BURG [79] has shown that the MVE is composed of N MEM estimators with increasing order. This follows from the following triangular decomposition of $\mathbf{R}^{-1}$ for $N = 3$

$$\mathbf{R}^{-1} = \begin{pmatrix} 1 & 0 & 0 \\ h_2 & 1 & 0 \\ h_3 & h_1 & 1 \end{pmatrix} \begin{pmatrix} P_3^{-1} & 0 & 0 \\ 0 & P_2^{-1} & 0 \\ 0 & 0 & P_1^{-1} \end{pmatrix} \begin{pmatrix} 1 & h_2^* & h_3^* \\ 0 & 1 & h_1^* \\ 0 & 0 & 1 \end{pmatrix} \quad (1.107)$$

The P_i denote the prediction error powers while the h_i in the triangular matrices are the coefficients of prediction error filters of decreasing order 3, 2, and 1. Therefore, the MVE power spectrum is a kind of average over several MEM spectra of different orders which have different resolution capabilities.

The response of the MUSIC technique at $F = 0$ is basically infinite but has been limited here so that an artificial sidelobe level can be noticed at about -57 dB. In practice there is always some mismatch due to errors in either the incoming wavefront received by the array or the steering vector. Therefore, infinitely high peaks are not obtained in practical applications.

An alternative superresolution technique is *Parametric Target Model Fitting* by NICKEL [479, 480]. A similar technique is the *Incremental Multi-Parameter (IMP) Algorithm* by MATHER [440]. It is a superresolution technique based on an optimisation by variation of an orthogonal projection matrix. It can be used for estimating the number, directions and amplitudes of sources and has even been proposed for jammer cancellation (NICKEL [482]).

In Chapter 3 we will illustrate the nature of airborne clutter by means of two-dimensional (φ − F) clutter spectra. The question is, which of the beforementioned spectral estimators is best suited for this purpose? It turns out that the signal match estimator suffers from its natural sidelobes which give a wrong impression of the true spectrum. Due to their high-resolution capability estimators such as MEM or MUSIC have been shown to produce the problem of 'graphical' undersampling. That means, due to the limited number of points in the plot the true signal component is not always met by the steering vector. This results in an extremely spiky spectrum. Increasing the number of points causes the 3-D plot to become entirely black. The best compromise between spikyness and sidelobe effects is given by the MVE which, therefore, will be used in Chapter 3.

1.4 Summary

Chapter 1 deals with an introduction into the problems of MTI processing, including the effect of radar platform motion. Reference is made to various authors who contributed to this field. The signal processing tools as required in subsequent chapters are briefly described. The major topics of Chapter 1 can be summarised as follows:

1. **Doppler clutter.** The platform motion causes clutter returns to be Doppler shifted. The Doppler shift is proportional to the platform velocity and the cosine of the angle between the flight axis and the direction of the individual scatterer. The total of clutter returns sums up in a clutter Doppler bandwidth. Slow targets may be buried in the clutter band.

2. The **optimum adaptive processor (OAP)** has been discussed repeatedly for jammer suppression (spatial processing) and clutter rejection purposes (temporal processing). It includes two linear components:
    - pre-whiten the received data with respect to the interference;
    - match to the desired signal vector.

    The properties of the OAP are discussed in some detail. Because of heavy computational load the OAP can hardly ever be used in practice, but it may serve as a reference to compare with suboptimum techniques.

3. The **orthogonal projection** technique is the zero noise approximation of the optimum processor. It forms exact nulls at the positions of the interference.

4. **Linear subspace transforms** (see Chapter 6) are used to reduce the signal vector space. Such transforms are a step towards suboptimum processing systems with *real-time* capability. A prominent example for a spatial pre-transform is the *sidelobe canceller*.

5. **Digital filters** (see Chapter 7) may yield some saving of operations in the time dimension if the data sequence is stationary, i.e., the covariance matrix is Toeplitz.

6. The optimum processor can be formulated in the **frequency (or angular) domain** without reduction of detection performance.

7. **Power spectral estimators** are based on the signal + interference + noise covariance matrix. Four different estimators have been mentioned:
    - signal matching (linear);
    - minimum variance estimator (linear; strongly related to the optimum processor);
    - maximum entropy (requires stationarity; related to the least squares FIR filter, see Chapter 7);
    - MUSIC (related to the orthogonal projection technique).

Chapter 2

Signal and interference models

In this chapter models for target, clutter, jammers and noise are derived which form the basis for the numerical evaluations carried out in the subsequent chapters. To some extent we follow the derivation by VAN TREES [662, pp. 238].

Based on such models ROMAN *et al.* [576] developed state variable models which indicate that multichannel models of *low* order can represent airborne ground clutter effectively. This will be confirmed by the results obtained in Chapters 5, 6, 7, and 9. These chapters deal with order-reducing subspace techniques for clutter rejection with real-time capability.

2.1 Transmit and receive process

Let us assume that the radar transmits a signal of the form

$$s_t(t) = A_t E(t) \cos \omega_c t = \Re[A_t E(t) e^{j\omega_c t}] \qquad (2.1)$$

where $\omega_c = 2\pi f_c$ denotes the carrier frequency, A_t is $\sqrt{2P_t}$ with P_t being the transmitted power, and $E(t)$ is the envelope of the transmitted waveform for which we assume

$$\int_{-\infty}^{\infty} |E(t)|^2 \, dt = 1 \qquad (2.2)$$

Assuming an ideal point-shaped reflector at distance R_t from the radar the received signal becomes

$$s_r(t) = \Re[A_r E(t-\tau) e^{j\omega_c(t-\tau)}] \qquad (2.3)$$

where $A_r = \sqrt{2P_r}$. P_r includes the transmit power and the two-way propagation and reflection processes according to the radar range equation (for details see SKOLNIK [609, pp. 2-1,...,2-73], [610, pp. 2-1,...,2-68]). τ is the round-trip delay of the transmitted wave between radar and target

$$\tau = \frac{2R_t}{c} \qquad (2.4)$$

where c is the velocity of light. The received signal then undergoes a complex demodulation which can be described simply by

$$\begin{aligned} s_{\mathrm{r}}(t) &= \mathrm{e}^{\mathrm{j}\omega_{\mathrm{c}}t}\Re[A_{\mathrm{r}}E(t-\tau)\mathrm{e}^{\mathrm{j}\omega_{\mathrm{c}}(t-\tau)}] \\ &= A_{\mathrm{r}}E(t-\tau)\frac{1}{2}[\mathrm{e}^{\mathrm{j}\omega_{\mathrm{c}}(t-\tau)}+\mathrm{e}^{-\mathrm{j}\omega_{\mathrm{c}}(t-\tau)}]\mathrm{e}^{\mathrm{j}\omega_{\mathrm{c}}t} \\ &= A_{\mathrm{r}}E(t-\tau)\frac{1}{2}[\mathrm{e}^{\mathrm{j}2\omega_{\mathrm{c}}(t-\tau)}+1] \end{aligned} \quad (2.5)$$

The high-frequency term will be suppressed by a low-pass filter and the factor of $\frac{1}{2}$ can be incorporated in the amplitude. Then the demodulated signal becomes

$$s_{\mathrm{r}}(t) = A_{\mathrm{r}}E(t-\tau) \quad (2.6)$$

The low-pass filter is usually followed by a band-limiting filter matched to the envelope function

$$s_{\mathrm{r}}(\tau) = A_{\mathrm{r}}\int_{t=-\infty}^{\infty} E(t-\tau)E^{*}(t-\tau_{\mathrm{m}})\mathrm{d}t \quad (2.7)$$

$\tau_{\mathrm{m}} = \tau$ is the time delay until the signal reaches the matched filter. The matched filter output becomes a maximum if $\tau_{\mathrm{m}} = \tau$. In this way the target range is estimated by measuring the echo delay. By use of (2.2) one then obtains

$$s_{\mathrm{r}}(\tau) = A_{\mathrm{r}}. \quad (2.8)$$

This is the response to a target at a range defined by $\tau_{\mathrm{m}} = \tau$. For a detailed theory of matched filtering see RIHACZEK [568].

2.2 The Doppler effect

We now consider a target moving at constant radial velocity v_{rad}. At $t = 0$ the target range is R_0. The target motion causes the target range to vary with t:

$$R(t) = R_0 - v_{\mathrm{rad}}t \quad (2.9)$$

A signal received at time t was reflected at $t - \tau(t)/2$. The associated target range is

$$R\left(t-\frac{\tau(t)}{2}\right) = R_0 - v_{\mathrm{rad}}\left(t-\frac{\tau(t)}{2}\right) \quad (2.10)$$

Inserting (2.4) into (2.10) and solving for $\tau(t)$ yields

$$\tau(t) = \frac{2R_0/c}{1-v_{\mathrm{rad}}/c} - \frac{(2v_{\mathrm{rad}}/c)t}{1-v_{\mathrm{rad}}/c} \quad (2.11)$$

In practice target velocities are small compared with the velocity of light, i.e. $v_{\mathrm{rad}}/c \ll 1$ so that (2.11) becomes

$$\tau(t) \approx \frac{2R_0}{c} - \frac{2v_{\mathrm{rad}}}{c}t = \tau_0 - \frac{2v_{\mathrm{rad}}}{c}t \quad (2.12)$$

where τ_0 is the delay associated with R_0. By inserting (2.12) into (2.3) the received target signal becomes

$$s_\mathrm{r}(t) = \Re\left[A_\mathrm{r} E\left(t - \tau_0 + \frac{2v_\mathrm{rad}}{c}t\right) \exp\left[\mathrm{j}\omega_\mathrm{c}\left(t - \tau_0 + \frac{2v_\mathrm{rad}}{c}t\right)\right]\right] \qquad (2.13)$$

The target motion has two effects on the received signal

- The envelope is stretched or compressed. That means that the matched filter (2.7) is mismatched to the received echo signal.[1] In radar this effect is negligible for realistic target velocities and signal time–bandwidth products (s. VAN TREES [662, p. 241]).

- The carrier frequency is shifted by

$$\omega_\mathrm{D} = 2\pi f_\mathrm{D} = 2\pi f_\mathrm{c} \frac{2v_\mathrm{rad}}{c} = 2\pi \frac{2v_\mathrm{rad}}{\lambda} \qquad (2.14)$$

with ω_D being the angular Doppler frequency. This Doppler shift causes a frequency mismatch with the matched filter. As long as the range of the Doppler frequencies is smaller than the signal bandwidth, i.e.,

$$\frac{2v_\mathrm{rad}}{c} f_\mathrm{c} \leq B_\mathrm{s} \qquad (2.15)$$

where B_s is the signal bandwidth, this mismatch effect can be neglected.

After demodulation and matched filtering (2.8) becomes for a single range increment

$$s_\mathrm{r}(t) = A_\mathrm{r} \exp\left(\mathrm{j}\omega_\mathrm{c} \frac{2v_\mathrm{rad}}{c} t\right) = A_\mathrm{r} \exp(\mathrm{j}\omega_\mathrm{D} t) \qquad (2.16)$$

2.3 Space-time signals

2.3.1 The spatial dimension: array geometry

Figure 2.1 shows the geometry of an airborne (or spaceborne) array radar. Without loss of generality we assume throughout this book that the radar platform moves in the x-direction. We further define the x-direction as zero azimuth. Instead of an elevation angle we use the depression angle θ which is more appropriate for the airborne configuration. R_s is the slant range, R_g the ground range, φ denotes azimuth, and v_p the platform velocity.

Two basic types of array antenna are depicted: the linear sideways and the linear forward looking array. Other configurations are possible and will be treated in Chapter 9. Most of the fundamental studies presented here are based on the two array configurations shown in Figure 2.1.

---

[1] In a paper by NAKHMANSON [474] this effect is taken into account for optimum reception of a broadband signal by an array.

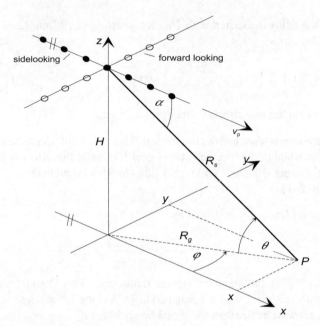

Figure 2.1: Geometry of airborne antenna arrays

Let us forget the platform motion for the time being. Consider a sensor with coordinates x_i, y_i, z_i relative to the array origin 'O'. The signal according to (2.3) received by this sensor from a stationary scatterer on the ground is phase shifted relative to the origin by

$$\Delta\varphi = j\frac{2\pi}{\lambda}((x_i \cos\varphi + y_i \sin\varphi)\cos\theta - z_i \sin\theta) \qquad i = 1,\ldots,N \qquad (2.17)$$

where N is the number of array elements. The received signal after demodulation and matched filtering according to (2.8) becomes

$$s_\mathrm{r}(R_\mathrm{s}, x_i, y_i, z_i) = A \exp\left[j\frac{2\pi}{\lambda}((x_i \cos\varphi + y_i \sin\varphi)\cos\theta - z_i \sin\theta)\right] \qquad (2.18)$$

where

$$\frac{H}{R_\mathrm{s}} = \sin\theta \qquad (2.19)$$

2.3.1.1 Sensor directivity patterns

For the sensors the following directivity patterns will be used in the sequel

$$D(\varphi) = 0.5(1 + \cos(2(\varphi - \varphi_0))) \qquad (2.20)$$
$$D(\theta) = 0.5(1 + \cos(2(\theta - \theta_0))) \qquad (2.21)$$

where the angle φ_0 denotes the direction of the maximum of the sensor pattern. For a forward looking array we have $\varphi_0 = 0°$, for the sidelooking configuration $\varphi_0 = 90°$. The vertical look angle is assumed to be $\theta_0 = 0$ throughout the book. It has some slight influence on the results only in the context of multiple-time-around clutter which will be treated in Chapter 16.

2.3.2 The temporal dimension: pulse trains

The frequency resolution of a single pulse is determined by the pulse length δ. For a rectangular pulse envelope the frequency response is

$$u(f) = \frac{\sin(\pi f \delta)}{\pi f \delta} \tag{2.22}$$

see SKOLNIK [609, p. 3–19]. As long as the range of Doppler frequencies is smaller than the signal bandwidth, i.e.,

$$\frac{2v_{\text{rad}}}{c} f_c \leq \frac{1}{\delta} \tag{2.23}$$

the Doppler information cannot be retrieved from a *single* pulse.

In pulse Doppler radar the Doppler frequency is measured by phase comparison between echo signals due to a transmitted coherent pulse train. The transmit signal is

$$s_t(t) = \Re[A_t E(t) \exp(j\omega_c (t + mT))] \qquad m = 1,\ldots,M \tag{2.24}$$

where T is the pulse repetition interval (PRI) and M is the number of transmitted pulses. The inverse of T

$$f_{\text{PR}} = \frac{1}{T} \tag{2.25}$$

is the pulse repetition frequency (PRF). Accordingly the receive signal due to a moving target becomes (after demodulation and matched filtering)

$$s_r = A_r \exp[j\omega_D (t + mT)] = A_r \exp\left[j\frac{2\pi}{\lambda} 2 v_{\text{rad}}(t + mT)\right] \qquad m = 1,\ldots,M \tag{2.26}$$

which reduces to

$$s_r = A_r \exp[j\omega_D (mT)] = A_r \exp\left[j\frac{2\pi}{\lambda} 2 v_{\text{rad}}(mT)\right] \qquad m = 1,\ldots,M \tag{2.27}$$

if the pulse length given by $E(t)$ is short compared with T.

2.3.2.1 *Response to a single clutter patch*

For the geometry shown in Figure 2.1, the signal received at the origin due to a stationary scatterer on the ground becomes for a certain range ring

$$s_r = A(\varphi) \exp\left[j\frac{2\pi}{\lambda} 2 v_p mT \cos\varphi \cos\theta\right] \qquad m = 1,\ldots,M \tag{2.28}$$

and for a sensor at x_i, y_i, z_i

$$s_\mathrm{r} = A(\varphi)\exp\left[j\frac{2\pi}{\lambda}\right. \tag{2.29}$$
$$\left.\times(((x_i+2v_\mathrm{p}mT)\cos\varphi+y_i\sin\varphi)\cos\theta-z_i\sin\theta)\right]$$
$$m=1,\ldots,M \qquad i=1,\ldots,N$$

since we assumed that the radar platform moves in the x-direction. For an equispaced array one obtains

$$s_\mathrm{r} = A(\varphi)\exp\left[j\frac{2\pi}{\lambda}\right. \tag{2.30}$$
$$\left.\times((id_x+2v_\mathrm{p}mT)\cos\varphi+kd_y\sin\varphi)\cos\theta-ld_z\sin\theta)\right]$$
$$m=1,\ldots,M \qquad i,k,l=1,\ldots,N$$

where d_x, d_y, d_z are the spacings in the x-, y-, z- directions.

2.3.2.2 Response to a moving target

For a moving target with radial velocity v_rad (2.29) becomes

$$s_\mathrm{r}(\varphi) = A\exp\left[j\frac{2\pi}{\lambda}\right. \tag{2.31}$$
$$\left.\times(2v_\mathrm{rad}mT+(x_i\cos\varphi_t+y_i\sin\varphi_t)\cos\theta-z_i\sin\theta)\right]$$
$$m=1,\ldots,M \qquad i=1,\ldots,N$$

where v_rad is the radial target velocity which includes the target–radar geometry as well as the target's velocity and cruise direction, and φ_t the target azimuth.

Basically there is an additional quadratic phase term due to the tangential component of the target velocity. For large range and small M (number of echoes) this term can be neglected. It will play a role in Chapter 13 where large values of M will be considered in connection with ISAR (Inverse Synthetic Aperture Radar).

As long as the azimuth angle interval passed by the target during the observation period is negligible the target velocity v_rad can be assumed to be constant. This is fulfilled for short pulse sequences. For longer pulse sequences as occur particularly in SAR applications the changing target–radar geometry due to radar and target motions has to be taken into account. In this case v_rad varies from pulse to pulse.

It should be noted that the space-time signal vector **s**(.) used for calculation of the improvement factor or power spectra is composed of elements given by (2.31). In these calculations we always assume that the processor is perfectly matched to the expected signal so as to obtain the maximum possible performance measure for the clutter supression operation. Therefore, **s**(.) will play both the roles of the expected signal and the space-time matched filter. The steering vector appears, for example, in (1.18), (1.19), (1.93), (1.99), (1.102), (1.104), and in most of the following chapters.

Notice that the space-time steering vector can be separated into a beamformer with look direction φ_L, θ_L

$$\mathbf{b}(\varphi) = \begin{pmatrix} \vdots \\ \exp[j\frac{2\pi}{\lambda}(x_i \cos \varphi_L + y_i \sin \varphi_L) \cos \theta_L - z_i \sin \theta_L)] \\ \vdots \end{pmatrix} \quad i = 1, \ldots, N \quad (2.32)$$

and a velocity (or Doppler) filter

$$\mathbf{d}(v_{\text{rad}}) = \begin{pmatrix} \vdots \\ \exp[j\frac{2\pi}{\lambda} 2v_{\text{rad}} mT \cos \varphi_L \cos \theta_L)] \\ \vdots \end{pmatrix} \quad m = 1, \ldots, M \quad (2.33)$$

$$= \begin{pmatrix} \vdots \\ \exp[j\omega_D mT] \\ \vdots \end{pmatrix} \quad m = 1, \ldots, M$$

2.4 Interference

2.4.1 Ground clutter

A model for airborne clutter has been presented by RINGEL [569]. In this model the clutter power contributions of individual range-Doppler cells of a single-channel radar are computed. Due to the non-linear boundaries of the range-Doppler cells this is a complex procedure. On the other hand the RINGEL model computes clutter *power* only. TOMLINSON [649] describes a STAP modelling computer code which can be used for monostatic and bistatic operation as well as for effects such as crabbing. Processing is done in the frequency domain (post-Doppler architecture, see Chapter 9). In the paper paper by ZEKAVAT and ABDI [744] patches of different roughness according to a priori information (maps etc.) are composed.

For the purpose of analysing space-time processing algorithms a simpler model is sufficient. Since space-time clutter rejection takes place on a pulse-to-pulse basis, computation of the clutter echoes is carried out for one range increment only. Instead of absolute power values we determine a clutter-to-noise ratio at the individual sensor and echo pulse. Contrary to the RINGEL model we have to generate *complex* clutter returns due to a train of coherent transmit pulses. Moreover, we have to consider an array antenna rather than a single channel radar.

For clutter echoes we make the following usual assumptions:

- The contributions of different scatterers to the clutter echo are statistically independent. Since the received clutter echoes are a sum over a large number of scatterers they are asymptotically gaussian.

- Temporal clutter fluctuations are slow compared with the observation time MT.

The total clutter echo is an integral over the various contributions from all ground scatterers in the visible range. Using (2.29) we obtain for a single range increment at the i-th sensor at the m-th instant of time

$$c_{im} = \int_{\varphi=0}^{2\pi} s_{\mathrm{r}}(\varphi) d\varphi \tag{2.34}$$

$$= \int_{\varphi=0}^{2\pi} A \exp\left[j\frac{2\pi}{\lambda}\right.$$

$$\times \left.(((x_i + 2v_{\mathrm{p}}mT)\cos\varphi + y_i \sin\varphi)\cos\theta - z_i \sin\theta)\right] d\varphi$$

where A is a circular complex gaussian distributed variable (gaussian amplitude and uniformly distributed phase). Like for the signal in (2.31), we omitted here the quadratic phase term due to the tangential component of the platform velocity. It plays a role only if the clutter range is not too large and the number of echo pulses is sufficiently large. It will play a role in the discussion of an ISAR problem in Chapter 13.

The Doppler frequency of the individual clutter arrival is

$$f_{\mathrm{D}} = \frac{2v_{\mathrm{p}}}{\lambda} \cos\varphi \cos\theta \tag{2.35}$$

This important relation was mentioned in the preface. It says that each direction is associated with an individual clutter Doppler frequency. As an abbreviation we write the temporal and spatial phase terms as follows

$$\Phi_m(v_{\mathrm{p}}, \varphi) = \exp\left[j\frac{2\pi}{\lambda} 2v_{\mathrm{p}}mT \cos\varphi \cos\theta\right]$$

$$\Psi_i(\varphi) = \exp\left[j\frac{2\pi}{\lambda} (x_i \cos\varphi + y_i \sin\varphi)\cos\theta - z_i \sin\theta\right] \tag{2.36}$$

Including the sensor pattern[2] $D(\varphi)$ (2.34) becomes

$$c_{im} = \int_{\varphi=0}^{2\pi} AD(\varphi)\Phi_m(v_{\mathrm{p}}, \varphi)\Psi_i(\varphi) d\varphi \tag{2.37}$$

The clutter model given by (2.37) can be slightly more refined by introducing two quantities $L(\varphi)$ and $G(\varphi)$

$$c_{im} = \int_{\varphi=0}^{2\pi} AD(\varphi)L(\varphi)G(\varphi, m)\Phi_m(v_{\mathrm{p}}, \varphi)\Psi_i(\varphi) d\varphi \tag{2.38}$$

$L(\varphi)$ stands for the reflectivity of the ground while $G(.)$ stands for the transmit directivity pattern. $L(\varphi)$ is in general range dependent. However, since we focus on one range increment we have to consider only the azimuth dependence. Notice that the

---

[2] Defined in (2.20).

parameter m in $G(\varphi, m)$ describes the instantaneous position of the antenna pattern at time m. The transmit directivity pattern of an array is defined as

$$G(\varphi) = \mathbf{b}(\varphi_L)^* \mathbf{b}(\varphi) \qquad (2.39)$$

with φ_L being the look direction and φ the angle of the individual clutter arrival. $\mathbf{b}$ is a beamformer vector with elements

$$b(\varphi, \theta) = \exp\left[j\frac{2\pi}{\lambda}((x_i \cos\varphi + y_i \sin\varphi)\cos\theta - z_i \sin\theta)\right] \quad i = 1, \ldots, N_t \qquad (2.40)$$

where N_t is the number of transmit array elements.

From the clutter echo (2.38) the elements of the clutter covariance matrix are calculated as follows

$$q_{ln}^{(c)} = E\{c_{im} c_{kp}^*\} \qquad (2.41)$$

where

$$l = (m-1)N + i \quad m = 1, \ldots, M; \quad i = 1, \ldots, N \qquad (2.42)$$

and

$$n = (p-1)N + k \quad p = 1, \ldots, M; \quad k = 1, \ldots, N \qquad (2.43)$$

are the indices of the $NM \times NM$ covariance matrix. i, k denote sensors and m, p echo pulses. $E\{\}$ means the mathematical expectation. We come back to the clutter covariance matrix in Section 3.2.1.

2.4.2 Moving clutter

We now consider a clutter background that moves at the radial clutter velocity. We assume that the clutter does not exceed a range increment during the observation time MT. This clutter model may be used to simulate different kinds of moving clutter, for instance sea, weather, chaff. In our model we introduce a radial clutter velocity v_c.

By addition of radial platform and clutter velocities the temporal phase term in (2.36) becomes

$$\Phi_m(\varphi, v_p, v_c) = \exp\left[j\frac{2\pi}{\lambda}(v_p \cos\varphi \cos\theta + v_c)2mT\right] \qquad (2.44)$$

The total clutter echo is given by replacing the temporal phase term $\Phi_m(\varphi, v_c, v_p)$ in (2.38) by (2.44).

Some calculations including moving clutter such as weather or chaff have been given by GHOUZ et al. [186, 187].

2.4.3 Jamming

In addition to clutter we will consider CW (continuous wave) noise jamming. The jammer signal received by the i-th sensor of the array at time m due to J jammers is

$$c_{im}^{(j)} = \sum_{j=1}^{J} A_j(m) \exp\left[j\frac{2\pi}{\lambda}\right. \qquad (2.45)$$
$$\left. \times (x_i \cos\varphi_j + y_i \sin\varphi_j)\cos\theta_j - z_i \sin\theta_j)\right]$$

where A_j are the jammer amplitudes and φ_j and θ_j are the angles that determine the positions of the jammers. We assume that the jammer signals are Doppler broadband. Therefore, the jammer amplitudes A_j are modelled as complex gaussian variables. We assume that

$$E\{A_j(m)A_j^*(p)\} = \begin{cases} P_j & : \; m = p \\ 0 & : \; m \neq p \end{cases} \quad (2.46)$$

which means that there is no temporal correlation between subsequent pulse repetition intervals. P_j is the power of the j-th jammer. We further assume that jammers are mutually uncorrelated

$$E\{A_j A_r^*\} = \begin{cases} P_j & : \; j = r \\ 0 & : \; j \neq r \end{cases} \quad (2.47)$$

However, the sensor output signals due to the j-th jammer are mutually fully correlated[3] according to

$$\begin{aligned} E\{c_i^{(j)}(j)c_k^{(j)*}(j)\} &= P_j \exp\Big[\mathrm{j}\frac{2\pi}{\lambda} \\ &\times ((x_i - x_k)\cos\varphi_j + (y_i - y_k)\sin\varphi_j)\cos\theta_j \\ &- (z_i - z_k)\sin\theta_j\Big) \Big] \end{aligned} \quad (2.48)$$

This expression includes the jammer power and the phase differences between different sensors.

2.4.4 Noise

The main limiting factor in radar detection is the noise generated by the radar receiver which is dominated by the first amplifier in the receiver chain. We assume the noise to be uncorrelated in time and space

$$E\{n_m n_p^*\} = \begin{cases} P_n & : \; m = p \\ 0 & : \; m \neq p \end{cases} \quad (2.49)$$

$$E\{n_i n_l^*\} = \begin{cases} P_n & : \; i = l \\ 0 & : \; i \neq l \end{cases} \quad (2.50)$$

2.5 Decorrelation effects

In this Section a few important effects are described which can degrade the performance of anti-clutter and anti-jamming measures. Due to the two-dimensional (space-time) nature of the received signals there are two different effects, one applying to the temporal, the other one to the spatial dimension.

---
[3] This is a narrowband assumption. For large radar bandwidth spatial decorrelation effects may occur.

2.5.1 Temporal decorrelation

Temporal decorrelation effects modify the space-time covariance matrix in the temporal dimension, i.e., with respect to the index m in equation (2.34). There are two temporal effects: internal clutter motion and range walk.

2.5.1.1 Clutter bandwidth

The statistical behaviour of clutter returns depends very much on the nature of the individual scatterers. Hard reflectors such as buildings tend to produce stable echo signals which are reflected in a high temporal correlation between successive echo samples. Equivalently, stable returns produce a narrow Doppler bandwidth. Soft scatterers such as vegetation produce less stable echoes with a larger Doppler bandwidth. This is even more true when the scattering background exhibits internal motion. Vegetation or weather under wind conditions are prominent examples for this effect.

It is common to model clutter fluctuations by means of a gaussian-shaped clutter spectrum.[4] We use here the gaussian model. A gaussian correlation function is related to a gaussian power spectrum by the equivalence (PAPOULIS, [518, pp. 25])

$$\exp\left(-\frac{\sigma^2}{2}\tau^2\right) \iff \sqrt{\frac{2\pi}{\sigma^2}} \exp\left(\frac{-\omega^2}{2\sigma^2}\right) \quad (2.51)$$

Let us define a clutter bandwidth $B = 2\sigma$. Then (2.51) becomes

$$\exp\left(-\frac{B^2}{8}\tau^2\right) \iff \sqrt{\frac{8\pi}{B^2}} \exp\left(\frac{-2\omega^2}{B^2}\right) \quad (2.52)$$

Defining a relative clutter bandwidth $B_c = B/f_{PR} = BT$ (clutter bandwidth normalised to the unambiguous Doppler band) one obtains with $\tau = (m-p)T$ the discrete normalised autocorrelation function

$$\rho_{mp}^{(c)} = \exp\left(-\frac{B_c^2(m-p)^2}{8}\right) \quad (2.53)$$

This temporal decorrelation applies to clutter echoes only. Jammer and noise signals have been supposed anyway to be uncorrelated in the time dimension. Target echoes are supposed to be deterministic and, hence, are fully correlated in time.

2.5.1.2 Range walk

Range walk can lead to temporal decorrelation of space-time clutter echoes. The influence of range walk on space-time clutter covariance matrices and the associated power spectra has been analysed by KREYENKAMP [368]. The principle of range

---

[4]Under windy conditions clutter may exhibit an exponential rather than a gaussian shape, see BILLINGSLEY et al. [48], TECHAU et al. [639].

62 Signal and interference models

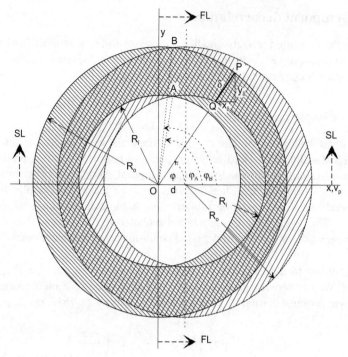

Figure 2.2: Geometry of range walk

walk due to radar platform motion is illustrated in Figure 2.2. It shows in essence two circular rings of width $R_o - R_i$ displaced by

$$d = v_p mT \qquad m = 1,\ldots,M \qquad (2.54)$$

which is the distance the radar passes during the time mT. Both rings denote range bins as seen by the radar at different instances of time.

With the usual assumption that the clutter background consists of a large number of scatterers producing spatially uncorrelated echoes we can conclude that the correlation between the two instances of time is given by the area where both range rings overlap. At a particular azimuth, φ, the correlation will be:

$$\rho_{mp}^{(\text{RW})}(\varphi) = \frac{\delta_{m-p}(\varphi)}{\Delta_R} \qquad m,p = 1,\ldots,M \qquad (2.55)$$

with $\Delta_R = R_o - R_i$ denoting the width of the range bin. Notice that, opposite to decorrelation by internal clutter motion, the decorrelation due to range walk is azimuth dependent. Obviously the correlation is a maximum in the broadsight direction of a sidelooking array (SL) and minimum in the forward looking direction (FL).

The distance $\delta(\varphi)$ can be calculated as follows

$$\delta(\varphi) = \sqrt{(x_P(\varphi) - x_Q(\varphi))^2 + (y_P(\varphi) - y_Q(\varphi))^2} \qquad (2.56)$$

where x_P, y_P and x_Q, y_Q are the coordinates of the points P and Q in Figure 2.2:

$0 \leq \varphi < \varphi_A$

$$x_P = R_o \cos \varphi; \quad y_P = R_o \sin \varphi$$
$$x_Q = [d \cos \varphi + \sqrt{R_i^2 - d^2 \sin^2 \varphi}] \cos \varphi; \quad y_Q = \sqrt{R_i^2 - (x_Q - d)^2}$$

$\varphi_A \leq \varphi < \varphi_B$

$$x_P = R_o \cos \varphi; \quad y_P = R_o \sin \varphi$$
$$x_Q = R_i \cos \varphi; \quad y_Q = R_i \sin \varphi \tag{2.57}$$

$\varphi_B \leq \varphi \leq \frac{\pi}{2}$

$$x_P = [d \cos \varphi + \sqrt{R_o^2 - d^2 \sin^2 \varphi}] \cos \varphi; \quad y_P = \sqrt{R_o^2 - (x_Q - d)^2}$$
$$x_Q = R_i \cos \varphi; \quad y_Q = R_i \sin \varphi$$

For azimuth angles $\varphi > \frac{\pi}{2}$ the correlation measure $\delta(\varphi)$ can be obtained via symmetry considerations.

It should be noted that in practice the temporal decorrelation effect due to range walk will be even stronger than described above. The receive signal normally passes a bandwidth limiting matched filter so that the output signal is the autocorrelation function of the transmit signal. In the case of range walk, the individual shape of the matched filter response causes additional decorrelation which depends on the shape of the transmitted waveform. In the subsequent discussion we omit this aspect and assume that the matched filter response is constant across the range bin.

2.5.2 Spatial decorrelation: effect of system bandwidth

The system bandwidth causes *spatial* decorrelation between signals appearing at the outputs of the antenna array. The decorrelation is determined by the signal bandwidth and by the travel time of the incoming wave between the sensor positions.

The travel time between the i-th and the k-th sensor of a wave due to a source at position (φ, θ) becomes for the configuration defined in Figure 2.1,

$$\tau_{ik} = \frac{1}{c}[((x_i - x_k) \cos \varphi + (y_i - y_k) \sin \varphi) \cos \theta - (z_i - z_k) \sin \theta] \tag{2.58}$$

2.5.2.1 *Rectangular impulse response*

The system bandwidth is determined by the spectrum or, equivalently, by the autocorrelation function of the transmitted pulse. In narrowband radar systems the modulation of the carrier is often a rectangular function

$$A(t) = \begin{cases} 1 & : \quad t = [-\delta/2, \delta/2] \\ 0 & : \quad \text{otherwise} \end{cases} \tag{2.59}$$

The autocorrelation function of such a rectangular impulse is

$$\rho_\tau = \begin{cases} 1 - \frac{|\tau|}{\delta} & : \quad t = [-\delta, \delta] \\ 0 & : \quad \text{otherwise} \end{cases} \tag{2.60}$$

Then the spatial decorrelation of a wave with complex amplitude A across the array is obtained by inserting (2.58) into (2.60)

$$\begin{aligned}\rho_{ik}^{(\text{BW})} &= E\left\{\frac{A_i A_k^*}{\sqrt{|A_i|^2 |A_k|^2}}\right\} \\ &= \begin{cases} 1 - \frac{|\tau_{ik}|}{\delta} & : \quad t = [-\delta, \delta] \\ 0 & : \quad \text{otherwise} \end{cases}\end{aligned} \tag{2.61}$$

Notice that the relative system bandwidth is related to the pulse length by

$$B_\text{s} = \frac{1}{\delta f_\text{c}} \tag{2.62}$$

2.5.2.2 Rectangular frequency response

In broadband radar applications such as SAR frequently a linear chirp modulation of the kind

$$u(t) = \frac{1}{\delta}\text{rect}\left(\frac{t}{\delta}\right) e^{j\pi\alpha t^2} \tag{2.63}$$

is used. δ is the pulse width as before, and α is the slope of the chirp. The frequency response of a chirp is approximately rectangular. The autocorrelation function of such a signal is (SKOLNIK [609, pp. 3-18])

$$\rho_{ik}^{(\text{BW})} = \text{rect}\left(\frac{\tau_{ik}}{2\delta}\right) \rho_\text{r}(\tau_{ik}) \frac{\sin(\pi\alpha\delta\tau_{ik}\rho_\text{r}(\tau_{ik}))}{\pi\alpha\delta\tau_{ik}\rho_\text{r}(\tau_{ik})} \tag{2.64}$$

with $\rho_\text{r}(\tau_{ik})$ being the autocorrelation function of a rectangular pulse as given by (2.60). The time-varying chirp frequency is $f_\text{ch} = \alpha t$ so that the chirp bandwidth is $B = \alpha\delta$. Then we can relate the slope factor with the relative system bandwidth

$$\alpha = \frac{B_\text{s} f_\text{c}}{\delta} \tag{2.65}$$

with f_c being the carrier frequency. As before τ_{ik} is given by (2.58).

2.5.3 Doppler spread within range gate

For certain applications a large range gate equivalent to low range resolution is desirable. Two examples are

- Target search by a coherent sequential test (WIRTH [719]) is more efficient if the range gates are chosen large.

Table 2.1 The standard parameter set

Parameter	Symbol	Value	Equ.	Fig.
System				
carrier frequency	f_c	10 GHz	2.1	
PRF	f_{PR}	12 KHz	2.25	
rel. system bandwidth	B_s	0	2.62, 2.65	
pulse length	δ	large influence negligible	2.63, 2.62	
Operational data				
platform speed	v_p	90 m/s		2.1
target velocity range	v_{rad}	-100...100 m/s	2.31	3.16, 4.5
look direction	φ_L	45°		2.1
range	R	10 km	2.19	2.1
altitude	H	3 km	2.19	2.1
flight path		horizontal over plane ground		2.1
Antenna				
array orientation		side/forward looking	2.30	
array type		linear equispaced	2.30	
no. of elements	N	24	2.17	
transmit array	$G(\varphi)$	24	2.38	
sensor spacing	d	0.015 m	2.30	
hor. sensor pattern	$D(\varphi)$	yes	2.20	
vert. sensor pattern	$D(\theta)$	yes	2.21	
Signal and interference and signal				
signal/noise ratio	SNR	-10 dB	Chapter 14	14.3 etc.
clutter/noise ratio	CNR	20 dB	3.26	
rel. clutter bandwidth	B_c	0	2.53	
reflectivity	$L(\varphi)$	1	2.38	
clutter power		uniform		
clutter phase		random		
jammer/noise ratio	JNR	20 dB	3.27	
receiver noise	$P_n\mathbf{I}$	white	2.49, 2.50	
multiple clutter		none	10.2	10.5
Signal Processing				
no. of temporal samples	M	24	2.24	
space-time channels	C	48	5.1	
no. of antenna channels	K	5	6.2	6.2
temp. dim. of FIR filter	L	5	7.3	
Doppler filter length	$M-L+1$	20	4. p.214	
tapering	$\mathbf{D}$	none		

- The omnidirectional low probability of intercept (OLPI) radar (WIRTH [720]) is based on a CW pseudonoise transmit code in conjunction with extremely long target signal integration periods. The range bins are chosen sufficiently large in order to prevent targets from range migration.

If the clutter Doppler is range dependent[5] a large range gate will produce a certain Doppler bandwidth instead of producing one Doppler frequency per direction. As this effect depends on the width of the range gate it depends in essence on the system bandwidth. Recall that decorrelation due to range walk increases as the range gate gets smaller. The effect of the Doppler spread behaves just the other way around because it increases as the range gate gets larger.

Let us consider the clutter return from a single scatterer at the position φ, θ as given by (2.29). The space-time correlation between sensors and echo pulses is

$$\rho_{mpik} = E\Big\{ A(\varphi, \theta) \exp\Big[j\frac{2\pi}{\lambda} \tag{2.66}$$
$$\times (((x_i + 2v_p mT)\cos\varphi + y_i \sin\varphi)\cos\theta - z_i \sin\theta)\Big]$$
$$\times A^*(\varphi, \theta) \exp\Big[-j\frac{2\pi}{\lambda}$$
$$\times (((x_k + 2v_p pT)\cos\varphi + y_k \sin\varphi)\cos\theta - z_k \sin\theta)\Big]\Big\}$$

Let us, for simplicity, consider linear arrays. Then $z_i = 0, z_k = 0$. Extending the single clutter patch to a range interval (or, equivalently, to an interval in $u = \cos\theta$) of width Δu gives

$$\rho_{mpik} = E\Big\{ \frac{1}{\Delta u} \int_{u_0 - \frac{\Delta u}{2}}^{u_0 + \frac{\Delta u}{2}} A(\varphi, u) \exp\Big[j\frac{2\pi}{\lambda} \tag{2.67}$$
$$\times ((x_i + 2v_p mT)\cos\varphi + y_i \sin\varphi)u\Big] du$$
$$\times \frac{1}{\Delta u} \int_{u_0 - \frac{\Delta u}{2}}^{u_0 + \frac{\Delta u}{2}} A^*(\varphi, v) \exp\Big[-j\frac{2\pi}{\lambda}$$
$$\times ((x_k + 2v_p pT)\cos\varphi + y_k \sin\varphi)v\Big] dv\Big\}$$

where $u, v = \cos\theta$ are directional cosines. Assuming that returns from different scatterers are statistically independent gives $E\{A(\varphi, u)A^*(\varphi, v)\} = 0$, $u \neq v$, and the clutter power is constant across the range gate so that $E\{A(\varphi, u)A^*(\varphi, v)\} = P_c$, $u = v$. Then (2.67) reduces to

$$\rho_{mpik} = E\Big\{ \frac{1}{\Delta u} \int_{u_0 - \frac{\Delta u}{2}}^{u_0 + \frac{\Delta u}{2}} A(\varphi, u) A^*(\varphi, u) \exp\Big[j\frac{2\pi}{\lambda} \tag{2.68}$$
$$\times ((x_i - x_k + 2v_p(m-p)T)\cos\varphi + (y_i - y_k)\sin\varphi)u\Big] du\Big\}$$

---

[5]For details on clutter spectra, range dependence of the clutter Doppler and its effect on the detection of moving targets see Chapters 3 and 4.

$$= \frac{\sin[\frac{2\pi}{\lambda}((x_i - x_k + 2v_\mathrm{p}(m-p)T)\cos\varphi + (y_i - y_k)\sin\varphi)\frac{\Delta u}{2}]}{\frac{2\pi}{\lambda}((x_i - x_k + 2v_\mathrm{p}(m-p)T)\cos\varphi + (y_i - y_k)\sin\varphi)\frac{\Delta u}{2}}$$
$$\times P_\mathrm{c}\exp\left[j\frac{2\pi}{\lambda}((x_i - x_k + 2v_\mathrm{p}(m-p)T)\cos\varphi + (y_i - y_k)\sin\varphi)u_0\right]$$
$$= \rho_{mpik}^{(\mathrm{RG})}$$
$$\times P_\mathrm{c}\exp\left[j\frac{2\pi}{\lambda}((x_i - x_k + 2v_\mathrm{p}(m-p)T)\cos\varphi + (y_i - y_k)\sin\varphi)u_0\right]$$

The term $\rho_{mpik}^{(\mathrm{RG})} = \operatorname{sinc}[\frac{2\pi}{\lambda}((x_i-x_k+2v_\mathrm{p}(m-p)T)\cos\varphi+(y_i-y_k)\sin\varphi)\frac{\Delta u}{2}]$ denotes the decorrelation due to the Doppler spread in a range gate of width Δu. Notice that this decorrelation is effective in space (i,k) and time (m,p). This decorrelation effect will occur whenever the range gate is large and the clutter Doppler is range dependent.

2.5.4 System Doppler spread

The clutter Doppler frequency is proportional to the carrier frequency as follows from (2.35)

$$f_\mathrm{D} = \frac{2v_\mathrm{p}f_c}{c}\cos\varphi\cos\theta \qquad (2.69)$$

If the carrier has a certain system bandwidth B_s then the Doppler exhibits a Doppler bandwidth proportional to the system bandwidth:

$$B_\mathrm{D} = \frac{2v_\mathrm{p}}{\lambda}B_s\cos\varphi\cos\theta \qquad (2.70)$$

with $B_s = B/f_c$ being the relative system bandwidth. Notice that the Doppler bandwidth B_D is direction dependent.

The Doppler bandwidth is equivalent to a temporal correlation. Assuming a rectangular system band the temporal correlation becomes

$$\rho_{mp}^{(\mathrm{SD})} = \frac{\sin(\pi B_\mathrm{D}(m-p)T)}{\pi B_\mathrm{D}(m-p)T} \qquad (2.71)$$

2.5.5 Total correlation model

Let us summarise the total correlation model for the clutter echo amplitude. Assuming that arrivals from different directions are uncorrelated we get by means of (2.53), (2.55), (2.64) or (2.68)

$$\rho_{ln} = \frac{E\{A(\varphi_r, i, m)A^*(\varphi_q, k, p)\}}{\sqrt{E\{|A(\varphi_r, i, m)|^2\}E\{|A(\varphi_q, k, p)|^2\}}} \qquad (2.72)$$
$$= \begin{cases} \rho_{mp}^{(\mathrm{C})}\rho_{mp}^{(\mathrm{RW})}\rho_{mp}^{(\mathrm{SD})}\rho_{ik}^{(\mathrm{BW})}\rho_{ikmp}^{(\mathrm{RG})} & : \varphi_r = \varphi_q \\ 0 & : \text{otherwise} \end{cases}$$

where i and k are spatial indices denoting the array elements, m and p are temporal indices denoting different echo pulses, and φ_r and φ_q are angles of arrival. The indices l and n have been defined in (2.42) and (2.43).

For jamming only the *spatial* correlation applies

$$\rho_{ln} = \begin{cases} \rho_{ik} & : \quad \varphi_r = \varphi_q, \quad m = p \\ 0 & : \quad \text{otherwise} \end{cases} \quad (2.73)$$

where now φ_r and φ_q denote the directions of the $r-$th and the $q-$th jammer.

2.6 The standard parameter set

As stated above the main objective of this book is to present results of a comprehensive numerical study on space-time adaptive processing with application to clutter suppression for air- and spaceborne radar. The main input parameters used in this study are listed in Table 2.5.2.1. All numerical examples shown further below are based on this set of parameters if no other values are specified.

The f_{PR} and d have been chosen so that the backscattered wavefield is sampled at the Nyquist rate in both the temporal and the spatial dimension. Notice that the Doppler bandwidth extends from $-2v_p/\lambda$ to $2v_p/\lambda$ so that

$$B_D = \frac{4v_p}{\lambda} = \frac{4 \times 90\text{m/s}}{0.03\text{m}} = 12\text{KHz} \quad (2.74)$$

It is well known that the resolution capability of a linear array is best for the broadside direction and decays with off-broadside look angles. Most of the subsequent discussions are based on linear equispaced arrays in either the sidelooking or forward looking configuration. The look direction has been chosen as $\varphi = 45°$ because this angle is the same for both sideways and forward looking array geometries. By this choice we avoid any beamwidth effects caused by different look directions when comparing these two array configurations.

2.6.1 Multiple-time around clutter

As indicated before, the PRF has been chosen so that the clutter Doppler band is sampled at the Nyquist rate. This means, however, that the radar operation is range ambiguous. The width of the unambiguous range interval is

$$\Delta R = \frac{c}{2f_{PR}} \quad (2.75)$$

For the chosen PRF of 12000 Hz the unambiguous range interval becomes 12.5 km while the line of sight is for this geometry at 187.5 km. For the sake of clarity we omitted the effect of multiple-time-around clutter in most of the following discussions. This important aspect will be discussed in Chapter 10.

2.6.2 Remark on image quality

In several chapters the reader will find graduated figures, in particular range-Doppler matrices. Some of them exhibit a coarse pixel raster. Notice that this is not a question of printing quality. These calculations, especially for range ambiguous radar, are extremely time consuming, so that a coarse pixel raster had to be chosen. These figures are a compromise between image quality and computing time. I do not think that these artifacts will significantly deteriorate the understandability of those graphics.

2.7 Summary

This book presents the results of a comprehensive model study on space-time adaptive processing, with particular emphasis on real-time processing techniques. Chapter 2 starts from the fundamentals of the radar transmit and receive process including the Doppler effect due to reflection at a moving scatterer. Then the mathematical *space-time* models for clutter, target signal, noise, and jamming are derived. In detail the following topics play an important role:

1. The **clutter Doppler frequency** of an individual scatterer is proportional to the angle between the flight axis and the direction of the scatterer. This relation suggests that

2. **Space-time** or **space-frequency** techniques have to be applied for clutter suppression.

3. **The elements** of an array antenna provide *spatial sampling* of the backscattered wavefield.

4. The **backscattered echo samples** contain the temporal (pulse-to-pulse) history of the reflecting background which can be interpreted in terms of Doppler frequency.

5. **Ground clutter** exhibits an azimuth-dependent Doppler colouring given by the platform velocity and direction.

6. **Targets** have their own velocity; therefore, there is no relation between Doppler frequency and direction.

7. For **jamming** we assume noise jammers whose bandwidth is large compared with the Doppler bandwidth (spatially correlated, uncorrelated in the time dimension).

8. **Receiver noise** is assumed to be white in space and time.

9. **Clutter fluctuations** are modelled by means of a gaussian correlation function.

10. **Range walk** is modelled by assuming uncorrelated scatterers and calculating the overlap of range rings shifted against each other.

70 Signal and interference models

11. The impact of the **system bandwidth** has been analysed for two different matched filters:
    - rectangular impulse response
    - rectangular frequency response

12. A model has been given for **Doppler spread** which may occur if the radar has low range resolution.

13. The **radar parameters** used throughout this book, unless denoted otherwise, are listed in Table 2.1.

Chapter 3

Properties of airborne clutter

In this chapter some basic properties of airborne clutter are analysed. These include space-Doppler characteristics, the space-time covariance matrix and the associated azimuth-Doppler.

3.1 Space-Doppler characteristics

3.1.1 Isodops

From (2.29), we can conclude that the ground is totally Doppler coloured since any pair of angles (φ, θ) denotes an individual clutter Doppler frequency. Curves of constant Doppler frequency on the ground are called *isodops*.

The clutter Doppler frequency due to a certain stationary scatterer on the ground is proportional to the radial velocity[1] (see Figure 2.1 and (2.14)):

$$f_D = \frac{2v_p}{\lambda} \cos \alpha \qquad (3.1)$$

In accordance with Figure 2.1, and (2.29), (3.1) becomes in ground coordinates

$$f_D = \frac{2v_p}{\lambda} \cos \varphi \cos \theta \qquad (3.2)$$

Notice that we assumed that the radar platform flies parallel to the ground and that the ground is planar. Clutter arrivals may come from all possible azimuth angles $\varphi = 0°, \ldots, 360°$. The Doppler frequency depends only on the platform velocity and the angle of arrival, but not on the array geometry. For scatterers at a particular range (elevation angle θ) the clutter Doppler bandwidth extends from

$$B_D = \left[-\frac{2v_p}{\lambda} \cos \theta, \frac{2v_p}{\lambda} \cos \theta \right] \qquad (3.3)$$

which means that the clutter bandwidth is larger at long range (small θ) than at close range.

72 Properties of airborne clutter

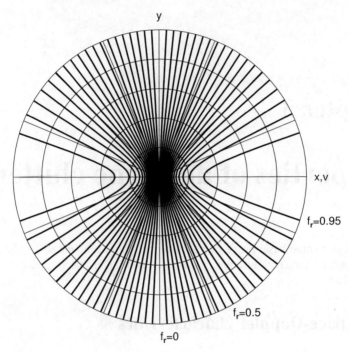

Figure 3.1: The isodops

For convenience we define a relative Doppler frequency

$$f_r = \frac{f_D \lambda}{2 v_p} = \cos\varphi \cos\theta \tag{3.4}$$

From Figure 2.1, we obtain the following relations:

$$\cos\varphi = \frac{x_P}{R_g} \qquad \cos\theta = \frac{R_g}{R_s} \qquad R_s^2 = H^2 + R_g^2 \tag{3.5}$$

so that (3.4) becomes

$$f_r = \frac{x_P}{R_s} = \frac{x_P}{\sqrt{H^2 + x_P^2 + y_P^2}} \tag{3.6}$$

which is a set of hyperbolae

$$\frac{x_P^2}{H^2 f_r^2/(1-f_r^2)} - \frac{y_P^2}{H^2} = 1 \qquad f_r = [-1, 1] \tag{3.7}$$

For $H = 0$ the hyperbolas turn into straight lines:

$$y_P = \frac{\sqrt{1-f_r^2}}{f_r} x_P \tag{3.8}$$

---
[1]This important relation has already been quoted in the preface.

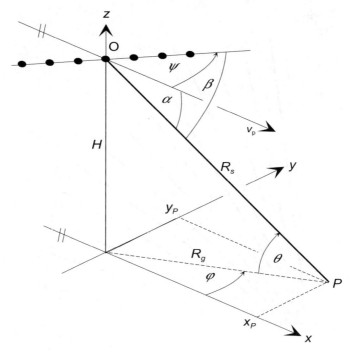

Figure 3.2: Geometry of a linear airborne array with crab angle

For $f_r = 1$ or $f_r = -1$ the isodop is a straight line in the flight direction, while for $f_r = 0$ its orientation is across the flight direction.

Figure 3.1 shows such set of hyperbolas on a R-φ grid. It should be noted that this symmetric set of hyperbolas is obtained only for a flight path parallel to the ground. For flight paths including a diving angle the hyperbolas will become non-symmetric with respect to the y-axis. Even parabolas or ellipses may be obtained, depending on the dive angle.

One can notice in Figure 3.1 that the Doppler frequency tends to become constant with range at larger distance and at angles close to 90°. There is a large Doppler gradient in the flight direction which decreases with range. For $H = 0$ there is no Doppler dependence with range.

3.1.2 Doppler-azimuth clutter trajectories

As a next step we derive the trajectories of clutter spectra in the Doppler-azimuth plane. Let us assume a linear array as depicted in Figure 3.2. The reason for considering a *linear* array is that the look direction can simply be expressed by one angle β. We focus for the time being on a single range increment, that means θ is constant. Let us

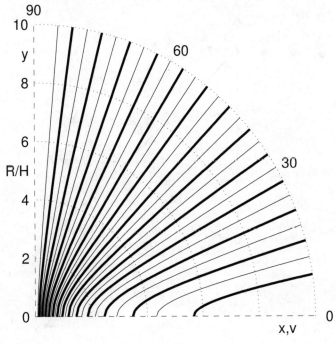

Figure 3.3: Beam traces (dark lines, only every second curve shown) and isodops for a sidelooking array

recall the relative clutter Doppler frequency (3.1) and (3.2)

$$f_r = \frac{f_D \lambda}{2v} = \cos \alpha \tag{3.9}$$

and

$$\cos \alpha = \cos \varphi \cos \theta \tag{3.10}$$

The look direction of the linear array is determined by

$$\cos \beta = \cos(\varphi - \psi) \cos \theta \tag{3.11}$$
$$= (\cos \varphi \cos \psi + \sqrt{1 - \cos^2 \varphi} \sin \psi) \cos \theta$$

Notice that β is the look direction relative to the array. This definition implies for example that $\beta = 0$ means the direction of the array axis.

Now we have to relate the relative Doppler frequency with the look direction. Solving (3.11) for $\cos \varphi$ gives

$$\cos \varphi = \frac{\cos \beta}{\cos \theta} \cos \psi \pm \sqrt{\left(\frac{\cos \beta}{\cos \theta}\right)^2 \cos^2 \psi - \left(\frac{\cos \beta}{\cos \theta}\right)^2 + \sin^2 \psi} \tag{3.12}$$

Inserting (3.12) into (3.9) and (3.10) one obtains after some manipulations

$$f_r^2 - 2 f_r \cos \beta \cos \psi + \cos^2 \beta = \sin^2 \psi \cos^2 \theta \tag{3.13}$$

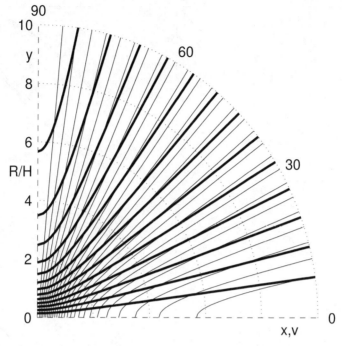

Figure 3.4: Beam traces (dark lines) and isodops for a forward looking array

Since the invariant

$$\delta = \begin{vmatrix} 1 & \cos\psi \\ \cos\psi & 1 \end{vmatrix} \qquad (3.14)$$

is greater than 0 (3.13) is a set of rotated ellipses in the f_r- $\cos\beta$ plane. The rotation angle is given by

$$\tan 2\gamma = \frac{2\cos\psi}{1-1} = \infty \qquad (3.15)$$

which means the ellipses are rotated by a constant angle of $45°$.

3.1.2.1 Sidelooking linear array

For a sidelooking array the crab angle becomes $\psi = 0$. Then (3.13) reduces to

$$f_r = \cos\beta \qquad (3.16)$$

which is a straight line in the f_r- $\cos\beta$ plane with a $45°$ slope.

3.1.2.2 Forward looking linear array

In the case of a forward looking configuration (3.13) becomes

$$f_r^2 + \cos^2\beta = \cos^2\theta \qquad (3.17)$$

76 Properties of airborne clutter

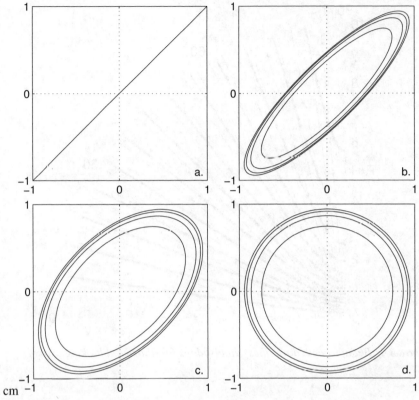

Figure 3.5: f_r-$\cos\varphi$ clutter trajectories for linear arrays: a. $\psi = 0°$; b. $\psi = 30°$; c. $\psi = 60°$; d. $\psi = 90°$; from inside to outside: $R/H = 1.5; 2; 2.5; 3$

which is a set of concentric circles with radii $\cos\theta$. Larger circles are associated with larger range.

In Figure 3.5 the f_r-$\cos\varphi$ trajectories of clutter Doppler spectra as given by (3.13) are depicted. The four plots have been calculated for different crab angles. The curves in each of the plots denote different range/altitude ratios R/H (increasing from inside to outside).

3.1.2.3 Range dependence of clutter Doppler

The Doppler frequency can be obtained by solving (3.12) for $f_r = \cos\varphi\cos\theta$:

$$f_r = \cos\psi\cos\beta \pm \sqrt{\cos^2\psi\cos^2\beta - \cos^2\beta + \sin^2\psi\cos^2\theta} \quad (3.18)$$

The sign before the square root denotes clutter Doppler frequencies left and right of the array axis.

For $\psi = 0°$ (sidelooking array) one gets

$$f_r = \cos\beta \quad (3.19)$$

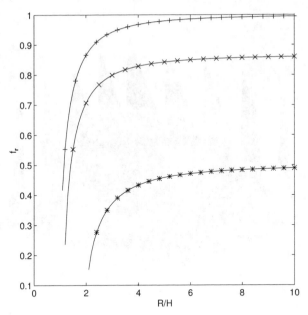

Figure 3.6: Range dependence of the clutter Doppler frequency for a forward looking array: $+\ \beta = 90°$ *(look direction = flight direction);* $\times\ \beta = 60°$*;* $*\ \beta = 30°$

which means that the Doppler frequency does not depend on the range which was stated already in (3.16). This is an expected result since the geometry of a sidelooking array coincides with the flight path. Therefore, beam traces on the ground coincide with isodops. This can be seen in Figure 3.3.

For $\psi = 90°$ (forward looking array) one gets

$$f_r = \pm\sqrt{\cos^2\theta - \cos^2\beta} \qquad (3.20)$$

By virtue of (3.11), we have $\cos\beta \leq \cos\theta$. The Doppler frequency becomes a maximum in the flight direction ($\beta = 90°$) and a minimum in the cross-flight direction ($\beta = 0°$). The range dependence of the clutter Doppler frequency is expressed in Figure 3.4, by the intersections between isodops and beam traces. A numerical evaluation of (3.20) leads to curves as shown in Figure 3.6. Notice that $\cos^2\theta$ has been replaced by $1 - H^2/R^2$. The plot shows the relative clutter frequency plotted versus range, normalised to the altitude of the radar above ground. As can be seen the major range dependence is in the area $R \leq 5H$. Due to the rotational symmetry of a linear array there are the same curves (but upside down) for negative Doppler frequencies. For $\beta = 0$ (endfire look direction) f_r is real only for $\theta = 0$ which means that the array looks at infinite range.

3.1.2.4 *Other array configurations*

While the isodops depend on the flight path the beam traces (dark curves in Figures 3.3 and 3.4) depend on the actual antenna configuration.

78 Properties of airborne clutter

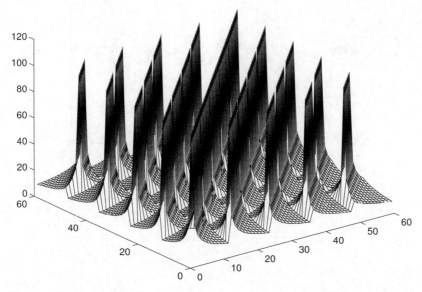

Figure 3.7: Modulus of the clutter covariance matrix vs. horizontal and vertical index (sidelooking linear array, $N = 12, M = 5$, omnidirectional sensors and transmission)

For planar arrays the total beam pattern is approximately the product of a horizontal and a vertical pattern (this is perfectly true for a rectangular array). Therefore, the results obtained for linear horizontal arrays concerning the range dependence of the Doppler frequency and the shape of the Doppler-azimuth trajectories apply approximately to planar arrays in the same way. The main difference is that clutter returns are weighted in range by the vertical beam pattern.

There are other possible array geometries, for example conformal arrays or volume arrays, see WILDEN and ENDER, [711].

ZATMAN [740] applies space-time adaptive processing to clutter data received by a circular array.

The Doppler-azimuth trajectories of these array configurations may be different from those of linear or planar antennas in that the unambiguous azimuth domain can be larger than $180°$ (in contrast to a linear array). Moreover, different beam shapes may cause different Doppler/range dependencies.

3.1.2.5 Historical note

Plotting the Doppler-azimuth trajectories means plotting a sine wave according to (3.9), (3.10) against another sine wave as given by (3.11), possibly with a different phase and frequency. As a result one obtains the so-called Lissajou graphs. They are well known from mechanics or other disciplines dealing with harmonic oscillations. Lissajou graphs can be generated by feeding one sine wave to the y- and another one to the x-input of an oscilloscope. The simplest graph is obtained by using the same

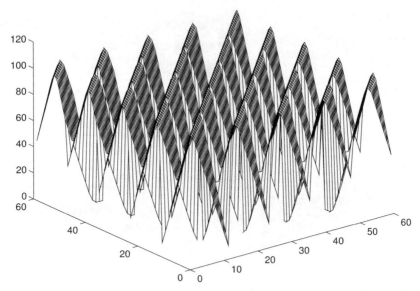

Figure 3.8: Modulus of the clutter covariance matrix vs. horizontal and vertical index (sidelooking linear array, $N = 12, M = 5$, directive sensors and transmission)

frequency and phase as shown in Figure 3.5a. With changing phase ψ (rotation of the array) one obtains ellipses or circles (Figure 3.5b–d).

3.1.2.6 Conclusions concerning the design of MTI systems

Sidelooking radar is used for surveillance, early warning and mapping purposes, especially in conjunction with SAR. The nose radar of a fighter aircraft is an example of forward looking MTI radar. The above conclusions on the range dependence of the clutter Doppler frequency have a serious impact on the design of adaptive MTI. For sidelooking radar training samples can be achieved from other range increments over all the visible range. For forward looking radar the use of data from different range increments is limited by the Doppler variation with range.

Alternatively, clutter data can be obtained in the time dimension for each individual range increment. That implies however some constraints on the operation of the radar because the antenna beam has to remain in a certain position as long as data for adapting the MTI are required.

3.2 The space-time covariance matrix

As carried out in Section 1.2 the covariance matrix of the interference (clutter, jamming, noise) plays a key role in interference rejection as well as for spectral representation.

80 Properties of airborne clutter

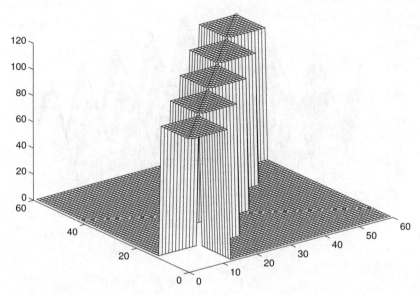

Figure 3.9: Jammer covariance matrix (sidelooking array, $N = 12$, $M = 5$, one jammer, absolute values)

Let us define the following vector quantities similar to (1.1):

$$\mathbf{c} = \begin{pmatrix} \mathbf{c}_1 \\ \mathbf{c}_2 \\ \vdots \\ \mathbf{c}_M \end{pmatrix} ; \quad \mathbf{s} = \begin{pmatrix} \mathbf{s}_1 \\ \mathbf{s}_2 \\ \vdots \\ \mathbf{s}_M \end{pmatrix} ; \quad \mathbf{n} = \begin{pmatrix} \mathbf{n}_1 \\ \mathbf{n}_2 \\ \vdots \\ \mathbf{n}_M \end{pmatrix} ; \quad \mathbf{j} = \begin{pmatrix} \mathbf{j}_1 \\ \mathbf{j}_2 \\ \vdots \\ \mathbf{j}_M \end{pmatrix} \quad (3.21)$$

where $\mathbf{c}, \mathbf{s}, \mathbf{n}$ and $\mathbf{j}$ denote the space-time vectors of clutter, target signal, noise and jamming. The elements of these vectors are defined by the following equations:

- vector of clutter echoes $\mathbf{c}$: (2.38)

- target signal vector $\mathbf{s}$: (2.31)

- noise vector $\mathbf{n}$: (2.49), (2.50)

- vector of jammer signals $\mathbf{j}$: (2.45)

Each subvector includes the spatial dimension, which means that the indices of the elements of the subvectors $i = 1, \ldots, N$ denote the array elements. The indices of the subvectors $m = 1, \ldots, M$ mean time, i.e. pulse repetition intervals.

Accordingly, the space-time covariance matrix of clutter, jammers and noise has

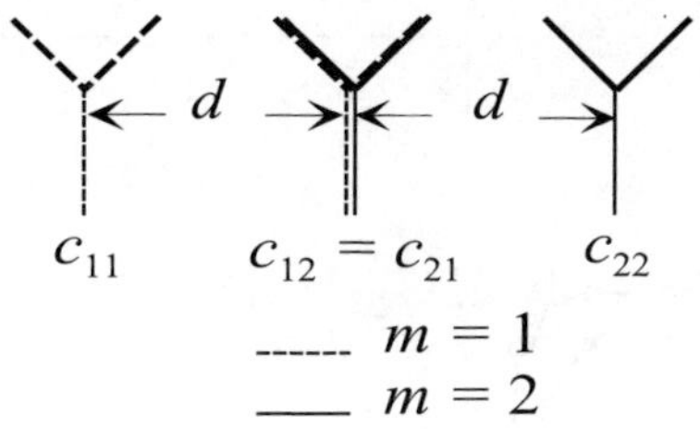

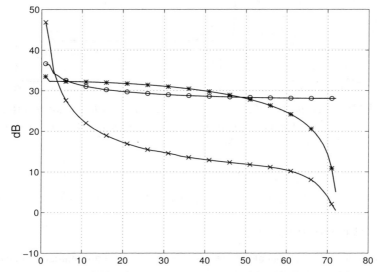

Figure 3.10: Principle of a two-pulse DPCA clutter canceller

Figure 3.11: Eigenspectra of a spatial covariance matrix. $N = 72, M = 1$; ○ omnidirectional sensors and transmission; ∗ directive sensors; × directive sensors and transmission

the form

$$\mathbf{Q} = E\{(\mathbf{c}+\mathbf{j}+\mathbf{n})(\mathbf{c}+\mathbf{j}+\mathbf{n})^*\} = \begin{pmatrix} \mathbf{Q}_{11} & \mathbf{Q}_{12} & \cdots & \mathbf{Q}_{1M} \\ \mathbf{Q}_{21} & \mathbf{Q}_{22} & \cdots & \mathbf{Q}_{2M} \\ \vdots & \vdots & \ddots & \vdots \\ \mathbf{Q}_{M1} & \mathbf{Q}_{M2} & \cdots & \mathbf{Q}_{MM} \end{pmatrix} \quad (3.22)$$

where again the indices of the submatrices m, p denote time (pulse repetition intervals) while the indices i, k run inside the submatrices and denote space (sensors).

Let $q_{ln}^{(c)} = E\{c_{im}c_{kp}^*\}$ denote an element of $\mathbf{Q}$. c_{im} is the clutter echo at the i-th sensor at time mT. The indices l, n are related to the sensor and echo indices through the relations (2.42) and (2.43):

$$l = (m-1)N + i \quad m = 1, \ldots, M; \quad i = 1, \ldots, N \quad (3.23)$$

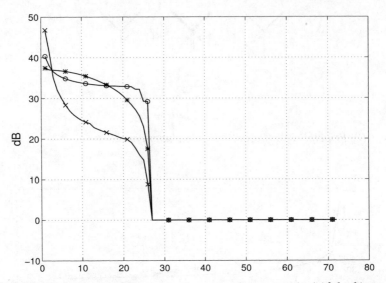

Figure 3.12: Eigenspectra of a space-time covariance matrix (sidelooking array). $N = 24$, $M = 3$; ○ omnidirectional sensors and transmission; ∗ directive sensors; × directive sensors and transmission

and
$$n = (p-1)N + k \quad p = 1,\ldots,M; \quad k = 1,\ldots,N \tag{3.24}$$

3.2.1 The components

We assume that all components of the covariance matrix $\mathbf{Q}$ are mutually independent so that $\mathbf{Q}$ is a sum of clutter, jammer and noise covariance matrices

$$\mathbf{Q} = \mathbf{Q}_c + \mathbf{Q}_j + \mathbf{Q}_n \tag{3.25}$$

The clutter-to-noise ratio is

$$\text{CNR} = \text{tr}\mathbf{Q}_c/\text{tr}\mathbf{Q}_n = \text{tr}\mathbf{Q}_c/(NM \cdot P_n) \tag{3.26}$$

and the jammer-to-noise ratio

$$\text{JNR} = \text{tr}\mathbf{Q}_j/\text{tr}\mathbf{Q}_n = \text{tr}\mathbf{Q}_j/(NM \cdot P_n) \tag{3.27}$$

3.2.1.1 Clutter

The elements of the space-time clutter covariance matrix are composed of the clutter echoes as defined in (2.38):

$$\begin{aligned} q_{ln}^{(c)} &= E\{c_{im}c_{kp}^*\} \\ &= E\left\{\int_{\varphi=0}^{2\pi} A(\varphi)D(\varphi)L(\varphi)G(\varphi,m)\Phi_m(\varphi,v_p)\Psi_i(\varphi)\mathrm{d}\varphi \right. \end{aligned} \tag{3.28}$$

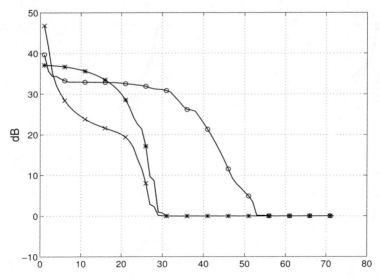

*Figure 3.13: Eigenspectra of a space-time covariance matrix (forward looking array). $N = 24$, $M = 3$; ○ omnidirectional sensors and transmission; * directive sensors; × directive sensors and transmission*

$$\times \int_{\phi=0}^{2\pi} [A(\phi)D(\phi)L(\phi)G(\phi,m)\Phi_p(\phi,v_{\mathrm{p}})\Psi_k(\phi)]^* \mathrm{d}\phi \Big\}$$

where $A(\varphi)$ and $A(\phi)$ are the complex amplitudes of clutter arrivals from angles φ and ϕ and the indices l, n depend on i, k, m, p as defined by (3.23) and (3.24). The phase terms $\Phi_p^{(\mathrm{t})}(\varphi, v_{\mathrm{p}})$ and $\Phi_k^{(\mathrm{s})}(\varphi)$ have been defined in (2.36).

We assume that the echoes from different clutter scatterers are mutually independent, i.e.,

$$E\{A(\varphi)A^*(\phi)\} = 0 \quad \varphi \neq \phi \qquad (3.29)$$

Then the cross-terms in azimuth vanish so that (3.28) becomes

$$\begin{aligned} q_{ln}^{(\mathrm{c})} &= E\{c_{im}c_{kp}^*\} \\ &= \int_{\varphi=0}^{2\pi} E\{AA^*\}D^2(\varphi)L^2(\varphi)G(\varphi,m)G^*(\varphi,p) \\ &\quad \times \Phi_m(\varphi,v_{\mathrm{p}})\Phi_p^*(\varphi,v_{\mathrm{p}})\Psi_i(\varphi)\Psi_k^*(\varphi)\mathrm{d}\varphi \end{aligned} \qquad (3.30)$$

Replacing $E\{AA^*\}$ by the correlation model (2.72), gives

$$\begin{aligned} q_{ln}^{(\mathrm{c})} &= E\{c_{im}c_{kp}^*\} \\ &= P_{\mathrm{c}}\rho(mp)\int_{\varphi=0}^{2\pi} \rho(\tau_{ik})D^2(\varphi)L^2(\varphi)G(\varphi,m)G^*(\varphi,p) \\ &\quad \times \Phi_m(\varphi,v_{\mathrm{p}})\Phi_p^*(\varphi,v_{\mathrm{p}})\Psi_i(\varphi)\Psi_k^*(\varphi)\mathrm{d}\varphi \end{aligned} \qquad (3.31)$$

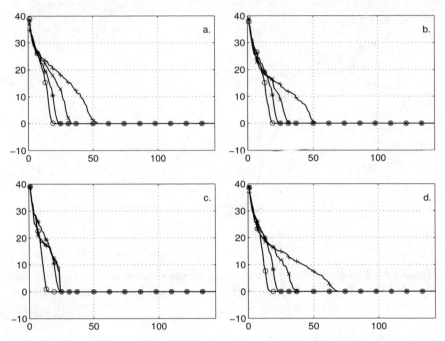

*Figure 3.14: Influence of PRF and sensor spacing on eigenspectrum (linear array, SL; numbers of eigenvalues in [ ]: N_{BS}, see (3.61)): a. $d/\lambda = 0.5$ (○ PRF = 24000 [18], * PRF = 12000 [23], × PRF = 6000 [34], + PRF = 3000 [56]); b. PRF = 12000 (○ $d/\lambda = 0.25$ [18], * $d/\lambda = 0.5$ [23], × $d/\lambda = 1$ [34], + $d/\lambda = 2$ [56]); c. $\gamma = 1$ [23] (○ PRF = 24000, * PRF = 12000, × PRF = 6000, + PRF = 3000; d. ○ PRF = 17000 $d/\lambda = 0.25$ [28], * PRF = 13000 $d/\lambda = 0.5$ [35], × PRF = 7400 $d/\lambda = 1$ [21], + PRF = 3700 $d/\lambda = 2$ [21]*

where

$$P_c = \sqrt{E\{|\,A(\varphi,i,m)\,|^2\}E\{|\,A(\varphi_r,k,p)\,|^2\}} = E\{|\,A(\varphi,i,m)\,|^2\}$$

is the clutter power at the individual sensor output.

For constant PRI ($T_m = T$) one gets for the temporal phase term

$$\Phi_m(\varphi, v_{\text{p}})\Phi_p^*(\varphi, v_{\text{p}}) = \Phi_{m-p}(\varphi, v_{\text{p}}) \qquad (3.32)$$

For a linear equispaced array (ULA) the spatial phase term becomes

$$\Psi_i(\varphi)\Psi_k^*(\varphi) = \Psi_{i-k}(\varphi) \qquad (3.33)$$

We assume for simplicity that the temporal clutter correlation due to internal clutter fluctuations is independent of the direction so it can be placed outside the integral. Notice that the spatial correlation due to the system bandwidth varies with the azimuth of the incoming wave which follows from (2.58). Therefore, $\rho(\tau_{ik})$ is a part of the integrand.

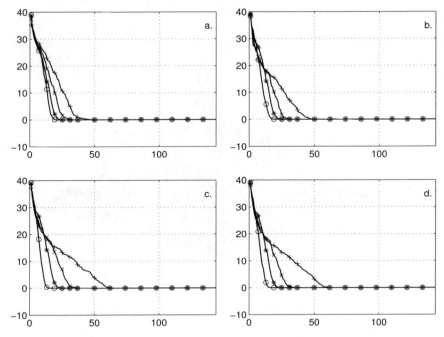

*Figure 3.15: Influence of PRF and sensor spacing on eigenspectrum (linear array, FL; numbers of eigenvalues in [ ]: $N_{\rm BS}$, see (3.61)): a. $d/\lambda = 0.5$ (○ PRF = 24000 [18], * PRF = 12000 [23], × PRF = 6000 [34], + PRF = 3000 [56]); b. PRF = 12000 (○ $d/\lambda = 0.25$ [18], * $d/\lambda = 0.5$ [23], × $d/\lambda = 1$ [34], + $d/\lambda = 2$ [56]); c. $\gamma = 1$ [23] (○ PRF = 24000, * PRF = 12000, × PRF = 6000, + PRF = 3000; d. ○ PRF = 17000 $d/\lambda = 0.25$ [28], * PRF = 13000 $d/\lambda = 0.5$ [35], × PRF = 7400 $d/\lambda = 1$ [21], + PRF = 3700 $d/\lambda = 2$ [21]*

3.2.1.2 Jamming

The elements of the spatial jammer covariance matrix $\mathbf{Q}_j$ are composed of the jammer signals following (2.48):

$$q_{ln} = \begin{cases} \sum_{j=1}^{J} P_j \exp[j\frac{2\pi}{\lambda} \\ \times((x_i - x_k)\cos\varphi_j + (y_i - y_k)\sin\varphi_j)\cos\theta_j \\ -(z_i - z_k)\sin\theta_j)] & m = p \\ 0 & m \neq p \end{cases} \quad (3.34)$$

where J is the number of jammers. The relations between the indices l, n of the covariance matrix and i, k, m, p have been defined in (3.23) and (3.24).

In (2.46) we assumed that the jammers are uncorrelated between different pulse repetition intervals. Therefore, the jammer space-time covariance matrix becomes a block diagonal matrix where the spatial submatrices include the spatial correlation across the array. A calculated jammer + noise covariance matrix (absolute values) is

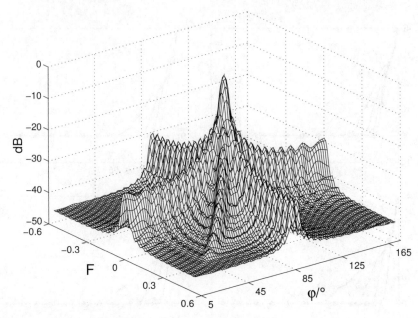

Figure 3.16: Fourier clutter spectrum (sidelooking array, $\varphi_L = 90°$)

shown in Figure 3.9. Notice the noise component (20 dB below jammer power) along the main diagonal.

3.2.1.3 Noise

The noise covariance matrix is, according to the assumptions made in (2.49) and (2.50), a $NM \times NM$ diagonal matrix

$$\mathbf{Q}_n = P_n \mathbf{I} \qquad (3.35)$$

3.2.2 The displaced phase centre antenna (DPCA) principle

The DPCA technique was the first method used for clutter rejection with platform motion compensation. The idea was to displace either physically or electronically the phase centres of two or more antennas in the flight direction. The principle is illustrated in Figure 3.10. The displacement of the antennas is such that during the flight the second antenna assumes the position of the first one after one PRI. At time $m = 1$ we have the dashed position of the antenna pair while the arrays position one PRI later ($m = 2$) is indicated by the solid lines. Notice that the signals c_{12} (first echo at second antenna) is received at the same position as c_{21} (second echo at first antenna). Clutter cancellation can simply be done by subtracting: $c_{12} - c_{21}$. Some detailed information on DPCA techniques can be found in SKOLNIK's *Radar Handbook*, 1st edition [609, Chapter 18], and 2nd edition [610, Chapter 16]. The interrelations between optimum

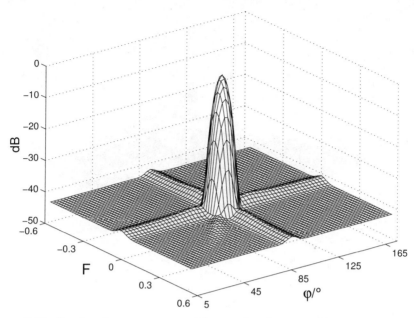

Figure 3.17: Fourier clutter spectrum with spatial and temporal Hamming weighting (sidelooking array, $\varphi_L = 90°$)

processing, DPCA and orthogonal projection have been illustrated by RICHARDSON [562].

MURRAY et al. [471] use airborne clutter data obtained at steep grazing angles to simulate spaceborne geometry. DPCA is used for clutter cancellation. NOHARA [503] describes the integration of a DPCA system in a signal processor for spaceborne surveillance radar. DPCA is applied also in the CARABAS UWB SAR (PETTERSON [535]).

Let us recall the phase terms (2.36)

$$\Phi_m(v_p, \varphi) = \exp\left[j\frac{2\pi}{\lambda}2v_p mT \cos\varphi \cos\theta\right]$$

$$\Psi_i(\varphi) = \exp\left[j\frac{2\pi}{\lambda}(x_i \cos\varphi + y_i \sin\varphi)\cos\theta - z_i \sin\theta\right] \quad (3.36)$$

Let us assume a sidelooking array with equidistant elements and transmission of pulses at constant rate. Then the complex phase terms in the clutter covariance (3.31) become

$$\Phi_{m-p}(v_p, \varphi) = \exp\left[j\frac{2\pi}{\lambda}2v_p(m-p)T \cos\varphi \cos\theta\right]$$

$$\Psi_{i-k}(\varphi) = \exp\left[j\frac{2\pi}{\lambda}(i-k)d \cos\varphi \cos\theta\right] \quad (3.37)$$

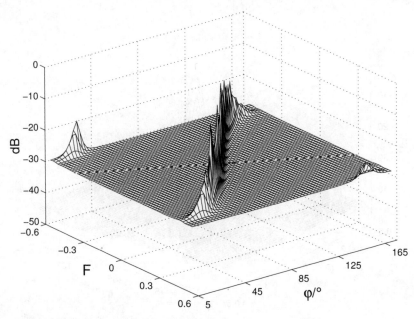

Figure 3.18: MV clutter spectrum (sidelooking array, $\varphi_L = 45°$)

The covariance (3.31) then contains the product

$$\Phi_{m-p}(v_{\mathrm{p}},\varphi)\Psi_{i-k}(\varphi) = $$
$$= \exp\left[j\frac{2\pi}{\lambda}(2v_{\mathrm{p}}(m-p)T + (i-k)d)\cos\varphi\cos\theta\right] \quad (3.38)$$

The DPCA condition is met whenever

$$(m-p)2v_{\mathrm{p}}T = -(i-k)d \quad (3.39)$$

In this case the phase term (3.38) is equal to 1 for all angles of arrival φ, θ so that the space-time covariance (3.31) becomes maximum. Such maxima show up as side-ridges in the covariance matrix (see Figure 3.7).

It should be pointed out that the DPCA principle is effective even if the system is undersampled, i.e., $d > \lambda/2$ (spatial undersampling) or $PRF < 4v_p/\lambda$ (PRF lower than clutter Doppler bandwidth, temporal undersampling) or both.

If either the spatial or the temporal dimension is sampled at or above Nyquist rate ($d \leq \lambda/2$, $PRF \geq 4v_p/\lambda$) then the interpolation property of Nyquist sampling provides high correlation among space-time samples even if the DPCA condition is not met.

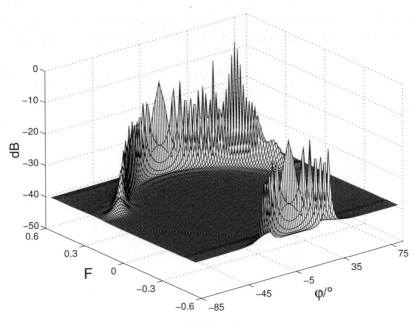

Figure 3.19: MV spectrum for forward looking linear array

3.2.2.1 A closer look at the clutter component of **Q**

Let us now have a closer look at the nature of space-time clutter covariance matrices. Consider a linear array in a sidelooking arrangement, which means that the array axis coincides with the flight axis. Assume a homogeneous clutter background with backscattered echoes coming from all directions φ. Then we obtain from (2.29) the response of an equispaced array to one arrival at angles (φ, θ)

$$c_{im}(\varphi) = A \exp\left[j\frac{2\pi}{\lambda}(di + 2v_{\mathrm{p}}mT)\cos\varphi\cos\theta\right] \tag{3.40}$$

with sensor spacing d and sensor index i. Now let us consider the so-called DPCA case

$$d = 2v_{\mathrm{p}}T \tag{3.41}$$

This means that the phase advance between any two positions of the array is equal to the phase difference between any two adjacent sensors. We call this assumption the *DPCA condition*. It means that each sensor assumes the same position as the preceding one after one PRI so that each sensor seems to be fixed in space for one (or more) PRIs. Strictly speaking, it does not assume the same position but the same phase. Notice that due to the factor of 2 in (3.41) the array moves only a distance of $d/2$ during one interval T (the factor of 2 orginates from the two-way delay between a moving receive array and a moving transmitter).

Properties of airborne clutter

Now let us generate a space-time signal vector of the form (3.21) with elements given by (3.40). It can be generated from a signal sequence of the following kind

$$g_i(\varphi) = A(\varphi)\exp\left[j\frac{2\pi}{\lambda}di\cos\varphi\cos\theta\right] \quad i=1,\ldots,N+M-1 \tag{3.42}$$

by using a shift operator

$$\mathbf{c} = \Sigma\mathbf{g} \tag{3.43}$$

where $\mathbf{g}$ includes elements $g_i(\varphi)$ and Σ is the shift operator which, for $N=4, M=6$, has the form

$$\Sigma = \begin{pmatrix} 1 & & & & & & & & \\ & 1 & & & & & & & \\ & & 1 & & & & \text{\Large 0} & & \\ & & & 1 & & & & & \\ 1 & & & & & & & & \\ & 1 & & & & & & & \\ & & 1 & & & & & & \\ & & & 1 & & & & & \\ & & & & 1 & & & & \\ & & & & & 1 & & & \\ & & & & 1 & & & & \\ & & & & & 1 & & & \\ & & & & & & 1 & & \\ & & & & & & & 1 & \\ & & & & & & 1 & & \\ & & & & & & & 1 & \\ & & & & & & & & 1 \\ & & & & & & & & 1 \\ & & \text{\Large 0} & & & & & 1 & \\ & & & & & & & & 1 \end{pmatrix} \tag{3.44}$$

Notice that Σ is a $(NM)\times(N+M-1)$ matrix. The shift operator Σ describes the motion of the array through the wavefield.

Integrating over φ leads to the covariance matrix of $\mathbf{c}$:

$$\mathbf{Q} = E\{\mathbf{cc}^*\} = \Sigma E\left\{\int_{\varphi=0}^{2\pi}\mathbf{g}(\varphi)d\varphi\int_{\psi=0}^{2\pi}\mathbf{g}^*(\psi)d\psi\right\}\Sigma^* \tag{3.45}$$

Assuming independent arrivals from different directions, i.e.,

$$E\{A(\varphi)A^*(\psi)\} = 0, \quad \varphi \neq \psi \tag{3.46}$$

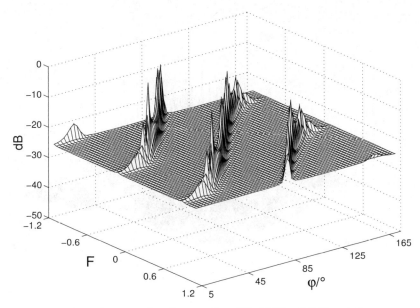

Figure 3.20: Effect of temporal undersampling (SL, PRF = 0.5 Nyquist)

and setting for simplicity $u = \cos\varphi$, the elements of the inner matrix become, using (3.42),

$$E\left\{\int_{u=-1}^{1} A(u)A^*(u)\mathbf{g}_i(u)\mathbf{g}_k^*(u)\mathrm{d}u\right\} = P_c \frac{\sin(\frac{2\pi}{\lambda}d(i-k)\sin\theta)}{\frac{2\pi}{\lambda}d(i-k)\sin\theta} \quad (3.47)$$

Let us consider the case of spatial Nyquist sampling $d = \lambda/2$ and assume $H = 0$ so that $\theta = 0$. The inner matrix of (3.45) is then proportional to the identity matrix $P_c \mathbf{I}$ and the covariance matrix becomes

$$\mathbf{Q} = P_c \boldsymbol{\Sigma}\mathbf{I}\boldsymbol{\Sigma}^* = P_c \boldsymbol{\Sigma}\boldsymbol{\Sigma}^* \quad (3.48)$$

where $P_c = E\{A(\varphi)A^*(\varphi)\}$ is the clutter power. In (3.49), an example of the matrix $\boldsymbol{\Sigma}\boldsymbol{\Sigma}^*$ can be seen. The dimensions $N = 5$ and $M = 3$ were chosen so that the matrix has the dimensions 15×15. There are single unity elements which denote full correlation while all other elements are zero. A simulated covariance matrix with $N = 12$, $M = 5$ is shown in Figure 3.7. The correspondence between (3.49) and Figure 3.7 is obvious.

92 *Properties of airborne clutter*

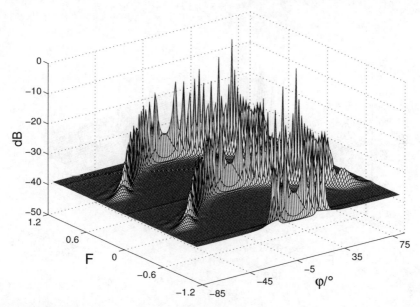

Figure 3.21: *Effect of temporal undersampling (FL, PRF = 0.5 Nyquist)*

$$\Sigma\Sigma^* = \begin{pmatrix} 1 & . & . & . & . & | & . & . & . & . & . & | & . & . & . & . & . \\ . & 1 & . & . & . & | & 1 & . & . & . & . & | & . & . & . & . & . \\ . & . & 1 & . & . & | & . & 1 & . & . & . & | & . & 1 & . & . & . \\ . & . & . & 1 & . & | & . & . & 1 & . & . & | & . & . & 1 & . & . \\ . & . & . & . & 1 & | & . & . & . & 1 & . & | & . & . & . & 1 & . \\ \hline . & 1 & . & . & . & | & 1 & . & . & . & . & | & . & . & . & . & . \\ . & . & 1 & . & . & | & . & 1 & . & . & . & | & 1 & . & . & . & . \\ . & . & . & 1 & . & | & . & . & 1 & . & . & | & . & 1 & . & . & . \\ . & . & . & . & 1 & | & . & . & . & 1 & . & | & . & . & 1 & . & . \\ . & . & . & . & . & | & . & . & . & . & 1 & | & . & . & . & 1 & . \\ \hline . & . & 1 & . & . & | & . & 1 & . & . & . & | & 1 & . & . & . & . \\ . & . & . & 1 & . & | & . & . & 1 & . & . & | & . & 1 & . & . & . \\ . & . & . & . & 1 & | & . & . & . & 1 & . & | & . & . & 1 & . & . \\ . & . & . & . & . & | & . & . & . & . & 1 & | & . & . & . & 1 & . \\ . & . & . & . & . & | & . & . & . & . & . & | & . & . & . & . & 1 \end{pmatrix} \quad (3.49)$$

As can be seen there is a correlation ridge along the main diagonal and there are several other high-correlation ridges in parallel to the main diagonal. These parallel ridges arise from the space-time signal concept. If only one dimension, for example the spatial dimension, were taken into account the covariance matrix would look like the first partial figure in the foreground of Figure 3.7, which corresponds to the upper left 5×5 submatrix in (3.49). This is just an identity matrix which does not give

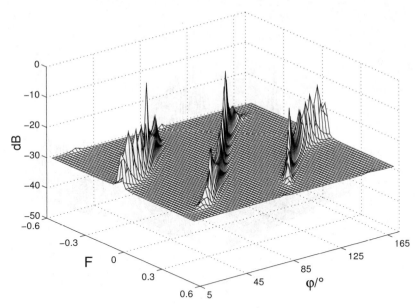

Figure 3.22: Effect of spatial undersampling (SL, $d/\lambda = 1$)

the chance for successful clutter rejection. The cross-correlation ridges are generated through the additional temporal dimension.

A look at the first off-diagonal in (3.49) reveals how the gaps in the individual off-diagonal ridges in Figures 3.7 and 3.8 are generated.

It should be noted that the clutter signal $g_i(\varphi)$ in (3.42) does not contain any temporal dimension (the time dimension enters our consideration through the motion of the array via the shift operator). Therefore the identity matrix in (3.48) means that the field is spatially white. If this is not satisfied due to some weighting in azimuth some off-diagonal elements will come up in this matrix. In Figure 3.8, the homogeneous clutter field has been weighted with sensor and array directivity patterns. As one can see some additional correlation values are coming up, resulting in less steep slopes of the correlation ridges.

3.2.2.2 Motion compensation through data selection strategy

In order to get more insight into the nature of air- and spaceborne radar clutter let us consider an interesting property brought up by LOMBARDO [414]. Under DPCA conditions the platform motion can be compensated for by a suitable choice of the received echo data. Let us select from (3.44) only those columns which are complete,

94 Properties of airborne clutter

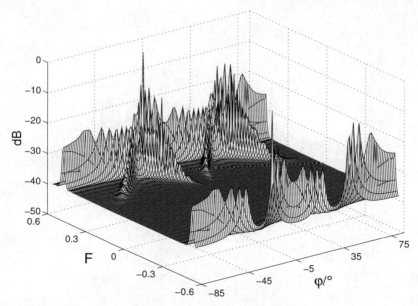

Figure 3.23: Effect of spatial undersampling (FL, $d/\lambda = 1$)

i.e., which contain N ones. This results in a reduced matrix as shown in (3.50):

$$\Sigma = \begin{pmatrix} 1 & & & & & & & & & & & \\ & 1 & & & & & & & & & & \\ & & 1 & & & & & & & & & \\ & & & 1 & & & & & & & & \\ & & & & 1 & & & & & & & \\ & & & & & 1 & & & & & & \\ & 1 & & & & & & & & & & \\ & & & & & & 1 & & & & & \\ & & & & & & & 1 & & & & \\ & & & & & & & & 1 & & & \\ & & & & & & & & & 1 & & \\ & & & & & & & & & & 1 & \end{pmatrix} \quad (3.50)$$

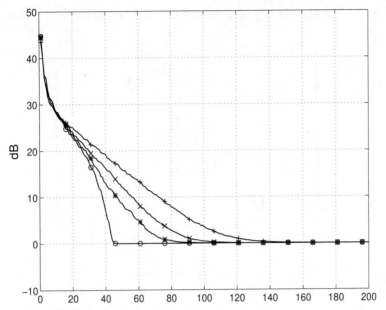

Figure 3.24: Eigenspectra for different clutter bandwidth (sidelooking, $N = M = 24$): $\circ$ $B_c = 0$; $*$ $B_c = 0.05$; $\times$ $B_c = 0.1$; $+$ $B_c = 0.2$

Let us, for illustration, interchange spatial (i, k) and temporal (m, p) indices[2] in (3.50). This results in a matrix of the form (3.51).

$$\Sigma = \begin{pmatrix} 1 & & & & & & & & & & & & \\ 1 & & & & & & & & & & & & \\ 1 & & & & & & & & & & & & \\ 1 & & & & & & & & & & & & \\ & 1 & & & & & & & & & & & \\ & 1 & & & & & & & & & & & \\ & 1 & & & & & & & & & & & \\ & 1 & & & & & & & & & & & \\ & & 1 & & & & & & & & & & \\ & & 1 & & & & & & & & & & \\ & & 1 & & & & & & & & & & \\ & & 1 & & & & & & & & & & \end{pmatrix} \qquad (3.51)$$

Then the clutter covariance matrix becomes a block diagonal matrix of the form shown in (3.52). As can be seen, the submatrices along the diagonal are dyadics, each of which contributes one clutter eigenvalue. So the total number of eigenvalues is

---

[2] See (3.23) (3.24).

$N_E = N$ where N is now the new spatial dimension of the reduced matrix (3.50).

$$\Sigma\Sigma^* = \begin{pmatrix} 1 & 1 & 1 & 1 & & & & & & & & \\ 1 & 1 & 1 & 1 & & & & & & & & \\ 1 & 1 & 1 & 1 & & & & & \mathbf{0} & & & \\ 1 & 1 & 1 & 1 & & & & & & & & \\ & & & & 1 & 1 & 1 & 1 & & & & \\ & & & & 1 & 1 & 1 & 1 & & & & \\ & & & & 1 & 1 & 1 & 1 & & & & \\ & & & & 1 & 1 & 1 & 1 & & & & \\ & & & & & & & & 1 & 1 & 1 & 1 \\ & & \mathbf{0} & & & & & & 1 & 1 & 1 & 1 \\ & & & & & & & & 1 & 1 & 1 & 1 \\ & & & & & & & & 1 & 1 & 1 & 1 \end{pmatrix} \quad (3.52)$$

Notice that (3.52) is similar[3] to the jammer covariance matrix illustrated in Figure 3.9. The main difference is that spatial and temporal indices are interchanged. In fact, jammers are discrete in space, but uncorrelated in Doppler whereas clutter is discrete in Doppler but uncorrelated in space. Jammers can be suppressed by a spatial only filter whereas clutter (after compensating for the platform motion) can be cancelled by a temporal only filter.

Practical advantages of this kind of motion compensation are not obvious because it relies on DPCA conditions which in practice are not fulfilled. In particular, the actual PRF might be dictated by other operational aspects rather than by requirements given by MTI design.

3.2.2.3 The DPCA two-pulse canceller

In the following example we will demonstrate the close connection between the DPCA principle and optimum processing. The reader is reminded that this connection exists only for sidelooking arrays.

Let us consider the special case $N = M = 2$. That means, we deal with a two-element array and two subsequent echo samples for clutter cancellation, see Figure 3.10. Then (3.49) reduces to

$$\mathbf{Q} = \begin{pmatrix} 1 & 0 & 0 & 0 \\ 0 & 1 & 1 & 0 \\ 0 & 1 & 1 & 0 \\ 0 & 0 & 0 & 1 \end{pmatrix} \quad (3.53)$$

---

[3] They are in general not identical because the submatrices of the jammer covariance matrix may contain more than one jammer which causes more than one eigenvalue. Also, the covariance matrix in Figure 3.9 contains additional receiver noise.

Adding the noise-to-clutter power ratio similar to (1.28), one gets

$$\mathbf{Q} = \begin{pmatrix} 1+a & 0 & 0 & 0 \\ 0 & 1+a & 1 & 0 \\ 0 & 1 & 1+a & 0 \\ 0 & 0 & 0 & 1+a \end{pmatrix} \quad (3.54)$$

The optimum clutter canceller is given by the inverse of $\mathbf{Q}$:

$$\mathbf{Q}^{-1} = \begin{pmatrix} \frac{1}{1+a} & 0 & 0 & 0 \\ 0 & \frac{1+a}{(2+a)a} & -\frac{1}{(2+a)a} & 0 \\ 0 & -\frac{1}{(2+a)a} & \frac{1+a}{(2+a)a} & 0 \\ 0 & 0 & 0 & \frac{1}{1+a} \end{pmatrix} \quad (3.55)$$

Normalising the inverse by $\frac{1+a}{(2+a)a}$ and setting $a = 0$ (noise-free case) one gets

$$\tilde{\mathbf{Q}}^{-1} = \begin{pmatrix} 0 & 0 & 0 & 0 \\ 0 & 1 & -1 & 0 \\ 0 & -1 & 1 & 0 \\ 0 & 0 & 0 & 0 \end{pmatrix} \quad (3.56)$$

Multiplying by a clutter vector gives

$$\mathbf{y} = \tilde{\mathbf{Q}}^{-1} \begin{pmatrix} c_{11} \\ c_{21} \\ c_{12} \\ c_{22} \end{pmatrix} = \mathbf{0} \quad (3.57)$$

since $c_{21} = c_{12}$, i.e., c_{21} and c_{12} are clutter echoes taken at the same position but at different instants of time. It is obvious that the operation (3.57) provides perfect clutter cancellation under the given conditions.

It has been found through simulation by FAUBERT et al. [161] that DPCA clutter cancellation may be an efficient solution as long as no mismatch in the multichannel antenna occurs. The performance is strongly affected by various kinds of mismatch (different subarray or sensor patterns, misaligned phase centres, aircraft crabbing, platform velocity errors). TSANDOULAS [653] used the aircraft's inertial navigation system to adjust the PRF to the instantaneous platform velocity. While most authors consider two-pulse DPCA, LIGHTSTONE et al. [400] compare the performance of optimum and suboptimum DPCA for different number of pulses.

3.2.2.4 The CPCT technique

CPCT (coincident phase center technique) is a straightforward generalisation of the DPCA principle to forward looking antennas. The antenna is illuminated in such a way that two phase centres are generated, one before and one behind the aperture. The distance between the two phase centres is chosen so that the phase of the rear phase centre due to the second pulse is equal to the phase of the front phase centre due to the first pulse. In this case one obtains DPCA conditions for the forward looking antenna. Some details on CPCT are given in SKOLNIK [610, p. 16–20]). To my knowledge no open literature is available on this technique.

Equivalently one may place two (or more) forward looking array antennas one behind the other to generate DPCA conditions for forward looking applications. However, as will be explained in Section 3.5.2.2, the DPCA principle is not necessary for space-time clutter suppression. Under idealised conditions, for range dependent adaptation, optimum clutter suppression can be obtained with a single forward looking array.

3.2.3 Eigenspectra

Eigenspectra for analysis of space-time covariance matrices have been introduced by KLEMM [298, 300]. In essence an eigenspectrum contains the rank-ordered eigenvalues of a covariance matrix (or power spectral matrix). The covariance matrix can be decomposed into eigenvectors and eigenvalues as follows

$$\mathbf{Q} = \mathbf{E}\mathbf{\Lambda}\mathbf{E}^* \tag{3.58}$$

$\mathbf{\Lambda}$ is the diagonal matrix of eigenvalues and $\mathbf{E}$ the unitary matrix of eigenvectors. Since $\mathbf{Q}$ is Hermitian the eigenvalues are real. Since $\mathbf{Q}$ is positive definite the eigenvalues are positive. The eigenvalue distribution of an interference + noise covariance matrix exhibits clearly how much vector space has been occupied by the interference and how much is left for the signal. The number of interference eigenvalues gives an indication of the number of degrees of freedom of the interference scenario. Any interference rejection operation should have at least this number of degrees of freedom.

In Figures 3.11, 3.12, and 3.13, the effect of directivity of the sensor elements and the transmit antenna on the clutter eigenspectrum is discussed.

In Figure 3.11, we consider a linear array assuming $N = 72$ sensors and $M = 1$ echo sample only. This is an entirely *spatial* situation. The circled curve has been calculated for omnidirectional sensors and omnidirectional transmission. As can be seen the curve is almost flat as in the case of white noise. The little 'colouring' on the left side is due to the depression angle θ. Introducing sensor directivity patterns leads to some additional colouring (asterisks) and even more for additional directivity on transmit. Nevertheless we notice that there is no clear distinction between clutter and noise eigenvalues. This, however, is the prerequisite for successful clutter suppression.

In Figure 3.12, we consider an array with $N = 24$ elements, but take $M = 3$ subsequent echoes into account. Notice that the size of the covariance matrix and, hence, the number of eigenvales is the same as in Figure 3.11. Now we see a clear distinction between clutter (on the left) and the floor of identical noise eigenvalues. As one can see the number of clutter eigenvalues is 26 which is $N + M - 1$. The relation

$$N_e = N + M - 1 \tag{3.59}$$

holds for a sidelooking array with Nyquist sampling in both dimensions, thus fulfilling the DPCA condition. This was first found heuristically by KLEMM[300].[4]

---
[4]Actually, in the quoted literature it reads $N_e = N + M$. The correct number is $N_e = N + M - 1$.

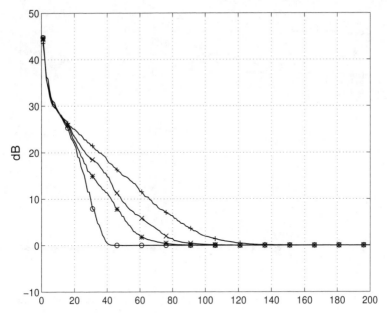

Figure 3.25: Eigenspectra for different clutter bandwidths (forward looking, N = M = 24): ○ $B_c = 0$; ∗ $B_c = 0.05$; × $B_c = 0.1$; + $B_c = 0.2$

We can observe that the matrix

$$\mathbf{Q} = P_c \mathbf{\Sigma I \Sigma}^* = P_c \mathbf{\Sigma \Sigma}^* \tag{3.60}$$

see (3.48), has exactly $N + M - 1$ eigenvalues since the shift operator $\mathbf{\Sigma}$ defined in (3.44) has $N + M - 1$ linearly independent columns.

Figure 3.13 shows the same curves as Figure 3.12, but for a forward looking array. We notice that for omnidirectional sensors and transmission (circles) the number of clutter eigenvalues is about twice as large as in case of a sidelooking array. This has to do with the left–right ambiguity of the sidelooking array. Both arrays receive the same spectrum, defined by frequencies from $-2v_p/\lambda$ to $+2v_p/\lambda$. However, the forward looking array is capable of distinguishing between left and right (relative to the flight axis) which means that for each Doppler frequency there are two directions of arrival. This results in twice the number of degrees of freedom, i.e. twice the number of eigenvalues than for the sidelooking array. Assuming horizontal sensor directivity patterns according to (2.20), all negative Doppler frequencies are cancelled so that only about one half of the eigenvalues remains. Notice that in both cases (SL and FL) the directivity of the transmit beam emphasises a very few eigenvalues.

In our model we assumed sensor patterns according to (2.20) and (2.21). That means there are no sidelobes in the back of the antenna. It has been shown by WARD [690, p. 43] that a certain backlobe level increases the number of eigenvalues for all antenna orientations other than *sidelooking* linear or planar arrays. This corresponds to the curve (○) in Figure 3.13, which has been calculated for a forward looking array with omnidirectional sensors and transmission.

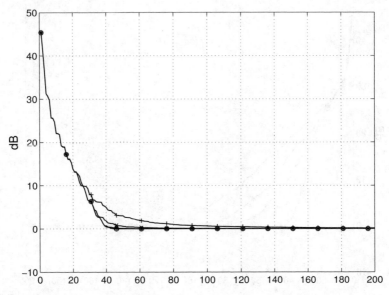

Figure 3.26: Effect of range walk, dependence of eigenspectra on range resolution (SL): ○ *no range walk;* ∗ $\Delta_R = 100m$; × $\Delta_R = 10m$; + $\Delta_R = 1m$

3.2.3.1 A generalised formula for the number of eigenvalues

BRENNAN and STAUDAHER [65] found the following rule for the number of clutter eigenvalues of the space-time clutter covariance matrix of a *sidelooking* linear array

$$N_{\text{BS}} \approx \text{int}\{N + (M-1)\gamma\} \tag{3.61}$$

where

$$\gamma = \frac{2v_{\text{p}}T}{d} \tag{3.62}$$

and int denotes the next integer number. Proofs of this rule have been published by WARD [690, p. 187], and ZHANG Q. and MIKHAEL [745].

γ denotes the array motion during one PRI related to the element spacing. For DPCA conditions one gets $\gamma = 1$. It is the ratio between spatial and Doppler clutter frequency. The normalised clutter Doppler frequency is

$$F_c = f_D T = \frac{2v_{\text{p}}T}{\lambda} \cos\varphi \cos\theta \tag{3.63}$$

while the spatial frequency is

$$f_s = \frac{d}{\lambda} \cos\varphi \cos\theta \tag{3.64}$$

$\gamma = F_c/f_s$ is the slope of the clutter trajectory in the $F_c - f_s$ plane. For $\gamma = 1$ spatial and temporal samples coincide in terms of phase (DPCA case).

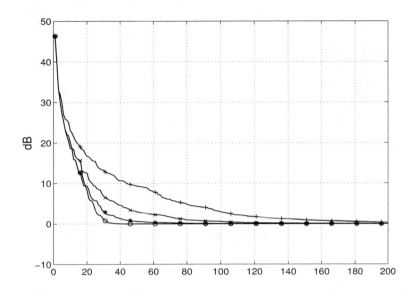

Figure 3.27: Effect of range walk, dependence of eigenspectra on range resolution (FL): ○ *no range walk;* ∗ $\Delta_R = 100m$; × $\Delta_R = 10m$; + $\Delta_R = 1m$

Figure 3.14, show some eigenspectra for a sidelooking array with $N = M = 12$ to verify Brennan's and Staudaher's rule (3.61). The numbers in brackets denote the number of eigenvalues according to (3.61). In Figure 3.14a γ was varied via variation of the PRF. It can be noticed that the eigenspectra are in good accordance with the figures given by (3.61). Similar results have been obtained if the variation of γ is caused by variation of the sensor spacing d (Figure 3.14b).

The curves in Figure 3.14c have been computed for $\gamma = 1$ with both the PRF and the spacing being varied. As can be seen the rule (3.61) is verified as long as the clutter echo field is spatially and temporally sampled at the Nyquist rate (∗) or undersampled (×, +). However, since spatial and temporal sampling are integer multiples spatial and temporal spectral ambiguities coincide.[5] Oversampling (○) results in a reduction of the number of eigenvalues below the number expected from (3.61).

In Figure 3.14d γ has been chosen so that spatial and temporal spectral ambiguities do not coincide. We notice that the undersampled curves (×, +), although computed for the same value of γ, differ considerably. It appears from this calculation that the rule (3.61) gives just a rough indication of the number of eigenvalues. It holds rigorously only for *integer* values of γ, see WARD [690, p. 187], which means that spatial and temporal spectral ambiguities coincide.

In WARD [697] the dependence of the clutter eigenspectrum is discussed for different kinds of thinned arrays.

It should be noted that in his proof WARD[690, p. 187] assumes that the spacing

---

[5]For spatial and temporal undersampling see Section 3.4.2.

102 Properties of airborne clutter

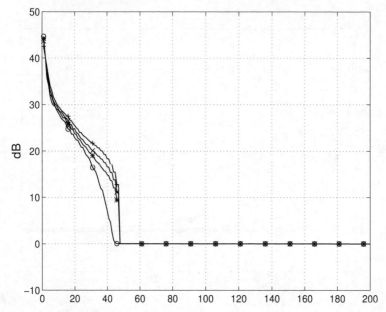

Figure 3.28: Impact of system bandwidth on clutter eigenspectra (rectangular impulse response, sidelooking): ○ $B_s = 0$; * $B_s = 0.05$; × $B_s = 0.1$; + $B_s = 0.2$

of the array is half wavelength. Therefore, a variation of γ in (3.62) is equivalent to a variation of T, *i.e.*, the *temporal* sampling while d is kept constant. Therefore, those curves in Figure 3.14, which have been calculated with variable element spacing do not necessarily satisfy BRENNAN'S rule (3.61).

Figure 3.15 shows the same eigenspectra as Figure 3.14, however for a forward looking array. It can easily be verified that there is no obvious relation to the rule (3.61). In fact, (3.61) has been created on the basis of a sidelooking linear array.

3.3 Power spectra

Once we have the clutter covariance matrix we can calculate a spectral representation of the covariance matrix using the techniques described in Section 1.3. The question is which power estimator is best suited for this purpose.

As carried out in Section 1.3 power spectra are obtained by multiplying a steering vector $s(\varphi, f_D)$, i.e., a variable signal replica, with some representation of the measured data, for instance the estimated clutter + noise covariance matrix or its inverse. As stated above the steering vector $s(\varphi, f_D)$ can be considered a beamformer cascaded with a Doppler filter. Steering $s(\varphi, f_D)$ over the whole range of azimuth and Doppler values is equivalent to cascading a set of beams with a Doppler filter bank. Notice that the transmit beam position is fixed. Therefore, only a cross-section of the spectra at $\varphi = \varphi_l$ (i.e., where transmit and receive beams coincide) reflects real radar operation.

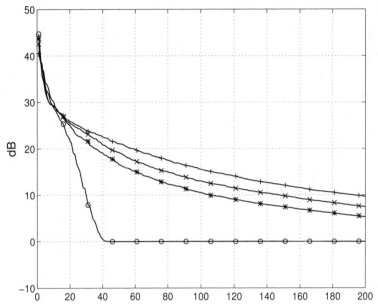

Figure 3.29: Impact of system bandwidth on clutter eigenspectra (rectangular impulse response, forward looking): ○ $B_s = 0$; ∗ $B_s = 0.05$; × $B_s = 0.1$; + $B_s = 0.2$

In all subsequent plots the Doppler frequency is normalised to the PRF

$$\mathsf{F} = \frac{f_D}{f_{PR}} \qquad (3.65)$$

3.3.1 Fourier spectra

The 'signal match' power estimator (1.102)

$$P_{SM}(\varphi, f_D) = \frac{\mathbf{s}^*(\varphi, f_D)\mathbf{Q}\mathbf{s}(\varphi, f_D)}{\mathbf{s}^*(\varphi, f_D)\mathbf{s}(\varphi, f_D)} \qquad (3.66)$$

becomes a two-dimensional Fourier transform if the array is equispaced and the PRF is constant. A one-dimensional example of a Fourier spectrum was shown in Figure 1.13 (circles). In Figure 3.16, a typical Fourier clutter spectrum for a sidelooking array has been plotted versus azimuth φ and the normalised Doppler frequency[6] F. The maximum at $\varphi = 90°$ appears at the position of the transmit beam. In projection on the Doppler axis it points at $\mathsf{F} = 0$ which is the associated clutter Doppler frequency.

The calculated clutter spectrum extends from the upper right to the lower left corner. It is modulated by the transmit beam pattern and by the sensor pattern which can be recognised by the decay of the spectrum close to $\varphi = 0$ and $\varphi = 180°$ (left and right corner).

---

[6]Normalised to the PRF f_{PR}.

104 *Properties of airborne clutter*

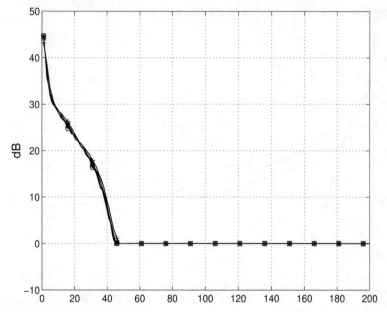

Figure 3.30: Impact of system bandwidth on clutter eigenspectra (rectangular frequency response, sidelooking): ○ $B_s = 0$; ∗ $B_s = 0.05$; × $B_s = 0.1$; + $B_s = 0.2$

There are two more 'sidelobe walls' one of which extends from $\varphi = 0, \ldots, 180°$ at F = 0, the other one from F = $-0.6, \ldots, 0.6$ at $\varphi = 90°$. Notice that these sidelobes are not a part of the clutter model in **Q** but just a spurious reaction of the 2-D Fourier estimator. Similar cross-shaped sidelobe patterns can be observed in SAR images as response to strong targets. As can be seen the spurious sidelobes hide most of the true clutter spectrum. Moreover, these sidelobes are not modulated by the transmit and sensor patterns which indicates that they are responses of the Fourier estimator to the main beam clutter power rather than antenna or Doppler filter sidelobes.

In Figure 3.17, the array has been tapered with a Hamming window $(\cos^2(.)$-weighting) on receive and transmit. Moreover a Hamming window has been applied in the time dimension. Now the main beam is clearly emphasised and broadened while all kinds of sidelobes are attenuated. The 'true space-time sidelobes' can hardly be noticed along the diagonal of the plot. The decay due to transmit and sensor pattern can be seen clearly. Nevertheless, the spurious 'sidelobe walls' are still dominant over the true sidelobe clutter.

In conclusion, the Fourier power estimator gives a wrong impression of the two-dimensional clutter spectrum because it exhibits sidelobes which have not been included in the clutter model. If the 2-D Fourier transform is used as a basis for processor design the spurious sidelobe effects have to be taken into account.

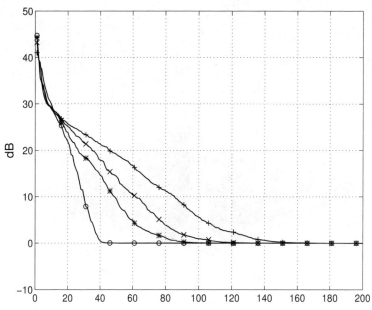

Figure 3.31: Impact of system bandwidth on clutter eigenspectra (rectangular frequency response, forward looking): ○ $B_s = 0$; ∗ $B_s = 0.05$; × $B_s = 0.1$; + $B_s = 0.2$

3.3.2 High-resolution spectra

In the following we discuss briefly the use of high-resolution power estimators for space-time clutter spectrum estimation.

3.3.2.1 *Minimum Variance Estimator*

The minimum variance estimator (MVE) was (1.104)

$$P_{\mathrm{MV}}(\varphi, f_{\mathrm{D}}) = (\mathbf{s}^*(\varphi, f_{\mathrm{D}}) \mathbf{Q}^{-1} \mathbf{s}(\varphi, f_{\mathrm{D}}))^{-1} \tag{3.67}$$

where $\mathbf{Q}$ is the clutter + noise covariance matrix and $\mathbf{s}(\varphi, f_{\mathrm{D}})$ is a signal replica or steering vector. An MV-clutter spectrum for a sidelooking linear array is shown in Figure 3.18. Like all high-resolution techniques the MVE tries to decompose a signal into single peaks. This leads to very realistic spectra if the signal is composed of single spectral lines (or point-shaped sources) but looks a bit awkward in the case of a continuous spectrum.

The advantage of this high-resolution spectrum is that it comes closest to the true clutter model. There are no spurious sidelobes, and the sidelobes of the transmit and receive beam can be noticed as clusters of peaks.[7]

---

[7]We made use of the fact that the shape of space-time MV spectra depends very much on the number of points used in the plot. There is always the problem of 'graphical undersampling' when applying such

106 Properties of airborne clutter

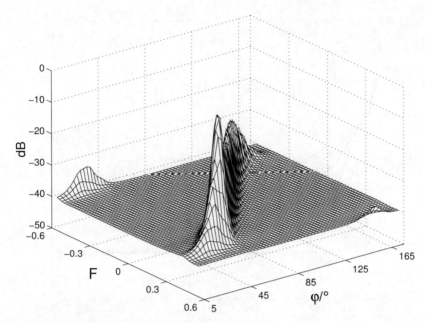

Figure 3.32: MV spectrum for a sidelooking array ($B_c = 0.2$)

Computations have shown (see Figure 1.13) that maximum entropy or MUSIC spectra exhibit even stronger peaks than the MV-spectrum. In some cases only one high peak in the look direction is visible and no realistic impression of the sidelobe structure is produced. The MVE is certainly the best compromise between suppression of spurious sidelobes and spikyness of the spectrum.

3.4 Effect of radar parameters on interference spectra

3.4.1 Array orientation

Figure 3.19 shows an MV clutter spectrum for a linear array in the forward looking orientation. It can clearly be seen that the footprint of the spectrum is a circle which is in accordance with the considerations made in Section 3.1.2. In this example we assumed that the sensors have forward looking directivity patterns according to (2.20) and (2.21). Therefore, the resulting spectrum is concentrated on a semicircle at positive Doppler frequencies. There is no clutter response at negative Doppler frequencies. However, targets with negative Doppler frequencies can be detected. The spectral components in the foreground are ambiguities due to the fact that $F = [-0.6, 0.6]$.

---

high-resolution techniques. However, the number of points is limited – not only by the computing time. If too many points are taken the plot becomes entirely black. Therefore some of the 3-D plots are plotted with 121×121 while others have 61×61 only. In any case, one has to be careful when drawing conclusions from details of such two-dimensional MV spectra.

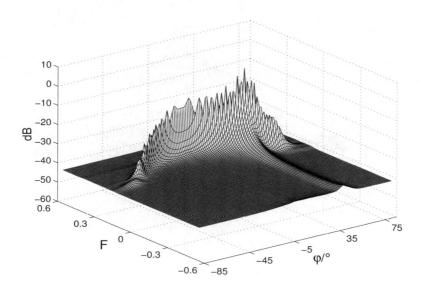

Figure 3.33: Clutter power spectrum (FL array) with range walk ($\Delta_R = 10$ m)

3.4.2 Temporal and spatial sampling

As pointed out in Chapter 2 we deal with two-dimensional signals $s(f_D, \varphi)$ which are sampled in both the temporal and spatial dimensions. By temporal sampling the succession of echoes as given by the PRF is understood while spatial sampling is done by the antenna array, specifically by the sensor positions. In Figures 3.20–3.23, the effect of undersampling in space and time on f_D–φ spectra is illustrated. In both cases (space and time) we assumed that the sampling frequency is half the Nyquist frequency. For details on space-time sampling see ENDER and KLEMM [134].

3.4.2.1 *Temporal undersampling*

In all previous examples the PRF was chosen to be 12 kHz which is the Nyquist frequency for the clutter bandwidth for the given parameters (see Table 2.1). In the following two examples we assumed PRF = 6 kHz. Figure 3.20 shows the clutter spectrum for a sidelooking linear array. As can be seen there are ambiguous clutter responses on both sides of the 'original' spectrum; these responses are 'copies' of the original spectrum shifted along the Doppler axis.

In Figure 3.21, the same situation has been plotted for a forward looking linear array. Again the temporal ambiguity produces a second 'replica' of the original spectrum. Now we notice that clutter portions appear at both positive and negative Doppler frequencies (compare with Figure 3.19).

Let us recall that each Doppler frequency is associated with a certain azimuth and

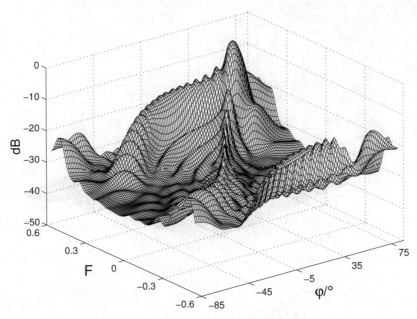

Figure 3.34: MV spectrum (FL array, rectangular impulse response, $B_s = 0.2$)

depression angle, see (2.35). It is obvious from this relation that any ambiguity in azimuth causes ambiguity in Doppler frequency and vice versa. This can be seen clearly in Figure 3.22.

3.4.2.2 Spatial undersampling

Figure 3.22 shows an MV clutter spectrum for sidelooking radar. The array spacing $d = x_{i+1} - x_i$ has been set to $d = \lambda$ in equations (2.30) and (2.36). As can be seen the spectrum is repeated every $90°$ instead of $180°$ for Nyquist spacing ($d = \lambda/2$).

Notice that for sidelooking radar the spatial ambiguities follow the same $f_D - \varphi$ trajectories as in case of *temporal* undersampling shown in Figure 3.22. The curvature of the trajectories is caused by the non-linear $\cos\varphi$-function in the steering vector (see 2.31). If the spectrum were plotted versus $\cos\varphi$ instead of φ (as indicated in Figure 3.18) the trajectories would become straight lines.

It is obvious that in Figure 3.20 the ambiguous responses are replicas of the original spectrum shifted in Doppler frequency while in Figure 3.22, the ambiguous responses are shifted in azimuth.

The spatial ambiguities in the case of a forward looking array give a quite confusing impression (Figure 3.23). There are three spectral components following semicircular trajectories which cross each other. The first one starts at $\varphi = -85°$ and ends at about $\varphi = -15°$. The second one starts at $\varphi = -75°$ and ends at about $\varphi = 60°$. The third one starts at $\varphi = 0°$ and ends at $\varphi = 75°$.

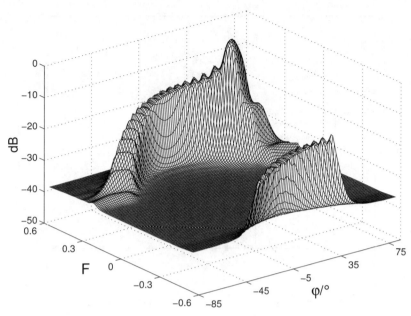

Figure 3.35: MV spectrum (FL array, rectangular frequency response, $B_s = 0.2$)

Let us recall that the improvement factor is just the inverse of the MV spectrum. In the discussion of the optimum processor in Chapter 4 we will come back to the problem of undersampling. The IF plots give a much clearer impression of the ambiguities than the 3D spectral plots.

3.4.3 Decorrelation effects

3.4.3.1 *Clutter bandwidth*

In Figures 3.24 and 3.25 we see the influence of clutter fluctuations due to internal motion on the eigenspectra of sideways and forward looking radar with a linear antenna array. Only the first 200 eigenvalues are shown. The total number amounts to $24 \times 24 = 576$. B_c is the Doppler bandwidth of clutter normalised by the unambiguous Doppler frequency range, see (2.53). As can be seen the effect of clutter bandwidth results in an increase of the number of clutter eigenvalues. The increase is independent of the array orientation which is clear because we deal here with a purely *temporal* effect.

As pointed out earlier the number of clutter eigenvalues is a measure for the complexity of a clutter suppression filter. We will come back to this point in the following chapters. Figure 3.32, shows the MV power spectrum for SL-radar with 20% relative clutter bandwidth. Obviously the increase in clutter eigenvalues is here reflected by a broadening of the spectrum (compare with 3.18) for the main beam as for the sidelobes.

110 *Properties of airborne clutter*

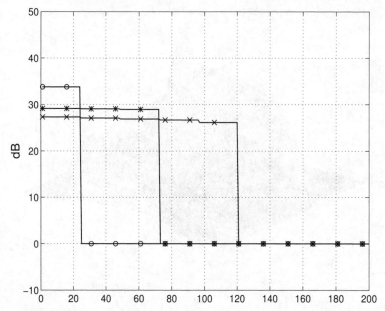

Figure 3.36: Space-time jammer eigenspectra (SL array): ○ $J = 1$; ∗ $J = 3$; × $J = 5$

3.4.3.2 *Range walk*

For a sideways looking array, under 'ideal' conditions (Nyquist sampling of the clutter Doppler band in space and time, no additional decorrelation effects) the number of clutter eigenvalues is $N_e = N + M - 1$, see (3.59). In the following we give some examples which illustrate the dependency of the eigenspectra on temporal decorrelation effects due to range walk as described in Section 2.5.1.2.

Only the first 200 eigenvalues are shown in Figures 3.26 and 3.27 to highlight the clutter portion of the spectra. The pulse-to-pulse correlation in the presence of range walk as defined by (2.55) is proportional to the overlap area in Figure 2.2. As can be seen in Figure 2.2 the correlation is strongest in the broadside direction of a sidelooking array and a minimum for a forward looking array. This is reflected by the associated space-time eigenspectra. Notice the difference in spreading of the eigenspectra in Figures 3.26 and 3.27.

An MV clutter spectrum illustrating the effect of range walk is shown in Figure 3.33. The width of the range bin was chosen to be 10 m. We notice that the spectrum is broadened mainly in the look direction, i.e., in the direction of the transmit beam. In contrast, internal motion causes a broadening of the entire spectrum (see Figure 3.32). Recall that the decorrelation caused by range walk is azimuth dependent. Obviously the transmit beam selects the individual decorrelation associated with the look direction.

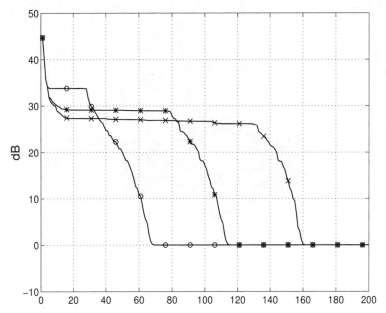

Figure 3.37: Space-time jammer + clutter eigenspectra (SL array)○ $J = 1$; ∗ $J = 3$; × $J = 5$

3.4.3.3 System bandwidth

In Figures 3.28 and 3.29 we find the effect of *system* bandwidth according to (2.63), on the eigenspectra of sideways and forward looking radar. In these examples we assumed a rectangular *time* response for the transmitted pulse and the band limiting matched filter, see (2.7).

The decorrelation due to travel delays across the array aperture was given by (2.64), for system characteristics with rectangular time response

$$\rho_{ik} = \begin{cases} 1 - \frac{|\tau_{ik}|}{\delta} & : \quad t = [-\delta, \delta] \\ 0 & : \quad \text{otherwise} \end{cases} \qquad (3.68)$$

For rectangular frequency response the correlation was given by (2.64).

A wave coming from direction (φ, θ) travelling across the array is delayed between any two sensors by (2.7):

$$\tau_{ik} = \frac{1}{c}[((x_i - x_k)\cos\varphi + (y_i - y_k)\sin\varphi)\cos\theta - (z_i - z_k)\sin\theta] \qquad (3.69)$$

If the array moves at speed v_p we get instead

$$\tau_{ik} = \frac{1}{c}[((x_i + 2mTv_p - x_k)\cos\varphi + (y_i - y_k)\sin\varphi)\cos\theta - (z_i - z_k)\sin\theta] \qquad (3.70)$$

Under DPCA conditions, i.e., for a sidelooking arrays and $x_{i+1} - x_i = 2Tv_p$ we get for an equispaced array certain combinations of x_i, x_k and $2mTv_p$ for which the

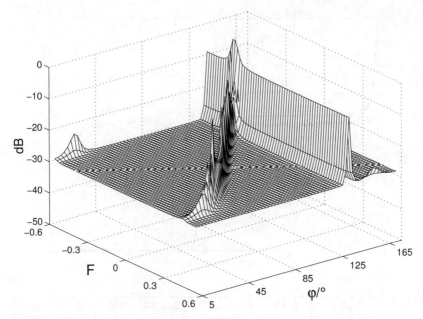

Figure 3.38: MV jammer + clutter spectrum (SL array, $J = 1$)

argument $x_i + 2mTv_\mathrm{p} - x_k$ in (3.70) becomes 0. In these cases of phase coincidence full correlation is obtained. The points of full correlation are identical to the unity elements in (3.49).

Transmit pulses with rectangular impulse response are very common in pulse Doppler radars. The frequency response of such a pulse is a $\sin x/x$-function. By 'bandwidth' we understand here the width of the main lobe (from null to null) of this function. However, the sidelobes of the $\sin x/x$-function are rather high so that the band limits are not very distinct.

In Figure 3.28, the effect of phase coincidence on the eigenspectra for a sidelooking array in DPCA mode is shown. As can be seen the number of clutter eigenvalues is not increased beyond $N + M - 1$. There are slight differences between eigenspectra below $N + M - 1$ which however is not important for clutter rejection.[8]

In the case of a forward looking array all sensors have the same x-coordinate and different y-coordinates. It follows from (3.70) that no DPCA effect (phase conincidences through platform motion) can occur for forward looking arrays. As a consequence the correlation between sensors is strongly effected by the system bandwidth (3.68). This is reflected in an increase of the number of clutter eigenvalues beyond the limit $N + M - 1$, see Figure 3.29.

Similar results are obtained for rectangular frequency response of the antenna channels. Again no increase of clutter eigenvalues can be noticed for the sidelooking array while for forward looking radar there is a considerable increase with increasing

---

[8] See Chapters 4 and 13.

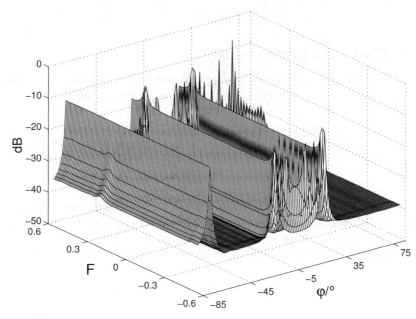

Figure 3.39: MV jammer + clutter spectrum (FL array, $J = 3$)

system bandwidth. However, compared with rectangular impulse response the increase of eigenvalues is much less severe because there are no significant sidelobes outside the bandwidth.

Figure 3.34, shows an MV spectrum for 20% relative system bandwidth with rectangular impulse response. As pointed out above the associated frequency response has quite high sidelobes which leads to a rather irregular spectrum.[9]

If instead the system bandwidth is strictly rectangular no significant sidelobes show up outside the clutter ridge. The example in Figure 3.35, reflects clearly the rectangular shape of the system bandpass. Notice that the clutter ridge is quite narrow at broadside ($\varphi = 0°$) and broadens for off-broadside angles on both sides. This has to do with the fact that the travel delays are large at endfire and small at the broadside direction.

It should be pointed out that these considerations cover only the *spatial* properties of STAP used with broadband arrays. There is an additional *spectral* effect. The signal bandwidth produces a Doppler bandwidth which is largest in the forward direction and smaller as the direction gets closer to endfire. This Doppler effect is large right where the spatial decorrelation effect is small, and vice versa.

For reasons of brevity we do not show the effect of system bandwidth on the clutter spectrum for sidelooking radar. As there is almost no system bandwidth effect on the MTI performance of sidelooking radar these spectra look like the one shown in Figure

---

[9]If such a spectrum were plotted by use of the two-dimensional Fourier transform (3.66) it would be difficult to verify which of the sidelobes come from the system bandpass and which of them are spurious responses of the Fourier estimator.

3.18.

3.4.4 Clutter and jammer spectra

Jammers are assumed to be Doppler broadband and discrete in space. The jammer covariance matrix is block diagonal with spatial submatrices (see Figure 3.9, and equations (2.45–2.48)). If only one jammer is present each of the submatrices is a dyadic and produces one eigenvalue of the covariance matrix. Accordingly, as we have M submatrices in the space-time matrix the total number of eigenvalues due to one jammer amounts to M. For J jammers the number of jammer eigenvalues becomes JM.

This can be seen in the numerical example Figure 3.36. The number of temporal samples is $M = 24$. Therefore, the number of jammer eigenvalues is 24 for one jammer, 72 for three jammers, and 120 for five jammers (the power was assumed to be equal for all jammers). If we have a superposition of clutter and jammers the number of interference eigenvalues becomes for a sidelooking array and DPCA conditions

$$N_e = N + (1 + J)(M - 1) \tag{3.71}$$

see Figure 3.37.

Equation 3.71 is a special case of

$$N_e = N + (\gamma + J)(M - 1) \tag{3.72}$$

(RICHARDSON [566]), where γ is the ratio of Doppler ambiguities to spatial ambiguities (see (3.62)).

The fact that a broadband jammer produces M eigenvalues of a space-time covariance matrix may have significant influence on the required number of degrees of freedom of the clutter canceller, in particular when suboptimum techniques based on the interference subspace are under discussion.

In Figures 3.38 and 3.39 two examples for a superposition of clutter and jammers are given. The jammers appear as thin walls along the F-axis.

It should be noted that the considerations concerning system bandwidth effects made above hold for jammers as well. Normally, even in ground-based radar systems, cancellation of off-broadside jammers is strongly degraded by a large system bandwidth. This can be compensated for by space-TIME processing where TIME means the time associated with wave propagation.[10]

3.5 Aspects of adaptive space-time clutter rejection

3.5.1 Illustration of the principle

The principle of space-time clutter rejection by space-time filtering is illustrated in Figure 3.40. In this example the sidelooking geometry was chosen. One recognises the clutter spectrum which runs along the diagonal of the azimuth-Doppler plane. The

---

[10] The fast radar time which corresponds to range.

clutter spectrum is modulated by the transmit directivity pattern and the sensor pattern. Let us consider some ways of suppressing this clutter spectrum.

3.5.1.1 *Optimum temporal filtering*

In the background, filter characteristics can be seen; this is referred to as inverse *temporal* clutter filter. This filter has been obtained by projecting the clutter spectrum on to the Doppler axis and taking the inverse. Such an inverse filter is the optimum *temporal* clutter canceller. Such temporal filters are used in ground-based radar where no motion-induced Doppler spread of the clutter spectrum occurs. As one can see there is a broad stop band which is determined by the transmit main beam. A fast target in the main beam may fall into the Doppler sidelobe domain of the inverse filter and, hence, may be detected. A slow target may fall into the stop band and will be rejected by the temporal filter.

3.5.1.2 *Optimum spatial filtering*

On the left we have a similar situation in the spatial dimension. Here we find again a filter based on to the inverse clutter spectrum projected on to the azimuth axis. Again the stop band is determined by the transmit beamwidth. However, as the radar beam looks at the clutter background this stop band is located in the beam look direction. The adaptive spatial clutter filter makes the radar blind!

3.5.1.3 *Space-time adaptive filtering*

In the left lower corner a cross-section of the transmission characteristics of a space-time clutter filter is depicted. A space-time filter operates in the whole φ-f_D plane and thereby recognises that the clutter spectrum is broad when looking from the lower right corner into the plot. It is quite narrow if one looks from the lower left side into the plot. Therefore, the filter forms a very narrow ditch along the trajectory of the clutter spectrum so that even slow targets may fall into the pass band and can be detected.

Another view of the problem is as follows: Clutter spectra of *stationary* radar are narrowband (let us forget about clutter fluctuations for the time being). The platform velocity causes a direction-dependent Doppler colouring of clutter echoes so that the spectrum is spread out along the diagonal of an azimuth-Doppler plane (instead of along a Doppler axis only). Since a space-time clutter suppression filter can operate in the φ-ω_D plane the motion-induced Doppler effect has no influence on the clutter rejection performance. This may be considered as the motion compensation capability of the space-time filter. In other words, if the right processing is applied there is no limitation in MTI performance caused by the radar platform motion.

3.5.2 Some conclusions

In the following some concluding remarks are summarised which may serve as guidelines for the further development of space-time processing concepts.

116 *Properties of airborne clutter*

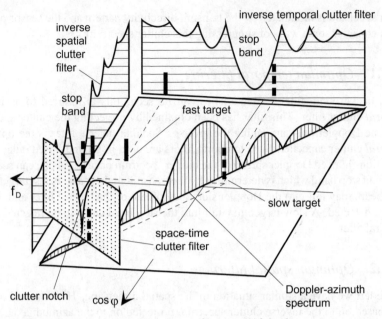

Figure 3.40: Principle of space-time clutter filtering (sidelooking array antenna)

3.5.2.1 Bandwidth limitations

Clutter echoes as defined by (2.38) are band limited in both the spatial and the Doppler domain. The Doppler frequency is defined by the platform velocity

$$f_D = \frac{2v_p \cos\beta}{\lambda} \qquad \beta = [0°, 180°] \qquad (3.73)$$

where β was the look direction relative to the array axis, see Figure 3.2. The clutter Doppler bandwidth is therefore

$$B_D = \frac{4v_p}{\lambda} \qquad (3.74)$$

The spatial frequency of clutter is

$$f_s = \frac{d \cos\beta}{\lambda} \qquad \beta = [0°, 180°] \qquad (3.75)$$

so that the spatial bandwidth becomes

$$B_s = \frac{2d}{\lambda} \qquad (3.76)$$

Let us make a general remark on the processing of space-time signal fields. As long as such clutter echoes are sampled properly in space and time, i.e., at the Nyquist rate or higher, no information loss through the sampling process will occur and therefore any kind of discrete signal processing (spectral analysis, filtering) can be principally applied to such data.

Historically the idea of motion compensation led to the development of DPCA systems. It should be noted, however, that the DPCA effect (motion compensation through displaced phase centres, RICHARDSON [562], RICHARDSON and HAYWARD [563]) *is not required* for explaining the principle of adaptive airborne clutter cancellation.

If the DPCA effect were necessary to explain the clutter cancellation operation then only sidelooking linear equispaced arrays would be suitable for airborne MTI radar. For example, there is no DPCA effect for a forward looking array. The spectra shown above demonstrate clearly that this is not the case. Moreover, we will show in Chapter 8 that the airborne MTI function can be combined with more complex antenna arrays for which no DPCA conditions can be defined.

3.5.2.2 The role of DPCA in space-time adaptive processing

The DPCA effect plays an important role in space-time radar signal processing. It makes the sidelooking array configuration a preferable choice, but not from the viewpoint of motion compensation!

The nature of the complex envelope $A(u)$ in (3.47) was not specified. We can distinguish between three cases in which arrivals from different directions are mutually uncorrelated, i.e., they fulfil the condition (3.46). The covariance matrix assumes the form (3.48).

- **Clutter suppression for narrowband radar**. We assume that by virtue of the reflection process the received amplitude is *slowly* varying so that subsequent echoes are coherent, however the arrivals from different directions are independent on a long-term basis. The DPCA effect is not required for space-time processing in narrowband systems. Since clutter echoes are band-limited in the spatial and temporal dimensions adaptive filtering will be capable of suppressing clutter returns for any kind of array geometry. Of course, clutter suppression is limited if the number of degrees of freedom of the radar (N, M) is smaller than the number of clutter eigenvalues of the space-time covariance matrix.

- **Clutter suppression for broadband radar**. The DPCA effect compensates for spatial decorrelation of clutter returns for broadband *sidelooking* radar. As the DPCA effect is based on the platform velocity in this respect the platform motion turns out to be an advantage compared with ground-based radar.

- **Forward looking radar**. Since for forward looking arrays no DPCA effect exists the effects of system bandwidth can be cancelled by space-TIME processing for jammer cancellation *and* space-time-TIME processing for clutter rejection. Alternatively, a CPCT configuration (see Section 3.2.2.4, and SKOLNIK [610, pp. 16–20])) may be used for compensation for system bandwidth effects in forward looking antennas. Basically no losses have to be encountered if the appropriate antenna configuration and processing is applied.

- **Clutter bandwidth**. There is no way of compensating for temporal decorrelation caused by fluctuations of the clutter background.

3.6 Summary

As an introduction, some general properties of airborne clutter such as isodops, azimuth-Doppler trajectories, and Doppler dependence of clutter are derived. The vector quantities for signal, clutter, jamming, and noise are extended to space-time vector quantities. By means of the models derived in Chapter 2 space-time clutter + noise covariance matrices can be developed. By use of the space-time models for target signals in Chapter 2 a steering vector can be defined by which spectral analysis of clutter echoes can be carried out. In detail the contents of this chapter can be summarised as listed below:

1. For horizontal flight and planar ground the **isodops** (curves of constant Doppler frequency on the ground) are hyperbolas.

2. For a **sidelooking linear array** the beam traces on the ground are hyperbolas which coincide with the isodops. Therefore, the clutter Doppler is range independent.

3. For a **forward looking linear array** the beam trace hyperbolas are rotated by 90° so that beam trace and isodop hyperbolas cross each other, especially at near range. Therefore, the clutter Doppler frequency is range dependent.

4. In practice the **clutter covariance matrix** is unknown and has to be estimated by use of secondary data. Secondary data are usually obtained from neighbouring range bins. If the clutter Doppler is range dependent (as is the case for all but sidelooking arrays) the clutter spectrum is broadened which results in degraded detectability of low Doppler targets (BORSARI [58]). Some improvement can be obtained by Doppler warping (compensation for the Doppler gradient) of the data, see BORSARI [58] and KREYENKAMP [369]).

5. **Space-time clutter covariance matrix**. If
    - the clutter background is homogeneous and
    - illuminated with an omnidirectional antenna and
    - the array has half-wavelength spacing between elements (spatial Nyquist sampling)

    clutter is spatially white, and the *spatial* clutter covariance matrix becomes a identity matrix. No clutter suppression is possible. Higher correlation components appear if a *space-time* covariance matrix is used.

6. The number of **clutter eigenvalues** of a linear sidelooking array is $N + M - 1$ under DPCA conditions. For a forward looking array the number is about $2(N + M - 1)$. If only one half-plane is illuminated (or directive sensors are used) the number of clutter eigenvalues is about $N + M - 1$ for both configurations.

7. A **2-D Fourier azimuth-Doppler clutter spectrum** exhibits high spurious sidelobes which might give a wrong impression of the true clutter spectrum.

8. The **minimum variance estimator** generates clutter spectra that are very close to the clutter model used.

9. For a **sidelooking** array the clutter power is concentrated on the diagonal across the $\cos\varphi$-F plane while for **forward** looking radar the clutter is on a circle. The $\cos\varphi$-F clutter trajectories are examples of the well-known Lissajou patterns.

10. **Spatial** and **temporal undersampling** lead to spatial and temporal ambiguities, respectively.

11. For sidelooking radar **DPCA** is a zero noise approximation of the optimum processor.

12. **Clutter fluctuations** lead to a broadening of the clutter spectrum that is constant for all Doppler frequencies.

13. Range walk as occurs mainly in high-resolution radar leads to temporal decorrelation and, hence, to a broadening of the clutter spectrum. The number of clutter eigenvalues increases. This effect is stronger for forward than for sideways looking radar.

14. The **system bandwidth** effect causes a broadening of the clutter spectrum proportional to the off-boresight look angle.

15. The **principles** of spatial, temporal and space-time clutter rejections are compared. Only space-time clutter filtering has the capability of slow target detection.

16. The **DPCA effect** is *not* the physical mechanism which makes space-time clutter filtering possible. Clutter echoes are two-dimensional (spatial-temporal) signals which are strictly band-limited in both dimensions. When sampled at the Nyquist rate these signals can be filtered in arbitrary ways. The potential of slow target detection lies in the fact that the clutter trajectory is a narrow line (straight, elliptic, circular).

17. DPCA is not the basis of space-time processing. It is merely a special case which for sidelooking linear arrays and appropriate synchronisation of PRF, v_p and spacing d, may achieve near-optimum clutter rejection.

18. In contrast to the DPCA techniques, space-time adaptive processing is almost independent of the PRF. Of course, if the PRF falls below the Nyquist rate ambiguous clutter notches occur.

Chapter 4

Fully adaptive space-time processors

4.1 Introduction

In this chapter we focus on two space-time processors which are fully adaptive. 'Fully' adaptive means that the number of degrees of freedom as given by the number of array elements and echo pulses will be preserved in the clutter rejection process. 'Adaptive' means that clutter suppression is based in some way on the received clutter data, for instance on an estimate of the clutter covariance matrix.

Why do we need *adaptive* clutter suppression? In principle the space-time clutter characteristics are well known a priori through the angle–Doppler relation. However, various kinds of errors in the receiving instrument or perturbations of the flight path through platform motion may degrade the performance of non-adaptive techniques such as DPCA. For example, a comparison between adaptive and non-adaptive DPCA processing (SHAW and MCAULEY [602]) in the presence of beam squint errors reveals that adaptivity can compensate for errors in the receiver mechanism. The use of *adaptive* space-time filtering to compensate for various kinds of radar inherent errors has been pronounced by SURESH BABU *et al.* [625].

In contrast to subspace techniques (see Chapters 5, 6, 7) which exploit the subspace properties of space-time clutter data the processors discussed in this chapter are based on the full space-time covariance matrix of the available data vector space. The application of the fully adaptive space-time processor for clutter rejection has been discussed by KLEMM [300] and WARD [690, p. 57]. For small data size N, M the optimum processor is easy to handle and has been frequently used in theoretical considerations, for instance by BARBAROSSA and FARINA [34].

Fully adaptive processing techniques are normally not useful for practical applications. The reason is the high computational complexity involved for:

- **Adaptation**. This involves the estimation of the inverse of the clutter + noise space-time covariance matrix. The number of data vector samples required for estimating the covariance matrix increases with the matrix dimension (for space-time processing at least $2NM$), see REED et al., [557]. Adaptation is a moderately fast operation which should be able to cope with any changes of the data statistics. Such changes may be caused by inhomogeneity of the clutter background and perturbations caused by platform motion.

- **Filtering**. Optimum filtering means multiplying the vectors of received data with the covariance matrix inverse, with a beamformer vector and a Doppler filter bank for all possible Doppler frequencies. This is a fast operation which has to be carried out sequentially for all range bins. Typical clock rate in the range direction is 1 to several MHz.

A realistic radar array antenna may include $N = 1000$ elements or more. Moreover, the number of coherent echoes involved in the detection process may vary from $M = 10,...,1000$. This results in dimensions of the space-time covariance matrix up to $NM = 10^6$! It is obvious that such huge vector spaces cannot be handled by reasons of computing time, numerical stability and cost.

In this chapter we analyse fully adaptive processors on the basis of a linear array with aperture of moderate size ($N = 24$) and a moderate number of echoes ($M = 24$). We have to keep in mind that linear arrays will rarely be used in practice. However, analysing the behaviour of linear arrays can give considerable insight into the problems with moderate computing time. Furthermore, on the basis of the fully adaptive processor we can study the various effects of radar parameters on space-time clutter rejection without caring about additional effects caused by suboptimum processing. In that sense Chapter 4 has some tutorial value. In the sequel the results obtained with the optimum processor will serve as a reference when we analyse suboptimum techniques for real-time processing, see Chapters 5, 6, 7. It should be noted that all numerical results are based on the assumption that the steering vector is perfectly matched in angle and Doppler to the target signal. Any mismatch between the steering vector and the target signal vector results in additional losses in SNIR (LUMINATI and HALE [421]).

4.2 General description

4.2.1 The optimum adaptive processor (OAP)

The optimum (likelihood ratio) processor was introduced in (1.3):

$$\mathbf{w}_{\text{opt}} = \gamma \mathbf{Q}^{-1} \mathbf{s} \qquad (4.1)$$

Now we deal with space-time vector quantities according to (3.21). The covariance matrix will have the form shown in (3.22). We assume that sequences of echo samples are stationary during the observation time. This assumption has been verified by numerous radar experiments and is valid as long as the radar transmits at constant PRF and the motion of the radar is such that the relative angle between the radar and incoming signals does not change significantly ($\sim 1/100^{\text{th}}$ of the beamwidth, see

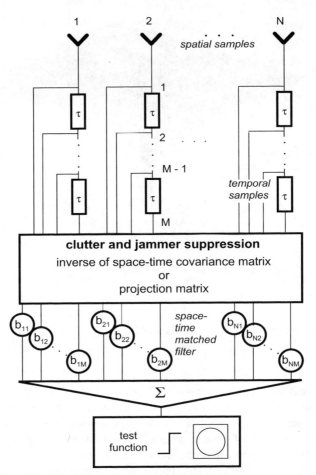

Figure 4.1: The optimum adaptive space-time processor

HAYWARD [256]). Then the clutter + noise covariance matrix $\mathbf{Q} = E\{\mathbf{qq}^*\}$ becomes block Toeplitz:

$$\mathbf{Q} = \begin{pmatrix} \mathbf{Q}_0 & \mathbf{Q}_1 & \cdot & \cdot & \mathbf{Q}_{M-2} & \mathbf{Q}_{M-1} \\ \mathbf{Q}_1^* & \mathbf{Q}_0 & \cdot & \cdot & \cdot & \mathbf{Q}_{M-2} \\ \cdot & \cdot & \cdot & \cdot & \cdot & \cdot \\ \cdot & \cdot & \cdot & \cdot & \cdot & \cdot \\ \mathbf{Q}_{M-2}^* & \cdot & \cdot & \cdot & \cdot & \mathbf{Q}_1 \\ \mathbf{Q}_{M-1}^* & \mathbf{Q}_{M-2}^* & \cdot & \cdot & \mathbf{Q}_1^* & \mathbf{Q}_0 \end{pmatrix} \quad (4.2)$$

The submatrices $\mathbf{Q}_m$ are *spatial* covariance matrices and have the dimensions $N \times N$. For instance, $\mathbf{Q}_0$ is the spatial covariance matrix measured at the same instant of time while $\mathbf{Q}_1$ is a spatial cross-covariance matrix between one echo and the next one. The *temporal* information is included in the relations between different submatrices.

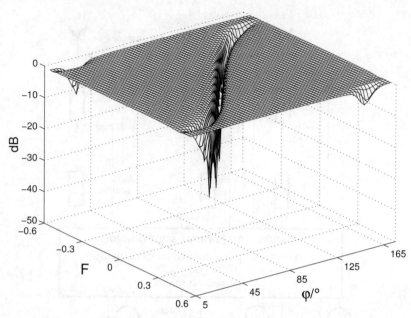

Figure 4.2: Improvement factor for a sidelooking array

A block diagram of the optimum processor is depicted in Figure 4.1. The N antenna elements provide *spatial sampling* of the backscattered wavefield. Each of the array channels includes amplification, complex demodulation and digitisation (not shown). Each channel is followed by a shift register to store subsequent echo samples. This is the temporal dimension of the space-time processor. All the spatial-temporal data are filtered by the inverse of the space-time covariance matrix. This is followed by a space-time matched filter including the coefficients of the signal reference (beamformer and Doppler filter coefficients) according to (2.31). In practice the space-time matched filter has to be implemented for all possible Doppler frequencies simultaneously (Doppler filter bank).

A test function is then calculated based on the actual output signals of the Doppler filter bank and is fed into a detection and indication device. One possible test function consists of a search for the maximum magnitude Doppler channel output

$$\max_m |\mathbf{x}^* \mathbf{Q}^{-1} \mathbf{s}_m| \begin{cases} > \eta & : \text{'target + noise'} \\ < \eta & : \text{otherwise} \end{cases} \quad (4.3)$$

This test function is called a *Doppler processor* and is widely used in practical applications. NAYEBI et al. [476] have shown that a discrete form of the likelihood function for gaussian noise and deterministic signals:

$$\sum_{m=1}^{\mu M} \exp(-\mathbf{s}_m^* \mathbf{Q}^{-1} \mathbf{s}_m) I_0(2 | \mathbf{x}^* \mathbf{Q}^{-1} \mathbf{s}_m |) \begin{cases} > \eta & : \text{'target + noise'} \\ < \eta & : \text{otherwise} \end{cases} \quad (4.4)$$

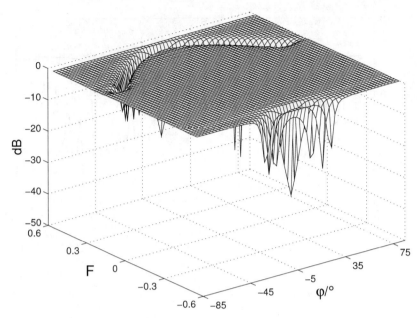

Figure 4.3: Improvement factor for a forward looking array

is near optimum for both white and coloured noise. I_0 is the modified Bessel function. This processor is called the *Constrained Averaged Likelihood Ratio* (CALC). μ determines the number of Doppler channels which usually assumes the values 1 or 2. As shown by NAYEBI [476] the Doppler processor (4.3) is near-optimum for white noise but suffers some considerable losses in the case of coloured noise. In his paper the author presents an overview of a large number of test functions (which in essence are different approximations of the optimum likelihood ratio processor).

As pointed out in Chapter 1 the efficiency of the processor can be jugded by calculating the improvement factor (IF). For the optimum processor (4.1) the IF is given by

$$\text{IF}_{\text{opt}} = \mathbf{s}^*(\varphi_L, f_D)\mathbf{Q}^{-1}\mathbf{s}(\varphi_L, f_D) \cdot \frac{\text{tr}\mathbf{Q}}{\mathbf{s}^*(\varphi_L, f_D)\mathbf{s}(\varphi_L, f_D)} \qquad (4.5)$$

where the elements of the signal reference vector (or steering vector) are given by (2.31). φ_L is the look direction while $f_D = 2v_{\text{rad}}/\lambda$ is the target Doppler frequency due to the radial target velocity v_{rad}. In contrast to the spectra shown in Chapter 3 the calculation of the optimum IF requires that the transmit beam is steered in the same direction as the receive beam (perfect signal match).

In Figures 4.2 and 4.3 the IF as given by (4.5) has been plotted versus the φ-F plane where F is again the target Doppler frequency normalised to the PRF. It can be noticed that along the φ-F clutter trajectories (see Figures 3.18 and 3.19), now a narrow trench has been formed by the optimum processor. In fact, as pointed out in Chapter 1 (compare (1.16) and (1.105)), the optimum IF is just the inverse of the minimum

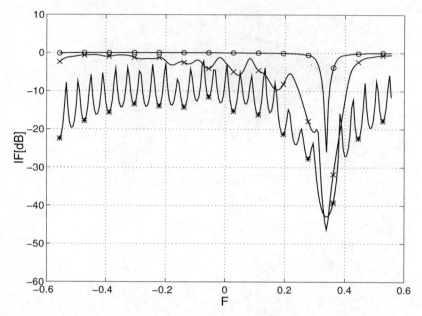

Figure 4.4: The potential of space-time adaptive processing: ○ *optimum processing;* ∗ *beamformer + Doppler filter;* × *beamformer + adaptive temporal filter*

variance spectrum. Everywhere outside the clutter trench we find an IF plateau where detection of moving targets is optimum.

4.2.1.1 The potential of space-time adaptive processing

In Figure 3.40 the principle of space-time adaptive clutter rejection was illustrated. In particular a comparison of space-time processing with pure temporal *or* spatial processing has been done. The comments given there are supported by the numerical example shown in Figure 4.4.

Like the examples given in Chapter 1 the plot shows the IF (improvement factor) in SCNR versus the normalised target Doppler frequency. The IF has been normalised by the theoretical maximum which is approximately $\text{CNR} \times N \times M$. Notice that this kind of plot is just a cross-section through a 3-D IF plot as shown for example in Figure 4.2. The cross-section runs along the Doppler axis in the look direction.

Figure 4.4 shows a comparison between space-time adaptive filtering, beamforming and Doppler filtering (but no clutter rejection), and beamforming cascaded with an optimum *temporal* clutter FIR filter as has been described by BUEHRING and KLEMM [75], and (1.90).

The curve for beamforming and Doppler filtering reflects the sidelobe response of the Doppler filter bank to clutter. Notice that there is a kind of inverse main lobe in the look direction. The losses compared with the optimum curve are dramatic.

A slightly better performance is achieved with a beamformer cascaded with an optimum *temporal* FIR filter, however there are still considerable losses, especially in

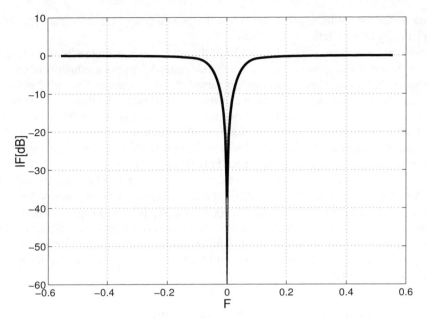

Figure 4.5: Comparison of optimum and orthogonal projection processing

the look direction ($F \approx 0.35$). These losses degrade particularly the detectability of slow targets.

We showed the comparison of these three ways of processing to demonstrate what the motivation for space-time adaptive processing is. In Chapters 5, 6 and 7 we try to approximate the performance of the optimum processor as much as possible by suboptimum processors with reduced computational expense and real-time capability. The IF curves of such processors will appear somewhere between the optimum space-time and the optimum temporal processor. Favourable solutions will have to be closer to the upper (o) than the lower (×) curve.

4.2.2 The orthogonal projection processor (OPP)

The orthogonal projection operator is given by (1.47)

$$\mathbf{P} = \mathbf{I} - \mathbf{E}(\mathbf{E}^*\mathbf{E})^{-1}\mathbf{E}^* \qquad (4.6)$$

where $\mathbf{E}$ is the matrix of eigenvectors belonging to the interference component in $\mathbf{Q}$. The IF becomes, according to (1.15),

$$\mathrm{IF} = \frac{\mathbf{s}^*\mathbf{P}^*\mathbf{s}\mathbf{s}^*\mathbf{P}\mathbf{s} \cdot \mathrm{tr}\mathbf{Q}}{\mathbf{s}^*\mathbf{P}^*\mathbf{Q}\mathbf{P}\mathbf{s} \cdot \mathbf{s}^*\mathbf{s}} \qquad (4.7)$$

where $\mathbf{s} = \mathbf{s}(\varphi_\mathrm{L}, f_\mathrm{D})$ is again the steering vector with elements given by (2.31), whose parameters assume all possible values of the $\varphi_\mathrm{L} - f_\mathrm{D}$ plane. The block diagram is the same as for the optimum processor except that the inverse of the covariance matrix is replaced by the projection matrix (4.6).

As we know from Figure 1.3, the IF of the orthogonal projection is very close to the optimum in one-dimensional applications. Figure 4.5 shows a similar example for airborne clutter cancellation (sidelooking array, $N = 12$ sensors, $M = 12$ echoes). In fact, these curves are cross-sections through 3-D IF plots like those in Figures 4.2 and 4.3. As can be seen the clutter rejection performance is almost the same for the optimum processor and the orthogonal projection method. The curves differ in that the clutter notch caused by the orthogonal projection method is an exact null while the optimum processor suppresses the clutter only down to the noise level.

TSAO et al. [654] discussed the choice of degrees of freedom in rank reduced STAP processors. It was found that a slight overestimate of the degrees of freedom does not degrade the performance. Underestimating is more critical, in particular when decorrelation effects such as clutter fluctuations come into play. GIERULL and BALAJI [193] decribe a class of fast projection techniques for suppression of airborne or spacebased ground clutter which are not based on eigendecomposition of the space-time covariance matrix.

4.2.2.1 Sidelooking array, DPCA conditions

LIU Q.-G. et al. [407] analysed the performance of an orthogonal projection processor based on a sidelooking array under DPCA conditions. In this case the clutter component of the space-time covariance matrix was $\mathbf{Q} = P_c \boldsymbol{\Sigma} \boldsymbol{\Sigma}^*$ (see (3.48)) where $\boldsymbol{\Sigma}$ is the shift operator defined by (3.44). Then the orthogonal projection operator is given by the pseudoinverse

$$\mathbf{P} = \mathbf{I} - \boldsymbol{\Sigma}(\boldsymbol{\Sigma}^*\boldsymbol{\Sigma})^{-1}\boldsymbol{\Sigma}^* \tag{4.8}$$

Since $\boldsymbol{\Sigma}$ is a sparse matrix the calculation of the projection matrix needs only a small amount of computation. Even under array errors near optimum performance is obtained, provided that DPCA conditions are satisfied.

The interrelation between optimum processing, DPCA and orthogonal projection have been exploited by RICHARDSON [562] for designing a suboptimum simple processor:

- The zero noise approximation of the space-time covariance matrix inverse $\mathbf{Q}^{-1}$ is a projection matrix of the form

$$\mathbf{P} = \mathbf{I} - \mathbf{B}(\mathbf{B}^*\mathbf{B})^{-1}\mathbf{B}^* \tag{4.9}$$

where the columns of $\mathbf{B}$ span the clutter subspace of $\mathbf{Q}$. For instance, $\mathbf{B}$ may include the clutter eigenvectors of $\mathbf{Q}$.

- Under DPCA conditions the matrix may be composed of identity vectors of the following kind ($N \geq M$)

$$\mathbf{b}_1 = (\mathbf{e}_N^T, \mathbf{0}^T, \ldots, \mathbf{0}^T)^T$$

$$\mathbf{b}_2 = (\mathbf{e}_{N-1}^{\mathrm{T}}, \ldots, \mathbf{e}_N^{\mathrm{T}}, \mathbf{0}^{\mathrm{T}}, \ldots, \mathbf{0}^{\mathrm{T}})^{\mathrm{T}}$$

$$\vdots$$

$$\mathbf{b}_M = (\mathbf{e}_{N-M+1}^{\mathrm{T}}, \mathbf{e}_{N-M+2}^{\mathrm{T}}, \ldots, \mathbf{e}_N^{\mathrm{T}})^{\mathrm{T}}$$
$$\mathbf{b}_{M+1} = (\mathbf{e}_{N-M}^{\mathrm{T}}, \mathbf{e}_{N-M+1}^{\mathrm{T}}, \ldots, \mathbf{e}_{N-1}^{\mathrm{T}})^{\mathrm{T}} \qquad (4.10)$$

$$\vdots$$

$$\mathbf{b}_{N+1} = (\mathbf{0}^{\mathrm{T}}, \mathbf{e}_1^{\mathrm{T}}, \mathbf{e}_2^{\mathrm{T}}, \ldots, \mathbf{e}_{M-1}^{\mathrm{T}})^{\mathrm{T}}$$

$$\vdots$$

$$\mathbf{b}_{N+M-1} = (\mathbf{0}^{\mathrm{T}}, \mathbf{0}^{\mathrm{T}}, \ldots, \mathbf{0}^{\mathrm{T}}, \mathbf{e}_1^{\mathrm{T}})^{\mathrm{T}}$$

where $\mathbf{e}_i$ is a $N \times 1$ identity vector with a '1' in the i-th position and 0 elsewhere. $\mathbf{0}$ is a $N \times 1$ zero vector.

A similar set of vectors can be found for $N < M$ [562]. It can be seen by inspection that Richardson's basis vector system (4.9), (4.10) is identical to (4.8), (3.44), i.e., it is in essence a multichannel/multipulse DPCA system. The use of DPCA for spaceborne radar has been proposed by TAM and FAUBERT [633] and WANG H. S. C. [666].

A procedure for adaptive calculation of the eigenvalues and eigenvectors belonging to a subspace of the covariance matrix (for instance, the clutter subspace) has been developed by MATHEW and REDDY [442].

Two techniques based on eigendecomposition of the sample covariance matrix have been compared with the 'joint domain' optimum processor ('JDO') by GOLDSTEIN *et al.* [197] on the basis of MOUNTAINTOP data (TITI [646]).

BALAJI and Gierull [27] have given an analytical expression for the SCNR achieved by the projection technique based on a finite number of training samples.

4.3 Optimum processing and motion compensation

In this Section we try to reveal the interrelation between motion compensation and optimum processing.

4.3.1 Principle of RF motion compensation

First let us recall the principle of platform motion compensation, see SKOLNIK [610, pp. 17-7]. Let us assume an antenna looking in the cross-flight direction (sidelooking). In Figure 4.6a, a phasor diagram for two subsequent clutter echoes $c(t)$ and $c(t + \tau)$ coming from a single clutter scatterer is depicted. Because of the platform motion there is a phase advance

$$\phi = 2\pi f_\mathrm{D} T = \frac{2\pi v_\mathrm{p} T \cos \varphi}{\lambda} \qquad (4.11)$$

between the two arrivals. φ denotes azimuth. The amplitude of $c(t)$ is proportional to the two-way antenna field intensity.

130 *Fully adaptive space-time processors*

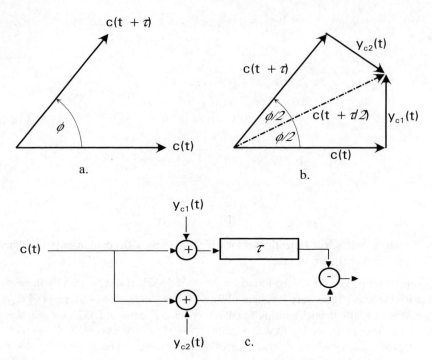

Figure 4.6: *Motion compensation with correction patterns*

The principle of platform motion compensation is shown in Figure 4.6b. Two so-called correction patterns $y_{c1}(t)$ and $y_{c2}(t)$ are added to the clutter signals in such a way that both echoes $c(t)$ and $c(t+\tau)$ are projected on a common phase angle $\phi/2$. Now both echoes are in phase so that they can easily be cancelled by subtraction, that means, by use of a conventional two-pulse RF clutter canceller which is available in any coherent radar.

This principle dates back to the times where digital technology was not yet available. Motion-compensated clutter cancellation was carried out in the RF domain. The correction patterns are given by

$$y_{c1}(t) = c(t)\tan(\phi/2) = \Sigma^2(\varphi)\tan\frac{\pi v_p T \cos\varphi}{\lambda} \tag{4.12}$$

$$y_{c2}(t) = -c(t+\tau)\tan(\phi/2) = -\Sigma^2(\varphi)\tan\frac{\pi v_p T \cos\varphi}{\lambda} \tag{4.13}$$

see SKOLNIK [610, pp. 17-7]. $\Sigma(\varphi)$ is the one-way antenna sum beam pattern. The correction pattern is very similar to the difference pattern of a monopulse tracking radar. Obviously no correction is applied in the look direction ($\varphi = 0$) while for any angle $\varphi \neq 0$ the difference pattern follows the sidelobe structure of the sum beam. Therefore, the main channel is associated with an auxiliary channel for clutter cancellation in such a way that the CNR is always the same in both the main and

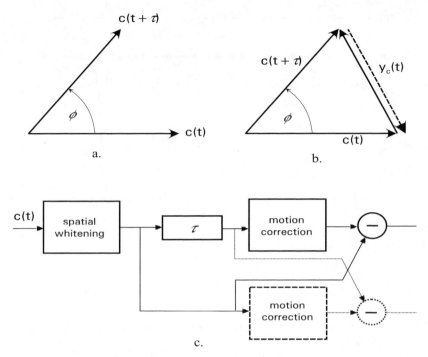

Figure 4.7: Motion compensation by the optimum processor

auxiliary channel.[1]

In fact, the output signal of a difference pattern is in quadrature with the sum beam. Therefore, with the two channels (sum and difference) of a standard monopulse antenna, motion-compensated airborne MTI can be realised.

It should be noted that this motion compensation technique works only in a sidelooking configuration where the main beam center Doppler frequency is zero. In the case where the antenna is horizontally tilted so as to look into a squint angle the clutter Doppler frequency in the beam center has to be compensated for because the standard two-pulse canceller is based on the subtraction of two equiphased signals. This compensation has been done by the TACCAR[2] loop, see SKOLNIK [610, pp. 17-32]. TACCAR is a control loop which adjusts the local oscillator frequency in such a way that the phases of subsequent echoes become equal.

4.3.2 Correction patterns

The above described motion compensation technique was developed for radar with reflector antenna. ANDREWS [19] proposed a correction pattern optimisation technique for use with antenna arrays which follows tightly the motion compensation principle

---

[1] Recall the considerations concerning the sidelobe canceller, Section 1.2.3 and Figure 1.6.
[2] Acronym for 'time-averaged-clutter coherent airborne radar'.

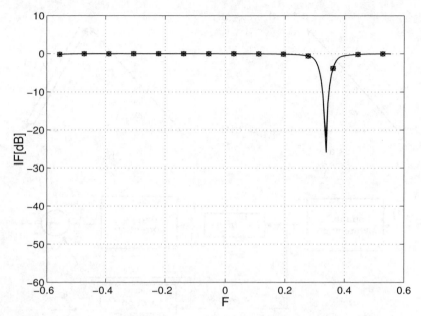

Figure 4.8: Influence of transmit beamwidth (SL): ○ $N_t = 1$; ∗ $N_t = 24$ *(like receive array);* × $N_t = 48$

illustrated before. The correction patterns $y_{c1}(t)$ and $y_{c2}(t)$ according to Figure 4.6 are optimised in such a way that the squared magnitude of the difference between the ideal and the actual *corrected* echoes is a minimum. The resulting optimised weightings for the array outputs are

$$\mathbf{y}(t+\tau/2) = (\mathbf{Q}_0^{-1}\mathbf{Q}_1(t+\tau/2) - \mathbf{I})\mathbf{b} \qquad (4.14)$$
$$\mathbf{y}(t-\tau/2) = (\mathbf{Q}_0^{-1}\mathbf{Q}_1(t-\tau/2) - \mathbf{I})\mathbf{b}$$

where **b** is a beamformer vector and $\mathbf{Q}_0$ and $\mathbf{Q}_1$ are spatial submatrices of the space-time covariance matrix, see (4.2), for the case $M = 2$.

Notice that the submatrix $\mathbf{Q}_1$ is a cross-variance matrix between the clutter echo vector **c** at time t and at time $t+\tau/2$ and $t-\tau/2$ respectively. They have to be calculated based on an analytical clutter model as has been done by ANDREWS [19]. An *adaptive* realisation is not easily achieved because at time $t + \tau/2$ no data are available.

This problem could be overcome by doubling the PRF which, however, might have unwanted effects on the radar operation. For instance, the unambiguous range is reduced by a factor of 2 when the PRF is doubled. The unambiguous Doppler range is doubled. However, if the number of pulses is kept constant the Doppler resolution is lowered by a factor of 2.

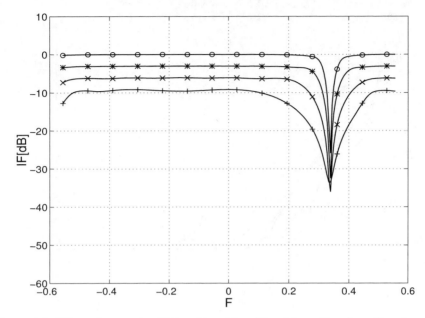

Figure 4.9: Effect of array size (SL): ∘ $N = 24$; ∗ $N = 12$; × $N = 6$; + $N = 3$

4.3.3 Interrelation with the optimum processor

4.3.3.1 *Example: the inverse of* $\mathbf{Q}$ *for* $M = 2$

Some interesting relations between motion compensation after ANDREWS [19] and optimum processing has been shown by KLEMM [301]. Similar to the considerations in Chapter 1 let us consider the space-time covariance matrix for the special case $M = 2$.

$$\mathbf{Q} = E\left\{ \begin{pmatrix} \mathbf{c}_1 \\ \mathbf{c}_2 \end{pmatrix} \begin{pmatrix} \mathbf{c}_1^* & \mathbf{c}_2^* \end{pmatrix} \right\} = \begin{pmatrix} \mathbf{Q}_0 & \mathbf{Q}_1 \\ \mathbf{Q}_1^* & \mathbf{Q}_0 \end{pmatrix} \tag{4.15}$$

where $\mathbf{c}_1 = \mathbf{c}(t)$ and $\mathbf{c}_2 = \mathbf{c}(t + \tau)$ denote the first and second echo vector. Using the Frobenius–Schur relation (BODEWIG [54, p. 217]) the inverse of $\mathbf{Q}$ becomes

$$\mathbf{Q}^{-1} = \begin{pmatrix} \mathbf{Q}_0^{-1} + \mathbf{Q}_0^{-1}\mathbf{Q}_1\mathbf{\Delta}_{21}^{-1}\mathbf{Q}_1^*\mathbf{Q}_0^{-1} & -\mathbf{Q}_0^{-1}\mathbf{Q}_1\mathbf{\Delta}_{21}^{-1} \\ -\mathbf{\Delta}_{21}^{-1}\mathbf{Q}_1^*\mathbf{Q}_0^{-1} & \mathbf{\Delta}_{21}^{-1} \end{pmatrix} \tag{4.16}$$

with

$$\mathbf{\Delta}_{21} = \mathbf{Q}_0 - \mathbf{Q}_1^*\mathbf{Q}_0^{-1}\mathbf{Q}_1 \tag{4.17}$$

being the partial covariance matrix of the clutter vector $\mathbf{c}_2$ given $\mathbf{c}_1$ (ANDERSON [15, pp. 27]). An equivalent version of (4.16) is given by (BODEWIG [54, p. 217])

$$\mathbf{Q}^{-1} = \begin{pmatrix} \mathbf{\Delta}_{12}^{-1} & -\mathbf{\Delta}_{12}^{-1}\mathbf{Q}_1^*\mathbf{Q}_0^{-1} \\ -\mathbf{Q}_0^{-1}\mathbf{Q}_1^*\mathbf{\Delta}_{12}^{-1} & \mathbf{Q}_0 + \mathbf{Q}_0^{-1}\mathbf{Q}_1^*\mathbf{\Delta}_{12}^{-1}\mathbf{Q}_1\mathbf{Q}_0^{-1} \end{pmatrix} \tag{4.18}$$

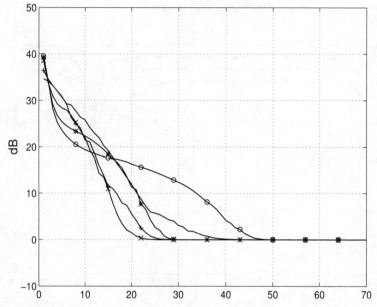

Figure 4.10: Array size versus sample size (FL), eigenspectra: ○ $N = 48, M = 3$; *
$N = 24, M = 6$; × $N = 12, M = 12$; + $N = 24, M = 6$; no symbol $N = 48, M = 3$

with

$$\mathbf{\Delta}_{12} = \mathbf{Q}_0 - \mathbf{Q}_1 \mathbf{Q}_0^{-1} \mathbf{Q}_1^* \qquad (4.19)$$

being the partial covariance matrix of the clutter vector $\mathbf{c}_1$ given $\mathbf{c}_2$. Comparing (4.16) with (4.19) one obtains

$$\mathbf{Q}^{-1} = \begin{pmatrix} \mathbf{\Delta}_{12}^{-1} & -\mathbf{Q}_0^{-1}\mathbf{Q}_1\mathbf{\Delta}_{21}^{-1} \\ -\mathbf{Q}_0^{-1}\mathbf{Q}_1^*\mathbf{\Delta}_{12}^{-1} & \mathbf{\Delta}_{21}^{-1} \end{pmatrix} \qquad (4.20)$$

4.3.3.2 Correction patterns

Let us now recall the principle of optimising correction patterns according to ANDREWS [19]. Instead of correcting for just half the phase advance $\phi/2$ (see Figure 4.6b) we now consider the case of correction for the total phase ϕ as depicted in Figure 4.7b. The beamformer responses to a clutter echo vector are

$$\begin{aligned} c_1 &= \mathbf{c}_1^* \mathbf{b} \\ c_2 &= \mathbf{c}_2^* \mathbf{b} \end{aligned} \qquad (4.21)$$

The signal at time t is to be phase corrected by adding a correction vector

$$\tilde{c}_1 = \mathbf{c}_1^*(\mathbf{b} + \mathbf{y}) \qquad (4.22)$$

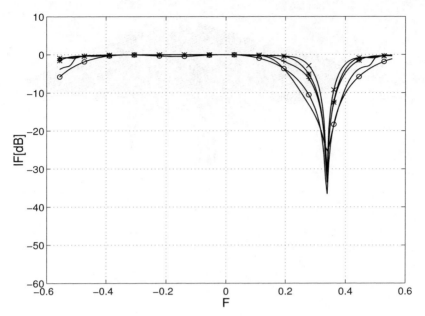

Figure 4.11: Array size versus sample size (FL), IF: ○ $N = 48, M = 3$; * $N = 24, M = 6$; × $N = 12, M = 12$; + $N = 24, M = 6$; *no symbol* $N = 48, M = 3$

where **y** is the vector of correction coefficients. The correction vector is to be chosen so that the difference between the corrected echo and the following echo becomes a minimum.

Hence we minimise the expression

$$Q = E\{|\, c_2 - \tilde{c}_1\,|^2\} = E\{c_2^* c_2 + \tilde{c}_1^* \tilde{c}_1 - c_2^* \tilde{c}_1 - \tilde{c}_1^* c_2\} \tag{4.23}$$

Using

$$\begin{aligned} \mathbf{Q}_0 &= E\{\mathbf{c}_1 \mathbf{c}_1^*\} = E\{\mathbf{c}_2 \mathbf{c}_2^*\} \\ \mathbf{Q}_1 &= E\{\mathbf{c}_1 \mathbf{c}_2^*\} \end{aligned}$$

(4.23) becomes

$$Q = \mathbf{b}^* \mathbf{Q}_0 \mathbf{b} + (\mathbf{b}+\mathbf{y})^* \mathbf{Q}_0 (\mathbf{b}+\mathbf{y}) - (\mathbf{b}+\mathbf{y})^* \mathbf{Q}_1 \mathbf{b} - \mathbf{b}^* \mathbf{Q}_1 (\mathbf{b}+\mathbf{y}) \tag{4.24}$$

The minimum is obtained by differentiation of (4.24) with respect to **y** and setting the result equal to the zero vector. The solution is

$$\mathbf{y} = (\mathbf{Q}_0^{-1} \mathbf{Q}_1 - \mathbf{I})\mathbf{b} \tag{4.25}$$

The corrected beamformer vector is then

$$\mathbf{b}_c = \mathbf{b} + \mathbf{y} = (\mathbf{Q}_0^{-1} \mathbf{Q}_1)\mathbf{b} \tag{4.26}$$

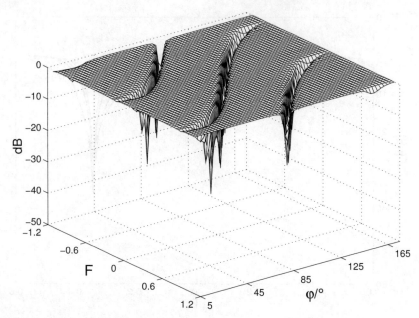

Figure 4.12: *Effect of temporal undersampling (SL, $f_{PR} = 0.5$ Nyquist)*

4.3.3.3 Comparison

Comparing (4.26) with (4.20) we find the correction factor $\mathbf{Q}_0^{-1}\mathbf{Q}_1$ in the off-diagonal submatrices. This indicates that in the second column of $\mathbf{Q}^{-1}$ the vector $\mathbf{c}(t)$ is corrected while the vector $\mathbf{c}(t + \tau)$ is not. The same happens in the first column, but the other way around. The minus signs in both the off-diagonal submatrices take care of the clutter cancellation like the minus sign occurring in the well-known two-pulse canceller.

The inverses of the partial covariance matrices $\mathbf{\Delta}_{12}^{-1}$ and $\mathbf{\Delta}_{21}^{-1}$ provide the spatial part of the space-time clutter cancellation.[3] Now we can summarise our insights into the nature of the optimum processor. It includes three components:

Motion compensation. The phase advance of one out of two subsequent clutter echoes due to the platform motion is corrected for by a correction matrix (4.26).

Spatial cancellation. The spatial part of the clutter cancellation is carried out by the inverses of the partial covariance matrices $\mathbf{\Delta}_{12}$ and $\mathbf{\Delta}_{21}$. Notice that this part of the optimum processor is not included in the DPCA techniques nor in ANDREWS' motion compensation technique.

It should be noted that in the DPCA case $\mathbf{Q}_0$ becomes the identity matrix and all off-diagonal submatrices $\mathbf{Q}_m$ are identity matrices with the main diagonal shifted by

---

[3]For details on partial correlation see MARDIA *et al.* [436, pp. 169-170] and ANDERSON [15, p. 29].

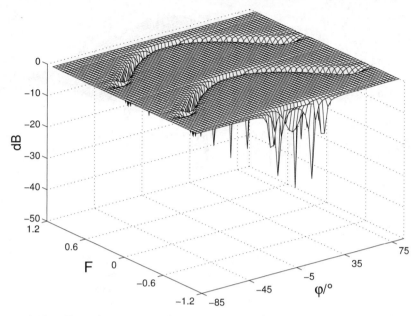

Figure 4.13: Effect of temporal undersampling (FL, $f_{PR} = 0.5$ Nyquist)

m steps in the cross-diagonal direction.[4] For example, for two sensors ($N = 2$) we get

$$\mathbf{Q}_0 = \begin{pmatrix} 1 & 0 \\ 0 & 1 \end{pmatrix} \quad \mathbf{Q}_1 = \begin{pmatrix} 0 & 0 \\ 1 & 0 \end{pmatrix} \quad (4.27)$$

Therefore, the partial covariance matrices $\mathbf{\Delta}_{12}$ and $\mathbf{\Delta}_{21}$ according to (4.19) and (4.19), respectively, become identity matrices. By looking at the covariance matrix (3.49), it becomes obvious that this holds for arbitrary values of N. Under ideal DPCA conditions no spatial decorrelation is necessary. This can be seen by inserting (4.27) into (4.19). In this case only motion compensation plus temporal cancellation is optimum.

If ideal DPCA conditions are not fulfilled, i.e., the zero elements in the covariance matrix (3.49)[5] assume values unequal to zero, the partial covariance matrices will play a role in spatial clutter rejection:

Non-ideal DPCA conditions may be caused by several reasons:

- The PRF is not precisely the Nyquist frequency of the clutter bandwidth as determined by the platform velocity.

- The PRF is not harmonised with the sensor spacing and the platform speed.

- The antenna array is different from an equispaced, linear or planar, array in the sidelooking configuration.

---

[4] Compare with (3.49).
[5] See also Figure 3.8.

138 *Fully adaptive space-time processors*

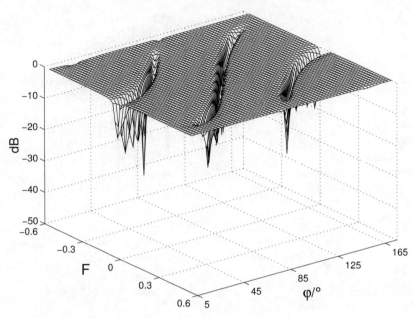

Figure 4.14: Effect of sensor spacing (SL, $d/\lambda = 1$)

- The clutter background is inhomogeneous. Singular peaks of high intensity may occur.
- Directive illumination through the transmit beam (compare the effect of directive illumination in Figures 3.7 and 3.8).
- The sensor directivity patterns of the receive array cause some additional directivity.

Under all these circumstances the optimum adaptive processor is superior to the motion compensation and subtraction method proposed by ANDREWS [19].

Temporal cancellation The temporal clutter cancellation component is given by subtraction of the clutter vectors after spatial decorrelation and motion compensation (multiplication with the correction matrix, see (4.26)).

A block diagram of the optimum processor is shown in Figure 4.7c. The processor has to be completed with a signal matching network (see Figure 4.1) of the form

$$\mathbf{Q} = \begin{pmatrix} \mathbf{b} \\ \mathbf{b}e^{-j\omega_t \tau} \end{pmatrix} \qquad (4.28)$$

where the factors attached to the beamformer vectors are the coefficients of a two-point Doppler filter and ω_t is the Doppler frequency of the target.[6]

---
[6]For $M = 2$ the length of the Doppler filter bank is only 2. In practice a MTI FIR filter as given by one

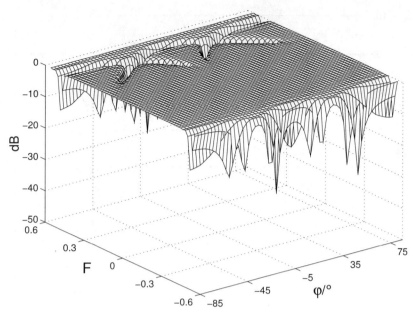

Figure 4.15: Effect of sensor spacing (FL, $d/\lambda = 1$)

4.4 Influence of radar parameters

In the following we discuss some properties of the optimum processor by variation of several radar parameters.

4.4.1 Transmit beamwidth

Let us consider first the influence of the directivity of the transmit antenna. For that purpose all clutter arrivals are weighted with a transmit directivity pattern as defined by (2.39) and (2.40).

In Figure 4.8 the improvement factor (IF) has been plotted versus the normalised Doppler frequency. The look direction has been kept constant ($45°$). Such curves are cross-sections through a three-dimensional plot as shown for example in Figures 4.2 and 4.3. We will use this kind of plot throughout the following Chapters.

The three curves in Figure 4.8 have been plotted for $N_t = 1$ sensor[7] (omnidirectional transmission), $N_t = 24$ (transmit array = receive array), $N_t = 48$

---

of the columns of $\mathbf{Q}^{-1}$ is cascaded with a Doppler filter bank whose length is chosen independent of the clutter filter length $M = 2$. The operation of the two columns is marked in Figure 4.7b by solid and dashed lines.

It should be noted that this way of processing (pre-filtering for clutter rejection followed by a Doppler filter bank) is a *suboptimum* approach which is based on a much larger data size than required for the adaptive clutter filter. The optimum processor for this data size would have the same temporal dimension, i.e. much larger than $M = 2$. We come back to this point in Chapter 7.

[7] N_t is the number of elements of the transmit array, see (2.39).

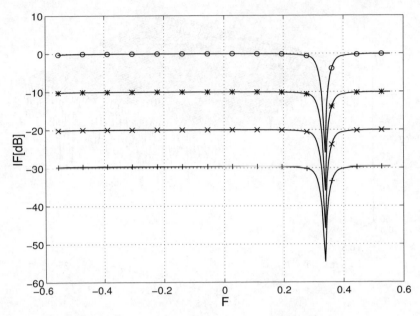

Figure 4.16: Effect of the CNR (SL): ○ *CNR = 40 dB;* ∗ *CNR = 30 dB;* × *CNR = 20 dB;* + *CNR = 10 dB*

(transmit array = 2 × receive array). The improvement factor has been normalised by the maximum IF which amounts approximately to

$$\text{IF}_{\max} = N \cdot M \cdot \text{CNR} \tag{4.29}$$

see (1.35).

We notice the following facts:

- The maximum IF is reached for all target Doppler frequencies except for the clutter notch. This notch appears at that clutter Doppler frequency which is associated with the look direction. In other words, the target Doppler equals the clutter Doppler. In the given example we have $\varphi_\text{L} = 45°$; $\cos 45° = 0.707$; therefore the notch appears at F = 0.35. Actually it is located slightly below F = 0.35 because of the depression angle given by the radar-ground geometry.[8]

- The width of the clutter notch is a measure for the detectability of slow targets.

- The IF is practically independent of the transmit beamwidth. The width of the clutter notch is merely determined by the clutter resolution capability of the adaptive receiver.

---

[8]See Table 2.1

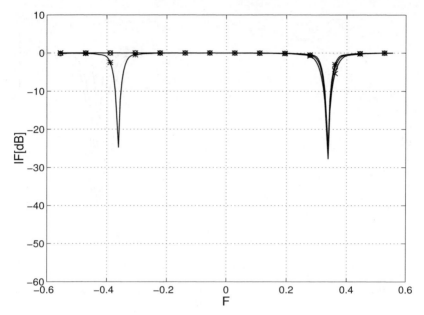

Figure 4.17: Effect of non-DPCA sampling (SL): ∘ f_{PR} =Nyquist; ∗ $f_{\text{PR}} = 0.7 \cdot$ Nyquist; × $f_{\text{PR}} = 1.43 \cdot$ Nyquist

4.4.2 Array and sample size

The effect of the array size (number of elements N) is illustrated in Figure 4.9. The four curves are plotted for $N = 3, 6, 12, 24$. In accordance with doubling the number of sensors the improvement factor increases in increments of 3 dB in the antenna sidelobe region, i.e., far away from the clutter notch.

Moreover, it can be noticed that the clutter notch becomes narrower as the array size increases. This fact is well known from one-dimensional processing (spatial *or* temporal), see, for example, Figure 1.2. Now, as we increased the number of array elements the *Doppler response* shows the same behaviour as the spatial response! This, however, follows directly from the equivalence between azimuth and Doppler, see (2.35). It is impossible to make a trench like the one in Figure 4.2 narrower in one dimension only without doing it in the other dimension as well.

Moreover, without giving further numerical examples, we can state that the clutter notch depends on the sample size M in a similar way. In summary, whichever dimension (N, M) is smaller limits the clutter resolution, i.e., the width of the clutter notch.

Let us recall that under DPCA conditions the number of clutter eigenvalues of the clutter covariance matrix for a sidelooking equidistant array is

$$N_e = N + M - 1$$

For a given total sample size NM the minimum number of eigenvalues is obtained if

$$N = M \tag{4.30}$$

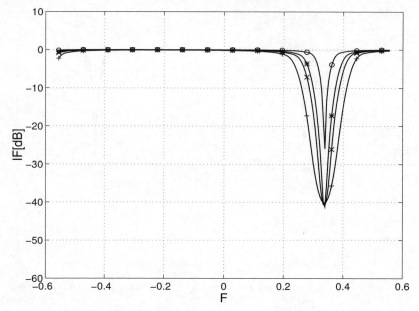

Figure 4.18: Influence of clutter bandwidth (SL): ○ $B_c = 0$; ∗ $B_c = 0.05$; × $B_c = 0.1$; + $B_c = 0.2$

The reader is reminded that the number of eigenvalues is a measure of the degrees of freedom of a clutter suppression filter.

In Figure 4.10 the total space-time sample size was assumed to be $NM = 144$. The plots show curves for different ratios N/M. It is clearly seen that for $N = M = 12$ the minimum number of clutter eigenvalues is obtained.[9]

Figure 4.11, shows the corresponding IF curves. In comparison with Figure 4.10 it can be noticed that the clutter notches are broadened as the number of clutter eigenvalues increases.

These examples were calculated for a forward looking array. Similar results can be obtained for a sidelooking array.

4.4.3 Sampling effects

As mentioned before the IF of the optimum processor is proportional to the inverse of the MV spectrum (compare (1.16) and (1.104)). Therefore, the following four plots correspond closely to Figures 3.20, 3.21, 3.22 and 3.23.

4.4.3.1 *Temporal undersampling*

Figure 4.12 shows the IF versus the azimuth angle φ and the normalised Doppler frequency F for a sidelooking linear array. The PRF has been set to one half of

---

[9]Notice that only 70 out of 144 eigenvalues are shown.

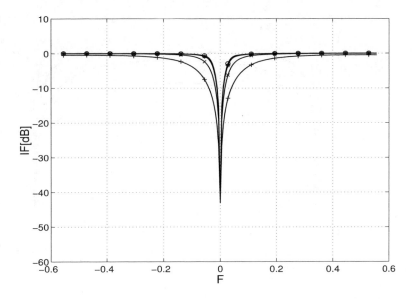

Figure 4.19: Effect of range walk (SL, $\varphi_L = 90°$): ○ *no range walk;* ∗ $\Delta_R = 100\ m$; × $\Delta_R = 10\ m$; + $\Delta_R = 1\ m$

the Nyquist frequency corresponding to the clutter bandwidth (6 KHz). Therefore, ambiguous responses can be noticed on both sides of the main clutter trench.

In Figure 4.13, we see an example for forward looking radar again with half the Nyquist frequency as PRF. The clutter trench, originally located only at positive Doppler frequencies, is now repeated at negative target Doppler frequencies.

4.4.3.2 Spatial undersampling

In the following two figures we assume that the sensor spacing is $d = \lambda$, i.e., twice the spatial Nyquist frequency. In Figure 4.14, we see again the clutter trench and the associated ambiguous responses on both sides for sidelooking radar. Notice that the clutter ambiguities run along the same trajectories as in the case of temporal undersampling (Figure 4.12). Therefore, if the clutter echoes are undersampled in both time and space in the same way the ambiguities fall onto the same trajectories. This has to do with the fact that for a sidelooking array the array motion is aligned with the array axis.

For a forward looking array with $d = \lambda$ the IF becomes as shown in Figure 4.15. Notice that there are multiples of the original semicircle which are shifted along the spatial φ-coordinate. Opposite to the sidelooking array case there is no coincidence of the trajectories for temporal and spatial undersampling (compare Figure 4.15 with Figure 4.13).

144 Fully adaptive space-time processors

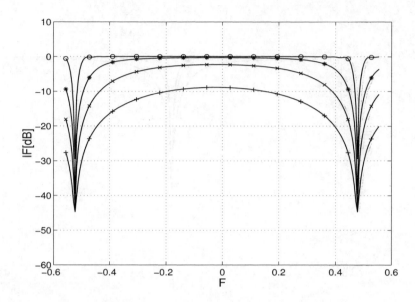

Figure 4.20: Effect of range walk (FL, $\varphi_L = 0°$): ○ *no range walk;* ∗ $\Delta_R = 100\ m$; × $\Delta_R = 10\ m$; + $\Delta_R = 1\ m$

4.4.3.3 Non-DPCA sampling

Let us recall the remarks made on the nature of the inverse of the clutter covariance matrix on page 136. It was found that the DPCA technique performs motion compensation and temporal clutter cancellation, but does not include partial spatial whitening as does the optimum processor. Let us further recall that the partial spatial covariance matrices in (4.17) and (4.19) become identity matrices only if the phase advance of subsequent echoes is identical to the phase advance due to sensor displacement, that means, the covariance matrix has the form (3.49) (DPCA condition). Therefore, the IF achieved by DPCA depends on the ratio of spatial (sensor displacement) and temporal (PRF) sampling. This effect is shown in SKOLNIK [609, p. 18-8] and SKOLNIK [610, p. 16-12].

In Figure 4.17, three IF curves are plotted. We chose PRF = Nyquist, PRF = 0.7· Nyquist, and PRF = 1.43· Nyquist. First of all we notice that for all three PRFs the maximum gain is achieved in the passband. There are very small differences in the width of the clutter notch which are due to the fact that for different PRFs the observation time and, hence, the Doppler resolution, is different. The additional clutter notch on the left is due to undersampling.

It appears from these results that the IF achieved by the optimum processor is independent of the ratio of spatial and temporal sampling, except for ambiguities and variations of the width of the clutter notch.

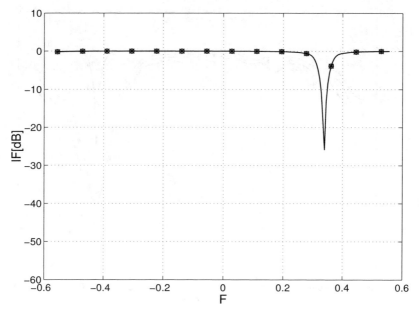

Figure 4.21: Influence of system bandwidth (SL, rectangular impulse response): ○ $B_s = 0$; ∗ $B_s = 0.05$; × $B_s = 0.1$; + $B_s = 0.2$

4.4.4 Influence of the CNR

Expressing the inverse of $\mathbf{Q}$ by eigenvalues and eigenvectors the optimum IF is

$$\text{IF}_{\text{opt}} = \mathbf{s}^*(\varphi_L, f_D)\mathbf{E}\mathbf{\Lambda}^{-1}\mathbf{E}^*\mathbf{s}(\varphi_L, f_D) \cdot \frac{\text{tr}\mathbf{Q}}{\mathbf{s}^*(\varphi_L, f_D)\mathbf{s}(\varphi_L, f_D)} \qquad (4.31)$$

where $\mathbf{\Lambda}$ is the diagonal matrix of eigenvalues of $\mathbf{Q}$ and $\mathbf{E}$ the unitary matrix of eigenvectors. The eigenvalues can be separated into clutter and noise eigenvalues so that (4.31) becomes

$$\text{IF}_{\text{opt}} = \mathbf{s}^*(\varphi_L, f_D)(\mathbf{E}\tilde{\mathbf{\Lambda}}_c\mathbf{E}^* + \frac{1}{P_w}\tilde{\mathbf{I}})\mathbf{s}(\varphi_L, f_D) \cdot \frac{\text{tr}\mathbf{Q}}{\mathbf{s}^*(\varphi_L, f_D)\mathbf{s}(\varphi_L, f_D)} \qquad (4.32)$$

where $\tilde{\mathbf{\Lambda}}_c$ is a diagonal matrix whose diagonal has the inverse clutter eigenvalues on the first N_c positions and zeroes elsewhere. $\tilde{\mathbf{I}}$ is a diagonal matrix with zeroes on the first N_c positions and unity elsewhere. N_c is the number of clutter eigenvalues. Except for the small contribution of the clutter subspace $\mathbf{E}\tilde{\mathbf{\Lambda}}_c\mathbf{E}^*$, i.e., for large CNR, IF$_{\text{opt}}$ becomes inversely proportional to the noise power P_w. This is illustrated in Figure 4.16. The four curves have been plotted for CNR = 10, 20, 30, 40 dB. Notice that the difference between the various curves is constant (10 dB), except for the two lower ones (low CNR) where the clutter subspace still has little influence.

146 Fully adaptive space-time processors

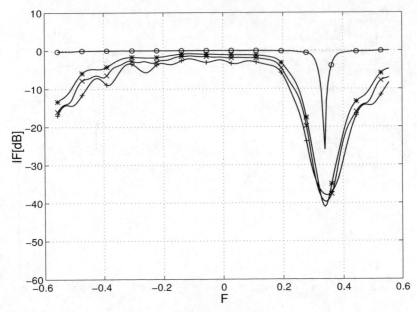

Figure 4.22: Influence of system bandwidth (FL, rectangular impulse response): ○ $B_s = 0$; ∗ $B_s = 0.05$; × $B_s = 0.1$; + $B_s = 0.2$

4.4.5 Bandwidth effects

4.4.5.1 *Clutter bandwidth*

The following results are based on the clutter fluctuation model (2.53). As we know from Figures 3.24 and 3.25, temporal decorrelation due to internal clutter motion is reflected by an increase in the number of clutter eigenvalues. This in turn has consequences for the width of the clutter notch and, hence, on the detection of slow targets.[10]

Figure 4.18, p. 142, shows the effect of clutter fluctuations on the detectability of slowly moving targets. As can be seen, a target whose Doppler frequency falls inside the clutter bandwidth will be attenuated by the clutter suppression process so that detection is degraded. Notice that outside the area covered by the clutter bandwidth the maximum gain is obtained.

According to BARILE *et al.* [35] internal clutter motion may lead to a strong decrease in improvement factor as well as to a rise of antenna sidelobes.

4.4.5.2 *Effect of system bandwidth: temporal decorrelation through range walk*

In Figures 4.19 and 4.20 the effect of range walk (for a model see Section 2.5.1.2) is illustrated for sideways and forward looking array configurations. Both arrays look into

---
[10]Compare, for example, Figures 4.10 and 4.11.

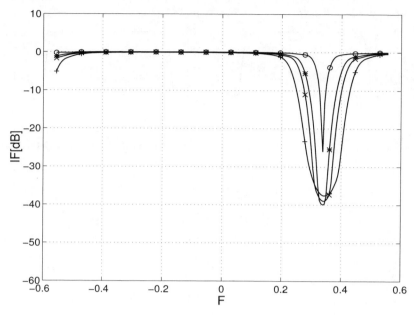

Figure 4.23: Influence of system bandwidth (FL, rectangular frequency response): ○ $B_s = 0$; ∗ $B_s = 0.05$; × $B_s = 0.1$; + $B_s = 0.2$

the broadside direction (90° and 0°, respectively). Both figures show that the higher the range resolution is, the larger the clutter notch of the IF curves becomes. A comparison of both figures confirms that range walk has more impact on the clutter notch in the 'in flight' direction than in the cross-flight direction.

These results indicate that the received data might have to be range corrected prior to adaptation and clutter filtering. This depends on parameters such as signal integration time (M) and range resolution. For details on range migration compensation techniques see CURLANDER and MCDONOUGH [109, pp. 189–196].

4.4.5.3 Effect of system bandwidth: spatial decorrelation through travel delay

The effect of system bandwidth on space-time clutter suppression has been discussed by KLEMM and ENDER [340], KLEMM [316], and HERBERT [257]. For the sidelooking array we know from the considerations in Chapter 3 (see the eigenspectra in Figure 3.28, p. 102 and Figure 3.30, p. 104) that the number of clutter eigenvalues is independent of the system bandwidth. We recall that the DPCA effect compensates for travel delays of incoming waves across the array, see Section 3.4.3. The following results are based on the system bandwidth models (2.60) and (2.64). Figure 4.21 shows the improvement factor of a sidelooking array under DPCA conditions (see Table 2.1). The curves have been calculated for different values of the relative system bandwidth B_s. A rectangular frequency response of the receive channels has been assumed. Recall that the associated eigenspectra in Figure 3.30, p. 104, are almost identical.

148 *Fully adaptive space-time processors*

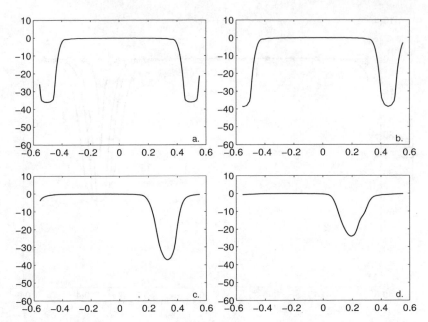

Figure 4.24: Influence of system bandwidth (FL, rectangular frequency response, $B_s = 0.2$): a. $\varphi_L = 0°$; b. $\varphi_L = 25°$; c. $\varphi_L = 50°$; d. $\varphi_L = 75°$

Accordingly, the IF curves coincide perfectly. There is no degradation in SCNR due to the system bandwidth.

The DPCA effect which compensates for system bandwidth effects in sidelooking arrays does not apply to forward looking arrays or any other array configuration. Therefore, as we know from the clutter spectra shown in Chapter 3 (Figure 3.34, p. 108 and Figure 3.35, p. 109) forward looking arrays are sensitive to spatial decorrelation caused by the system bandwidth.

In Figure 4.22, IF curves for different system bandwidth (rectangular impulse response according to (2.60)) are shown. It can be noticed that the clutter notch is severely broadened. Even in the pass band the optimum gain is not reached. This can be explained with the relatively high sidelobes of the $\sin x/x$ autocorrelation function of a rectangular impulse.

For a channel response with rectangular spectrum according to (2.64), p. 64, the bandwidth effect on the improvement factor is mitigated in that outside the clutter frequency region the optimum IF is reached (Figure 4.23, p. 147). However, some broadening of the clutter notch (loss in SCNR for slow targets) has to be taken into account.

In Figure 4.24 the dependence of the clutter notch on the look angle φ_L is shown. As can be seen even for $\varphi_L = 0°$ the clutter notch is broadened significantly. This happens because all clutter arrivals whose angle of arrival is different from $\varphi = 0°$ (even those within the beamwidth) are subject to decorrelation. It can also be noticed that the width of the clutter notch increases with increasing angle of arrival (i.e., with increasing travel delay across the array aperture).

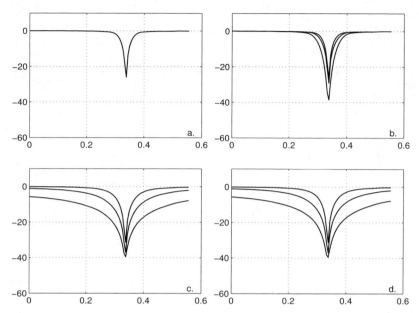

Figure 4.25: Superposition of system bandwidth and range walk effect (IF vs F, FL, rectangular frequency response): a. No bandwidth effect; b. system bandwidth effect only; c. range walk effect only; d. superposition of both bandwidth induced effects. Upper to lower curves: $B_s = 0.0001$ ($\Delta_R = 150$ m); $B_s = 0.001$ ($\Delta_R = 15$ m); $B_s = 0.01$ ($\Delta_R = 1.5$ m)

If required system bandwidth effects can be compensated for by space-time-TIME or space-time-FREQUENCY processing (by TIME is meant the echo delay time equivalent to range). FREQUENCY is the associated frequency after Fourier transform. It should be noted that the temporal decorrelation due to clutter motion cannot be compensated for.

In Figure 4.30, the effect of system bandwidth (rectangular frequency response) for a sidelooking antenna configuration is shown. The three curves have been plotted for different PRFs. This plot corresponds to Figure 4.17, p. 141, however with $B_s = 0.1$. The figure is to illustrate that the optimum space-time processor compensates for spatial decorrelation effects caused by the system bandwidth even under non-DPCA sampling conditions (that means, the array elements do not assume the positions of their predecessors while the array is moving). Obviously the system bandwidth has no effect on the clutter notch.

The spatial decorrelation due to the system bandwidth occurs only when narrowband array steering is applied. ZATMAN and BARANOSKI [739] have shown that the effect of system bandwidth can be compensated by using time-delay steering in angle and Doppler. TECHAU [638] analyses the effect of different characteristics of the receive filter on the hot clutter mitigation performance. RABIDEAU and KOGON [549] discuss the impact of the system bandwidth in the context of space-based STAP radar. They propose to use sufficiently narrow subband filters to reduce the system bandwidth

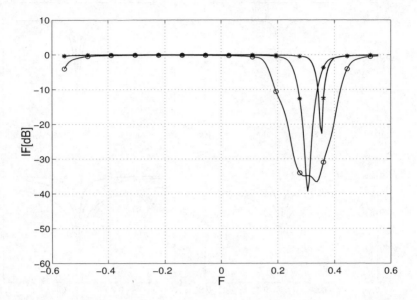

Figure 4.26: Doppler spread within range gate (FL, $H = 1$ km): ○ $R_s = 2$ km, $\Delta_R = 1500$ m; * $R_s = 2$ km, $\Delta_R = 150$ m; × $R_s = 20$ km, $\Delta_R = 1500$ m; + $R_s = 20$ km, $\Delta_R = 150$ m

effect. Some considerations on STAP with wideband radar have been examined by HERBERT [257]. He concludes that by using true delay lines rather than phase shifters for beamsteering the performance of the STAP processor can be improved even for forward looking arrays. For application in communications, BOCHE and SCHUBERT [53] give an upper bound for signal errors caused by the system bandwidth.

4.4.5.4 Superposition of range walk and travel delay effects

Both of the two above discussed effects (temporal decorrelation due to range walk, spatial decorrelation due to travel delays of incoming waves) have the same origin, namely the bandwidth of the radar system. Although both effects have been treated separately in the previous Sections they occur in practice simultaneously.

This principle is illustrated by Figure 4.25. In Figure 4.25a we see the optimum IF curve without any decorrelation effects. Figure 4.25b shows the system bandwidth effect as would occur in a stationary radar with the associated bandwidths (range resolutions Δ_R). The impact of platform velocity is shown in Figure 4.25c while d shows a superposition of both effects. As can be seen the curves in c and d look identical. Obviously range walk is the dominating effect.

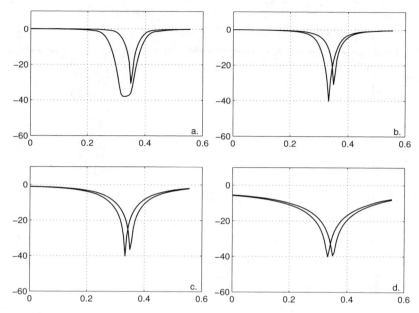

Figure 4.27: Doppler spread and range walk (IF vs. F, FL, $H = 1$ km; *left clutter notch:* $R_s = 3$ km, *right clutter notch:* $R_s = 10$ km*): a.* $\Delta_R = 1500$ m; *b.* $\Delta_R = 150$ m; *c.* $\Delta_R = 15$ m; *d.* $\Delta_R = 1.5$ m

4.4.5.5 Doppler spread within range gate

It was shown in the previous subsections that temporal decorrelation due to range walk becomes stronger if the range gate is getting narrower (high range resolution). For large range gates some Doppler spread within the range gate may occur, provided the clutter Doppler frequency is range dependent. As was demonstrated in Chapter 3 this applies to all array configurations other than sidelooking.[11] A model for the Doppler spread was given in Section 2.5.3.

Figure 4.26 shows a numerical example for the effect of Doppler spread on the IF. It can be noticed that a significant degradation of the improvement factor takes place only at short range (i.e., R_s is of the same order of magnitude as H), and if the width of the range gate is of the order of magnitude of R_s and H.[12] These are quite exceptional conditions. We can conclude that Doppler spread within the range gate does not limit the performance of the adaptive space-time processor in most applications.

There is obviously a trade-off between the Doppler spread and the range walk effect. In the example given in Figure 4.27 both effects have been superimposed. The two curves in each of the subplots have been calculated for $R_s = 3$ km and for $R_s = 10$ km, respectively. Notice that the clutter notches for the two distances are Doppler shifted due to different depression angles ($\sin \theta = H/R_s$).

---

[11] For bistatic radar operation even sidelooking arrays may exhibit range dependence of the clutter Doppler, see Chapter 12.

[12] For the definitions of R_s and H see Figure 2.1.

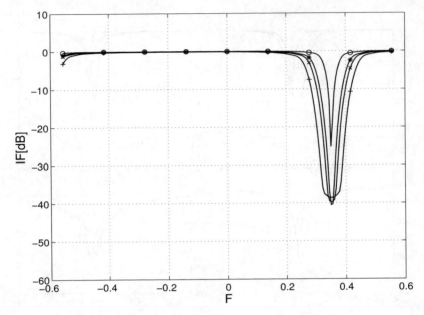

Figure 4.28: Influence of bandwidth induced Doppler spread (SL, rectangular frequency response): ∘ $B_s = 0$; ∗ $B_s = 0.05$; × $B_s = 0.1$; + $B_s = 0.2$

For a large range bin ($\Delta_R = 1500$ m, Figure 4.27a) the Doppler spread dominates. Notice that at short range the clutter notch is considerably broadened whereas at far range the effect is not as significant. As the range bin gets smaller ($\Delta_R = 150$ m, Figure 4.27b) the clutter notch becomes narrower even for short range. In this case the impact of both Doppler spread and range walk is rather weak. For even smaller range gates (Figures 4.27c and d) decorrelation through range walk dominates. This effect is nearly independent of range.

4.4.5.6 System Doppler spread

The Doppler frequency is proportional to the carrier frequency, see (2.70). Therefore, if the received echo signal has a certain system bandwidth, the clutter Doppler frequency has a certain Doppler bandwidth too. Since the clutter Doppler frequency depends on the direction of arrival the Doppler bandwidth does as well (see (2.70)). Since the decorrelation effect is purely temporal it is independent of array orientation and shape.

In Figure 4.28 the effect of system Doppler spread is illustrated. In contrast to Figure 4.23 we assumed here a sidelooking array. The broadening of the clutter notch with increasing system bandwidth can be seen clearly. For a forward looking array similar results would be obtained.

Figure 4.29 shows the dependence of the clutter notch on the look direction. The shift of the clutter notch as well as the variation of the width of the clutter notch with the look angle can be noticed.

Like all decorrelation effects induced by the system bandwith the effect of system

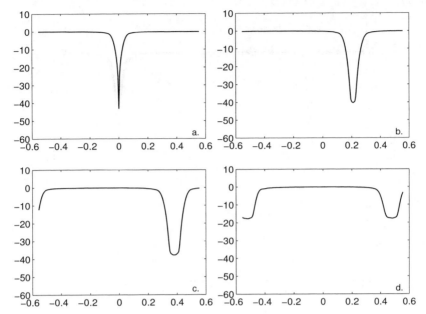

Figure 4.29: Influence of bandwidth induced Doppler spread (SL, rectangular frequency response): a. $\varphi_L = 90°$; b. $\varphi_L = 65°$; c. $\varphi_L = 40°$; d. $\varphi_L = 15°$

Doppler spread can be mitigated by adding the fast (range equivalent) time dimension to the signal vector space so that the STAP processor operates in space-time-TIME. Equivalently, the system bandwidth may be subdivided into subbands. Then space-time processing has to be conducted for all subbands separately.

4.4.6 Moving clutter

The mathematical description of the phase relations for Doppler clutter was given in (2.44), p. 59. It is obvious from (2.44) that the radial clutter velocity causes just a Doppler shift of the clutter spectrum. That means in turn that the resulting IF plots are identical to those shown in Figure 4.2, p. 124, or Figure 4.3, p. 125, but shifted by a Doppler term corresponding to the clutter velocity v_c. Adding a clutter velocity component results in a shift of the clutter notch according to the radial clutter velocity. Since this is obvious we omit any numerical example.

4.5 Range-Doppler IF matrix

A common means of illustrating the reactions of a radar to clutter, interference and targets is the range-Doppler matrix. The range-Doppler matrix normally shows the Fourier transforms of radar echo sequences for various range increments. It can, however, be used as well to illustrate the performance of signal processing techniques.

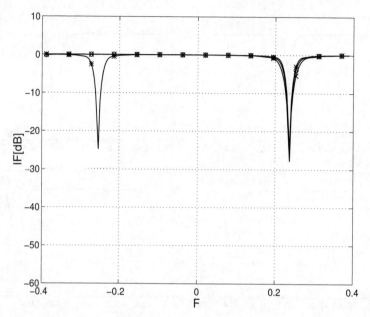

Figure 4.30: Effect of non-DPCA sampling (SL, system bandwidth $B_s = 0.1$): ○ f_{PR} =Nyquist; * $f_{PR} = 0.7 \cdot$Nyquist; × $f_{PR} = 1.43 \cdot$Nyquist

The principle of clutter suppression by use of the optimum processor is illustrated in Figures 4.31, p. 155, and 4.32, p. 156. The figures show the improvement factor according to (4.5), p. 125, versus range and normalised Doppler frequency. We assumed that the CNR is 20 dB at 15000 m range.

In Figure 4.31 the clutter notch appears as a vertical straight line in black. It indicates that the clutter Doppler frequency is range independent for a sidelooking array (compare with (3.19), p. 76). Notice that the improvement factor decreases with range. This is in accordance with the radar range equation which states that the backscattered radar power due to a single point scatterer is $P_c \propto \frac{1}{R^4}$.

For the forward looking radar case[13] the clutter notch runs according to the range law of the clutter Doppler frequency given by (3.20), p. 77. Since we assumed a radar platform altitude of $H = 3000$ m, $R_s = 3000$ m means a look perpendicular onto the ground. Therefore there is no radial velocity component between radar and ground so that the clutter Doppler frequency becomes zero. For larger range the depression angle decreases and, hence, the clutter Doppler frequency approaches the Doppler frequency associated with the platform velocity.[14]

---

[13] For the parameter configuration used in Figure 4.32 the clutter is range ambiguous. For simplification this aspect has been neglected here. We will come back to this aspect in Chapter 16.

[14] Recall that the choice PRF=12000 is the clutter Nyquist frequency. Therefore, the platform velocity induced Doppler frequency appears at $F = 0.5$.

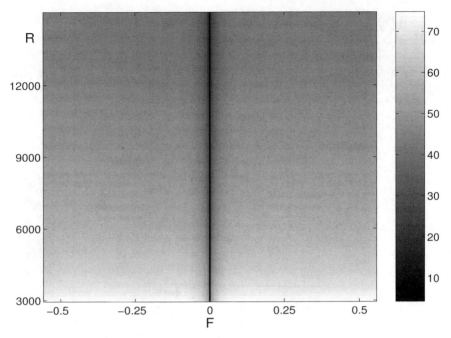

Figure 4.31: Range-Doppler matrix (greytones denote IF/dB, R = range/m, SL, φ_L = 90°)

4.6 Summary

This Chapter deals with fully adaptive processors, i.e., space-time processors that operate adaptively on the full sample size as given by the number of sensors × the number of echoes ($N \times M$). In detail the following observations can be made:

1. The **improvement factor** is just the inverse of the MV spectrum (see Chapter 3).

2. The **optimum adaptive space-time** processor (OAP) is far superior to one-dimensional (spatial or temporal) processing.

3. The **orthogonal projection** processor (OPP) approximates the optimum processor almost perfectly. The only difference is that the OPP forms exact nulls at the clutter Doppler frequency while the OAP reduces the improvement factor down to the minimum eigenvalue of the clutter + noise of the inverse of the covariance matrix.

4. The **principle of clutter rejection** by the optimum processor can be interpreted as a three-step function:

    - compensation for motion-induced phase advances between subsequent clutter echoes;

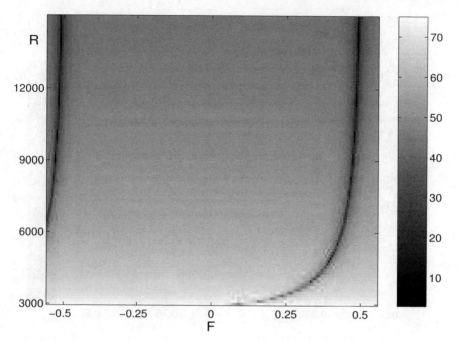

Figure 4.32: Range-Doppler matrix (greytones denote IF/dB, R = range/m, FL, $\varphi_L = 0°$)

- spatial clutter cancellation through the inverses of partial covariance matrices;
- temporal clutter cancellation through subtraction of subsequent echo vectors.

5. The clutter rejection performance of the optimum processor depends on the **sampling rates** as follows:

- Spatial undersampling (sensor spacing larger than $\lambda/2$) leads to spatial ambiguities.
- Temporal undersampling (PRF below the Nyquist rate of clutter bandwidth) leads to temporal ambiguities.
- The optimum adaptive processor functions optimally even if the PRF is not chosen at the DPCA (Nyquist) rate.

6. Decorrelation effects degrade the clutter rejection performance in the following way:

- The clutter bandwidth due to internal clutter fluctuation causes a *temporal* decorrelation which results in a broadening of the clutter notch (equivalent to degradation in slow target detection). These losses cannot be compensated for.

- Range walk is another temporal decorrelation effect which causes broadening of the clutter notch.
- The system bandwidth causes *spatial decorrelation* of clutter echoes and signals because of delays of arriving waves travelling over the antenna aperture. This decorrelation results also in a broadening of the clutter notch. This effect can be compensated for by space-time-TIME processing.
- Temporal decorrelation due to range walk and spatial decorrelation due to travel delays across the aperture have the same origin (the system bandwidth) and, therefore, occur simultaneously. It appears that the range walk effect dominates over the spatial decorrelation.
- Decorrelation may occur due to Doppler spread within the individual range gate if the clutter Doppler is range dependent. This does not apply to sidelooking arrays. The effect of Doppler spread becomes significant only for large range gates at short distance.
- Temporal decorrelation due to bandwidth induced Doppler spread is proportional to the system bandwidth and to the sine of the angle of arrival.

7. The **range-Doppler matrix** is a useful way of illustrating the range dependence of the clutter Doppler frequency.

Chapter 5

Space-time subspace techniques

The optimum space-time processor as analysed in Chapter 4 suffers from a high degree of computational complexity so that its use in practical applications, especially for real-time operation, is unlikely.

In Chapters 5, 6, 7 and 9 we discuss several ways of reducing the signal vector space and, hence, the computational workload associated with space-time adaptive MTI processing. All of these techniques are based on linear subspace transforms as has been described already in Section 1.2.3.

The principle of linear transforms to reduce the signal subspace[1] has been addressed by several authors (KLEMM [302], WARD [690, p. 81], HAIMOVICH et al. [239], SELIKTAR et al. [599], BARANOSKI [31]). WANG Y. and PENG [681] present a unified approach for various kinds of transform processors (element-pulse, beamspace-pulse, element-Doppler, beamspace-Doppler). A similar overview of these techniques has been given by WARD [691, 690]. DE GRÈVE et al. [113] propose a canonical framework for a large class of suboptimum STAP processors. CALDWELL et al. [81] has pointed out that a reduction of the system dimensions (N, M) is essential for adaptive processing in clutter with range dependent Doppler which, for instance, occurs in forward looking arrays, see Chapter 3.

Transforming the covariance matrix by the discrete wavelet transform (DWT) may be a way of reducing the computational load for matrix inversion, exploiting the sparsening property of the DWT (BRAUNREITER et al. [61], KADAMBE and OWECHKO [284]).

BOJANCZYK and MELVIN [55] analyse a least squares STAP technique based on the preconditioned conjugate gradients iterative method.

HIMED and MELVIN [262] give a brief overview of different subspace processors: factored time-space (frequency-dependent spatial processing), extended factored time-space (frequency dependent spatial processing, with additional auxiliary Doppler channels involved), adaptive displaced phase centre (ADPCA), eigencanceller (orthogonal projection), and eigen-based cross-spectral metric (CSM, GOLDSTEIN and REED [198], GUERCI et al. [227]). Some of them are compared using a set of clutter

---

[1] Referred to as rank deflation by some authors, e.g., BARANOSKI [31].

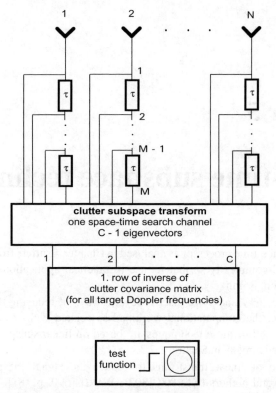

Figure 5.1: The auxiliary eigenvector processor (AEP)

data measured with the MCARM system, and with an artificial target signal inserted. It is remarkable that processing based on a subspace order $N_e = N + M - 1$ does not appear to be sufficient to detect the target in clutter.

5.1 Principle of space-time subspace transforms

In this chapter we consider space-time transforms, i.e., transforms that are effective in both the time and space dimensions. Such a transform $\mathbf{T}$ is a $NM \times C$ matrix, where C is the dimension of the reduced vector space.[2]

$$\mathbf{T} = \begin{pmatrix} \mathbf{a}_1 & \mathbf{a}_2 & \cdots & \mathbf{a}_C \end{pmatrix} = \begin{pmatrix} \mathbf{a}_{11} & \mathbf{a}_{12} & \cdots & \mathbf{a}_{1C} \\ \mathbf{a}_{21} & \mathbf{a}_{22} & \cdots & \mathbf{a}_{2C} \\ \vdots & \vdots & \ddots & \vdots \\ \mathbf{a}_{M1} & \mathbf{a}_{M2} & \cdots & \mathbf{a}_{MC} \end{pmatrix} \tag{5.1}$$

Notice that each of the spatial subvectors $\mathbf{a}_{ml}$ has the dimension $N \times 1$ of the array.

---
[2]Technically speaking, C is the number of channels.

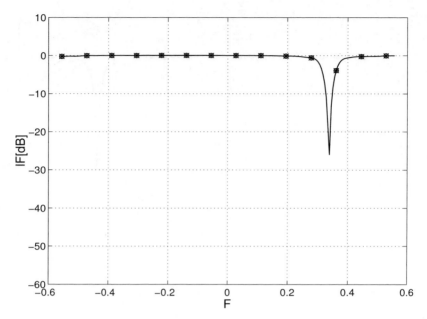

Figure 5.2: Comparison of eigenvector and optimum processor ($N = M = 24$; $C = 48$, FL): ○ *auxiliary eigenvector processor;* ∗ *optimum processor*

The transform has to be in accordance with the criteria presented in Chapter 1 in section 1.2.3. Following criterion 1, one of the columns of $\mathbf{T}$ should be matched to the expected signal in space and time. We call this the search channel. The associated space-time vector $\mathbf{a}(\varphi_L, v_{\mathrm{rad}})$ includes beamformer and Doppler filter coefficients as given by (2.31).

The remaining $C - 1$ space-time vectors describe auxiliary channels for measuring the clutter covariance matrix in the reduced vector space. Then the space-time transform becomes

$$\mathbf{T} = \begin{pmatrix} \mathbf{s}(\varphi_L, f_D) & \mathbf{a}_2 & \ldots & \mathbf{a}_L \end{pmatrix} = \begin{pmatrix} \mathbf{s}(\varphi_L, f_D) & \mathbf{A} \end{pmatrix} \quad (5.2)$$

After transforming the data according to (1.53),

$$\mathbf{q}_T = \mathbf{T}^*\mathbf{q}; \quad \mathbf{s}_T = \mathbf{T}^*\mathbf{s}; \quad \mathbf{x}_T = \mathbf{T}^*\mathbf{x}; \quad \mathbf{Q}_T = \mathbf{T}^*\mathbf{Q}\mathbf{T} \quad (5.3)$$

the optimum processor in the transformed domain becomes, according to (1.54),

$$\mathbf{w}_T = \gamma \mathbf{Q}_T^{-1} \mathbf{s}_T \quad (5.4)$$

and the improvement factor according to (1.55) becomes

$$\mathrm{IF} = \frac{\mathbf{w}_T^* \mathbf{s}_T \mathbf{s}_T^* \mathbf{w}_T \cdot \mathrm{tr}(\mathbf{Q})}{\mathbf{w}_T^* \mathbf{Q}_T \mathbf{w}_T \cdot \mathbf{s}^* \mathbf{s}} \quad (5.5)$$

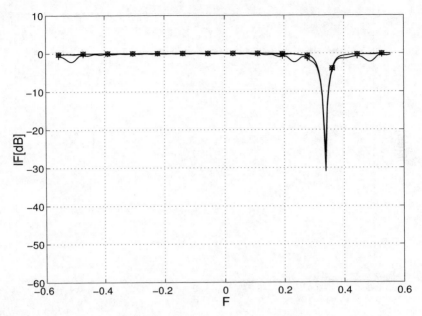

Figure 5.3: Reduction of the number of channels (FL, AFP): ∘ $C = 48$; ✶ $C = 40$; × $C = 32$; + $C = 24$

The transformed signal vector is

$$\mathbf{s}_T = \mathbf{T}^*\mathbf{s} = \begin{pmatrix} \mathbf{s}^* \\ \mathbf{a}_2^* \\ \vdots \\ \mathbf{a}_C^* \end{pmatrix} \mathbf{s} \tag{5.6}$$

Notice that the first row of $\mathbf{T}^*$ is matched to the signal reference while the auxiliary channels have to be matched in some way to the interference. Therefore we can write

$$\mathbf{s}^*\mathbf{s} \gg \mathbf{a}^*\mathbf{s} \tag{5.7}$$

so that the transformed signal reference becomes approximately

$$\mathbf{s}_T = \mathbf{T}^*\mathbf{s} \approx NM \begin{pmatrix} 1 \\ 0 \\ \vdots \\ 0 \end{pmatrix} = NM\mathbf{e}_1 \tag{5.8}$$

Then the subspace processor becomes

$$\mathbf{w}_T = \gamma \mathbf{Q}_T^{-1} \mathbf{e}_1 \tag{5.9}$$

In this processor only the signal component in the search channel $\mathbf{s}$ is evaluated while the signal contributions of the auxiliary channels are neglected. In (5.4) the signal

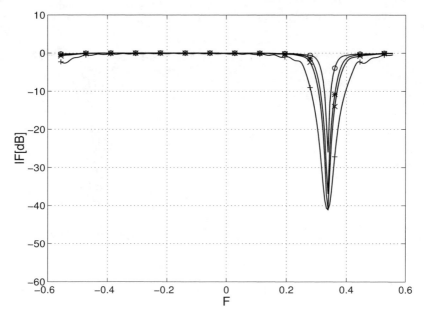

Figure 5.4: Influence of clutter bandwidth (FL, AEP): ○ $B_c = 0$; ∗ $B_c = 0.01$; × $B_c = 0.03$; + $B_c = 0.1$

contributions of all channels (search and auxiliary) are included. Notice that (5.9) is a further simplification of (5.4). The processor (5.9) approximates the one in (5.4) well if the condition (5.7) is satisfied.

The auxiliary channels can be chosen in several ways. In any case they should be matched as well as possible to the clutter so as to produce reference signals with high CNR. In the following two Sections we discuss two techniques which lead basically to near-optimum clutter rejection performance.

5.2 The auxiliary eigenvector processor (AEP)

One way of focusing the auxiliary channels $\mathbf{a}_2 \ldots \mathbf{a}_C$ on the clutter echoes is to use the clutter eigenvectors as auxiliary channels (KLEMM [303]). This principle was discussed in Section 1.2.3. It was shown (see also NICKEL, [490]) that this auxiliary eigenchannel processor is identical to the orthogonal projection processor (OPP) described in Section 4.2.2, if there is a perfect match between the eigenvectors of the covariance matrix and the channels of the sidelobe canceller transform. The sidelobe canceller transform matrix is in this case

$$\mathbf{T} = (\; \mathbf{s}(\varphi_L, f_D) \quad \mathbf{E} \;) \quad (5.10)$$

where $\mathbf{E}$ is the $NM \times (C-1)$ matrix of clutter eigenvectors of $\mathbf{Q}$. The number of auxiliary channels $C - 1$ must be chosen so that the total clutter subspace is included.

164 Space-time subspace techniques

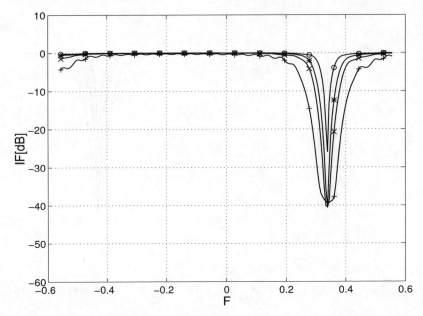

Figure 5.5: Influence of system bandwidth (FL, AEP, rectangular frequency response): ○ $B_s = 0$; ∗ $B_s = 0.01$; × $B_s = 0.03$; + $B_s = 0.1$

For a sidelooking linear array we have to take (3.59) into account and postulate

$$C \geq N + M \tag{5.11}$$

According to GUERCI et al. [227] the inclusion of the signal vector **s** in the transform leads to higher compression of the subspace of the cross-spectral metric based auxiliary eigencanceller technique which is a similar kind of processor as the AEP. The transform (5.10) belongs to the class of data dependent transforms (PECKHAM et al. [533]).

A block diagram of the auxiliary eigenvector processor (AEP) based on (5.9) and (5.10) is given in Figure 5.1. The signals received by N antenna elements are demodulated and digitised (not shown) and stored in shift registers of length M. The NM space-time samples are then transformed by the auxiliary eigenvector transform according to (5.10).

After the transform the $C \times C$ clutter covariance matrix is estimated (not shown). The remaining C channels are multiplied with the first column[3] of $\mathbf{Q}_T^{-1}$ for clutter cancellation. This operation has to be carried out for all range increments of the visible range.

Notice that the search channel includes both a beamformer and a Doppler filter. That means, beamforming and Doppler filtering is carried out *before* clutter rejection. Therefore, clutter rejection in the processor in Figure 5.1 has to be carried out for all

---
[3] We assume that the first column denotes the search channel as in (5.10).

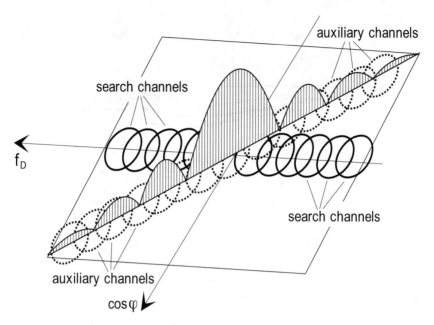

Figure 5.6: Principle of the auxiliary channel processor (ACP)

target Doppler frequencies of interest. The Doppler channel with the maximum power is selected for detection and Doppler estimation.

In the following we analyse the performance of the auxiliary eigenvector processor by calculating the improvement factor (IF) as a function of the normalised target Doppler frequency F. For the numerical examples we assumed a forward looking linear array. Similar results can be obtained for a sidelooking array configuration.

5.2.1 Comparison with the optimum adaptive processor (OAP)

The performance of the auxiliary eigenvector processor in comparison with the optimum fully adaptive processor described in Section 4.2.1 is illustrated in Figure 5.2. The IF has been plotted versus the normalised Doppler frequency. The look direction was again $45°$ as stated in the parameter list on page 65. This is reflected in the off-zero Doppler shift of the clutter notch. The number of channels was chosen to be $C = 48$ which is in accordance with the number of eigenvalues since $N = M = 24$.

In Figure 5.2 ideal conditions (identical receive channels, no bandwidth effects) have been assumed. It can be noticed that under these conditions the two curves for the AEP and the OAP coincide perfectly. Using the full subspace of clutter eigenvectors is obviously an optimum choice.

166 *Space-time subspace techniques*

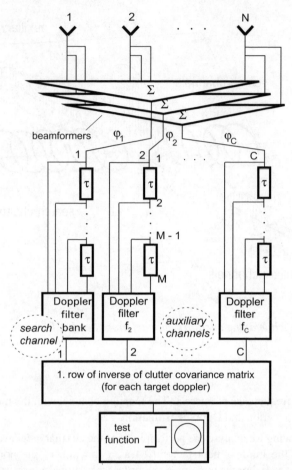

Figure 5.7: Block diagram of the auxiliary channel processor (ACP)

5.2.2 Reduction of the number of channels

Now we try to answer the question of how much further the number of channels can be reduced. The motivation for further reduction of the degrees of freedom of the clutter filter (5.9) lies in the fact that usually some of the clutter eigenvalues are relatively small so that the associated eigenvectors do not play a significant role.

Figure 5.3 shows four curves with different numbers of channels C. It can be seen that for $C = 32$ still the optimum IF curve is obtained. For $C = 24$ some small losses can be noticed. Obviously, under the conditions assumed, the number of channels can be divided by 2 which means that the computational expense for calculating the matrix inverse[4] is reduced by a factor of 8.

It should be noted that the required number of channels depends heavily on the

---
[4]Matrix inversion needs $\propto C^3$ complex operations.

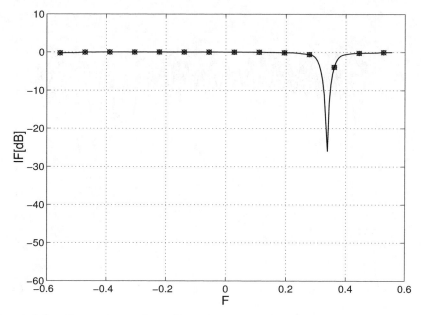

Figure 5.8: Comparison of auxiliary channel and optimum processor (forward looking): ○ *auxiliary channel receiver ($N = M = 24$; $N_c = 48$);* ∗ *optimum processor*

actual number of dominant eigenvalues which in turn depends on parameters such as transmit pattern, sensor pattern, system and clutter bandwidth, etc. Let us keep in mind that the auxiliary eigenvector processor tends to be tolerant with increase in the number of eigenvalues.

5.2.3 Bandwidth effects

In the following two examples we consider again the case $C = N + M = 48$, which means that the number of auxiliary channels has been matched to the number of eigenvalues. Let us recall that the number of eigenvalues can be increased by bandwidth effects (see Chapter 3, Figures 3.24–3.31).

The increase in the number of eigenvalues may have two effects on the clutter rejection performance of the space-time processor:

- the clutter notch is broadened;

- losses in IF may occur due to lack of degrees of freedom of the clutter filter.

While the first effect occurs mainly close to the clutter Doppler frequency the second one may also influence the IF at Doppler frequencies far away from the clutter notch. Experience has shown that a lack in degrees of freedom results in some sidelobe ripple in the IF curves due to Doppler filter sidelobes.

168 Space-time subspace techniques

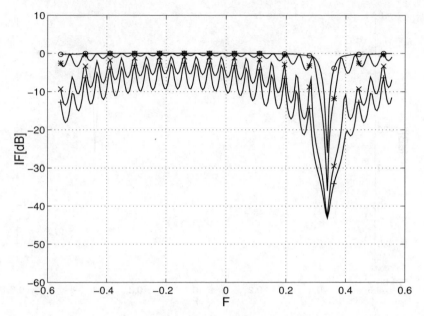

Figure 5.9: Reduction of the number of channels (FL, ACP): ○ $C = 48$; ∗ $C = 40$; × $C = 32$; + $C = 24$

5.2.3.1 Clutter bandwidth

In Figure 5.4 the effect of clutter Doppler bandwidth due to clutter fluctuations (for the clutter fluctuation model see (2.53)) is shown. It can be seen that for $B_c = 0, \ldots, 0.03$ the IF curves run smoothly while for $B_c = 0.1$ some ripple can be noticed. Obviously the number of degrees of freedom $C = 48$ is sufficient for a relative clutter bandwidth up to $B_c = 0.03$.

5.2.3.2 System bandwidth

In Figure 5.5, the effect of spatial decorrelation due to the system bandwidth is illustrated for a forward looking array (for sidelooking arrays there is no spatial decorrelation due to the system bandwidth). A rectangular frequency response according to (2.64) was assumed. As can be seen the spatial decorrelation due to the system bandwidth leads to similar degradation in clutter rejection performance as the temporal decorrelation due to internal clutter motion. For $B_s = 0.1$ some slight ripple can be recognised: This indicates that the processor is short of degrees of freedom, which means that the number of channels C has been chosen too small. Some considerations on the impact of the system bandwidth on reduced rank space-time processing have been made by ZATMAN [737].

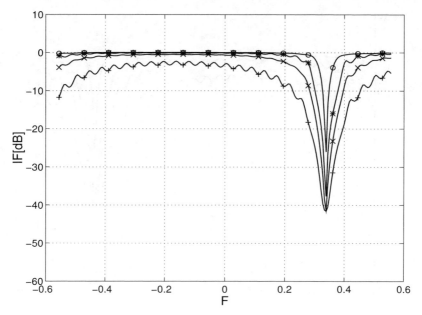

Figure 5.10: Influence of clutter bandwidth (FL, ACP): ○ $B_c = 0$; ∗ $B_c = 0.01$; × $B_c = 0.03$; + $B_c = 0.1$

5.2.3.3 Related techniques

In the previous discussion we assumed that the space-time transform matrix contains $N_E = N + M - 1$ or fewer eigenvectors associated with the *largest* eigenvalues of the clutter covariance as auxiliary channels plus an additional search channel. GOLDSTEIN and REED [198] propose a generalised sidelobe canceller in which those eigenvectors are selected as auxiliary channels which maximise the *cross-spectral metric* (CSM)

$$\frac{|\mathbf{v}_i \mathbf{r}_{xs}|^2}{\lambda_i} \quad (5.12)$$

where $\mathbf{v}_i$ are the eigenvectors of the auxiliary subspace, λ_i the associated eigenvalues, and $\mathbf{r}_{xs}$ is the cross-correlation between search channel and auxiliary channels. As has been shown by BERGER and WELSH [40, 41] this criterion does not necessarily select those eigenvectors associated with the largest eigenvalues. The effect of limited secondary data support for updating the clutter covariance in the CSM algorithm has been analysed by HALE [242]. The performance of the CSM and related techniques has been analysed by simulations and MOUNTAINTOP data (TITI [646]) by GUERCI *et al.* [227].

Starting from the above-mentioned CSM technique, BERGER and WELSH [40] propose a SINR (signal-to-interference + noise ratio) metric

$$\frac{|\mathbf{f}_i \mathbf{s}|^2}{\lambda_i} \quad (5.13)$$

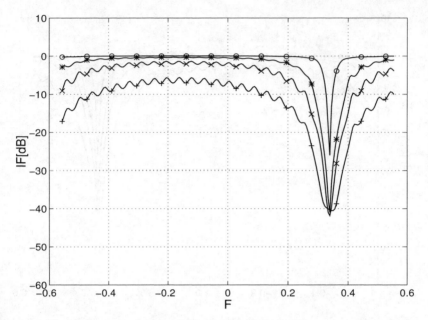

Figure 5.11: Influence of system bandwidth (FL, ACP): rectangular frequency response): ∘ $B_s = 0$; ∗ $B_s = 0.01$; × $B_s - 0.03$; + $B_s - 0.1$

which allows the selection of eigenvectors for a space-time transform based on maximisation of the SINR (f_i are eigenvectors to be selected). It is shown that this techniques provides graceful degradation when the number of channels is reduced below the number of eigenvalues of the clutter covariance matrix. In this respect the SINR method is superior to the CSM method. It should be noted that the straightforward method analysed above (one search channel, eigenvectors belonging to the largest eigenvalues) is quite tolerant against reduction of the number of channels (see Figure 5.3).

5.3 Auxiliary channel processor (ACP)

The auxiliary eigenvector processor described above has the disadvantage that the eigenvectors of the covariance matrix are not known but have to be calculated from an estimate of the clutter covariance matrix, in other words, the transform is data dependent.

A more intuitive concept for a space-time auxiliary channel processor which circumvents these problems by being data independent (PECKHAM *et al.* [533]) has been proposed by KLEMM [302] and is described in the following. This concept is based on forming auxiliary beams cascaded with Doppler filters. CERUTTI-MAORI [88] uses auxiliary beams in a space-based multi-static GMTI concept.

Let us consider first the one-dimensional problem of jammer cancellation by spatial nulling. Suppose a number of jamming sources are radiating on a sensor array and assume that the jammer positions are known. Under these conditions a subspace processor can be designed by steering one beam on each of the jammers and having one more beam for target search. It has been shown numerically that such a processor is practically optimal (KLEMM [290]). NICKEL [490] has shown analytically that this kind of sidelobe canceller is identical to the orthogonal projection processor (see Section 4.2.2). The problem with this concept is that the positions of the jammers are normally unknown.[5] For some more details on multibeam configurations see Section 1.2.3.2.

In this respect airborne clutter suppression is easier because the positions of clutter in space (normally homogeneous) as well as the direction dependant clutter Doppler frequencies are known. We can, therefore, design a space-time processor according to (5.9) with a transform matrix of the form

$$\mathbf{T} = (\ \mathbf{s}(\varphi_L, f_D) \quad \mathbf{a}_2 \ldots \mathbf{a}_C\ ) = (\ \mathbf{s}(\varphi_L, f_D) \quad \mathbf{A}\ ) \qquad (5.14)$$

The column $\mathbf{s}(\varphi_L, f_D)$ denotes again the space-time search channel while $\mathbf{a}_2 \ldots \mathbf{a}_C$ are auxiliary space-time channels matched to the clutter in direction and Doppler frequency.

The principle of the auxiliary channel processor[6] is illustrated in Figure 5.6. We consider the case of a sidelooking array configuration as in Figure 3.40. One recognises first of all the clutter power spectrum extending along the diagonal of the plot.

The ellipses in the f_D-cos φ plane denote footprints of space-time receive channels. There are auxiliary channels (dotted ellipses) covering the whole clutter azimuth range, each of them matched to the Doppler frequency associated with the individual direction.

In the look direction we have a number of search channels for all possible target Doppler frequencies (solid ellipses). Notice that there must be an auxiliary channel in the look direction in order to receive the transmit main beam clutter. No search channel should be matched exactly to the clutter Doppler frequency (centre of plot) because then the transform matrix would become singular.[7]

Figure 5.7 shows a block diagram of the auxiliary channel processor. The output signals of the N antenna channels are transformed by a multiple beamformer network. Each of the auxiliary beams points into a different direction so as to cover the whole azimuth range. For instance, this multiple beam network has been implemented in the OLPI radar (WIRTH [720, 718]) and is used frequently in sonar systems.

Each of the auxiliary beams is followed by a Doppler filter matched to the Doppler frequency associated with the look direction of the individual beam (dotted ellipses in Figure 5.6). The search beam is cascaded with Doppler filter bank (solid ellipses). The output signals are multiplied with the first column of the inverse covariance matrix and fed into a detection device.

---

[5]In fact, it is the task of an *adaptive* processor to cope with unknown jammer positions.

[6]In KLEMM [302] this processor was referred to as an auxiliary channel receiver (ACR).

[7]This problem can be avoided by adding 'artificial noise' to the covariance matrix. If no transform is applied one adds $\mathbf{Q}^{(L)} = \mathbf{Q} + \mu \mathbf{I}$ (diagonal loading). This is a well-known method to improve the condition of the covariance matrix. In the transformed domain one has to add instead $\mathbf{Q}_T^{(L)} = \mathbf{Q}_T + \mu \mathbf{T}^*\mathbf{T}$.

172 Space-time subspace techniques

5.3.1 Comparison with optimum processor

First the performance of the auxiliary channel processor is compared with the optimum processor treated in Chapter 4. Figure 5.8 shows that the IF curves of both processors coincide perfectly, as in case of the auxiliary eigenvector processor, see Figure 5.2.

5.3.2 Reduction of the number of channels

A substantial difference between the auxiliary clutter channel processor and the auxiliary eigenvector processor (AEP) described in Section 5.2 is that the AEP has important and less important channels, depending on the magnitude of the associated eigenvalue. In the auxiliary channel processor all channels have the same priority so that reducing the number of channels is more critical.

Figure 5.9 shows the effect of reducing the number of auxiliary channels on the performance of the auxiliary channel processor. Comparing Figure 5.9 with Figure 5.3, it is obvious that the ACP is much more sensitive to a reduction of channels. We notice strong ripple even for a slight reduction by eight channels (asterisks). As stated in the previous paragraph this is an expected result.

It can be noticed that on the one hand there is a degradation close to the clutter Doppler frequency (clutter notch). On the other hand, significant losses show up in the pass band. The fact that the sidelobe structure of the Doppler filter bank becomes apparent indicates a lack of degrees of freedom (or channels).

5.3.3 Bandwidth effects

In the following examples we assumed again that the number of auxiliary channels is matched to the number of eigenvalues of the ideal clutter covariance matrix: $C = N + M = 48$. As in Section 5.2.3 we want to find out what the effect of additional eigenvalues of $\mathbf{Q}$ caused by bandwidth effects is.

5.3.3.1 Clutter bandwidth

Again we use the clutter fluctuation model given by (2.53). The IF curves are shown in Figure 5.10. Comparing the curve $B_c = 0$ with the others shows that the auxiliary channel processor is short of degrees of freedom because the actual number of eigenvalues of $\mathbf{Q}$ has been increased through the clutter fluctuation model. The clutter notch is broadened which results in degraded detection capability of slow targets. Even in the pass band some significant IF losses occur for larger clutter bandwidth. Comparing this plot with Figure 5.4 shows that the auxiliary eigenvector processor is much more tolerant against an increase in the number of clutter eigenvalues of $\mathbf{Q}$ than the ACP.

5.3.3.2 System bandwidth

Similar effects are caused by the system bandwidth. Like in Figure 5.5, a rectangular frequency response according to (2.64) was assumed. Again a forward looking array geometry was assumed. Figure 5.11 shows some numerical results. Basically the effect of system bandwidth (spatial decorrelation) on the clutter suppression performance is similar to the influence of the clutter bandwidth (temporal decorrelation).

5.4 Other space-time transforms

In the preceding sections two techniques (AEP and ACP) have been described which promise to approximate the clutter rejection performance optimally because their auxiliary channels are well matched to the clutter. It was found that under ideal conditions (no bandwidth decorrelation effects) the improvement factor achieved by the optimum processor (OAP, see Section 4.2.1) was reached perfectly by both of the space-time auxiliary channel processors.

Instead of using auxiliary eigenvectors (AEP) or auxiliary beams and Doppler filters (ACP) one might think of a variety other space-time auxiliary channel configurations.

5.4.1 Single auxiliary elements and echo samples transform

One possiblity is to choose single array elements at single instants of time as auxiliary channels. The advantage over the previously described techniques is that no auxiliary beams and Doppler filters have to be formed. This concept is a space-time analogue of the usual sidelobe canceller commonly used for jammer suppression.

There are disadvantages, however. As stated in Chapter 1, Section 1.6, the clutter rejection performance of the sidelobe canceller is degraded whenever the interference-to-noise ratio in the auxiliary channels is smaller than that in the search channel. This can happen if the sidelobe level of the search beam is higher than the gain of the auxiliary element. This effect was shown in the numerical example Figure 1.6.

5.4.2 Space-time sample subgroups

The idea of this technique is to subdivide the total of spatial[8] and temporal[9] samples into space-time sample subgroups (KLEMM, [304]). The samples of each space-time subgroups are combined by a primary beamformer and a primary Doppler filter so that all sample subgroups are focused in the same direction and on the same Doppler frequency. Then the space-time covariance matrix is estimated and inverted at the subgroup level.

In a second stage a secondary beamformer can be steered inside the primary beam pattern, and a secondary Doppler filter bank can be designed to perform Doppler analysis in the limits of the primary Doppler filter main lobe. The shapes of the primary beamformer and the primary Doppler filter determine the required number of primary φ-f_D positions.

There is a variety of possibilities of designing the space-time subgroups. They may be different in the spatial and temporal dimensions, they may overlap or not, adjacent or

---

[8] Array elements.
[9] Echoes.

displaced samples can be chosen. The appropriate design of such processors is a wide playground for system designers. It should be noted that not all transforms perform equally well. We confine our considerations to the two examples treated above (AEP, ACP) whose performance is near optimum.

5.4.3 Space-time blocking matrices

All adaptive processors are based on the assumption that the clutter covariance matrix does not contain any portion of the desired signal. That means that the covariance matrix should be estimated from signal-free data. This can be a problem, especially if the amount of data available is small. In some applications (SAR, passive sonar) the signal (if there is any) is *continuously* present.

Inclusion of the desired signal in the adaption leads to severe signal suppression because the processor treats the signal as interference. No degradation in signal power may be obtained by the *optimum* processor (Chapter 4) if the processor (i.e., the steering vector) is perfectly matched to the signal (COX [106]). In a radar search mode the angular (Doppler) cells are determined by the beamwidth (Doppler filter) so that a mismatch between steering vector and signal happens intentionally.

A way to mitigate the effect of signal inclusion in sidelobe canceller types of processors (such as the AEP, ACP) is to apply a *blocking matrix* so that the transformed ('blocked') auxiliary channels

$$\tilde{\mathbf{A}} = \mathbf{B}\mathbf{A} \tag{5.15}$$

become insensitive to the signal. $\mathbf{B}$ is called a *space-time* blocking matrix. It can be composed for example of some columns of a space-time projection matrix orthogonal to the signal vector

$$\mathbf{P} = \mathbf{I} - \frac{\mathbf{s}(\varphi_L, f_D)\mathbf{s}^*(\varphi_L, f_D)}{\mathbf{s}^*(\varphi_L, f_D)\mathbf{s}(\varphi_L, f_D)} \tag{5.16}$$

so that

$$\mathbf{s}^*\tilde{\mathbf{A}} = \mathbf{0} \tag{5.17}$$

Notice that the column rank of $\tilde{\mathbf{A}}$ is reduced by 1. Therefore, the number of auxiliary channels has to be reduced at least by 1. Details on the use of blocking matrices can be found in the papers by SCOTT and MULGREW [594], SU and ZHOU [620], GOLDSTEIN and REED [199], and GOLDSTEIN *et al.* [200].

5.4.4 The JDL-GLR

The *Joint Domain Localised Generalised Likelihood Ratio Detector* (JDL-GLR) by WANG H. and CAI belongs also to the class of space-time transform processors. The auxiliary space-time vectors can be generated by using the FFT this processor will be described in Chapter 9.

5.5 Aspects of implementation

5.5.1 General properties

5.5.1.1 *Number of operations*

Space-time auxiliary channel processors can reach the performance of the optimum receiver (see Chapter 4) at greatly reduced computational expense. A comparison of all processors treated in this book in terms of computational complexity will be given in Chapter 16.

The number of space-time auxiliary channels depends on the number of clutter eigenvalues which is approximately $N + M - 1$. Therefore, the complexity of space-time auxiliary channel receivers increases with increasing number of antenna elements and the length of the pulse burst. Such solutions are useful for small numbers of N and M. In practice inhomogeneity of the clutter background (dominating clutter discretes) may lead to a reduced number of clutter eigenvalues and, hence, to less complex receiver structures.

5.5.1.2 *System bandwidth*

The system bandwidth effect does not occur for linear or planar sidelooking arrays, see the comments at the end of Chapter 3, page 115.

For other than linear or planar sidelooking arrays the system bandwidth effects can be compensated for by means of space-time-TIME or space-time-FREQUENCY processing (see the remark on page 117).

5.5.1.3 *Calculation of the matrix inverse*

As mentioned earlier the search channel in the transform $\mathbf{T}$ includes a Doppler filter. That means that the clutter filter (first column of $\mathbf{Q}_T^{-1}$) has to be calculated for all possible target Doppler frequency. The inverse covariance matrix can be calculated efficiently by exploiting the matrix inversion lemma in the following way:

The transformed covariance matrix assumes the form

$$\mathbf{QT} = \mathbf{T}^*\mathbf{QT} = \begin{pmatrix} P(m) & \mathbf{h}^*(m) \\ \mathbf{h}(m) & \mathbf{D} \end{pmatrix} \tag{5.18}$$

where m is the Doppler frequency index, $P(m) = \mathbf{s}(m)^*\mathbf{Qs}(m)$ is the clutter power in the search channel, $\mathbf{h}(m) = \mathbf{A}^*\mathbf{Qs}(m)$ is the cross-variance between search channel and auxiliary channels ($\mathbf{A}$ is the matrix of auxiliary channels, see (5.2)), and $\mathbf{D} = \mathbf{A}^*\mathbf{QA}$ is the covariance matrix of auxiliary channels. Notice that $\mathbf{D}$ does not depend on the Doppler frequency.

Then the inverse becomes

$$\mathbf{Q}_T^{-1} = (\mathbf{T}^*\mathbf{QT})^{-1} = \begin{pmatrix} P(m) & \mathbf{h}^*(m) \\ \mathbf{h}(m) & \mathbf{D} \end{pmatrix}^{-1} \tag{5.19}$$

$$= \begin{pmatrix} \frac{1}{P(m)} & \mathbf{o}^* \\ \mathbf{o} & \mathbf{0} \end{pmatrix} + \begin{pmatrix} \frac{1}{P^2(m)}\mathbf{h}^*(m)\mathbf{C}^{-1}\mathbf{h}(m) & -\frac{1}{P(m)}\mathbf{h}^*(m)\mathbf{C}^{-1} \\ -\frac{1}{P(m)}\mathbf{C}^{-1}\mathbf{h}(m) & \mathbf{C}^{-1} \end{pmatrix}$$

where $\mathbf{C} = \mathbf{D} - \frac{1}{P(m)}\mathbf{hh}^*(m)$, and $\mathbf{o}$ and $\mathbf{0}$ denote the zero vector and matrix, respectively. If $\mathbf{D}^{-1}$ is known $\mathbf{C}^{-1}$ is obtained as follows:

$$\mathbf{C}^{-1} = (\mathbf{D} - \frac{1}{P(m)}\mathbf{hh}^*(m))^{-1} = \mathbf{D}^{-1} + \frac{\mathbf{D}^{-1}\mathbf{h}(m)\mathbf{h}^*(m)\mathbf{D}^{-1}}{P(m) - \mathbf{h}^*\mathbf{D}^{-1}\mathbf{h}(m)} \quad (5.20)$$

The following steps have to be carried out:

1. Estimate $\mathbf{D}, P(m), \mathbf{h}(m)$ for $m = 1, \ldots, M$
2. Compute $\mathbf{D}^{-1}$
3. Compute $\mathbf{C}^{-1}(m)$ for $m = 1, \ldots, M$ (5.20)
4. Compute $\mathbf{Q}_T^{-1}$ for $m = 1, \ldots, M$ (5.19).

The computational expense for inverting the matrix is about $(C-1)^3 + 2C^2 M$.

5.5.2 Auxiliary eigenvector processor

5.5.2.1 *Generation of auxiliary channels*

The eigendecomposition of the clutter covariance matrix requires more computations than taking the inverse of the matrix. In this view the auxiliary eigenvector processor has only theoretical value. However, this problem might be circumvented by selecting the appropriate transform from a set of 'typical'[10] pre-calculated eigenvector matrices according to the actual flight conditions (KLEMM [303]).

For the actual eigenvalues of $\mathbf{Q}$ the transformed clutter covariance matrix assumes the form

$$\mathbf{Q}_T = \mathbf{T}^*\mathbf{Q}\mathbf{T} = \begin{pmatrix} P_c & \rho_2 & \cdots & \rho_C \\ \rho_2^* & \lambda_2 & 0 & \\ \vdots & & 0 & \ddots \\ \rho_C^* & & & \lambda_C \end{pmatrix} \quad (5.21)$$

$P_c = \mathbf{s}^*\mathbf{Q}\mathbf{s}$ is the clutter power in the search channel. $\rho_i = \mathbf{s}^*\mathbf{e}_i\lambda_i$ are the cross-variances between search and auxiliary channels, $\mathbf{e}_i$ and λ_i are clutter eigenvectors and eigenvalues of $\mathbf{Q}$, respectively. The special form of (5.21) may offer some simplification in calculating the matrix inverse. However, if pre-calculated transform matrices are used there will always be some mismatch between the transform and the actual covariance matrix so that the lower right $(C-1) \times (C-1)$ submatrix is no longer diagonal.

---

[10]For example, for homogeneous clutter and, if necessary, for various flight velocities.

5.5.3 Auxiliary channel processor

5.5.3.1 Generation of auxiliary channels

Compared with the auxiliary eigenvector prosessor the auxiliary channel processor has the advantage that the pre-transform consisting of clutter beams and Doppler filters are known a priori while the eigenvectors of the AEP have to be calculated beforehand.

5.5.3.2 Number of channels

It was shown that the number of channels of the ACP must not be reduced below $N + M$. On the other hand it might be of interest to increase the number of channels beyond $N + M$ in order to cope with decorrelation effects caused by system or clutter bandwidth, see Section 5.3.3, or by channel errors, see Chapter 16. It should be noted that for DPCA conditions the pre-transform $\mathbf{T}$ cannot have more than $C = N + M - 1$ auxiliary space-time vectors. This follows from the fact that a DPCA covariance matrix without corruption by stochastic effects has only $C = N + M - 1$ eigenvalues. Therefore, more than $C = N + M - 1$ space-time clutter matched vectors become linearly dependent so that $\mathbf{Q}$ becomes singular. It is not possible to match the number of channels to the actual number of eigenvalues of $\mathbf{Q}$. Diagonal loading of $\mathbf{QT}$ with 'artificial noise' (see footnote on p. 171) might be a way out of this dilemma.

5.5.3.3 Related concept

WANG Y. and PENG [678, 680] propose a similar processor, however with *tapered* Doppler filter weights so as to reduce the filter sidelobes. This results in a reduced number of degrees of freedom of the space-time clutter covariance matrix. Also LIU Q.-G. *et al.* [408] found that the *auxiliary channel processor* is quite sensitive to array channel errors. They propose a modified version by applying temporal and spatial taper weights[11] to the data before estimating the space-time clutter covariance matrix. Such weightings reduce the sidelobe level and, hence, the number of significant clutter eigenvalues. This results in less sensitivity to channel errors.

5.6 Summary

1. **Space-time transform processors** basically offer a possibility of clutter suppression in a reduced signal vector space. They are based on a linear transform whose columns are space-time vectors. One of these vectors is a signal matched *search* channel while the others serve as *clutter reference* channels.

2. The **number of channels** (dimension of the subspace) has to be about $N + M$. Therefore, these techniques have the disadvantage that the computational workload increases with increasing dimension of the signal vector space, that is, number of sensors and the number of coherent echoes M.

---
[11] 40 dB Dolph–Chebychev.

3. The **auxiliary eigenvector transform** achieves near-optimum clutter rejection performance without degradation in slow target detection. In detail this technique is characterised by the following properties:

   - The transform requires the eigendecomposition of the clutter covariance matrix. This needs more arithmetic operations than the inversion involved in the optimum processor.
   - The transform may be pre-calculated off-line. Since the transform is followed by an *adaptive* processor in the clutter subspace some mismatch between the transform and the actual clutter subspace may be tolerated.
   - This processor technique is relatively robust against bandwidth effects.
   - The AEP requires a fully digitised array, at least in the horizontal dimension.

4. The CSM technique by GOLDSTEIN and REED [198] and the SINR metric technique by BERGER and WELSH [40] belong to the class of eigenvector transform techniques.

5. The **auxiliary channel processor (ACP)** transform uses a bunch of parallel beamformers which cover the entire angular domain. Each of the beamformers is cascaded with a receiver chain and a Doppler filter which is matched to the clutter Doppler frequency of the individual beam. This processor has the following properties:

   - The total number of channels must not exceed $N + M$, otherwise the associated space-time vectors become linearly dependent which results in a singular transformed covariance matrix.
   - The ACP is sensitive to bandwidth effects and to channel errors. The number of degrees of freedom cannot increased beyond $N + M$, see above.
   - The problem associated with the linear dependence of channel might be circumvented by adding 'artificial' noise in the transformed domain (see the footnote on p. 171).
   - The multibeam auxiliary channels can be implemented in the RF domain. However, each beam has to be followed by a receiver chain. Therefore, the number of required digital array channels is about the same as for the AEP.

6. A variety of other transforms may be used, for instance forming space-time sample subgroups. It is up to the fantasy of the designer to create different kinds of sample subgroups. It should be noted, however, that not all transforms show satisfactory performance.

A comparison of all techniques in terms of computational complexity is presented in Chapter 16.

Chapter 6

Spatial transforms for linear arrays

It was pointed out in Chapter 5 that linear *space-time* transforms can be used to reduce the signal vector space down to the clutter subspace which leads to a reduction of the computational expense in the processing.

However, we noticed a dependency between the required number of degrees of freedom of the processor and the dimension of the signal vector space (number of antenna elements N, echo sample size M). Therefore, these techniques are useful mainly for small antenna arrays and small echo sample size.

It was found furthermore that such systems may suffer from a lack of degrees of freedom in the case of additional eigenvalues due to bandwidth effects or channel errors.

In this chapter we analyse the effect of *spatial* transforms in the context of space-time adaptive MTI filters. Such transforms have been widely used in adaptive jammer nulling. One prominent example is the sidelobe canceller, see Section 1.2.3.1. For some details of spatial transforms see Section 1.2.3.

While the space-time transforms treated in the previous chapter[1] reduce the signal vector space in both the spatial and the temporal dimension simultaneously spatial transforms reduce the spatial dimension only. There are many ways of designing a spatial order reducing transform. One possible solution is the partner filter approach by WIRTH [714]. In Chapter 7 we will discuss a way of simplifying the adaptive processor in the time dimension. We will concentrate in the sequel on techniques using subarrays or auxiliary sensors.[2]

A *spatial* transform of *space-time* vectors and matrices can be described by a

---

[1] (5.10), and (5.14).
[2] Sidelobe canceller type.

$NM \times KM$ matrix of the following form

$$\mathbf{T} = \begin{pmatrix} \mathbf{T}_s & & & \\ & \mathbf{T}_s & 0 & \\ & 0 & \ddots & \\ & & & \mathbf{T}_s \end{pmatrix} \qquad (6.1)$$

where the $N \times K$ submatrix $\mathbf{T}_s$ is the spatial transform. If $K \ll N$ this transform can strongly reduce the computational expense for matrix inversion and data filtering: The inversion of the transformed covariance matrix takes $\propto (KM)^3$ operations; filtering needs $(KM)^2$ operations per range increment.

Comparing (6.1) with (5.1), it is obvious that (5.1) shows a space-time transform while (6.1) operates in the spatial dimension only.

The vectors of signal, clutter and noise are transformed according to (5.3), the optimum processor in the transformed domain is given by (5.4), and the improvement factor by (5.5).

The simplification of processing obtained by the spatial transform is in part based on the properties of linear arrays with equidistant sensor spacing. As will be shown these techniques can be extended to the use of planar (rectangular, circular, elliptic) arrays with equidistant sensor spacing.

In most of the following numerical examples a forward looking array is assumed. Similar results are obtained for sidelooking arrays.[3]

6.1 Subarrays

Subdividing a large antenna into subarrays is a common way of reducing the signal vector space. Such techniques have been discussed by FERRIER et al. [163], JOHANNISSON and STEEN [282], and NICKEL [484] in the context of jammer suppression. Some examples for forming subarrays have been given in Figure 1.5. In contrast to the beam manifold used in the auxiliary channel processor (Section 5.3) here all the subarray elements are weighted in such a way that all subarray beams are steered in the same look direction.

SWINGLER [630] demonstrates that reduction of the rank of the covariance matrix by subarray beamforming reduces the amount of training data required for adaptation. GAFFNEY et al. [174] discuss STAP performance at subarray level with application to fast-scan radar operation.

---

[3]Except for the effect of system bandwidth, see Section 3.4.3.3. Here the sidelooking array offers some advantages over other sensor configurations.

6.1.1 Overlapping uniform subarrays (OUS)

The transform matrix that forms overlapping subarrays has already been presented in Chapter 1 (1.70):

$$\mathbf{T}_s = \begin{pmatrix} b_1 & & & & \\ \vdots & b_2 & & 0 & \\ b_g & \vdots & & & \\ & b_{g+1} & & & \\ & & \ddots & & \\ & & & b_{N-g+1} & \\ & 0 & & \vdots & \\ & & & b_N & \end{pmatrix} \qquad (6.2)$$

The beamformer coefficients b_i include the spatial phase terms of the expected signal and have been defined in (2.32). The related space-time transform is obtained by inserting (6.2) into (6.1). $\mathbf{T}_s$ is a $N \times K$ rectangular matrix which transforms the N-dimensional signal vector space into a subspace of dimension K. If $K \ll N$ this may result in strong savings of computing cost, provided that the resulting subspace processor works properly. This depends very much on the individual spatial transform chosen.

In (6.2) the overlap includes all subarray elements but one. Therefore, the phase centres of the subarrays are displaced by the same amount as the sensors (normally $d = \lambda/2$). In the transformed domain one obtains an array whose sensors

- are displaced according to the spatial Nyquist frequency,

- are directive according to the subarray size, and therefore,

- have a clutter response with reduced Doppler bandwidth, or equivalently,

- have a reduced number of clutter eigenvalues.

WARD and STEINHARDT [692] apply the concept of overlapping subarrays in the *time domain*. Overlapping data segments are used to generate temporal degrees of freedom for a post-Doppler processor. The authors proved that a transform according to (6.2) changes Brennan's rule in the following way. The number of clutter eigenvalues of the space-time covariance matrix are[4] (see (3.61)))

$$N_e \approx \text{int}\{N + (M-1)\gamma\} \qquad (6.3)$$

where

$$\gamma = \frac{2v_\text{p}T}{d} \qquad (6.4)$$

---
[4] For sidelooking linear eqispaced array, no bandwidth effects or channel errors.

After applying a subarray transform like (6.2) in the *time domain* the number of eigenvalues changes into

$$N_e \approx \text{int}\{N + (L-1)\gamma\} \tag{6.5}$$

where L is the number of subarrays and $\text{int}\{\}$ denotes the next higher integer number. For DPCA conditions ($\gamma = 1$) we have

$$N + L - 1 < NL \tag{6.6}$$

i.e., the dimension of the clutter subspace of the transformed covariance matrix is smaller than the order of the matrix which offers the potential of effective clutter rejection.

ZHANG and MIKHAEL [745] have shown for a sidelooking array that a *spatial* transform according to (6.2), together with (6.1), yields a transformed space-time covariance matrix of reduced order KM (instead of NM) with roughly

$$N_e \approx \text{int}\{N - N_{es} + 1 + (M-1)\gamma\} \tag{6.7}$$

clutter eigenvalues. N_{es} is the number of elements in each of the subarrays. For realistic values of γ, N_e is again smaller than the total vector space KM.[5] Therefore, transforms like (6.2) promise high clutter suppression performance. The number of subarrays, subarray size and array size are related as follows:

$$K = \frac{N - N_{\text{lap}}}{N_{es} - N_{\text{lap}}} \tag{6.8}$$

where N_{lap} is the amount of overlap between adjacent subarrays. In their proof the authors exploit similarities between moving array radars and the theory of band-limited signals. Based on the idea that for a sidelooking array space and time are interchangeble, they apply a theorem by LANDAU and POLLAK [375] which says that a signal which is approximately limited in time and bandwidth[6] can be decomposed into $N_e = 2(BT + 1)$ orthogonal signal components. By applying this theorem to space-time covariance matrices for sidelooking radar the formulas (6.3) and (6.7) can be found (ZHANG and MIKHAEL [745]).

6.1.1.1 *Comparison with optimum processing*

Considering a transform according to (6.2), we are still faced with properly sampled space-time clutter echo signals so that near-optimum clutter filtering performance can be expected. The spatial dimension of the processor is reduced from N down to K. Nevertheless, the clutter resolution (width of the clutter notch) which is a measure for detectability of slow targets is determined by the directivity of the subarray patterns. We can, therefore, expect about the same resolution as for the optimum processor.

In Figure 6.1 the IF has been calculated for the optimum processor (OAP, see Chapter 4) and the overlapping subarray processor. The number of channels was chosen to be $K = 5$, the subarray displacement is $d_s = \lambda/2$. As can be seen the overlapping subarray processor approaches the optimum improvement factor very well. This has been confirmed by GRIFFITHS *et al.* [214].

---

[5]Notice that this formula holds for overlapping and disjoint subarrays as well.
[6]Strictly speaking, this is a contradiction.

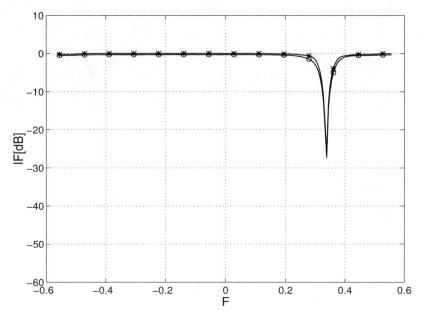

Figure 6.1: Subarray concept (FL, $K = 5$): ○ *overlapping uniform subarrays;* ∗ *optimum processor*

6.1.1.2 Number of channels

In the following example the dependence of the improvement factor on the number of channels K after the spatial transform is illustrated. Figure 6.2 shows IF curves for $K = 2, 4, 6, 8$. For $K = 2$ we notice some degradation in the neighbourhood of the clutter notch. For $K = 4, 6, 8$ almost the same near-optimum curves are obtained.

6.1.1.3 The overlapping subarray processor

A block diagram of the overlapping subarray processor is given in Figure 6.13. The sensor array is followed by a subarray beamformer network. All subarray beamformers are steered in the same look direction. The beamformer network is followed by shift registers for storing M successive echoes, and by the inverse of the transformed clutter covariance matrix $\mathbf{Q}_T$. Then the spatial dimension is eliminated by combining the clutter-free subarray outputs by a secondary beamformer. The remaining M temporal samples are then fed into a Doppler filter bank whose output is used for target detection and indication in the usual way.

6.1.2 Effect of subarray displacement

In Section 6.1.1, overlapping subarrays were considered which are displaced by just one sensor spacing of the original array. In this way the subarray outputs form an array with fewer channels than the original array, however with directive sensors,

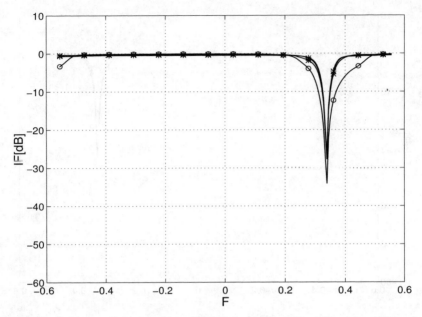

Figure 6.2: Influence of the number of channels (OUS, FL): ∘ $K = 2$; ∗ $K = 4$; × $K = 6$; + $K = 8$

the directivities being given by the subarray beampatterns, and with phase centres displaced at the foot distance $d_s = \lambda/2$.

Now we want to investigate the effect of larger displacement of the subarrays. In this case one has to take into account that the subarray phase centres are displaced by more than $\lambda/2$. This may lead to angular ambiguities (grating lobes in the case of beamforming, grating nulls in the case of interference rejection).

For example, the transform

$$\mathbf{T}_s = \begin{pmatrix} b_1 & & 0 \\ \vdots & b_1 & \\ b_g & \vdots & b_1 \\ & b_g & \vdots \\ 0 & & b_g \end{pmatrix} \quad (6.9)$$

leads to a subarray subspace with three channels. The beamformer coefficients b_i have been defined in (2.32). The displacement of subarrays is $3d$ where d is the spacing between the array sensors. An extreme case of displacement of subarrays are disjoint subarrays. Disjoint subarrays play an important role for practical reasons: each sensor output is used only once. The associated spatial transform becomes[7]

---

[7]Compare (1.69).

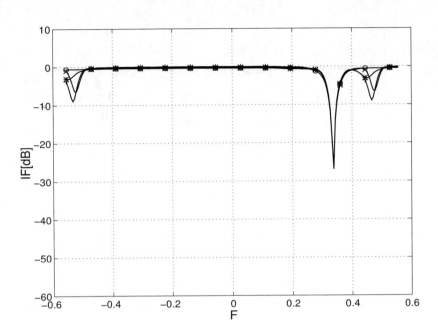

Figure 6.3: Subarray processing: Effect of subarray displacement (FL, $K = 6$): $\circ\ d_s = \lambda/2$; $*\ d_s = \lambda$; $\times\ d_s = 1.5 \cdot \lambda$; $+\ d_s = 2 \cdot \lambda$

$$\mathbf{T}_s = \begin{pmatrix} b_1 & & & & \\ \vdots & & & \mathbf{0} & \\ b_g & & & & \\ & b_1 & & & \\ & \vdots & & & \\ & b_g & & & \\ & & \ddots & & \\ & & & b_1 & \\ & \mathbf{0} & & \vdots & \\ & & & b_g & \end{pmatrix} \quad (6.10)$$

with beamformer coefficients as defined in (2.32). For a linear array all subarrays can be made equal so that the resulting subarray beam patterns become identical. In this case one obtains an array with a reduced number of receive channels with directive sensors. Since the phase centres of the subarrays are displaced by more than $\lambda/2$ grating lobes may occur. This effect can be avoided by designing non-uniform subarrays. A practical example for irregular disjoint subarrays is the ELRA antenna (see GROEGER [215], GROEGER *et al.* [216]).

In Figure 6.3 the effect of subgroup displacement is shown. The number of channels was chosen to be $K = 6$. The four curves have been calculated for sensor displacements $d_s = 0.5\lambda, \lambda, 1.5\lambda, 2\lambda$. Accordingly, the number of elements in each

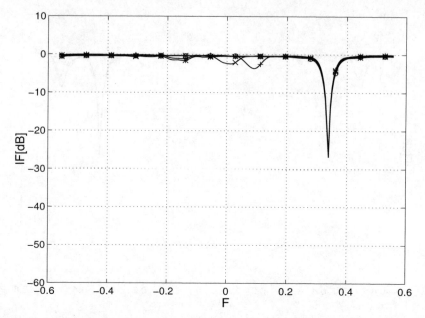

Figure 6.4: Subarray processing: Effect of subarray displacement (SL, K = 6): ○ $d_s = 0.5 \cdot \lambda$; ∗ $d_s = \lambda$; × $d_s = 1.5 \cdot \lambda$; + $d_s = 2 \cdot \lambda$

subarray are 19, 13, 7, 4. Notice that $d_s = 2\lambda$ means disjoint subarrays according to (6.10).

The curves show that there is almost no effect of the sensor displacement in the major part of the pass band and even in the clutter notch area. Some small losses are encountered in the broadside direction which in the case of a forward looking array is the flight direction, i.e., the direction with the maximum clutter Doppler frequency (F ≈ 0.5). In this direction the clutter power is a maximum because the sensor directivity patterns have their maximum in the flight direction for a forward looking array.

Let us recall that a displacement of array *sensors* by more than $\lambda/2$ leads to spatial ambiguities (see Figures 4.14 and 4.15, and Section 3.4.2.2). If the sensors have a directivity pattern as determined by the subarrays all directions other than the look directions are modulated by the subarray sidelobe pattern. The ambiguous responses (grating lobes) of the secondary beamformer (Figure 6.13) fall into the subarray sidelobe area and are attenuated. Therefore, the ambiguous clutter responses of the subarray processor are relatively small.

Similar results are obtained for the sidelooking array. In Figure 6.4 the clutter notch is nearly independent of the subgroup displacement. Some small losses in the sidelobe area can be noticed. They appear again close to the broadside direction where the sensor directivity patterns have their maximum. Recall that for a sidelooking array broadside means zero Doppler frequency.

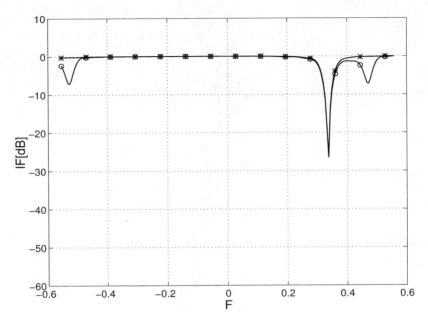

Figure 6.5: Non-uniform subarrays (FL, $K = 5$): ∘ overlapping uniform subarrays; ∗ optimum processor

6.1.3 Non-uniform subarrays

So far we discussed only uniform subarrays. Let us now consider an example for non-uniform subarrays. As mentioned earlier non-uniform subarrays may be useful for several reasons, especially for reduction of grating lobes. This principle has been used in the ELRA experimental radar system (see GROEGER et al. [216]).

6.1.3.1 Some background

The effect of non-uniform subarrays is that the spatial sampling of the backscattered wave field is done with non-uniform directivity patterns. Each of these directivity patterns cuts its individual Doppler spectrum out of the Doppler coloured clutter background. The effect of non-uniform subarrays on the performance of the space-time clutter covariance matrix can be explained by recalling (3.28). We have to modify the integrands in that the transmit directivity patterns $D(\varphi)$ and $D(\phi)$ have to be replaced by two different receive patterns for the i-th and k-th subarray and a transmit patterns $D_t(\varphi)$

$$\begin{aligned} q_{ln}^{(c)} &= E\{c_{i,m} c_{k,p}^*\} \\ &= E\Big\{ \int_{\varphi=0}^{2\pi} A(\varphi) D_t(\varphi) D_i(\varphi) L(\varphi) G(\varphi, m) \Phi_m^t(\varphi, v_p) \Phi_i^s(\varphi) \mathrm{d}\varphi \\ &\quad \times \int_{\phi=0}^{2\pi} A(\phi) D_t(\phi) D_k(\phi) L(\phi) G(\phi, m) \Phi_p^t(\phi, v_p) \Phi_k^s(\phi) \mathrm{d}\phi \Big\} \end{aligned} \quad (6.11)$$

188 *Spatial transforms for linear arrays*

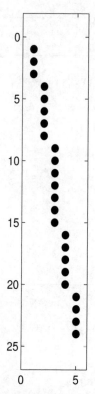

Figure 6.6: Disjoint non-uniform subarrays: The transform matrix

where $D_t(\varphi)$ is the transmit pattern and

$$D_i(\varphi) = \mathbf{b}_i^*(\varphi_L)\mathbf{b}_i(\varphi) \qquad (6.12)$$
$$D_k(\phi) = \mathbf{b}_k^*(\phi_L)\mathbf{b}_k(\phi)$$

are the complex directivity patterns of the i-th and k-th subarray, with $\mathbf{b}_i(\varphi)$ and $\mathbf{b}_k(\varphi)$ being beamformer vectors according to (2.32). φ_L and ϕ_L denote the look direction. Recall that the indices l, n of the covariance matrix are related with i, k, m, p through (3.23), (3.24).

The i-th subarray produces a noise power

$$P_n = \mathbf{b}_i^*(\varphi)\mathbf{b}_i(\varphi) = N_i \qquad (6.13)$$

where N_i is the number of subarray elements. Normalising the covariance coefficient (6.11) to the individual subarray noise levels means using normalised directivity patterns

$$\tilde{D}_i(\varphi) = \mathbf{b}_i^*(\varphi_L)\mathbf{b}_i(\varphi)/N_i \qquad (6.14)$$
$$\tilde{D}_k(\varphi) = \mathbf{b}_k^*(\varphi_L)\mathbf{b}_k(\varphi)/N_k$$

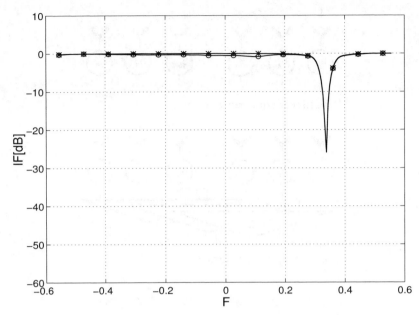

Figure 6.7: Non-uniform subarrays (SL, $K = 5$): ○ *overlapping uniform subarrays;* ∗ *optimum processor*

Assuming again independent clutter arrivals from different directions leads to

$$E\{A(\varphi)A^*(\phi)\} = 0 \qquad (6.15)$$

Then the cross-terms in azimuth vanish so that the covariance (6.11) becomes

$$\begin{aligned} q_{ln}^{(c)} &= E\{c_{im}c_{kp}^*\} \qquad (6.16) \\ &= \int_{\varphi=0}^{2\pi} E\{AA^*\} D_t^* D_t \tilde{D}_i^*(\varphi) \tilde{D}_k(\varphi) L^2(\varphi) G(\varphi, m) G^*(\varphi, p) \\ &\quad \times \Phi_{m-p}^t(\varphi, v_p) \Phi_{i-k}^s(\varphi) \mathrm{d}\varphi \end{aligned}$$

where the indices of the covariance matrix l, n have been defined in (3.23) and (3.24). (6.16) is a scalar product whose absolute value becomes a maximum when $\tilde{D}_i(\varphi) = \tilde{D}_k(\varphi)$. We conclude that *uniform* subarrays are the optimum choice for adaptive processing because they maximise the off-diagonal terms of the clutter covariance matrix.[8]

6.1.3.2 Examples

The subarray structure chosen for a numerical example is depicted in Figure 6.6. In Figure 6.5, the subarray transform according to Figure 6.6 has been applied to a

---

[8]It should be noted here that, on the other hand, uniform subarrays whose phase centres are displaced by more than $\lambda/2$ tend to create grating lobes.

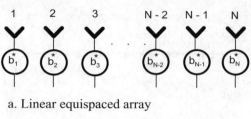

a. Linear equispaced array

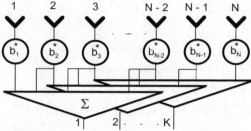

b. Overlapping subarrays

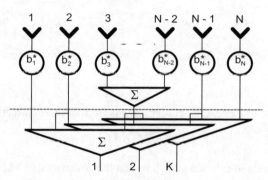

c. Summation of central elements

Figure 6.8: From subarrays to auxiliary elements

forward looking array. In comparison with the optimum processor some losses can be recognised at the clutter Doppler frequency in the look direction (F ≈ 0.5). The result is similar to the uniform disjoint subarray processor (+ curve in Figure 6.3). Obviously the heterogeneous subarray structure in Figure 6.6 does not have significant influence on the clutter rejection performance.

Figure 6.7 shows for comparison the same array configuration, however in sidelooking arrangement. Again some very small losses in the broadside direction of the array (F = 0) can be noticed. However, opposite to Figure 6.4, the losses are smoothed which is caused by the heterogeneous subarray structure. It turns out that an irregular subarray structure may offer some advantages in supression of ambiguous clutter response.

It should be noted that in all examples shown the number of echo samples in the space-time covariance matrix was $M = 24$, i.e., as large as the original number of

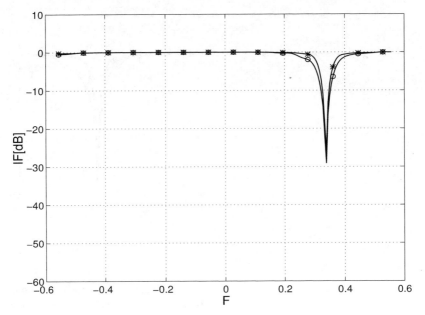

Figure 6.9: Auxiliary sensor configuration (FL, $K = 5$): ○ *symmetric auxiliary sensors;* ∗ *optimum processor*

antenna elements. Therefore, the clutter rejection operation $\mathbf{Q}^{-1}$ has a large number of temporal degrees of freedom. This may compensate for bandwidth effects, subarray displacement, and irregular subarray size. This aspect will be revisited in Chapter 7. There we will try to reduce the number of degrees of freedom in the temporal dimension.

6.2 Auxiliary sensor techniques

6.2.1 Symmetric auxiliary sensor configuration (SAS)

It can be shown that a symmetric auxiliary sensor configuration is strongly related to the overlapping uniform subarray technique. This relation is illustrated in Figure 6.8. Consider a linear equispaced array with beamformer weights. A beam can be formed by simply summing all the output signals in Figure 6.8a.

An alternative way of forming a beam is given in Figure 6.8b. Primary beams are formed for all subarrays like in Section 6.1.1. Combination of subarrays is then done by a secondary beamformer. Notice that both concepts are identical as far as the output is concerned, but lead to different processing schemes.

The antenna configuration in Figure 6.8c is essentially the same as that in Figure 6.8b. The only difference is that all those central elements which are connected with *all* subarrays are pre-summed. Notice again that the K output channels in 6.8c are equivalent to those in Figure 6.8b.

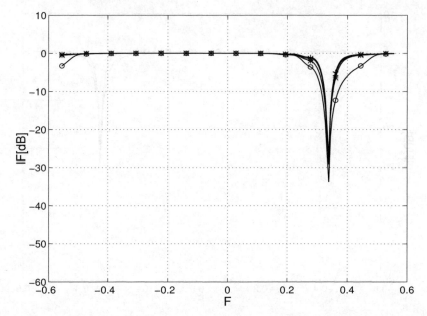

Figure 6.10: Influence of the number of channels (SAS, FL): ○ $K = 3$; ∗ $K = 5$; × $K = 7$; + $K = 9$

The sums below the dotted line constitute a linear transform of the vector space above the line and the K outputs. Notice that we have generated a symmetric auxiliary sensor-beam configuration above the dotted line. Since the vector space above the line is larger than that below by two channels we can conclude that a symmetric auxiliary sensor configuration is equivalent to uniform overlapping subarrays.

The auxiliary sensor transform matrix becomes

$$\mathbf{T}_s = \begin{pmatrix} 1 & & & & & & \\ & \ddots & & & & & \\ & & 1 & & 0 & & \\ & & & b_1 & & & \\ & & & \vdots & & & \\ & & & b_B & & & \\ & & 0 & & 1 & & \\ & & & & & \ddots & \\ & & & & & & 1 \end{pmatrix} \qquad (6.17)$$

where the b_i are beamformer weights according to (2.32). The weighting coefficients of the auxiliary channels have been set to 1. Any other phase coefficient can be chosen as well.

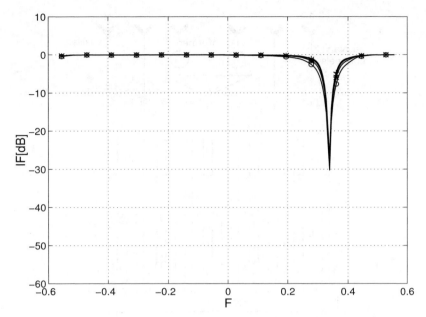

Figure 6.11: Influence of the number of channels (SAS, SL): ○ $K = 3$; ∗ $K = 5$; × $K = 7$; + $K = 9$

6.2.1.1 Comparison with optimum processing

A comparison of the symmetric auxiliary sensor processor with the optimum fully adaptive processor is shown in Figure 6.9. The number of channels was chosen to be $K = 5$ as in the previous examples. It can be seen that both IF curves coincide almost perfectly. Some slight losses can be noticed close to the look direction.

At first glance it appears that the overlapping subarray processor (Section 6.1.1) performs even slightly better for the same number of output channels[9] $K = 5$. To explain this let us have a look again at Figure 6.8c. Notice that the number of channels of the symmetric auxiliary sensor part above the dotted line is larger by two channels. Therefore, by keeping $K = 5$ constant the auxiliary sensor receiver has two channels less than the overlapping subarray processor. This explains the slight difference in the IF curves.

6.2.1.2 Number of channels

The dependence of the improvement factor on the number of channels K is shown in Figure 6.10, for a forward looking, and Figure 6.11, for a sidelooking linear array. The curves have been calculated for $K = 3, 5, 7, 9$.

Obviously the improvement factor is almost independent of the number of antenna channels. In the case of the forward looking array, however, we find some losses for $K = 3$ while in the sidelooking configuration no significant losses appear.

---

[9]Notice that the number of output channels is the spatial dimension of the adaptive processor, see (5.9).

194 *Spatial transforms for linear arrays*

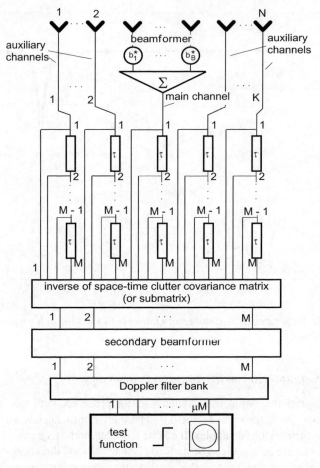

Figure 6.12: The symmetric auxiliary sensor processor

To explain this effect let us recall (3.2)

$$f_\mathrm{D} = \frac{2v_\mathrm{p}}{\lambda} \cos\varphi \cos\theta \qquad (6.18)$$

which states that due to the motion of the radar platform in the direction $\varphi = 0$ each point on the stationary background is associated with one Doppler frequency. Since the cosine function is symmetric with respect to $\varphi = 0$ each individual Doppler frequency appears at two different azimuth angles.

Suppose we have omnidirectional sensor patterns. The geometry of a *sidelooking* array coincides with the flight path so that beam patterns are symmetrical about $\varphi = 0$. Therefore, for one Doppler frequency the sidelooking array perceives signals arriving from one direction only. Concerning the clutter covariance matrix this means that one single frequency causes one clutter eigenvalue.

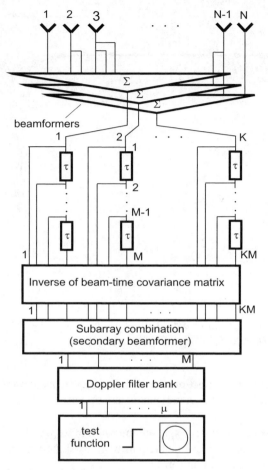

Figure 6.13: Overlapping subarray processor

The *forward looking* array can distinguish between left ($\varphi < 0$) and right ($\varphi > 0$). Therefore, the forward looking array observes two different directions with the same clutter Doppler frequency. This means that for the forward looking array one Doppler frequency causes two different eigenvalues of the space-time clutter covariance matrix. Eigenspectra of sidelooking and forward looking arrays have been presented in Figures 3.12 and 3.13. Compare the ○ curves: The number of clutter eigenvalues for the forward looking array is twice the number of eigenvalues for the sidelooking configuration.

Using directive sensors (∗ curves in Figures 3.12 and 3.13) leads to about[10] the same number of eigenvalues for sidelooking and forward looking arrays. However, for a sidelooking array one gets one direction for each Doppler frequency. For the forward

---

[10]The exact number of clutter eigenvalues is determined to some extent by the sensor.

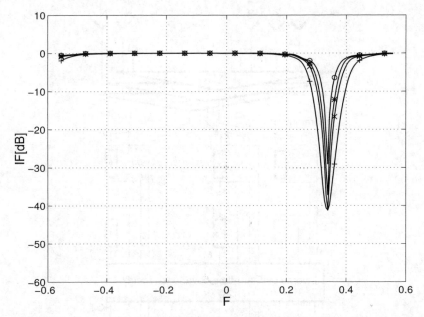

Figure 6.14: Influence of clutter bandwidth (SAS, $K = 5$, FL): ○ $B_c = 0$; ∗ $B_c = 0.01$; × $B_c = 0.03$; + $B_c = 0.1$

looking configuration two directions for each Doppler frequency are obtained, but one deals only with one half of the clutter bandwidth because no clutter echoes come from the backward direction.

Therefore, when reducing the spatial dimension of a space-time processor one has to keep in mind that the forward looking array needs slightly more spatial degrees of freedom (K) than the sidelooking array.

6.2.1.3 The processor

A block diagram of the symmetric auxiliary sensor processor (SAS) is shown in Figure 6.12. A beam is formed by using the centre elements of the array in such a way that the same number of auxiliary sensors is left on both sides. The next stage is the shift register array for storing successive echoes. At this level the inverse of the $KM \times KM$ clutter covariance matrix has to be calculated. The processor is completed with a space-time signal matching network which in essence is a secondary beamformer cascaded with a Doppler filter bank.

The secondary beamformer is a weighted sum of the K antenna channel output signals. If the number of elements in the pre-beamformer is large compared with the number of auxiliary elements then the processor can be simplified by omitting the target signal contributions of the auxiliary sensors.[11] In this case the $KM \times 1$ signal reference vector s_T reduces to a $M \times 1$ vector which contains Doppler filter weights.

---

[11] A similar simplification has been described in the context of the space-time transform (5.8).

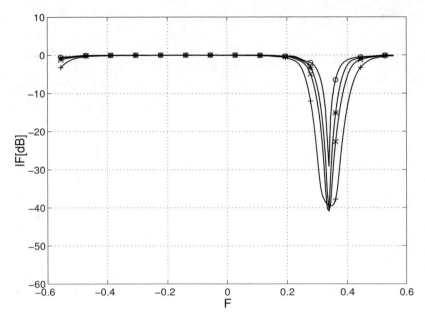

Figure 6.15: Influence of system bandwidth (SAS, FL): ○ $B_s = 0$; ∗ $B_s = 0.01$; × $B_s = 0.03$; + $B_s = 0.1$

For filtering the received echo data only a submatrix of $\mathbf{Q}^{-1}$ is required. In (6.19) those elements of $\mathbf{Q}^{-1}$ related to the beamformer are denoted as column vectors $\mathbf{q}_{nm}$ while the dots indicates those elements due to the auxiliary elements which may be omitted:

$$\mathbf{Q}^{-1} = \begin{pmatrix} \cdot & \cdot & \mathbf{q}_{nm} & \cdot & \cdot & | & \cdot & \cdot & \mathbf{q}_{nm} & \cdot & \cdot & | & \cdots & | & \cdot & \cdot & \mathbf{q}_{nm} & \cdot & \cdot \\ \cdot & \cdot & \mathbf{q}_{nm} & \cdot & \cdot & | & \cdot & \cdot & \mathbf{q}_{nm} & \cdot & \cdot & | & \cdots & | & \cdot & \cdot & \mathbf{q}_{nm} & \cdot & \cdot \\ & & \cdots & & & | & & & \cdots & & & | & \cdots & | & & & \cdots & & \\ & & \cdots & & & | & & & \cdots & & & | & \cdots & | & & & \cdots & & \\ \cdot & \cdot & \mathbf{q}_{nm} & \cdot & \cdot & | & \cdot & \cdot & \mathbf{q}_{nm} & \cdot & \cdot & | & \cdots & | & \cdot & \cdot & \mathbf{q}_{nm} & \cdot & \cdot \end{pmatrix} \tag{6.19}$$

where $n = 1, \ldots, M$ and $m = (N+1)/2 + r(M-1)$, with r being the temporal index which denotes the spatial submatrices. By omitting the target signal contributions of the auxiliary elements the filter matrix becomes even smaller ($KM \times M$) than the reduced order covariance matrix $\mathbf{Q}^{-1}$ ($KM \times KM$). Notice that this simplification is possible because the rejection operation takes place *after* beamforming. This is not possible for the overlapping subarray processor where all subarray outputs have the same priority.

6.2.1.4 *Further reduction of the signal vector space*

As has been noted above the advantage of the symmetric auxiliary sensor concept is that it can easily be implemented in the RF domain. Due to its symmetric structure the

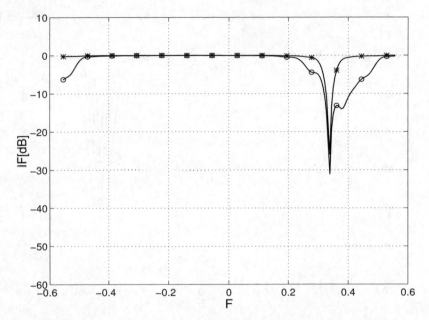

Figure 6.16: Auxiliary sensor configuration (FL): ○ *asymmetric auxiliary sensors;* ∗ *optimum processor*

number of channels is an odd number, with three channels being the absolute possible minimum.

The number of channels can be reduced even further by a second transform of the following kind

$$\mathbf{T}_2 = \begin{pmatrix} 1 & & & & & \\ \vdots & 1 & & \mathbf{0} & & \\ 1 & \vdots & & & & \\ & 1 & & & & \\ & & \ddots & & & \\ & & & & 1 & \\ & \mathbf{0} & & & \vdots & \\ & & & & 1 & \end{pmatrix} \quad (6.20)$$

By this transform the outputs of the auxiliary sensor scheme are summed up so as to form overlapping subarrays. This summation has been illustrated in Figure 6.8 below the dotted line. It can be carried out in the digital domain. The formation of subarrays can be easily seen by multiplying

$$\mathbf{T}_{\text{tot}} = \mathbf{T}_s \mathbf{T}_2 \quad (6.21)$$

where $\mathbf{T}_s$ was defined by (6.17). The performance of the overlapping subarray processor has been discussed in Section 6.1.1. The resulting number of channels is reduced from $2K - 1$ down to K so that the minimum possible number of channels is two instead of three. As can be seen from Figure 6.2, some degradation in performance has to be tolerated if the number of channels is minimised.

6.2.2 Bandwidth effects

In this section we discuss again the influence of bandwidth effects on the performance of the symmetric auxiliary sensor processor. As carried out in Chapter 2 such effects may occur due to either clutter fluctuations or wide system bandwidth. Again we consider a 24-element forward looking array.[12] As the concept of symmetric auxiliary sensors is equivalent to overlapping subarrays we discuss bandwidth effects only for the symmetric auxiliary sensor processor.

6.2.2.1 Clutter bandwidth

As pointed out in Section 2.5.1.1 internal motion of the clutter background may occur due to wind or sea state effects. Such fluctuations lead to *temporal* decorrelation of the clutter background effects which result in a broadening of the clutter spectrum and the notch of the clutter filter.

In Figure 6.14, the influence of the clutter bandwidth on the performance of the symmetric auxiliary sensor processor is demonstrated. As can be seen there is a broadening of the clutter notch which results in degraded detectability of slow targets.

However, there is no additional effect on the IF in the pass band, i.e., far away from the clutter notch, as in the case of the auxiliary channel processor, see Figure 5.10. This behaviour can be expected since the number of degrees of freedom chosen for the processing vector space is larger than the number of clutter eigenvalues. The number of eigenvalues of the covariance matrix used in Figure 6.14 is slightly above $N + M = 24 + 24 = 48$ while the dimension of the vector space is $KM = 120$.

6.2.2.2 System bandwidth

As carried out in Section 2.5.2 the bandwidth of the radar system which is usually matched to the transmitted waveform may cause a degradation of the cross-correlation between the output signal of different array channels (*spatial* decorrelation).

The effect of the relative system bandwidth ($B_s = 0, 0.01, 0.03, 0.1$) is demonstrated in Figure 6.15. As one can see the clutter notch is broadened with increasing system bandwidth. However, no additional effects such as a ripple in the pass band due to the lack of degrees of freedom can be noticed (compare with Figure 5.11). The curves look similar to those obtained for the optimum processor, see Figure 4.23. Obviously the increase in the number of eigenvalues due to the system bandwidth is not in conflict with the reduced number of antenna channels.

---

[12] The reader is reminded that sidelooking arrays do not suffer from the system bandwidth effect.

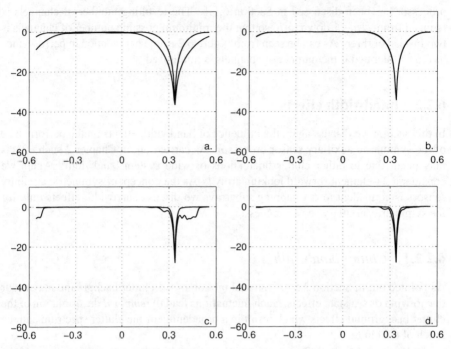

Figure 6.17: Optimum and auxiliary sensor processing (FL, $K = 5$): Dependence on temporal dimension M. a. asymmetric auxiliary sensors ($M = 5$); b. symmetric auxiliary sensors ($M = 5$); c. asymmetric auxiliary sensors ($M = 48$); d. symmetric auxiliary sensors ($M = 48$)

6.2.3 Asymmetric auxiliary sensor configuration

After we found that the symmetric auxiliary sensor configuration (SAS) yields near optimum clutter rejection performance we raise the question of what the performance of asymmetric auxiliary sensors may be. Consider for example a configuration with one main beam and some auxiliary sensors on *one* side. The corresponding transform matrix is, see (1.56),

$$\mathbf{T}_s = \begin{pmatrix} b_1 & & & & \\ \vdots & & & 0 & \\ b_B & & & & \\ & 1 & & & \\ & & \ddots & & \\ & 0 & & & 1 \end{pmatrix} \quad (6.22)$$

with beamformer weights b_i according to (2.32). The space-time transform is then given by inserting (6.22) into (6.1).

Figure 6.16, shows the MTI performance of such asymmetric antenna configuration compared with the optimum (fully array) adaptive processor. As can be seen the

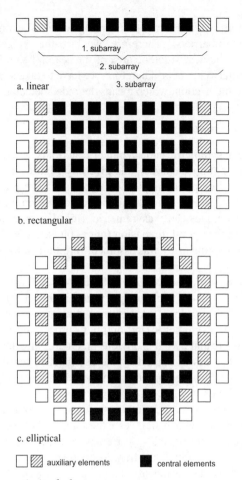

Figure 6.18: Clutter optimised planar array antennas

asymmetric auxiliary sensor processor performs well in the pass band, i.e., far away from the clutter notch. Close to the clutter notch some considerable losses can be noticed. In other words, detection of slow targets is quite poor compared with the symmetric auxiliary sensor configuration (Figure 6.9).

This degradation can be explained as follows. The symmetric auxiliary sensor array is identical to the overlapping subarray configuration. The concept of overlapping subarrays is equivalent to equidistant Nyquist sampling in the spatial dimension, however, with the subarrays being directive sensors. If the directivity patterns including the look directions of all subarrays are the same then the clutter spectra received by each of the output signals of the subarrays are identical except for a phase shift. Consequently the cross-correlation between identical channels is a maximum. In other words, the CNR is the same for all subarrays, see the criteria for subspace transforms, Section 1.2.3.

The asymmetric array configuration according to (6.22) cannot be interpreted as an array of identical subarrays. Therefore, we face the typical sidelobe canceller problem: the CNR in the near sidelobe area of the search beam is higher than in the omnidirectional auxiliary channels. Therefore, we have to take some losses in IF close to the look direction into account which means degraded detectability of slow targets.

6.2.3.1 *Effect of temporal dimension*

A comparison between symmetric and asymmetric auxiliary sensor configurations with the optimum processor (upper curves) was given by Figures 6.9 and 6.16. In these examples the number of temporal samples was chosen as large as the number of sensors ($M = 24$). Figure 6.17 shows again a comparison of the optimum processor (upper curves) with both techniques (left: asymmetric, right: symmetric auxiliary) for two different choices of the temporal dimension (upper plots: $M = 5$; lower plots: $M = 48$). Again a forward looking array was assumed. Similar results can be obtained for a sidelooking array.

For small temporal dimension ($M = 5$) the IF achieved by the asymmetric processor is considerably lower than the optimum (Figure 6.17a) while the symmetric processor (Figure 6.17b) reaches the optimum almost perfectly. For $M = 48$ the differences between asymmetric and optimum processing are mitigated (6.17c). Comparing Figures 6.17c and 6.16 it is obvious that increasing the temporal dimension equalises to some extent losses caused by the asymmetric sensor arrangement. The symmetric processor reaches the optimum performance almost perfectly even for large temporal dimension (Figure 6.17d).

As stated before, the symmetric auxiliary sensor configuration is equivalent to overlapping equal subarrays which all receive the same clutter spectrum. This property leads to near-optimum clutter rejection performance. If the subarrays are different (so that each subarray channel produces a different clutter spectrum) optimum performance can be approximated by increasing the temporal dimension of the adaptive space-time processor. These additional temporal degrees of freedom compensate for the differences in the clutter spectra received by the individual antenna channels.

6.2.4 Optimum planar antennas

The symmetric auxiliary sensor configuration, described in Section 6.2.1, has been shown to be a very good approximation of the optimum processor. In particular it has been shown that the number of channels required is almost independant of the antenna size N. This means that the computational load for the processor is independant of the antenna dimension.

This configuration has another attractive feature: The pre-beamforming operation can easily be done in the RF domain. There is no need to implement a fully digitised array. Only the K output channels have to be equipped with amplifiers, demodulators, filters, and A/D converters. Implementation of an overlapping subarray configuration in the RF domain is much more diffcult.

Figure 6.18 shows the evolution of auxiliary sensor configurations for implementation in patch array technology. The upper figure shows again the linear array with

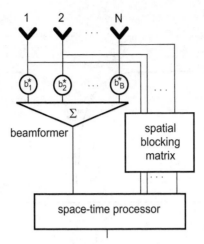

Figure 6.19: Adaptive sidelobe canceller with blocking matrix

two auxiliary sensors on either side. The brackets indicate the connection between auxiliary sensors and subarray configurations. The planar rectangular antenna array follows immediately from the linear configuration (Figure 6.18b). In Figure 6.18c an elliptical antenna with main beam and auxiliary subarrays is shown. Notice that the auxiliary subgroups have been chosen so that three *identical* overlapping subarrays can be formed.

The class of arrays which allow the design of equal subarrays arranged in the horizontal is even wider than that suggested by the above examples. In particular all kinds of cylindrical surfaces with the cylinder axis parallel to the ground may be used to design an array antenna with overlapping equal subarrays. The orientation of the array axis is arbitrary as long as it is parallel to the ground. Conformal arrays fixed on the surface of a cylindrical air vehicle are a prominent example.

It should be noted that these statements are based on the assumption that the PRF is chosen to be greater than or equal to the Nyquist frequency of the clutter bandwidth. If a lower PRF has to be used, for instance because of operational requirements, the displacement of the subarrays has to be matched to the PRF so that the DPCA condition is roughly met.

In the considerations on overlapping subarrays and symmetric sidelobe cancellers we assumed for simplicity that the PRF is always the Nyquist frequency of the clutter band. If a lower PRF is used the spacing of the array processing architecture should be matched to the PRF so as to obtain roughly DPCA conditions. That means that overlapping subarrays are no longer displaced by half the wavelength but at longer distance. For a related sidelobe canceller scheme the spacing of the auxiliary sensors should be matched roughly to the PRF, for instance, by forming little auxiliary subarrays instead of single auxiliary sensors.

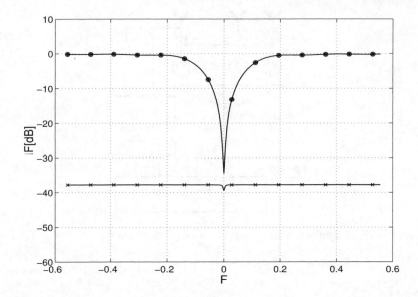

Figure 6.20: Effect of signal inclusion ($N = 12$, $M = 5$, number of blocking channel coefficients: 4): ○ *optimum processor, no signal present;* ∗ *sidelobe canceller with blocking matrix, no signal present;* × *sidelobe canceller with blocking matrix, signal present (SNR = CNR = 20 dB)*

6.3 Other techniques

6.3.1 Spatial blocking matrix transform

The use of a blocking matrix to form auxiliary channels in a sidelobe canceller configuration has been suggested by several authors for jammer nulling (for example, SCOTT and MULGREW [594], SU and ZHOU [620], GOLDSTEIN and REED [199]). The principle is shown in Figure 6.19. A beamformer is formed to maintain the signal energy in a desired direction. The channels (columns) in the auxiliary blocking matrix are determined so that they are all orthogonal to the look direction. The effect of such channels is that any desired signal in the look direction will not enter the auxiliary channels and, hence, will not contribute to the adaptive filter. Therefore, suppression of the desired signal by the adaptive filter is avoided.

Both the beamformer output and the auxiliary channels are connected with an adaptive processor for interference suppression. in the case of jammer suppression the blocking matrix is spatial only. In the following we want to discuss the use of a *spatial* blocking matrix for adaptive clutter rejection by a *space-time* sidelobe canceller. The rationale for using a *spatial* blocking matrix is that only one blocking matrix has to be designed for all target Doppler frequencies.

The spatial transform becomes

$$\mathbf{T}_s = \begin{pmatrix} b_1 & a_{12} & \cdots & a_{1K} \\ \vdots & \vdots & \cdots & \vdots \\ b_N & a_{N2} & \cdots & a_{NK} \end{pmatrix} = (\mathbf{b}(\varphi_L), \mathbf{A}) \qquad (6.23)$$

The auxiliary channel coefficients may be obtained by taking columns of a projection matrix

$$\mathbf{P} = \mathbf{I} - \frac{\mathbf{b}(\varphi_L)\mathbf{b}^*(\varphi_L)}{\mathbf{b}^*(\varphi_L)\mathbf{b}(\varphi_L)} \qquad (6.24)$$

where $\mathbf{b}(\varphi_L)$ is a beamformer vector in the look direction.

The width of the signal notch can be modified by weighting the projection matrix (6.24) with the coefficients

$$\frac{\sin(\pi\Delta(i-k))}{\pi\Delta(i-k)} \qquad (6.25)$$

where $0 < \Delta < 1$ is to adjust the width of an approximately rectangular signal notch.

If special forms of the auxiliary channels are required (for example, auxiliary *beams*) the auxiliary matrix $\mathbf{A}$ can be replaced by a projection orthogonal to the look direction

$$\tilde{\mathbf{A}} = \mathbf{P}\mathbf{A} \qquad (6.26)$$

so that

$$\mathbf{b}^*\tilde{\mathbf{A}} = \mathbf{0} \qquad (6.27)$$

For linear or rectangular arrays one can calculate a projection matrix after (6.24) for a subarray with C elements. Then one column of the projection matrix of reduced order can be applied in the fashion of a spatial FIR filter. The associated transform matrix becomes

$$\mathbf{T}_s = \begin{pmatrix} b_1 & a_{12} & & & 0 \\ \cdot & \cdot & a_{12} & & \\ \cdot & \cdot & \cdot & a_{12} & \\ \cdot & a_{C2} & \cdot & \cdot & a_{12} \\ \cdot & & a_{C2} & \cdot & \cdot \\ \cdot & & & a_{C2} & \cdot \\ b_N & 0 & & & a_{C2} \end{pmatrix} \qquad (6.28)$$

As before, the spatial transform matrix $\mathbf{T}_s$ has to be inserted into (6.1) to obtain a space-time version of the spatial transform.

6.3.1.1 *Numerical example*

Figure 6.20 shows a numerical example for the application of a sidelobe canceller with spatial blocking matrix in comparison with the optimum processor. The upper curves have been calculated for the optimum processor based on a signal-free space-time covariance matrix and the sidelobe canceller with no target signal present. Both curves coincide.

The third curve has been plotted for the case that the target signal is included in the adaptation of the sidelobe canceller. The signal power was chosen equal to the clutter power. Although the auxiliary channels are spatially blocked in the look direction we notice a dramatic signal cancellation effect. The reason for this effect is that the signal is coherent from pulse to pulse so that after blocking the signal spatially some temporal correlation remains. This can be illustrated by a simple example.

Consider for simplicity a signal coming from broadside and having zero Doppler frequency. For $N = 3, M = 2$ the received space-time signal vector is

$$\mathbf{s}^* = (1\ 1\ 1\quad 1\ 1\ 1) \tag{6.29}$$

A typical transform with blocking matrix becomes for this signal

$$\mathbf{T}_s = \begin{pmatrix} 1 & 1 & & & & \\ & -1 & 1 & & 0 & \\ & & -1 & 1 & & \\ & & & -1 & 1 & \\ & 0 & & & -1 & 1 \\ 1 & & & & & -1 \end{pmatrix} \tag{6.30}$$

Notice that the auxiliary channel forms dipole diagrams with the null steered in the signal direction. Inserting this matrix into (6.1) and multiplying the transform with the signal (6.29) gives

$$(\mathbf{Ts})^* = (\ 3\ \ 0\ \ 0\ \ 3\ \ 0\ \ 0\ ) \tag{6.31}$$

so that the covariance matrix becomes

$$(\mathbf{Ts})^*(\mathbf{sT}) = \begin{pmatrix} 9 & 0 & 0 & 9 & 0 & 0 \\ 0 & 0 & 0 & 0 & 0 & 0 \\ 0 & 0 & 0 & 0 & 0 & 0 \\ 9 & 0 & 0 & 9 & 0 & 0 \\ 0 & 0 & 0 & 0 & 0 & 0 \\ 0 & 0 & 0 & 0 & 0 & 0 \end{pmatrix} \tag{6.32}$$

The off-diagonal terms denote the temporal correlation which causes temporal signal cancellation by a space-time adaptive filter.

The correct way of handling this problem would involve a space-time blocking matrix. This matrix depends, however, on the target Doppler so that for each frequency a separate adaptive filter is required. This is not attractive if clutter cancellation is to be carried out before Doppler filtering (pre-Doppler processing, see Chapter 7). For post-Doppler techniques (see Chapter 9) the concept of space-time blocking matrices to generate signal-free auxiliary channels may be applicable. For further remarks on blocking matrices see Section 16.3.4 in Chapter 16.

6.3.2 Σ-Δ-processing

The concept of Σ-Δ-processing introduced by WANG H. *et al.* [671] and ZHANG Y. and WANG H. [746] is strongly related to the traditional DPCA techniques described

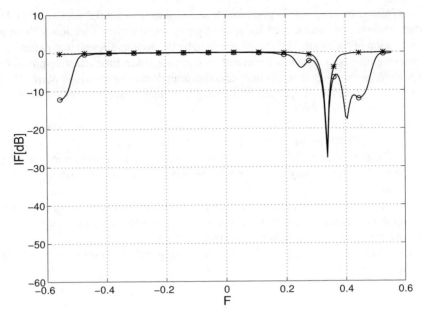

Figure 6.21: Comparison of processing techniques (FL): ○ Σ-Δ *processing;* ∗ *optimum processor*

in SKOLNIK [609, Chapter 18, p. 7] which uses a difference pattern to compensate for the platform motion induced phase advances of clutter returns in the sum beam. Some numerical results on Σ-Δ-processing are also given by BAO *et al.* [30].

NOHARA *et al.* [504] present results on the use of Σ-Δ-processing for non-sidelooking antennas. Comparisons with a two-subarray antenna are made.

Some favourable properties are summarised in the paper by BROWN and WICKS [69]:

- **Affordability.** Σ-Δ-channels are standard in any modern tracking radar.

- **Small training data sets.** If a two-pulse STAP canceller is implemented the dimension of the space-time covariance matrix is just 4×4. For adaptation less than 20 data sets are required. This has advantages in non-homogeneous clutter.

- **The small dimension** of the vector space guarantees the capability of real-time processing.

- **Channel calibration** is greatly simplified since there are only two channels.

It should be noted, however, that Σ-Δ clutter suppression fails if additional jamming is present. Another disadvantage is that by using the difference pattern for clutter rejection the capability of monopulse position finding is lost.

A comprehensive analysis of the performance of Σ-Δ-STAP based on MCARM data (SURESH BABU *et al.* [626, 627]) can be found in BROWN *et al.* [72]. The effect of array tapering is discussed in some detail.

As WANG H. *et al.* [671] point out Σ and Δ channels are frequently available in radar antennas and can be used for space-time adaptive clutter rejection without any modification of the radar. The results obtained by the authors suggest that in fact Σ-Δ-processing may be a cheap and efficient solution for space-time adaptive processing. They combine the Σ-Δ channels with adaptive joint domain processing (WANG H. and CAI [668]). This is a suboptimum processing technique which reduces the number of degrees of freedom in the Doppler domain.

ZHANG and WANG [747] analysed the effect of using elevation difference patterns instead of or in conjunction with azimuth patterns. It turns out that an additional elevation difference channel Δ_e offers some improvement in clutter rejection. However, if only two channels are allowed, the azimuth pattern Δ_a is the better choice.

MAHER *et al.* [426] combine the Σ-Δ processor with some auxiliary elements so as to form an additional sidelobe canceller for jammer rejection. Good results in clutter and jammer suppression have been reported.

The spatial transform matrix for Σ-Δ processing includes a sum and a difference channel

$$\mathbf{T}_s = \begin{pmatrix} b_1 & b_1 \\ \vdots & \vdots \\ b_{N/2} & b_{N/2} \\ b_{N/2+1} & b_{N/2+1} \\ \vdots & \vdots \\ b_N & -b_N \end{pmatrix} \qquad (6.33)$$

with b_i being the beamformer weights according to (2.32). In the numerical example we use the sum and difference channels and $M = 24$ echo pulses in the time domain. The associated clutter covariance matrix has dimensions $2M \times 2M = 48 \times 48$, which means that no reduction of the vector space in the time or frequency domain has been done. Therefore the result presented is the best one can obtain with a Σ-Δ processor and a given Doppler filter length.

A comparison with the optimum gain can be seen in Figure 6.21. As can be seen the optimum IF is reached perfectly in the pass band (off-look direction). We notice some significant losses in the neighbourhood of the clutter notch so that the detectability of slow targets is degraded.[13]

It may be that for other look directions and for sidelooking arrays better results can be obtained. But it should not be overlooked that the beam pattern of the Δ channel is quite different from that of the Σ channel. These differences can only be compensated for by a sufficient number of *temporal* degrees of freedom.

However, channels with different directivity patterns are generally unfavourable for interference cancellation, see the remarks on sidelobe canceller in Section 1.2.3.1. Notice that in the case of the Σ-Δ technique the auxiliary (Δ) channel receives very low clutter power in the direction where the sum channel (Σ) receives maximum clutter power. We come back to the Σ-Δ technique in the context of circular arrays in Chapter 8.

---

[13] Keep in mind that slow target detection is the primary goal of space-time MTI processing.

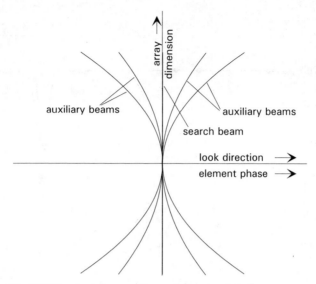

Figure 6.22: The CPCT principle: auxiliary channel weightings

6.3.3 CPCT Processing

CPCT (Coincident Phase Centre Technique, SKOLNIK[610, Chapter 16, p. 20]) is an attempt to apply the DPCA principle (Chapter 3) to forward looking arrays. The principle of CPCT was already mentioned in Section 3.2.2.4. A number of antenna channels are created by applying different weightings to the outputs of an array. These auxiliary weighting vectors $\mathbf{a}_k$, together with a beamformer vector $\mathbf{b}$ for spatial signal match, have to be inserted into a spatial transform matrix of the form

$$\mathbf{T}_s = \begin{pmatrix} \mathbf{a}_1 & \cdots & \mathbf{a}_{K-1} & \mathbf{b} \end{pmatrix} \tag{6.34}$$

which has to be inserted into (6.1) in order to form a space-time transform.

The weightings $\mathbf{a}_k$ have to be chosen in such a way that each of them has a different phase centre, and all phase centres are aligned with the flight path, just as in the case of DPCA. Unfortunately, the reference quoted in SKOLNIK [610, Chapter 16, p. 20] is not available so that the exact knowledge of how the displaced phase centres are to be generated is missing.

To give an example we have chosen a heuristic approach which is illustrated in Figure 6.22. In addition to a planar beamformer in the centre, a number of parabolic wavefronts ('auxiliary beams') are generated. All of these wavefronts have different phase centres which, because of symmetry, lie on the flight axis.

Figure 6.23 shows the result of CPCT processing in comparison with the optimum processor. Notice that the temporal dimension is as large as the array size ($M = 24$). It can be seen that there is no significant difference in the IF curves. The CPCT seems to be a reasonable approximation to the optimum processor.

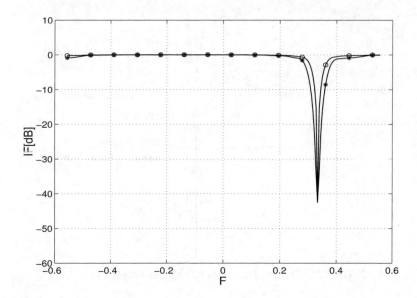

Figure 6.23: The CPCT technique ($N = M = 24$, $K = 5$): ∘ *optimum processing;* ∗ *CPCT processing*

In Figure 6.24 a similar example is shown, however with a reduced temporal dimension ($M = 6$). As a consequence the clutter notches of both the optimum and the CPCT processors are broadened. Moreover, we notice that the difference between the two IF curves has become much larger than in Figure 6.23. This is an indication that the number of temporal degrees of freedom ($M = 6$) is not sufficient.[14]

This can be explained by the following consideration. Near-optimum performance with a small number of echo samples M is obtained only if the auxiliary channels have all the same directivity patterns and are all steered in the same direction. Identity of channels means that each of the channel outputs produces the same clutter spectrum. Subtracting spectra from each other[15] leads to zero clutter if the spectra are identical. If they are not identical some equalisation of the different spectra can be obtained by use of a larger amount of temporal degrees of freedom. These aspects have been addressed already in the context of symmetric and asymmetric sensor configurations (Section 6.2.3.1).

Obviously the auxiliary channels $\mathbf{a}_k$ are all different and, therefore, produce different clutter spectra. Therefore, the CPCT approximates the optimum processor better if the number of temporal samples is large (Figure 6.23). If the number of temporal samples is very large as for instance in synthetic aperture radar space-frequency techniques can applied (see Section 9.5). In this case the number of temporal (or spectral) degrees of freedom is large enough for STAP processing to be

---

[14] These results based on CPCT with parabolic auxiliary wavefronts have been reported in KLEMM [328].
[15] This is what the inverse of the covariance matrix in essence is doing.

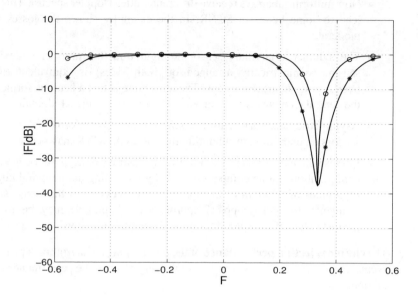

Figure 6.24: The CPCT technique ($N = 24$, $M = 6$, $K = 5$): ○ *optimum processing;* * *CPCT processing*

accomplished with different kinds of auxiliary channels.

6.4 Summary

Chapter 6 deals with the concept of *spatial* subspace transforms. Such transforms are a way of reducing the signal vector space in the spatial dimension. This leads to a reduced requirement in processor capacity. Moreover, only a few channels need to be equipped with complete receiver chains (pre-amplifier, mixer, A/D converters).

1. The transform techniques analysed in this chapter are based on **equispaced linear** (or planar rectangular) antenna arrays.

2. A variety of **linear subspace transforms** is possible:

   - Overlapping uniform subarrays, altogether steered in the look direction, preserve the original spacing, that is, the phase centres of the subarrays are at the same distance as the sensors. Practical implementation in the RF domain may be difficult. The uniform size of subarrays means that the clutter spectra received by the individual subarrays are equal.

   - Increasing the subarray displacement beyond $\lambda/2$ may result in losses in SCNR.

- Non-uniform subarrays receive different clutter Doppler spectra. This may lead to some losses in SCNR. In the examples shown the losses were moderate.
- The symmetric auxiliary sensor configuration (beamformer + auxiliary sensors symmetrically arranged on both sides) is equivalent to the overlapping subarray concept. The advantage is that a beam is formed for the central elements which can easily be realised in the RF domain.
- Asymmetric auxiliary sensor configurations (beamformer + auxiliary sensors on one side) are suboptimum. Losses in SCNR may occur.
- Losses which occur due to non-uniform subarrays or asymmetric auxiliary configuration can be compensated for by increasing the temporal number of degrees of freedom of the space-time processor. In this sense space-frequency (or post-Doppler) approaches (Chapter 9) may be tolerant against non-uniform spaced arrays and unequal subarrays.

3. The **clutter rejection performance** of the *overlapping subarray* concept and the *symmetric auxiliary sensors* concept comes very close to the performance of the optimum processor.

4. The clutter rejection performance is almost independent of the **number of channels**. $K = 2$ or $K = 3$ may be sufficient. This is independent of the number of clutter eigenvalues of the clutter covariance matrix.

5. **Bandwidth effects** (system bandwidth, clutter bandwidth) have the same consequences for slow target detection as in the case of the optimum processor: the clutter notch is broadened.

6. A *spatial only* blocking matrix is not an appropriate solution for preventing signal cancellation due to signal inclusion into the clutter covariance matrix. The space-time filter would cancel the target signal in the time dimension.

7. Σ-Δ **processing** uses the sum and difference outputs of a monopulse antenna as the spatial basis for space-time processing. This is a very efficient technique because these two channels are available anyway. However, the performance is suboptimum. Moreover, the monopulse capability of the antenna is lost.

8. **CPCT** processing is an analogy of DPCA for forward looking arrays. It works well only if the number of temporal degrees of freedom is sufficiently high.

A comparison of all techniques in terms of computational complexity is presented in Chapter 16.

Chapter 7

Adaptive space-time digital filters

In Chapter 6 some transforms to reduce the signal vector space in the *spatial* dimension have been analysed. In this chapter we investigate the use of digital filters for space-time clutter suppression. In this way we try to reduce the signal vector space in the *temporal* dimension. Temporal (i.e., pulse-to-pulse) digital filters have been used for clutter rejection in ground-based radar systems. For example, the usual two-pulse or three-pulse clutter cancellers are typical temporal anti-clutter pre-filters.

An adaptive clutter filter based on the prediction error filter has been described by BÜHRING and KLEMM [75]. It has been applied to a surveillance radar with rotating antenna.

BARBAROSSA and PICARDI [33] discuss the use of predictive techniques for adaptive clutter rejection. The superposition of two types of clutter is addressed. The LUD algorithm suggested by MAO et al. [435] allows one to compute the coefficients of the prediction error filter more efficiently than by using the BURG algorithm [79, 80]. GOLDSTEIN et al. [202, 201, 203, 204] propose a multistage STAP Wiener filter with CFAR capability. Simulations (GOLDSTEIN et al. [204], GUERCI et al. [227]) and experiments with MCARM data (GUERCI et al. [225]) have demonstrated outstanding clutter rejection performance at high computational efficiency. An extension of this concept to knowledge-aided implementation has been proposed by HIEMSTRA et al. [261].

WEIPPERT et al. [704, 703] present an efficient method for calculating a large number of multistage Wiener filters required for operation with a Doppler filter bank or with multiple beams. HIMED and MICHELS [264]achieved good suppression of hot[1] and cold clutter by using 3D STAP (space-time-TIME) based on a multistage Wiener filter. JIANG et al. [281] and LI et al. [393] analyse the 'VAR filter' (a least squares FIR filter) in comparison with DPCA with respect to clutter rejection performance and target parameter estimation. It is shown that the VAR filter is superior to the DPCA technique because it is adaptive. The VAR filter has been proposed by KIM and ILTIS

---
[1] Terrain scattered jamming.

[288] to GPS receiver synchronisation in presence of jamming.

The concept of space-time pre-whitening for suppression of airborne clutter has been illustrated by WANG Z. and BAO [685]. GRIFFITHS [213] used a pre-Doppler STAP architecture with MOUNTAINTOP data. The parametric clutter rejection technique by SWINDLEHURST and PARKER [629] and the vector autoregressive (VAR) filtering technique by LIU and LI [410] as well as PARKER and SWINDLEHURST [529], and RUSS et al. [580] are strongly related to the least squares FIR filter described below. PARKER proved that the space-time AR filter performs well in non-homogeneous clutter [531].

There are close relations between the optimum processor and the space-time FIR filter. LOMBARDO [412] and in more detail in [413] has shown that for special conditions:

- sidelooking linear equispaced array
- DPCA conditions
- exponential temporal clutter correlation[2] $\rho_t(i) = \rho_t^{|i|}$
- exponential spatial clutter correlation[3] $\rho_s(i) = \rho_s^{|i|}$

the inverse of the space-time covariance matrix becomes a tridiagonal symmetric matrix whose block rows describe a space-time FIR filter with temporal and spatial filter length $L = 3, K = 3$. Under the given conditions this means a two-dimensional space-time FIR filter which operates in the time (pulse-to-pulse) domain as well as the spatial domain (across the array).[4] It can be expected that for different clutter models adaptive FIR filters approximate well the optimum processor (see WANG Z. and BAO [686]).

From experience with ground-based radar several aspects of digital filters are known:

1. In phased array radar the beam steering direction is kept constant during the detection phase. If the PRF is constant clutter echoes can be considered to be *stationary* because no effect due to inhomogeneity of the clutter background nor through irregular sampling will occur. Digital filters applied to stationary signals have constant coefficients independent of time.

2. In the case of staggered PRF the signal sequence is no longer stationary. Then the filter has to be trained anew with every PRI. Training has to be done on the basis of data coming from range. As a result one obtains a FIR filter whose coefficients vary during filtering.

3. The number of filter coefficients may be chosen independently of the Doppler filter length. This is of high importance for radar with changing modes of operation (low, medium, and high PRF).

---
[2]In contrast to the gaussian correlation assumed in our model (2.53).
[3]Which differs considerably from our models (2.61) and (2.64).
[4]Such a filter is depicted in Figure 7.17.

4. The number of coefficients of a digital clutter filter may be small compared with the Doppler filter length. This property is different from all processors treated in the previous chapters. It may cause a dramatic reduction of the instantaneous computational load per pulse repetition interval during filtering the data through all range increments. Also the amount of data required for updating the filter coefficients will be reduced.

5. The filter can operate properly only if entirely filled with data. Therefore, it must not slide over the edges of the data record. Consequently, the resulting data length after filtering is $M - L + 1$ where L is the filter length. That means, the length of the Doppler filters is shortened by $L - 1$. This shortening results in some signal loss. The trade-off between clutter filter length and signal integration length has been discussed by DILLARD [116]. It is therefore essential that the clutter filter length is small compared with the total length of the data record.

6. If the clutter echoes are stationary the covariance matrix is Toeplitz and the prediction error filter can be calculated via the LEVINSON algorithm (BURG [79, 80], BÜHRING and KLEMM [75]) from one row of the covariance matrix. That means for the adaptation that instead of a covariance matrix only a vector of covariance coefficients has to be estimated.

7.1 Least squares FIR filters

The principle of least squares FIR filtering of stationary signals and its use for clutter cancellation was briefly addressed in Section 1.2.3.2, and in [75]. In this chapter we want to extend this principle to space-time processing i.e., we have to add the spatial dimension.

7.1.1 Principle of space-time least squares FIR filters

In the following a multichannel space-time least squares FIR filter for adaptive suppression of airborne clutter is derived. It is based on the principle of prediction error filtering and is, therefore, closely connected with the maximum entropy method (BURG [79, 80]). The *parametric adaptive matched filter* (PAMF) presented by ROMAN et al. [577] and further discussed by MICHELS et al. [465] also belongs to this category of filters.

7.1.1.1 Some remarks on prediction error filters

It was shown in (1.91) that the prediction error filter is defined by

$$\begin{pmatrix} P_{t+1} & \mathbf{q}_{xx}^*(1) \\ \mathbf{q}_{xx}(1) & \mathbf{Q}_{xx}(0) \end{pmatrix} \mathbf{h}_{\mathrm{PE}} = \begin{pmatrix} P_{\mathrm{PE}} \\ 0 \\ \vdots \\ 0 \end{pmatrix} \qquad (7.1)$$

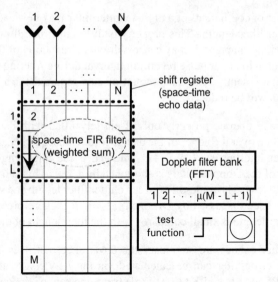

Figure 7.1: Space-time FIR filter processor

It can easily be verified that the matrix on the left side is the clutter covariance matrix of order $L+1$ so that the prediction error filter becomes

$$\mathbf{h}_{PE} = \mathbf{Q}^{-1} \begin{pmatrix} P_{PE} \\ 0 \\ \vdots \\ 0 \end{pmatrix} \qquad (7.2)$$

i.e., the prediction error filter is proportional to the first column of the inverse of the clutter + noise covariance matrix.

It should be mentioned that in this case a *future* value of the data sequence is predicted, see (1.87). The prediction filter as defined by (1.89) can also be used to estimate a value *inside* the data segment, that means prediction of a value $\hat{x}_i$ with $i = t + L - 1, \ldots, t$. Then the i-th column of $\mathbf{Q}^{-1}$ will play the role of a prediction error filter. This case is called *interpolation* rather than prediction. If a value in the past ($\hat{x}_{t-L}$) is to be predicted the last column of $\mathbf{Q}^{-1}$ represents the prediction error filter. This application is called *smoothing*.

We can conclude that each column of $\mathbf{Q}^{-1}$ represents a least squares filter, that means, a filter that minimises the output power of a stationary process defined by $\mathbf{Q}$. Therefore basically each of the columns of $\mathbf{Q}^{-1}$ can be used as a clutter rejection filter. In the following we use the first column of $\mathbf{Q}^{-1}$ while keeping in mind that the others would work in a similar way. Since the outer columns are based on more correlation values than the inner columns the width of the clutter notch will be narrower. This is in accordance with the MEM resolution achievable with outer or inner columns of the Toeplitz matrix inverse, respectively.

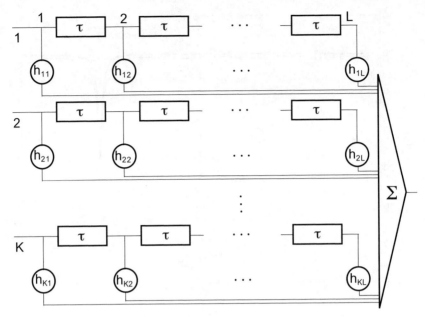

Figure 7.2: Detailed block diagram of the space-time FIR filter

7.1.1.2 Extension to space-time radar data

The extension of the prediction error filter to space-time filtering is straightforward. The space-time clutter + noise covariance matrix of a data segment of length L has the form (compare with (3.22))

$$\mathbf{Q} = \begin{pmatrix} \mathbf{Q}_{11} & \mathbf{Q}_{12} & \cdots & \mathbf{Q}_{1L} \\ \mathbf{Q}_{21} & \mathbf{Q}_{22} & \cdots & \mathbf{Q}_{2L} \\ \vdots & \vdots & \ddots & \vdots \\ \mathbf{Q}_{L1} & \mathbf{Q}_{L2} & \cdots & \mathbf{Q}_{LL} \end{pmatrix} \qquad (7.3)$$

Since we assumed that the clutter echoes are stationary in the temporal dimension the covariance matrix is block Toeplitz. Recall that the submatrices $\mathbf{Q}_{mp}$ have dimension $N \times N$, i.e., the spatial dimension of the array.

The inverse of $\mathbf{Q}$ has the same form:

$$\mathbf{K} = \mathbf{Q}^{-1} = \begin{pmatrix} \mathbf{K}_{11} & \mathbf{K}_{12} & \cdots & \mathbf{K}_{1L} \\ \mathbf{K}_{21} & \mathbf{K}_{22} & \cdots & \mathbf{K}_{2L} \\ \vdots & \vdots & \ddots & \vdots \\ \mathbf{K}_{L1} & \mathbf{K}_{L2} & \cdots & \mathbf{K}_{LL} \end{pmatrix} \qquad (7.4)$$

In analogy with the one-dimensional prediction error filters described by (1.91), a space-time FIR filter operating in the *temporal* dimension only is a matrix given by

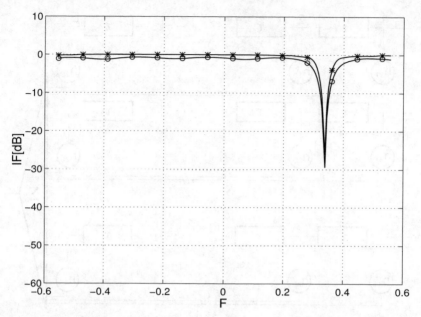

Figure 7.3: The FIR filter processor (full array, FL): ∘ *FIR filter processor* ($L = 5$); ∗ *fully adaptive optimum processor*

the first block column of $\mathbf{K}$:

$$\tilde{\mathbf{K}} = \mathbf{K} \begin{pmatrix} \mathbf{I} \\ \mathbf{O} \\ \vdots \\ \mathbf{O} \end{pmatrix} = \begin{pmatrix} \mathbf{K}_{11} \\ \mathbf{K}_{21} \\ \vdots \\ \mathbf{K}_{L1} \end{pmatrix} \quad (7.5)$$

The filtering operation can be formulated as follows

$$\mathbf{y}(\tau) = \tilde{\mathbf{K}}^* \mathbf{x}(\tau) \quad (7.6)$$

where the $\mathbf{x}(\tau)$ are vectors containing segments of the data sequence shifted by τ

$$\mathbf{x}(\tau) = \begin{pmatrix} \mathbf{x}_{1+\tau} \\ \mathbf{x}_{2+\tau} \\ \vdots \\ \mathbf{x}_{L+\tau} \end{pmatrix} \quad \tau = 0 \ldots M - L \quad (7.7)$$

where the dimension of the subvectors is N.

Including all subvectors of $\mathbf{x}(\tau)$ in a matrix $\mathbf{X}$ leads to

$$\mathbf{Y} = \tilde{\mathbf{K}}^* \mathbf{X} = \tilde{\mathbf{K}}^* \begin{pmatrix} \mathbf{x}_1 & \mathbf{x}_2 & \mathbf{x}_3 & \ldots & \mathbf{x}_{M-L+1} \\ \mathbf{x}_2 & \mathbf{x}_3 & \mathbf{x}_4 & \ldots & \mathbf{x}_{M-L+2} \\ \vdots & \vdots & \vdots & \ldots & \vdots \\ \mathbf{x}_L & \mathbf{x}_{L+1} & \mathbf{x}_{L+2} & \ldots & \mathbf{x}_M \end{pmatrix} \quad (7.8)$$

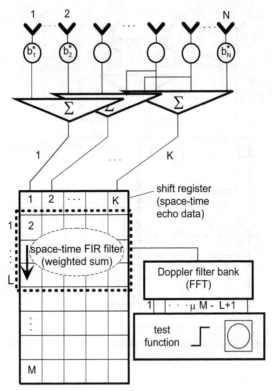

Figure 7.4: Overlapping subarray processor with space-time FIR filter

The filtered output data matrix $\mathbf{Y}$ has the spatial dimension $K \times (M - L + 1)$, i.e., the spatial dimension which is either given by the array size $K = N$ or by any of the pre-transforms described in Chapter 6.

Equation (7.8) describes the space-time data $\tilde{\mathbf{K}}$ moving over the filter matrix. An equivalent version of this equation can be obtained by moving the filter matrix relative to the data:

$$\mathbf{y} = \tilde{\mathbf{H}}^* \begin{pmatrix} \mathbf{x}_1 \\ \mathbf{x}_2 \\ \vdots \\ \mathbf{x}_M \end{pmatrix} \tag{7.9}$$

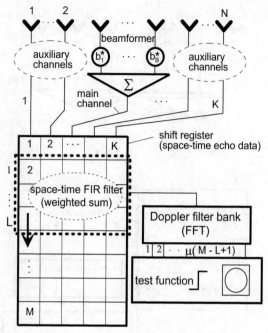

Figure 7.5: *Symmetric auxiliary sensor processor with space-time FIR filter (ASFF)*

where

$$\tilde{\mathbf{H}}^* = \begin{pmatrix} \mathbf{K}_{11}^* & \mathbf{K}_{21}^* & \cdots & \mathbf{K}_{L1}^* & & & & \\ & \mathbf{K}_{11}^* & \mathbf{K}_{21}^* & \cdots & \mathbf{K}_{L1}^* & & \mathbf{0} & \\ & & \mathbf{K}_{11}^* & \mathbf{K}_{21}^* & \cdots & \mathbf{K}_{L1}^* & & \\ & \mathbf{0} & & & \ddots & & & \\ & & & & & \mathbf{K}_{11}^* & \mathbf{K}_{21}^* & \cdots & \mathbf{K}_{L1}^* \end{pmatrix}$$
(7.10)

The submatrices $\mathbf{K}_{m1}^*$ have been defined in (7.4). The filter operator $\tilde{\mathbf{H}}^*$ has dimension $N(M-L+1) \times NM$ and the dimension of the vector of output signals $\mathbf{y}$ is $N(M-L+1)$.

7.1.1.3 Pre-beamforming

The filter operation (7.8) or (7.9) has to be followed by signal match in both the spatial and Doppler dimension. By spatial matching the spatial dimension is removed. This is done by multiplying the filtered data matrix with a beamformer vector

$$\mathbf{z} = \mathbf{Y}^*\mathbf{b} \tag{7.11}$$

The same result is obtained by pre-multiplying the filter matrix with the beamformer

$$\mathbf{h} = \tilde{\mathbf{K}}\mathbf{b} \tag{7.12}$$

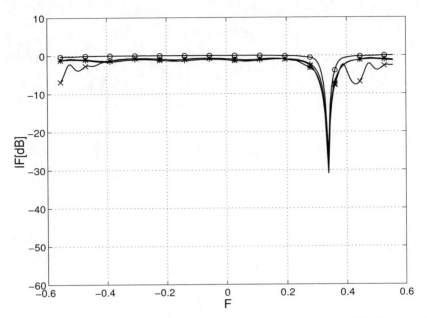

Figure 7.6: Space-time FIR filtering and spatial transforms (FL, $L = 5$): ○ optimum processor; ∗ symmetric auxiliary sensors $K = 5$; × disjoint subgroups $K = 6$; + overlapping subarrays $K = 5$

so that the space-time FIR filter is just a vector with dimension KL. This filter was first proposed by KLEMM and ENDER [339, 341, 342] and later on by BARANOSKI [31]. A mathematical derivation of this processor has been given by JAFFER [277]. The numerical results presented in [277] agree well with those given below.

The filtering operation is given by

$$z^* = b^* \tilde{K}^* X = h^* \begin{pmatrix} x_1 & x_2 & x_3 & \cdots & x_{M-L+1} \\ x_2 & x_3 & x_4 & \cdots & x_{M-L+2} \\ \vdots & \vdots & \vdots & \cdots & \vdots \\ x_L & x_{L+1} & x_{L+2} & \cdots & x_M \end{pmatrix} \qquad (7.13)$$

The vector z contains the output signal sequence which has length $M - L + 1$. Usually the clutter filter is followed by a Doppler filter bank which has to be applied to the shortened data sequence

$$\mathbf{f} = \mathbf{Fz} \qquad (7.14)$$

where $\mathbf{F}$ is a matrix whose rows are Doppler filter vectors according to (2.33). In practice the Doppler filter bank is usually a DFT or FFT. The elements of $\mathbf{f}$ are fed into a detection device.

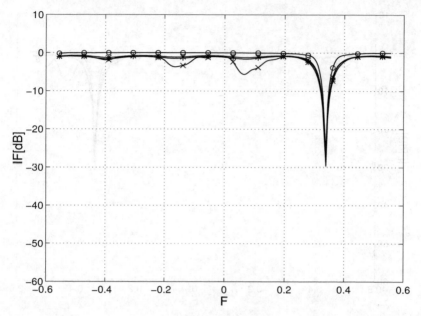

Figure 7.7: Space-time FIR filtering and spatial transforms (SL, $L = 5$): ○ *optimum processor;* * *symmetric auxiliary sensors $K = 5$;* × *disjoint subgroups $K = 6$;* + *overlapping subarrays $K = 5$*

7.1.1.4 Improvement factor

The efficiency of the filter technique described above can be judged by calculating the improvement factor as before. An equivalent way of formulating the FIR filter operation is as follows

$$z^* = H^* x = H^* \begin{pmatrix} x_1 \\ x_2 \\ \vdots \\ x_M \end{pmatrix} \quad (7.15)$$

where

$$H^* = \begin{pmatrix} h_1^* & h_2^* & \cdots & h_L^* & & & & 0 \\ & h_1^* & h_2^* & \cdots & h_L^* & & & \\ & & h_1^* & h_2^* & \cdots & h_L^* & & \\ & & & & & \ddots & & \\ 0 & & & & & h_1^* & h_2^* & \cdots & h_L^* \end{pmatrix} \quad (7.16)$$

Notice that the subvectors h_m of h have the spatial dimension N, with N being the number of antenna elements. The matrix H contains replicas of the filter vector h that are shifted by N between any two rows of H.

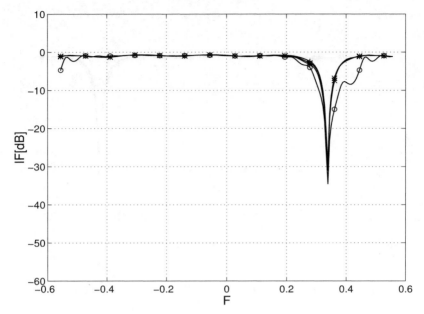

Figure 7.8: Effect of spatial filter length (ASFF, FL, $L = 5$): ○ $K = 3$; ∗ $K = 5$; × $K = 7$; + $K = 9$

Then the improvement factor becomes in accordance with (1.93),

$$\text{IF}(\omega_D) = \frac{\mathbf{s}^*(\omega_D)\mathbf{HH}^*\mathbf{s}(\omega_D)\mathbf{s}^*(\omega_D)\mathbf{HH}^*\mathbf{s}(\omega_D) \cdot \text{tr}(\mathbf{Q})}{\mathbf{s}^*(\omega_D)\mathbf{HH}^*\mathbf{QHH}^*\mathbf{s}(\omega_D) \cdot \mathbf{s}^*(\omega_D)\mathbf{s}(\omega_D)} \tag{7.17}$$

where we assumed once again that the processor is matched to the expected target signal vector $\mathbf{s}(\omega_D)$. Notice that the filter matrix $\mathbf{H}$ always appears twice, once for whitening the received signal, and once for equalising the target signal reference (or matched filter). This is in accordance with (1.71) and Figure 1.7b.

7.1.1.5 Filter length and eigenvalues

From the considerations in Chapter 3 we know that under ideal conditions (PRF = Nyquist frequency for clutter bandwidth, no additional bandwidth effects, clutter echoes received from one semiplane only) the number of eigenvalues of $\mathbf{Q}$ is approximately equal to $N + M - 1$, see Figures 3.12 and 3.13 and equation (3.63).

The length of the FIR filter L has not yet been determined. It would be desirable to choose $L \ll M$ in order to save arithmetic operations. This is of particular importance because the filtering operation has to be conducted in real-time for all range increments. How does a short filter length L cope with the number of eigenvalues of $\mathbf{Q}$ determined by M? Does the reduction M down to L cause lack of degrees of freedom? There are two ways to look at this problem:

- The FIR filter is calculated from a $NL \times NL$ space-time covariance matrix according to (7.12), where L is the length of a N-channel data segment of the

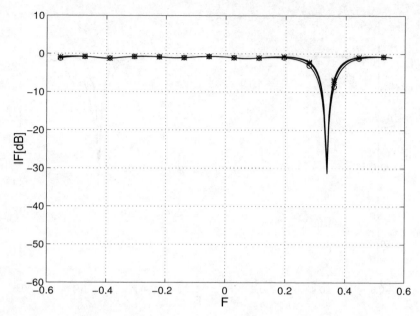

Figure 7.9: Effect of spatial filter length (ASFF, SL, $L = 5$): ○ $K = 3$; ∗ $K = 5$; × $K = 7$; + $K = 9$

received echo data sequence of length M. While sliding over the data sequence the filter operates at a certain instant of time only on an instantaneous data segment of length L. Opposite to all processing schemes described in Chapters 4–6 the FIR filter does not have to cope simultaneously with the complete data sequence of length M. Therefore, there is no lack of degrees of freedom which otherwise would cause serious degradation of the clutter rejection performance.

- The full filter operation on the data sequence of length M has been described in (7.15) by a matrix $\mathbf{H}$. The column rank of $\mathbf{H}$ is $N(M - L + 1)$. Since $N(M - L + 1)$ is larger than the number of eigenvalues $N + M - 1$ if L is sufficiently small no lack of degrees of freedom has to be taken into account.

7.1.2 Full antenna array

Let us start with the performance of a space-time FIR filter processor that uses all output signals of an array antenna. A block diagram is depicted in Figure 7.1. The N output signals of the antenna channels are stored in a $N \times M$ memory. The FIR filter is illustrated by a dashed rectangle. Each of the data inside this rectangle is weighted and summed up according to (7.13). The filter moves through the memory as indicated by the arrow. If the shift inherent in the filter operation is synchronised with the PRF, i.e., the data flow is pipelined, then a shift register of size $N \times L$ may replace the $N \times M$ memory.[5] A more detailed view of the space-time FIR filter is given in Figure 7.2. In

---
[5] Of course, this operation has to be carried out for all range increments.

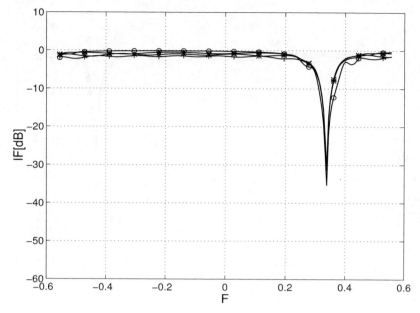

Figure 7.10: Effect of temporal filter length (ASFF, $K = 5$, FL): ○ $L = 2$; ∗ $L = 4$; × $L = 6$; + $L = 8$

comparison with Figure 7.1 we replaced N with K so as to suggest that such filter can as well be used after a spatial transform as treated in Chapter 6.

The numerical example in Figure 7.3 shows a comparison of the FIR filter processor with the optimum processor. As before the plot shows the normalised improvement factor in the signal-to-clutter + noise ratio versus the normalised target Doppler frequency. The IF is normalised by its theoretical maximum while the target Doppler frequency is normalised by half the PRF.

It can be noticed that the IF achieved by the FIR filter processor runs closely to the optimum curve. The difference between both curves of about 1–2 dB is caused by the loss in target signal energy due to shortening of the data sequence by the FIR filter. Recall that the final data length after FIR filtering is $M - L + 1$ instead of M because the FIR filter should work only *inside* the data record. The correct design of the filters length is a compromise between losses in signal power due to reduced integration time and losses due to imperfect clutter cancellation. If the data length M is large compared with the temporal filter length L the FIR filter processor approximates the optimum sufficiently.

PARK *et al.* [527, 528] optimise a linear predictor under the constraint that the target Doppler frequency is unknown. Hence, the Doppler part of the signal covariance matrix is a unit matrix. The clutter cancellation performance achieved shows considerable degradation compared with the optimum processor.

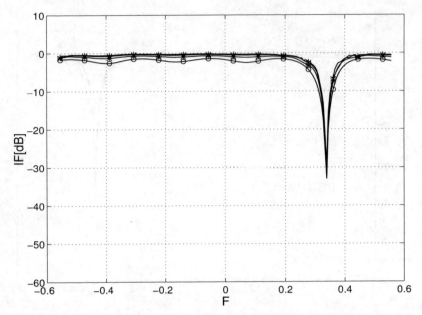

Figure 7.11: Effect of Doppler filter length (symmetric auxiliary sensors, FL, $K = L = 5$): ○ $M = 12$; ∗ $M = 24$; × $M = 48$; + $M = 96$

7.1.3 Spatial transforms and FIR filtering

Now we combine the spatial transforms treated in Chapter 6 with the concept of a space-time FIR filter. In this way two different kinds of simplification of the processor scheme (spatial and temporal) are achieved. The mathematical description of such pre-transform/FIR filter processors is in essence the same as above. All vectors and matrices have to be replaced by quantities transformed with a spatial transform according to (6.1), and the number of antenna elements N has to be replaced by the reduced number of channels K.

7.1.3.1 *Uniform overlapping subarrays*

Figure 7.4 shows a block diagram for a processor in which the overlapping subarray configuration has been combined with a FIR filter processor. The subarray transform was given by (6.1), with the spatial submatrices being defined by (6.2).

7.1.3.2 *Symmetric auxiliary sensors*

The symmetric auxiliary sensor configuration has been described by (6.1), where the spatial submatrices have the form (6.17). A block diagram of an auxiliary sensor FIR filter processor (ASFF) is given in Figure 7.5.

This processor has two beneficial properties. On the one hand the spatial transform (beamforming of centre elements) can be realised in the RF-domain as was noted in

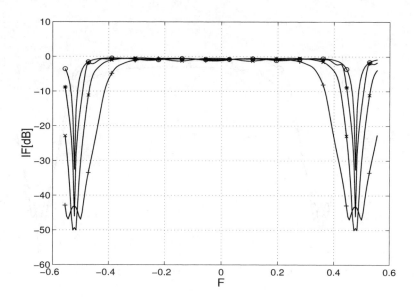

*Figure 7.12: STAP FIR filter, effect of internal clutter motion (FL, $\varphi_L = 0°$): ○ $B_c = 0$; * $B_c = 0.01$; × $B_c = 0.1$; + $B_c = 0.3$*

the previous chapter. On the other hand, the product

$$\mathbf{h} = \tilde{\mathbf{K}}\mathbf{b}$$

(7.12), which incorporates the beamformer into the space-time FIR filter, can be omitted for the auxiliary sensor processor and can be replaced by just selecting the column of $\tilde{\mathbf{K}}$ which belongs to the beamformer. The simplified processor becomes

$$\mathbf{h} = \tilde{\mathbf{K}}\mathbf{e}_{(K+1)/2} \qquad (7.18)$$

where $\mathbf{e}_{(K+1)/2}$ is the centre column of the $K \times K$ unit matrix.

7.1.3.3 Disjoint subarrays

The matrix formulation for a processor with disjoint subarrays was again given by (6.1), but with submatrices of the form given by (6.10). The corresponding FIR filter processor is similar to the one shown in Figure 7.4 but with the antenna array subdivided into non-overlapping subarrays.

7.1.3.4 Comparison of processors

The performance of the three processors is compared with the optimum processor (Chapter 4) in Figure 7.6, for a linear array in the forward looking configuration, and in Figure 7.7, for a sidelooking array. The filter length (data segment) was chosen to be

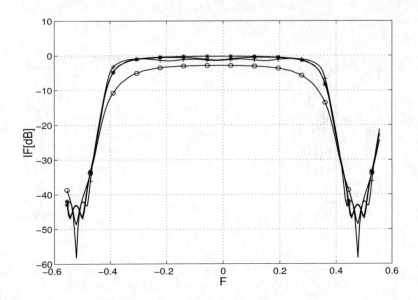

Figure 7.13: STAP FIR filter, filter length and clutter fluctuations (FL, $\varphi_L = 0°$): ○ $L = 2$; * $L = 3$; × $L = 5$; + $L = 7$

$L = 5$ while the total length of the data record is $M = 24$. The length of the resulting data sequence is then $M - L + 1 = 20$.

The following observations can be made:

- The auxiliary sensor/FIR filter and the overlapping subarray/FIR filter approximate the optimum IF very well.

- The difference in IF between the optimum and the two suboptimum curves is due to the shortening of the data record caused by the FIR filter of length L ($M - L + 1 = 20$). Therefore, the target signal energy is reduced by a factor of $20/24$.

- The disjoint subarray/FIR filter processor shows some losses at those frequencies which are associated with the maximum of the sensor directivity pattern (F = 0.5 for the forward looking array, F = 0 for the sidelooking array). These losses are due to spatial ambiguities caused by the displacement of the subarray phase centres. This effect was already observed in Figures 6.3 and 6.4.

7.1.3.5 Computational complexity

At this point we arrive at a space-time processing scheme which by virtue of a spatial transform combined with a space-time FIR filter offers a dramatic reduction in arithmetic operations while the performance is very close to the optimum. The amount of computations required for the various tasks associated with clutter rejection

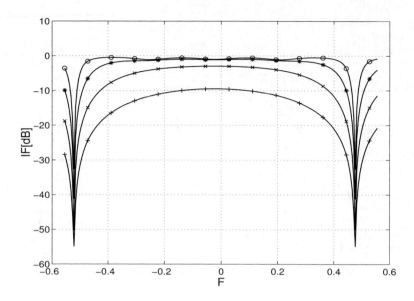

Figure 7.14: STAP FIR filter, effect of range walk (FL, $\varphi_L = 0°$): ○ no range walk; ∗ $\Delta_R = 100\ m$; × $\Delta_R = 10\ m$; + $\Delta_R = 1\ m$

Table 7.1 Comparison of OAP and ASFF

Processor	Adaptation	Inversion	Filtering/range increment
OAP	$2N^3M^2$	$8N^3M$	$(NM)^2$
ASFF	$2(K^3L^2)$	$8K^3L$	KL/PRI, $KL(M-L+1)$ total

is given in Tables 7.1 and 7.2. As can be seen the saving in arithmetic operations is extraordinary.

It should be noted that filtering has to be carried out for all range increments and is, therefore, the most critical operation. It is remarkable how little arithmetic effort is required for the FIR filter to accomplish this task. Notice that the total memory size required is KLR if the filter operation is synchronised with the PRF (R is the number of range increments). For the calculation of the FIR filter coefficients we assumed that the filter is calculated via the Akaike algorithm for block Toeplitz matrix inversion (see AKAIKE [12]). A comparison of the computational load of all techniques discussed in this book is given in Tables 12.1 – 12.4.

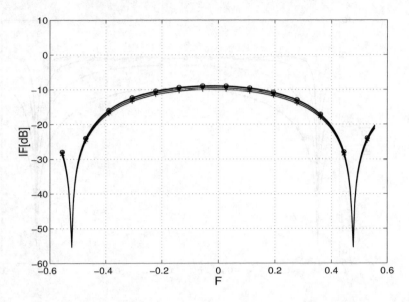

Figure 7.15: STAP FIR filter, filter length and range walk (1 m range resolution, FL, $\varphi_L = 0°$): $\circ\ L = 2$; $*\ L = 3$; $\times\ L = 5$; $+\ L = 7$

Table 7.2 Numerical example: $N = M = 24$, $K = L = 5$

Processor	Adaptation	Inversion	Filtering/range increment
OAP	$16 \cdot 10^6$	$2.6 \cdot 10^6$	331776
ASFF	6250	5000	25/PRI, 500 total

7.2 Impact of radar parameters

7.2.1 Sample size

In this section we briefly discuss the effect of spatial and temporal sample size, i.e., antenna aperture and length of the echo sequence, on the performance of the symmetric auxiliary FIR filter processor.

7.2.1.1 *Spatial filter dimension*

The influence of the number of channels K of the symmetric auxiliary sensor processor was already addressed in Chapter 6, Section 6.2.1, Figures 6.10 and 6.11. It was found that the performance of the processor was almost independent of the number of antenna channels, which is a highly desirable property. This means, in contrast to the processors discussed in Chapter 5, that the spatial dimension K of the space-time processor can be chosen independently of the antenna array size N.

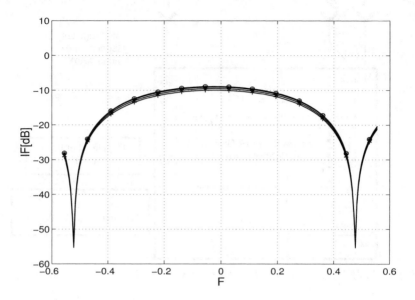

Figure 7.16: Symmetric auxiliary sensor processor with STAP FIR filter, effect of filter length (1 m range resolution, FL, $\varphi_L = 0°$, $K = 5$): ○ $L = 2$; ∗ $L = 3$; × $L = 5$; + $L = 7$

The two subsequent figures are to confirm that this property is preserved even if the pre-transform is followed by a suboptimum FIR filter instead of a fully adaptive processor. Figure 7.8 shows IF curves for the forward looking linear array, Figure 7.9, for the sidelooking configuration. As can be seen the sidelooking array is almost independent of the number of antenna channels.

For the forward looking array we encounter some loss close to the clutter notch for $K = 3$ which is the minimum possible number of channels for this kind of processor. As was already discussed in Section 6.1.1.2, these losses have to do with the fact that for a forward looking antenna configuration there are always two clutter arrivals for each Doppler frequency symmetrical to $\varphi = 0°$. Since the forward looking array can distinguish between left and right of the flight direction these symmetric clutter arrivals determine the number of degrees of freedom required. For two arrivals (for each Doppler frequency) the choice of $K = 3$ channels is the absolute minimum. As can be seen for $K = 5$ we obtain an almost optimum IF curve.

This problem does not occur for a sidelooking array. A sidelooking array cannot distinguish between left and right of the flight direction and therefore perceives both symmetric arrivals due to one Doppler frequency as *one* arrival only. Therefore, even $K = 3$ leads to satisfactory performance.

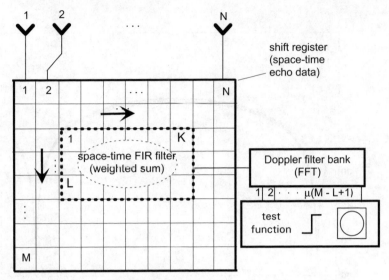

Figure 7.17: 2-D space-time FIR filter processor

7.2.1.2 Temporal filter dimension

In Figure 7.10 the dependence of the IF on the temporal dimension L of the space-time FIR filter is illustrated. A forward looking array was assumed. It can be noticed that the performance is almost independent of the filter length. Some slight losses can be recognised in the pass band that are due to the shortening of the data record through the FIR filter. Recall that the filtered data record has length $M - L + 1$. Shortening of the data record leads to losses in target signal energy. The shorter the filter is the closer the IF curve runs along the theoretical optimum. The uppermost curve has been calculated for $L = 2$. However, $L = 2$ shows some slight deviations from the other curves in the clutter notch region.

The results obtained for a sidelooking array are very similar and have therefore been omitted.

7.2.1.3 Doppler filter length

Practical radar systems have different operational modes which may imply different signal integration lengths or, equivalently, different Doppler resolution. The length of the Doppler filter bank may vary considerably. The structure of all processors discussed in the previous chapters depends heavily on the length of the Doppler bank.

In Figure 7.11 the clutter filter dimension was kept constant ($L = 5$) while the Doppler filter length was varied. As the IF is proportional to the Doppler filter length each curve has been normalised to its individual maximum. As can be seen, the longer the Doppler filter is, the closer the IF approaches the maximum. For $M \gg L$ the shortened length $M - L + 1$ approaches M. Furthermore, it can be noticed that the

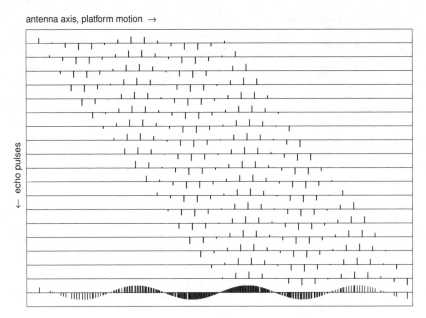

Figure 7.18: Projection of space-time echo samples onto a common time axis: clutter

width of the clutter notch is practically independent of the Doppler filter length. Some slight losses can be recognised for very short Doppler filter lengths (in our example $M = 8$ which means that the effective Doppler filter length is only $M - L + 1 = 4$).

The fact that the temporal and spatial dimensions of the auxiliary sensor FIR filter are almost independent of the antenna size as well as of the Doppler filter length makes this processor very attractive. The filter operation requires KL operations per PRI and range gate which can easily be accomplished in real-time with current digital technology.

7.2.2 Decorrelation effects

7.2.2.1 *System bandwidth*

As explained in Chapter 2, Section 2.5.2, the system bandwidth causes *spatial* (i.e., sensor-to-sensor) decorrelation of clutter arrivals. This decorrelation leads to an increase of the number of eigenvalues of the space-time clutter covariance matrix. Equivalently, we encounter a broadening of the clutter spectrum, see for example Figure 3.35, and a broadening of the clutter notch of the optimum clutter filter, see Figure 4.23.

Similar results have been obtained for the suboptimum symmetric auxiliary sensor processor, see Figure 6.15. It should be emphasised that in the pass band no losses occur, which indicates that the number of degrees of freedom of the reduced vector space is still sufficient to cope with the increased number of clutter eigenvalues.

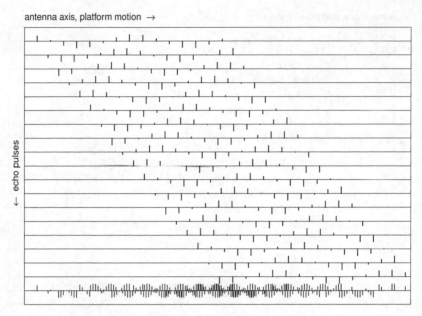

Figure 7.19: Projection of space-time echo samples onto a common time axis: Doppler target

The spatial side of the FIR filter processors discussed in this chapter is the same as described in Chapters 4 and 6. Therefore, we can expect that the results illustrated by the above-quoted figures apply as well to the FIR filter processors of this chapter. A numerical evaluation has, therefore, been omitted.

7.2.2.2   Clutter bandwidth

Clutter bandwidth owing to internal clutter motion causes a *temporal* decorrelation of subsequent clutter echoes which can also be expressed by an increase in the number of clutter eigenvalues of the space-time clutter covariance matrix (see for example Figure 3.25). The processors discussed in this chapter are characterised by their *short* temporal filter dimension which has been obtained by exploiting the temporal stationarity of clutter echoes. Therefore some effect of the clutter bandwidth (losses due to a lack of degrees of freedom) on the performance of this kind of processor can be expected.

The question to be answered is how far the increase of the number of clutter eigenvalues leads to a degradation in the clutter suppression performance of the space-time FIR filter. It should be kept in mind that normally an increase in the number of degrees of freedom in the interference covariance matrix can be compensated for by an increase in degrees of freedom of the clutter filter. Since we are talking about the temporal dimension of the problem (temporal decorrelation versus filter length) we can expect that the filter length for obtaining the best possible clutter rejection has to be increased.

Figure 7.12 shows the improvement factor versus the normalised Doppler frequency for different clutter bandwidths, corresponding to different strengths of internal clutter motion. As can be seen the clutter notch is broadened according to the clutter bandwidth. In the pass band, however, the optimum improvement is almost reached. There is a slight ripple in the pass band which is typical for space-time FIR filters. The loss compared with the theoretical optimum (0 dB) of the upper curve is due to the fact that the coherent signal integration is shortened by $L - 1$ because the FIR filter must not move outside the echo sequence.

Figure 7.13 shows a comparison of different filter lengths of the space-time FIR filter when applied to clutter with strong internal motion ($B_c = 0.3$). The broadened clutter notch can be recognised clearly. For $L = 2$ some losses in the pass band can be noticed. For larger values of L the improvement factor becomes nearly independent of the filter length.

We refer here to the results presented by LIU and PENG [409] who analysed such 'short-time processing' schemes. Their results indicate that the clutter notch of FIR filter processors is broadened by the clutter bandwidth. However, there is no effect in the remaining pass band due to a lack of temporal degrees of freedom of the clutter filter. This is confirmed by our results (KLEMM [329]).

7.2.2.3 Range walk

Range walk is another temporal decorrelation effect (see Section 2.5.1.2). Figure 7.14 shows the effect of range walk on the clutter rejection performance of the space-time adaptive FIR filter. As pointed out earlier the forward looking case is the configuration which is most sensitive to decorrelation by range walk. Therefore, we focus in the following on this case. The curves shown in Figure 7.14 are very similar to those of the optimum processor as shown in Figure 4.20 except for pass band effects which have been explained before (loss in signal integration due to shortening of the data record from M down to $M - L + 1$ by the FIR filter).

For strong range walk (1 m resolution) it is even more obvious that the filter length does not really play a significant role (Figure 7.15).

7.2.2.4 Spatial transforms and space-time FIR filters

Both the effects of internal motion as well as range walk are temporal by nature. We can expect that reducing the number of spatial degrees of freedom (as shown in the block diagrams Figures 7.5 and 7.4) of the processor will not have any influence on the performance of the FIR filter because the FIR filter length is temporal. This expectation is verified by the results shown in Figure 7.16 (calculated for a processor according to Figure 7.4) which look identical to the curves in Figure 7.15. For the processor the number of antenna channels was chosen to be $K = 5$ (one main beam and four auxiliary elements).

It should be noted that this processor has 5×5 adaptive coefficients (the number could be reduced even further) whereas the inverse covariance matrix in the optimum processor (1.3) would include $(24 \cdot 24) \times (24 \cdot 24) = 576 \times 576$ elements without offering significant advantages in performance.

7.2.3 Depth of the clutter notch

The optimum processor is strongly related to the minimum variance (MV) spectral estimator, see Chapters 1 and 4. In a similar way the least squares FIR filter is related to the maximum entropy method (MEM). It is known that the MV technique produces lower spectral peaks than the MEM. Inversely, the underlying least squares FIR filter is supposed to produce deeper nulls than the optimum processor. Deep nulls may be an advantage in heterogeneous clutter environments and in presence of strong clutter discretes. BÜRGER et al. [78] have shown that this is not fulfilled in the case of limited sample size.

7.2.4 Computation of the filter coefficients

The use of digital space-time filters is justified by the *temporal* stationarity of echo sequences. Because of the temporal stationarity space-time covariance matrices are block Toeplitz. An effective recursive algorithm for block Toeplitz matrix inversion has been presented by AKAIKE [12]. The Akaike algorithm has been extended by GOVER and BARNETT [209] to block Toeplitz matrices that are not strongly non-singular. This variant may be useful in cases where the clutter-to-noise ratio is large.

7.3 Other filter techniques

7.3.1 FIR filters for spatial and temporal dimension

In the case of a linear or rectangular equidistant antenna array the FIR filter principle so far used in the temporal dimension can be used as well in the spatial domain. In this case the FIR filter moves in both the temporal and spatial domain. This can be considered an alternative way of reducing the instantaneous vector space in the spatial dimension (KLEMM [341]).

The space-time subarray technique described by PILLAI et al. [540] is strongly related to two-dimensional space-time FIR filtering. It is shown that by forming subarrays the required sample support can be significantly reduced. In fact, to adapt a FIR filter with KL coefficients requires about $2KL$ data vectors which is in general much less than $2NM$ samples.[6]

Figure 7.17 shows a block diagram of a space-time FIR filter processor, with the FIR filter operating in both the temporal and spatial domain. This requires a form of the FIR filter with no beamformer incorporated because the beamforming has to take place after moving the FIR filter over the antenna aperture.

In comparison with the other processors discussed in this chapter the concept of a 2-D FIR filter processor has some disadvantages:

- The beamforming operation should be done simultaneously in order not to waste operation time. For temporal processing FIR filtering is particularly attractive because the data flow is determined by the PRF. In this case FIR filtering can

---

[6]For the number of data samples required for adaptation see REED et al. [557].

be carried out by pipelining the data. This option does not exist for the antenna array.

- Spatial FIR filtering is restricted to linear or rectangular equidistant arrays. Many operational antennas are circular or elliptical.

- The array has to be fully digitised. This is expensive and should be avoided.

- Since the FIR filter dimensions are small compared with the total number of space-time samples (array elements, echo pulses) a lot of remaining data are available for estimation of the small space-time covariance matrix. The data of one echo snapshot may be sufficient.

7.3.2 The projection technique

In this section we describe briefly a space-time filtering technique which makes use of strict synchronisation of PRF, displacement of the sensors of a linear array, and the flight velocity. This idea has been developed by ENDER and KLEMM [133]. The use of this technique for detecting and imaging moving targets with SAR has been proposed by ZHOU and FENG [750].

The technique is based on the following conditions:

- A linear array moves at constant speed in the direction of the array axis (sidelooking configuration).

- The sensor spacing is subdivided into N subintervals.

- The PRF is adjusted to the platform velocity so that the spatial advance of the array's position between any two echoes is an entire multiple of half[7] the subinterval.

- The number of subintervals between two neighbouring sensors (identical to the number of sensors) and the number of half-subintervals travelled by the array during one PRI have to be mutually prime.

Under these conditions all space-time echo samples can be projected onto a common time (or space) axis in such a way that an *equally spaced* signal sequence is generated. This principle is illustrated in Figure 7.18. The various lines show the echo samples of an array (horizontal) at different instants of time (vertical). Only one component of the complex signal is shown.

The signal in Figure 7.18 is a clutter echo coming from a point reflector at a certain angle relative to the array. Notice the spatial frequency which is a measure for the angle of arrival. After each PRI (from line to line) the array has moved by a certain amount. We chose PRF and the velocity such that the array moves 13 out of $N = 15$ half-subintervals. Interpreting the half-subintervals in terms of phase (or time shift) leads to a shift pattern as given in Figure 7.18. Notice that the sine wave is fixed in space while the array is moving.

---

[7] Due to two-way propagation.

If all of the space-time samples as depicted on the various lines are projected on one common time axis then we obtain an oversampled sine wave as can be recognised in the centre of the lower line. The continuous sine wave is obtained when steady state is reached. The gaps on the left and on the right are due to transient effects owing to the limited space of the drawing. If we have realistic clutter coming from all directions we obtain an oversampled signal which is band-limited by the platform velocity.

In the case of a moving target both the sine wave and the array are moving (Figure 7.19). After projection on the common time axis (lower line) the resulting signal now contains higher-frequency components which are due to the target Doppler. Clutter rejection can easily be carried out by a simple (temporal) high pass filter operating on the projected signal sequence, with the clutter bandwidth as cut-off frequency. A low pass filter instead provides the echo components of the stationary background and may be exploited for SAR imaging.

The beauty of this technique is the simplicity of clutter filtering which is enabled just by data reordering. However, it can work properly only if the above mentioned conditions are precisely fulfilled. This is doubtful for airborne applications because the aircraft motion will perturbate the projected data sequence so that the samples of the projected signal sequence are no longer equidistant. For more details of this technique the reader is referred to ENDER and KLEMM [133].

7.3.3 Space-time IIR filters

Basically IIR (infinite impulse response) filters may be designed for space-time applications. IIR filters can be used for designing steep cut-off frequency slopes with a relatively small number of coefficients. However, IIR filters have long transient times which may degrade the clutter suppression performance close to clutter edges or at the limits of a data window.

IIR filters have been used in the past for MTI purposes in radar systems with rotating antennas. Such radar produces very long data sequences so that no transient problems occur as far as limited data size is concerned. Application to phased array radar systems seems to be inappropriate because such radar uses normally pulse bursts of limited duration.

Finally, we have seen that even very short FIR filters accomplish almost optimum clutter suppression. Short FIR filters have short impulse responses so that transient effects at clutter edges will be moderate. Furthermore, we have already taken in the above analysis the limits of the pulse burst into account in that the filter moved only inside the data record.

7.3.4 Adaptive DPCA (ADPCA)

BLUM et al. [50] propose a suboptimum techniques which they call ADPCA (adaptive DPCA).[8] This technique is similar to FIR filtering in that the echo sequence is subdivided into segments. Then space-time adaptive processing is carried out for the individual data segments. It has been shown by the authors (GU et al. [219]) that these

---

[8]This technique has been referred to by WARD [690] as *element space pre-Doppler STAP*.

suboptimum techniques may outperform the optimum (SMI) processor in the case of non-homogeneous clutter because the ADPCA converges more rapidly than the fully adaptive SMI technique.

Other pre-Doppler techniques as described by WARD [690] and improved by BARANOSKI [31] use full submatrices according to (7.3) sliding in the time dimension in the same way as the vectors in (7.16). The clutter notch produced by this technique, however, appears to be broader than the one generated by the FIR filter.

7.4 Summary

Space-time digital filters can be used to reduce the vector space of the clutter rejection function in the time dimension. By using digital filters it is assumed that the echo sequences are stationary so that the space-time covariance matrix is block Toeplitz. The FIR filter is calculated from a short segment of length L of the received data record.

1. The **space-time least squares FIR filter** is given by the first block column of the inverse of the $N \times L$ space-time covariance matrix. This matrix is a submatrix of the $N \times M$ covariance matrix of the total data record.

2. Further **simplification** of the filter can be obtained by pre-multiplying the first block column of $\mathbf{Q}^{-1}$ with a beamformer vector.

3. The **space-time FIR filter** approximates the optimum processor very well.

4. The **clutter rejection performance** is almost independent of the temporal filter length. $L = 3, \ldots, 5$ is sufficient. The filter length is independent of the number of clutter eigenvalues of the space-time clutter covariance matrix.

5. The aforementioned property is important especially when the number of coherent echoes is varied during the radar operation.

6. In conjunction with the spatial transforms the dimension clutter covariance matrix reduces to $K \times L$ which leads to very **cost efficient** clutter filters with the capability of real-time processing.

7. Temporal decorrelation effects (internal clutter motion, range walk) lead to a broadened clutter notch, however, no additional losses owing to a lack of degrees of freedom of the FIR filter can be observed.

8. **Two-dimensional FIR filters** can designed which operate in the fashion of a FIR filter in both the temporal and spatial dimension. This requires, however, that the antenna array is

    - linear or planar rectangular;
    - equispaced;
    - fully digitised.

The last requirement is the most prohibitive one with current technology. An array whose channels are all fully equipped with receive channels including A/D converters is not attractive for practical reasons (cost, power consumption, heat, etc.).

A comparison of all techniques in terms of computational complexity is presented in Chapter 16.

Chapter 8

Antenna related aspects

8.1 Introduction

In this chapter some antenna specific aspects are discussed. Chapter 6 dealt in some detail with spatial transforms to reduce the signal vector space in the spatial dimension. Some specific properties of linear arrays were identified. It was shown how to use identical subarrays to reduce the number of antenna channels in order to reduce the number of degrees of freedom of the space-time processor. The concept of overlapping subarrays led directly to the concept of symmetric auxiliary channels (Figure 6.8). Planar array configurations for near-optimum space-time MTI processing were derived (Figure 6.18). In this chapter we try to find out how these or similar techniques can be applied to realistic (also non-linear) antenna arrays.

While in the previous chapters all results were based on linear arrays, in this chapter we consider array configurations which have more aptitude to practical applications. In particular the MTI capability of circular planar arrays as used for example in aircraft nose radar will be analysed. Different subarray configurations will be compared. For instance, random subarrays have been implemented in the ELRA ground-based phased array radar system (SANDER [582]).

Other geometries such as randomly spaced arrays, conformal arrays and planar horizontal array antennas will be discussed as well. Moreover, we will consider antennas with a realistic number of sensors. Most of the results presented in earlier chapters are based on a linear array with 24 elements which is not a realistic number. The antennas used in the following will have 480 or 1204 elements.

Tapering the array aperture is an important point of interest. It is desirable to control the antenna sidelobes. In this context the question arises of how far space-time processing may increase the sidelobe level. We further analyse the synthesis of difference patterns at subarray level and the effect of space-time filtering on the sidelobes.

The second aspect is the fact that aperture tapering attenuates the sidelobe clutter spectrum. We will discuss the question of how far tapering can lead to simplified processing. In essence we compare space-time adaptive processing with conventional

MTI processing (beamformer + temporal clutter filter). The issue of tapering clutter data in both the spatial and temporal dimension and its impact on STAP performance has been discussed by SMITH and HALE [612].

8.2 Non-linear array configurations

All the previous results have been based on a very special kind of antenna array: linear arrays with equidistant sensor spacing. There are several good reasons to consider this array type. First it is the simplest array configuration which appears to be suitable for fundamental research. Secondly, as has been shown in Chapter 6, linear (or rectangular planar) arrays have certain benign properties which lead to near-optimum processor schemes operating at very low computational expense, thus making real-time operation possible.

A special class of planar arrays was already addressed in Figure 6.18. These subarray configurations were particularly optimised for airborne MTI processing in that subgrouping is done only in the *horizontal*. However, MTI is normally not the only function of an array antenna so that other functions, in particular jammer suppression, will dictate the kind of subarray. Since jammers may be located in all three dimensions (different from clutter) the subgrouping will be done in such a way that there will also be *vertical* degrees of freedom.[1] Such array configurations will be discussed in some detail in the following. In Chapter 16 we will show that vertical degrees of freedom can be used to mitigate ambiguous clutter returns.

8.2.1 Circular planar arrays

Circular planar arrays play an important role particularly in aircraft nose radars, for example in the concept of the AMSAR radar (ALBAREL *et al.* [13], GRUENER *et al.* [218]).

The number of elements is typically about one to several thousand. In the following numerical analysis we assume an array with 1204 elements. Some results on STAP with circular forward looking antenna have been reported by KLEMM [320, 321].

Obviously it does not make sense to design the *optimum* adaptive processor according to (4.1) and Figure 4.1. One ends up with covariance matrices of dimension 6000×6000 which is of no use for practical applications and even causes trouble for simulation on a general-purpose computer.

8.2.1.1 *Space-time processor with aperture tapering*

A block diagram of the space-time processor used in this analysis is shown in Figure 8.4. The N sensor outputs are followed by taper coefficients and phase shifters for beamforming. Tapering of the receive array may be required to maintain low sidelobes of the directivity pattern. No tapering is normally applied in the transmit path of the radar in order not to lose signal energy.

---

[1] Aspects of manufacturing may have an impact on subarray forming as well.

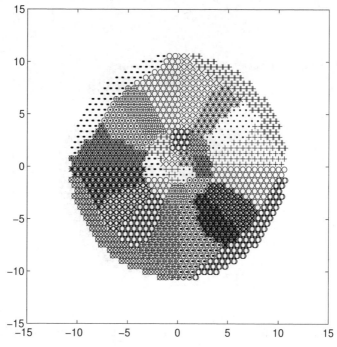

Figure 8.1: Irregular subarrays

In the next stage K subarrays are formed whose beams are steered altogether in the same direction. The subarray outputs are normalised in such a way that the white noise level is the same for all subarrays. This is necessary because the inverse of the clutter covariance matrix in the adaptive processor tries to normalise all outputs down to uniform noise levels. This, however, would just negate the tapering which is an undesirable effect (NICKEL [489]). Other technique for maintaining low sidelobes in adaptive arrays is given by colored diagonal loading (HIEMSTRA [260]).). MADURASINGHE and CAPON [423] propose a subspace technique for interference nulling while maintaining a desired sidelobe pattern.

After the noise normalisation we find the inverse of the space-time covariance matrix or the space-time adaptive FIR filter according to (7.12). The following numerical results are based on a space-time FIR filter with temporal filter length $L = 5$.

The clutter filter is followed by some additional coefficients for forming sum or difference patterns. Before entering the secondary beamformer the signals have to be weighted again with the inverse noise normalisation factors to compensate for the normalisation applied to the subarray outputs before adaptive processing. The output signal of the secondary beamformer is then fed into a Doppler filter bank according to (1.12) which in turn is connected with a detection and indication device.

The whole processing chain can be described by a linear transform as in (6.1) and

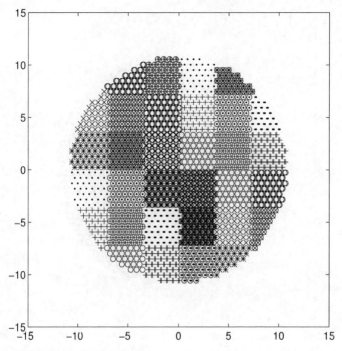

Figure 8.2: Checkerboard subarrays

(6.10). The received space-time signal vector at subarray level is given by

$$\mathbf{x}_T = \mathbf{T}^* \mathbf{x} \qquad (8.1)$$

where

$$\mathbf{T} = \begin{pmatrix} \mathbf{T}_s & & & \\ & \mathbf{T}_s & & 0 \\ & 0 & \ddots & \\ & & & \mathbf{T}_s \end{pmatrix} \qquad (8.2)$$

is the space-time transform. The spatial subvectors of the received space-time signal vector are first multiplied with taper weights, phase coefficients and normalisation factors so that the spatial transform at subarray level becomes

$$\mathbf{T}_s = \boldsymbol{\Delta}_t \mathbf{T}_{sa} \boldsymbol{\Delta}_n \qquad (8.3)$$

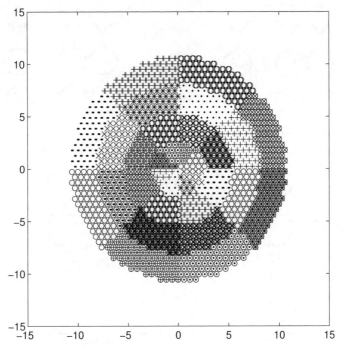

Figure 8.3: Dartboard subarrays

where

$$\mathbf{T}_{\mathrm{sa}} = \begin{pmatrix} 1 & & & & & & 0 \\ \vdots & & & & & & \\ 1 & & & & & & \\ & 1 & & & & & \\ & \vdots & & & & & \\ & 1 & & & & & \\ & & \ddots & & & & \\ & & & & 1 & & \\ & & & & \vdots & & \\ 0 & & & & 1 & & \end{pmatrix} \qquad (8.4)$$

describes the subarray forming. $\mathbf{\Delta}_{\mathrm{t}}$ is an $N \times N$ diagonal matrix containing the N taper weights and the beamsteering phase coefficients. $\mathbf{\Delta}_{\mathrm{n}}$ is a $K \times K$ diagonal matrix containing the subarray normalisation coefficients. The normalisation coefficients are given by

$$h_k = \sqrt{\sum_{i=1}^{N_{\mathrm{sa}}} w_i^2} \qquad (8.5)$$

246 Antenna related aspects

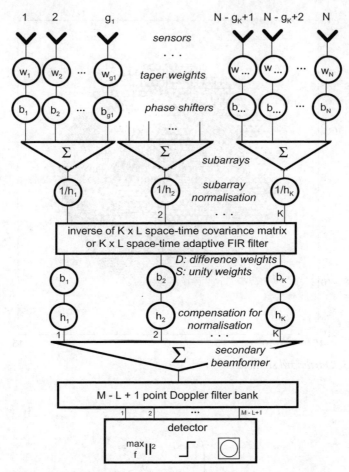

Figure 8.4: The subarray FIR filter processor with tapered antenna array

where N_{sa} is the number of antenna elements of the k-th subarray while w_i is the taper weight associated with the i-th antenna element. This normalisation provides a constant receiver noise power at all subarray outputs.

As in Chapters 5 and 6 the vector quantities of clutter, noise, target signal and received echoes become

$$\mathbf{q}_T = \mathbf{T}^*\mathbf{q}; \quad \mathbf{n}_T = \mathbf{T}^*\mathbf{n}; \quad \mathbf{s}_T = \mathbf{T}^*\mathbf{s}; \quad \mathbf{x}_T = \mathbf{T}^*\mathbf{x} \tag{8.6}$$

and the clutter covariance matrix is

$$\mathbf{Q}_T = \mathbf{T}^*\mathbf{Q}\mathbf{T} \tag{8.7}$$

where $\mathbf{T}$ is defined by (8.2), (8.4) and (8.3).

The total processor in the transformed domain then becomes

$$\mathbf{w}_T = \gamma \mathbf{Q}_T^{-1} \mathbf{\Delta}_\mathrm{n}^{-1} \mathbf{s}_T \tag{8.8}$$

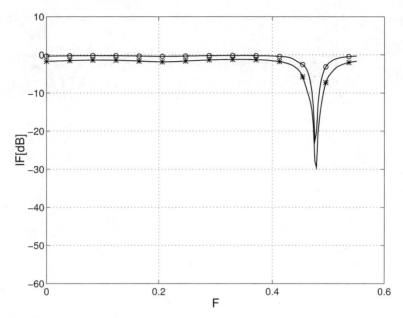

Figure 8.5: Improvement factor for irregular subarray configuration ($\varphi_L = 0°$): ○ no tapering; ∗ Taylor weighting and subarray normalisation

where $\mathbf{\Delta}_n^{-1}$ compensates for the subarray normalisation and $\mathbf{s}_T$ is the transformed signal replica including a Doppler filter and the transformed (or secondary) beamformer. In practice the Doppler filter will be replaced by a Doppler filter bank which is followed by a detection device.

The improvement factor becomes according to (1.55),

$$\mathrm{IF} = \frac{\mathbf{w}_T^* \mathbf{s}_T \mathbf{s}_T^* \mathbf{w}_T \cdot \mathrm{tr}(\mathbf{Q})}{\mathbf{w}_T^* \mathbf{Q}_T \mathbf{w}_T \cdot \mathbf{s}^* \mathbf{s}} \tag{8.9}$$

This processor has been analysed by the author [320] with respect to directivity patterns, especially for low sidelobe design. As any space-time processor has by definition spatial dimension some impact of the clutter rejection on the resulting directivity patterns can be expected. It has been shown in [320] that thanks to the subarray normalisation almost no effect on the sidelobe level can be noticed. Even for the difference patterns formed at subarray level relatively good sidelobe behaviour was achieved.

For the subsequent numerical evaluation a 40 dB Taylor weighting has been used. For Taylor array pattern synthesis for circular arrays see TAYLOR [636] and MAILLOUX [428, pp. 157–160]. The Bayliss synthesis for creating difference patterns with low sidelobes is given in BAYLISS [38] and MAILLOUX [428, pp. 160-162].

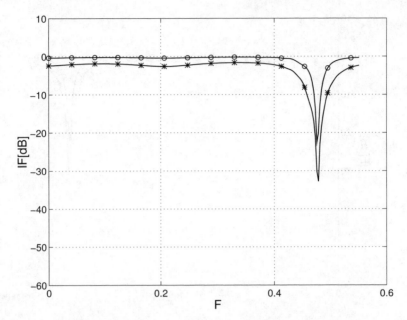

Figure 8.6: Improvement factor for checkerboard subarray configuration ($\varphi_L = 0°$): ○ *no tapering;* ∗ *Taylor weighting and subarray normalisation*

8.2.1.2 Clutter rejection performance

In Figures 8.5, 8.6 and 8.7 improvement factor curves are presented for the three different subarray configurations shown in Figures 8.1 (irregular subarrays),[2] 8.2 (checkerboard), and 8.3 (dartboard). The upper curves show the improvement factor in SCNR for the case that no tapering was applied. The arrays were assumed to be in the forward looking orientation, with look angle $\varphi = 0°$. For clutter filtering the space-time FIR filter as described in Chapter 7 was used.

First of all we notice the clutter notch which appears at nearly $F = 0.5$. The reader is reminded that the PRF is chosen so that the maximum clutter Doppler frequency is sampled at the Nyquist rate. The maximum clutter Doppler frequency occurs in the forward looking direction ($\varphi = 0°$). The slight deviation from $F = 0.5$ is due to the elevation of the radar (see Table 2.1).

The upper curves (no tapering, i.e., w_i=1 in Figure 8.4) coincide well with the theoretical optimum. There are no losses as in the case of linear arrays and disjoint subarray processing; compare with (6.3) and (6.4). The IF curves are quite independent of the subarray forming.

The lower IF curves show the case of 40 dB Taylor weighting. In the pass band some 2 dB losses have to be taken into account. This tapering loss is the penalty for achieving low sidelobes. Also, the clutter notch is slightly broadened which degrades the detectability of slow targets. Notice that the checkerboard subarray

---

[2]This subarray structure has been optimised so that the sidelobes of a difference beam formed at subarray level are minimised (NICKEL [489]).

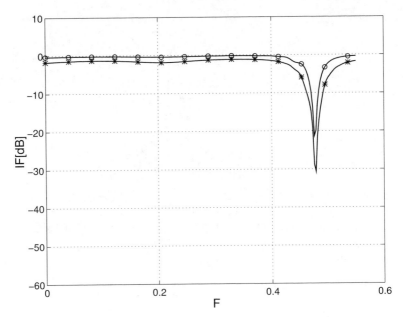

Figure 8.7: Improvement factor for dartboard subarray configuration ($\varphi_L = 0°$): ∘ *no tapering;* ∗ *Taylor weighting and subarray normalisation*

antenna performs worst. This has to do with the regular subarray structure which has a tendency for high grating sidelobes. However, for reasons of manufacturing this antenna might be particularly attractive for practical use.

8.2.1.3 *Impact of clutter rejection on beampatterns*

Now let us find out how the processor shown in Figure 8.4 performs in terms of antenna directivity patterns. First we have to answer the fundamental question: why do we need low sidelobes after adaptive clutter rejection? We can assume that jamming has been cancelled, either by an additional spatial anti-jamming pre-filter or by the space-time filter simultaneously with the clutter. WU R. et al. [730] have developed a technique for controlling the sidelobes of an array in the presence of adaptive jammer cancellation.

Clutter is by nature strongly heterogeneous. When adapting a clutter rejection filter a certain amount of data is needed to estimate the covariance matrix, for example by averaging space-time clutter echo dyadics. Such averaging can basically be done in the time or range dimension or both. In both cases data are averaged that stem from different locations on the ground and, due to the heterogeneous nature of the clutter background, have different clutter statistics. The resulting clutter rejection filter is therefore matched only to the *average* clutter and might not be able to suppress strong clutter discretes.[3] To cope with this effect low sidelobes are desirable.

---

[3]Projection type processors circumvent the problems of dynamic range of clutter echoes because they

250 Antenna related aspects

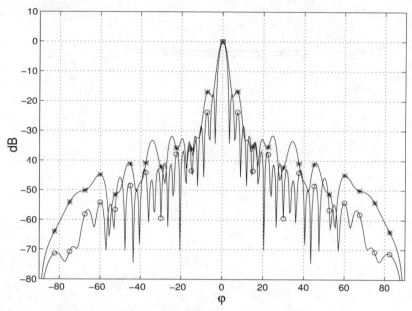

Figure 8.8: Irregular subarrays, no tapering ($\varphi_L = 0°$): ○ *azimuth beampattern, no clutter rejection;* ∗ *azimuth beampattern after space-time processing*

There is another interesting aspect of aperture tapering. The reduction of antenna sidelobes results in attenuation of sidelobe clutter Doppler frequencies. This might lead to simplified clutter rejection architectures. We will discuss this question in Section 8.4.

Figures 8.8, 8.9 and 8.10, show three different modes of operation of this processor. The antenna directivity pattern is in general calculated as

$$D(\varphi) = |\, \mathbf{b}^*(\varphi)\mathbf{b}(\varphi_0)\,| \tag{8.10}$$

where $\mathbf{b}(\varphi)$ is a beamformer vector (or the spatial part of the target signal replica), while $\mathbf{b}(\varphi_0)$ plays the role of a source located at angle φ_0.

The directivity pattern of the processor shown in Figure 8.4 is accordingly

$$D(\varphi) = |\, \mathbf{b}_T^*(\varphi)\mathbf{K}_{11}\mathbf{b}_T(\varphi_0)\,| \tag{8.11}$$

where $\mathbf{K}_{11}$ is the upper left (spatial) submatrix of $\mathbf{Q}_T^{-1}$.

It should be noted that (8.11) represents the directivity pattern before Doppler filtering. After Doppler filtering the directivity pattern becomes

$$D(\varphi, f_D) = |\, \mathbf{s}_T^*(\varphi, f_D)\mathbf{Q}^{-1}\mathbf{s}_T(\varphi_0, f_D)\,| \tag{8.12}$$

which means that basically the antenna pattern varies with Doppler. For simplicity we use (8.11) in the following analysis. However, some deviations from the results shown below can be expected if Doppler filtering is included in the consideration.

---

form a clutter notch that ideally is an exact null, see the remarks in Chapter 16.

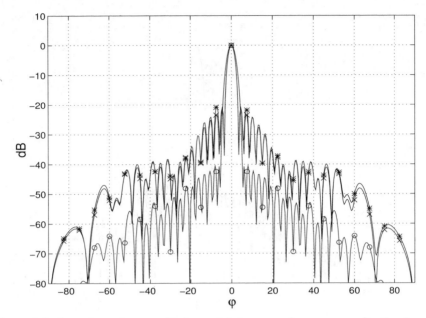

Figure 8.9: Irregular subarrays, Taylor weighting, no subarray normalisation ($\varphi_L = 0°$): ∘ *azimuth beampattern, no clutter rejection;* ∗ *azimuth beampattern after space-time processing, white noise case;* × *azimuth beampattern after space-time clutter rejection*

Figure 8.8 shows the beampatterns for an array after Figure 8.1 (irregular subarrays). No tapering has been applied. Therefore, the maximum sidelobe of the beampattern without processing (circles) is about −18 dB. As can be noticed from the second curve (asterisks) additional space-time clutter suppression leads to some increase of the sidelobes and slight broadening of the main beam. Also, the nulls in the beampattern are flattened out.

Figure 8.9 shows the same situation as before but with 40 dB Taylor weighting and without subarray normalisation. We notice first of all a broadening of the main lobe which is a natural consequence of the tapering. In a way the taper function reduces the effective aperture size.

It can be noticed that in the circled curve (no clutter rejection) all sidelobes of the antenna are suppressed below 40 dB. When space-time adaptive processing is applied, either for white receiver noise only (∗), or for clutter rejection (×), the effect of tapering is negated so that the sidelobe come up on the same level as in Figure 8.8. The adaptive processor has the property of normalising the subarray outputs in such a way that the white noise power is uniform in all channels. Therefore the effect of tapering is cancelled.

In Figure 8.10 we finally show the effect of additional subarray output normalisation. As can be seen the sidelobe level is well below 40 dB, for both cases (with and without clutter rejection). It can be noticed that there is little effect of the clutter rejection filter on the resulting beam pattern.

252 Antenna related aspects

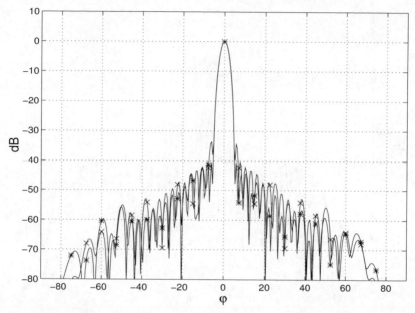

Figure 8.10: Irregular subarrays, Taylor weighting with subarray normalisation ($\varphi_L = 0°$): ∗ *azimuth beampattern, no clutter rejection;* × *azimuth beampattern after space-time clutter rejection*

It can be seen from Figures 8.11 and 8.12 that similar beampatterns are obtained for the checkerboard and dartboard subarray configurations. The beampattern is obviously quite independent of the kind of subarray formation.

8.2.1.4 Difference pattern generation at subarray level

Difference patterns are the basis of monopulse techniques (SKOLNIK [610, Chapter 18]). Monopulse techniques are used for location of detected targets with high accuracy, i.e., accuracy beyond the beamwidth determined by the antenna aperture. Monopulse techniques play a major role in tracking radar.

There are many possibilities of forming difference patterns. One prominent example is the Bayliss weighting, see BAYLISS [38] and MAILLOUX [428, p. 160].

The monopulse angle estimator is activated only if a target has been detected. In order to save radar operation time it is desirable to use the same data as obtained already for detection by the sum beam. Therefore, the Bayliss weighting cannot be applied directly to the sensor outputs. Instead we have to use data which have been gathered by use of a *tapered* antenna array at the subarray outputs. The difference weighting at the subarray outputs is approximately (NICKEL [489])

$$\mathbf{g}_{\mathrm{sa}} = \mathbf{T}_{\mathrm{sa}}^{*}\mathbf{g} \tag{8.13}$$

where **g** is the vector of Bayliss coefficients at the sensor level while $\mathbf{g}_{\mathrm{sa}}$ includes the

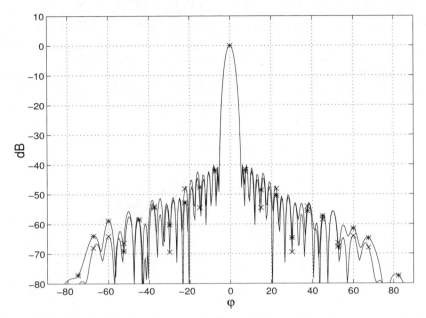

Figure 8.11: Checkerboard subarrays, Taylor weighting with subarray normalisation ($\varphi_L = 0°$): ∗ azimuth beampattern, no clutter rejection; × azimuth beampattern after space-time clutter rejection

transformed Bayliss coefficients at the subarray level. The subarray transform $\mathbf{T}_{sa}$ was defined in (8.4).

Figures 8.13–8.16 show some numerical examples for difference pattern synthesis with the irregular subarray structure given by Figure 8.1. In Figure 8.13 the case of Bayliss weighted *sensor* outputs can be seen, however without subarray normalisation.

As can be seen the sidelobe level is below 40 dB if no adaptive processing is applied (◦). The effect of adaptive processing, either against white noise (×) or clutter (∗), is dramatic. Obviously the adaptive processor compensates for the Bayliss weighting (except for the null in look direction) so that the sidelobes come up tremendously. The same effect was observed for the Taylor weighting applied to sum pattern synthesis in Figure 8.9. Including subarray normalisation according to (8.5) keeps the sidelobe level below 40 dB (Figure 8.14).

In Figure 8.15 the difference pattern was formed at the *subarray level* using the transformed Bayliss weighting according to (8.13). No subarray normalisation was used. As can be seen one sidelobe is at -24 dB, the others are below 30 dB.

As a result we find that additional subarray normalisation gives only very little further reduction of the sidelobe level (Figure 8.16). In Figures 8.17 and 8.18 the difference patterns of the checkerboard and dartboard arrays are calculated for the same conditions as in Figure 8.16. In particular, the checkerboard array tends to produce a high sidelobe level. It should be noted that the irregular structure shown in Figure 8.1 has been optimised to achieve low sidelobe levels (NICKEL [489]).

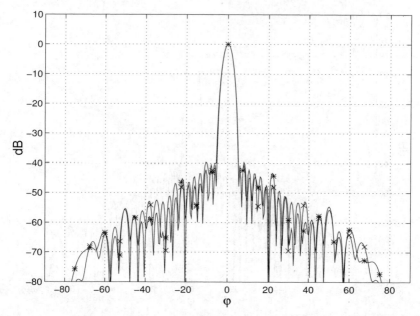

*Figure 8.12: Dartboard subarrays, Taylor weighting with subarray normalisation ($\varphi_L = 0°$): * azimuth beampattern, no clutter rejection; × azimuth beampattern after space-time clutter rejection*

8.2.1.5 Reduction of the number of subarrays

So far we considered circular apertures partitioned into 32 subarrays. The number 32 looks fairly high for the purpose of clutter rejection only. This high number of channels might be required when clutter suppression has to be carried out under jamming conditions (see Chapter 11).

In the following we try to reduce the number of output channels by combining several of the subarrays into one. For the look direction we assumed $\varphi_L = 45°$. Recall that at this angle the clutter returns are Doppler broadband while in the flight direction ($\varphi_L = 0°$) clutter echoes are narrowband.[4] Therefore, clutter rejection at $\varphi_L = 45°$ is a more difficult task than in the flight direction.

Consider for example Figure 8.19a. It shows the 32 subarrays of the checkerboard array which was presented in Figure 8.2. The other three pictures in Figure 8.19 show different ways of combining the square subarrays into larger ones.

Figure 8.20 shows the corresponding IF curves. As can be seen the clutter suppression performance is only slightly degraded if the number of subarrays is reduced (c and d). With $K = 6$ channels reasonable clutter rejection is still obtained.

In Figure 8.21 we combine subarrays of the dartboard array so as to cut down the number of subarrays to $K = 4$, 8 or 16. In Figure 8.22 we find that for only four subarrays (case b in Figure 8.21) the adaptive filter does not perform sufficiently well

---

[4]Notice that the IF curves in Figures 8.5, 8.6 and 8.7 have been calculated for the forward look direction ($\varphi_L = 0°$) because the associated beampatterns have been calculated for broadside steering.

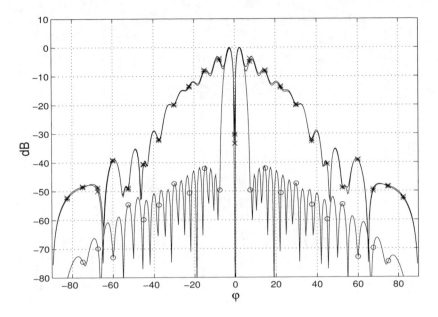

Figure 8.13: Irregular subarrays, Bayliss weighting, no subarray normalisation ($\varphi_L = 0°$): ○ *azimuth difference pattern, no clutter rejection;* × *azimuth difference pattern after space-time processing, white noise case;* ∗ *azimuth difference pattern after space-time clutter rejection*

(∗ curve). Notice that we reduced the number of degrees of freedom down to $K = 4$. In the horizontal dimension we have practically only two degrees of freedom. As the flight direction is horizontal the effective number of degrees of freedom is reduced down to two which is obviously too small, particularly in view of the fact that the two subarrays are not equal and, hence, produce different clutter spectra. Some losses can be noticed even for eight subarrays. These may be mitigated, however, by applying a larger filter length, of course at the expense of higher computational load. In the examples shown we used $L = 5$.

Even for an irregular subarray structure (Figures 8.23 and 8.24), forming of a smaller number of subarrays (in this example $K = 7$) has no significant detrimental effect on the clutter rejection performance.

We should keep in mind (see the considerations in Section 6.1.3) that optimum performance was obtained for *uniform* overlapping subarrays in the case of a linear array. Accordingly, optimum planar arrays were proposed in Figure 6.18. While the checkerboard array still has some similarity with a linear array the dartboard and the irregular configuration have no relation to a linear array.

It should further be noted that the losses in SNIR exhibited by Figures 8.20, 8.22 and 8.24 may be mitigated by increasing the temporal filter length L. In all shown examples $L = 5$ was chosen (according to the list of standard parameters in Chapter 2, p. 65).

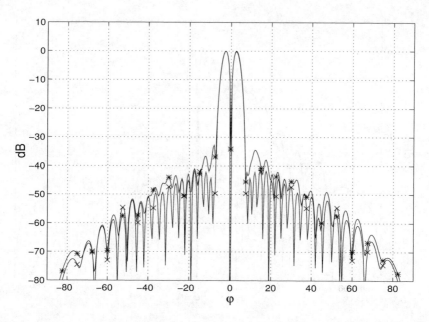

Figure 8.14: Irregular subarrays, Bayliss weighting with subarray normalisation ($\varphi_L = 0°$): × *azimuth difference pattern after space-time processing, white noise case;* ∗ *azimuth difference pattern after space-time clutter rejection*

8.2.1.6 Σ-Δ-processing

In the following we come briefly back to Σ-Δ-processing already addressed in Section 6.3.2. This techniques uses two channels only as a spatial basis, a sum beam and a difference beam, which both are available in many commercial radar antennas. Recall that WANG H. *et al.* [671] have found that this technique may be a cheap solution for space-time clutter rejection.[5] MOO [467] has found that Σ-Δ-STAP may be useful even in forward looking radar. Examples for linear arrays have been presented in Section 6.3.2. In the following we apply the same technique to a circular planar array.

Σ-Δ-processing can be described by a spatial transform as given by (6.33). The first column denotes the coefficients of the sum beamformer while the second column includes the weights of a difference beam. This spatial transform has to be inserted into (6.1), in order to obtain a space-time notation.

In the following examples the temporal filter length has been assumed to be $L = 2$ which is the minimum possible filter length. In this way the space-time FIR filter according to (7.12) will have only four coefficients.

Figure 8.25 shows the performance of Σ-Δ-processing for a sidelooking circular planar array (look directions 90° and 45°). It can be noticed that the performance is quite poor if no additional tapering is applied (Figure 8.25a and 8.25c). Tapering alleviated the space-time processing task considerably as can be seen from Figures 8.25b and 8.25d.

---

[5]The authors used MCARM data which have been generated by a linear sidelooking array.

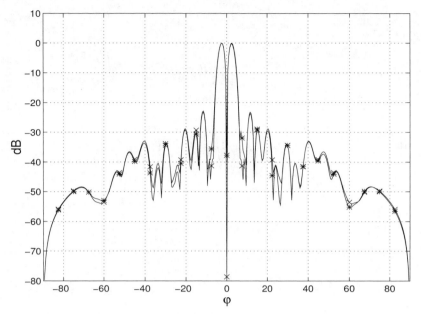

*Figure 8.15: Irregular subarrays, transformed Bayliss weighting at subarray level, no subarray normalisation ($\varphi_L = 0°$): × azimuth difference pattern after space-time processing, white noise case; * azimuth difference pattern after space-time clutter rejection*

Similar results for a forward looking array are shown in Figure 8.26 (look directions 0° and 45°). For look direction 0° we have the Doppler narrowband case because in the flight direction the isodops are run fairly tangential through the azimuth-range cells. Therefore, each cell contains only one Doppler frequency. As can be seen from Figures 8.26a and 8.26b this case is easy to handle for the Σ-Δ-processor. The single clutter frequency in the range cell leads to a narrow clutter notch, with and without tapering.

In the look direction 45° we encounter Doppler broadband clutter returns. Obviously, the Σ-Δ-processor does not cancel the clutter echoes sufficiently (Figure 8.26c). As in the sidelooking case, tapering reduces the number of degrees of freedom and leads to satisfactory clutter rejection performance (Figure 8.26d). This works only as long as the sidelobe level of the array is of the order of magnitude of the CNR. For large CNR (40 dB in Figure 8.27) severe losses at off-broadside directions occur even with 40 dB Taylor tapering applied (Figure 8.27d).

8.2.2 Randomly spaced arrays

In the following we consider briefly two kinds of array antennas with random distribution of the antenna elements. The aptitude of such antenna configurations for space-time clutter rejection is discussed.

258 *Antenna related aspects*

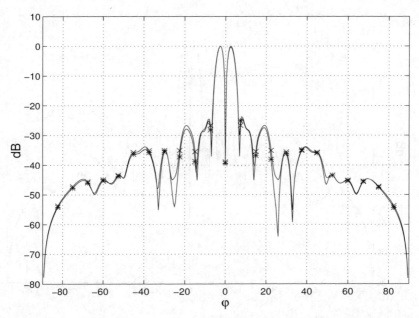

Figure 8.16: Irregular subarrays, transformed Bayliss weighting at subarray level and subarray normalisation ($\varphi_L = 0°$): × *azimuth difference pattern after space-time processing, white noise case;* ∗ *azimuth difference pattern after space-time clutter rejection*

8.2.2.1 Example for planar arrays: the ELRA antenna

The ELRA system is an experimental multifunction ground-based phased array radar (GROEGER et al. [216], SANDER [583, 582]). The phased array antenna is the functional basis for multifunction operation, which in essence means search and track. The radar includes various auxiliary functions such as MTI, jammer cancellation, sequential detection, power management, etc.

The receive antenna has a planar circular surface on which 768 sensors are distributed in a pseudo-random fashion. The operating frequency is 2.7 GHz, the array diameter is about 5 m. This means that the array is strongly thinned compared with a $\lambda/2$ element raster. However, thanks to the random element distribution in both dimensions ambiguities in either azimuth or elevation are largely suppressed. The width of the main beam is determined by the diameter of the aperture; the sidelobe level, however, is determined by the number of elements. The element density decreases from the center of the aperture to the edge so as to provide some density tapering.

The array is subdivided into 48 subarrays with 16 elements each. Subarray beamforming is formed in the RF domain. Due to the random positions of the elements all subarrays have different shapes which in turn leads to suppression of ambiguities (grating lobes, grating nulls) due to secondary beamforming (combination of subarrays).

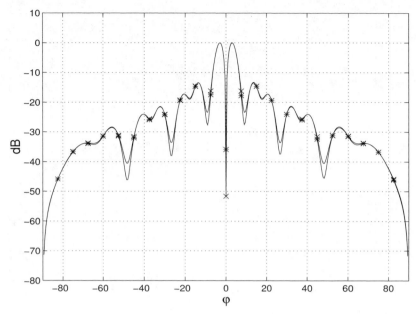

*Figure 8.17: Checkerboard subarrays, transformed Bayliss weighting at subarray level and subarray normalisation ($\varphi_L = 0°$): × azimuth difference pattern after space-time processing, white noise case; * azimuth difference pattern after space-time clutter rejection*

As far as space-time clutter filtering is concerned the ELRA receive array is comparable to the irregular subarray structure in Figure 8.1. In both cases we have irregular subarrays which all receive different clutter spectra and different clutter-to-noise ratios. Thanks to the large numbers of subarrays in both cases (32 and 48, respectively) space-time adaptive filtering can be expected to work properly. We omit a numerical analysis for this type of antenna.

8.2.2.2 Volume arrays

The crow's nest antenna[6] (WILDEN and ENDER [711]) is a spherical aperture randomly filled with horizontal ring dipoles. This kind of array has the property of 360° azimuthal coverage. In contrast to spherical *conformal* antennas *all* array elements contribute to the beamforming process for *all* look directions.

This kind of array has a natural density tapering because the sphere is thicker in the middle than at the outer edges. The positions of the elements have been chosen at random, however under the constraint that no spacing between neighbouring elements must be smaller than half a wavelength. This random spacing is to provide a constant beampattern independent of the look angle. The problem of space-time processing

---

[6]In fact, this antenna looks like a crow's nest in a tree. A 'crow's nest' is also a lookout platform on top of the mast of a sailing boat from where an overview of 360° is possible.

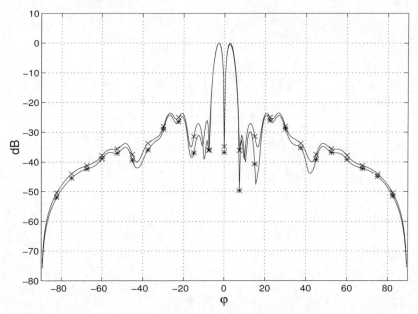

*Figure 8.18: Dartboard subarrays, transformed Bayliss weighting at subarray level and subarray normalisation ($\varphi_\mathrm{L} = 0°$): × azimuth difference pattern after space-time processing, white noise case; * azimuth difference pattern after space-time clutter rejection*

with such an array has been addressed by KLEMM [309], and GOODMAN and STILES [206].

For space-time clutter rejection only the x-coordinate of the individual sensor positions is relevant because this is the flight direction (see Figure 2.1). Figure 8.28a shows schematically a projection of the sensor positions of a randomly spaced volume array in the xz-plane.

The upper curve in Figure 8.30 shows the gain curve for a fully adaptive space-time FIR filter. The number of elements was chosen to be $N = 480$. The temporal dimension was limited[7] to $L = 2$.

As can be seen a reasonable performance is basically possible with such exotic antenna. The question now is how to reduce the signal subspace in such a way that this kind of antenna could be used in practical systems.

For this purpose let us come back to the considerations on subarrays made in Chapter 6. We stated that a prerequisite for efficient reduction of the number of channels through forming subarrays is that all subarrays are identical. Only then do they have identical beampatterns and, therefore, receive all the same clutter spectrum. This is the condition under which the number of antenna channels can strongly be reduced without significant losses in clutter rejection.

To illustrate this in the case of the crow's nest antenna we subdivide the spherical

---

[7]Larger values of L would have exceeded the capacity of the computer used for the analysis.

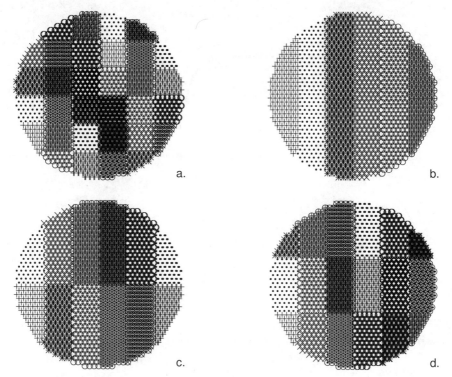

Figure 8.19: Subarray combination based on checkerboard structure: a. 32 square subarrays; b. 6 columns; c. 12 subarrays; d. 18 subarrays

array into subarrays in various ways and calculated the improvement factor (see Figure 8.29). For curve a the sphere has been subdivided into eight octants as is required for monopulse with a spherical antenna anyway. For curve b the sphere has been cut into five slices, each with identical numbers of sensors. Therefore, the thickness of the subarrays (extension in the flight direction) is different. In curve c five slices of equal thickness have been formed. Finally, curve d shows overlapping subarrays in the flight direction.

All four concepts have one property in common: The shapes of the individual subarrays are different. This means that the subarray beam shapes are different so that the received clutter Doppler spectra are different. Therefore the spectra of clutter echoes between different subarray outputs are decorrelated. No near-optimum clutter rejection performance can be expected. In fact, the four IF curves in Figure 8.29 indicate quite bad performance.

We have to look for some way to achieve identical subarrays. One possible solution may be found by modifying the antenna array according to Figure 8.28. Each of the sensors is replaced by a sensor doublet. The doublets are arranged in the flight direction. Now we have two identical arrays displaced by a certain distance in the flight direction. After beamforming for both subarrays in the same look direction we are left with two channels only which both receive the same clutter spectrum. There is

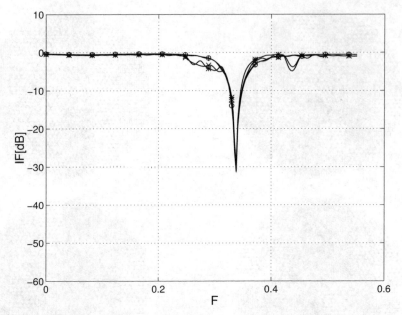

Figure 8.20: Number of subarrays (checkerboard, FL, $\varphi_L = 45°$): ○ *32 square subarrays;* ∗ *6 columns;* + *12 subarrays;* × *18 subarrays*

just a phase difference due to the displacement of the subarrays.

The lower curve (∗) in Figure 8.30 shows the improvement factor for the doublet sensor array. Since the temporal dimension of the clutter FIR filter was chosen to be $L = 2$ the total number of space-time filter coefficients is only $KL = 4$ which is the absolute possible minimum! There are some losses close to the clutter Doppler frequency in the look direction, however, we notice that this curves approximates the optimum (upper) curve quite well. Replacing the doublet sensors by triplets yields even further improvement (×).

8.2.3 Conformal arrays

Conformal antenna arrays will play a major role in future air- and spaceborne radar systems. The advantage of conformal arrays is that the hull of an air or space vehicle can be used as the supporting construction. Application of space-time adaptive processing to conformal arrays has been described by HERSEY et al. [259] and TANG et al. [634].

THOMPSON and PASALA [642] analysed linear arrays with a curved geometry. They found that the number of degrees of freedom increases due to the curvature, and that clutter suppression is degraded.

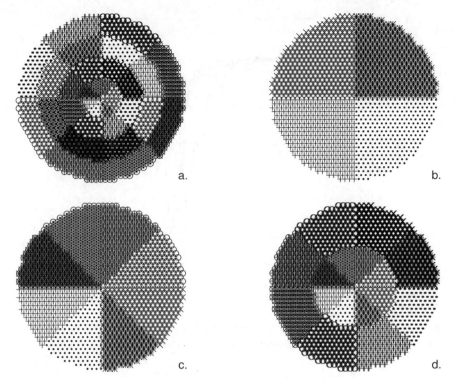

Figure 8.21: Subarray combination based on dartboard structure: a. 32 subarrays; b. 4 subarrays; c. 8 subarrays; d. 16 subarrays

8.2.3.1 Cylindrical arrays

A lot of air vehicles have a cylindrical shape with the cylinder axis coinciding with the flight direction. We are not talking necessarily about *circular* cylinders. By 'cylinder' we mean any surface whose yz-dimensions are constant with the flight direction x (for the coordinates see Figure 2.1). There are a lot of air vehicles such as airplanes, missiles, RPVs, etc., which have a cylindrical shape (or parts of which are cylindrical).

For all kinds of cylindrical antenna arrays all the considerations that have been made in Chapter 6 for linear or rectangular arrays are valid. That means, formation of identical subarrays or the implementation of symmetric auxiliary sensor configurations after Figure 6.12 is straightforward. Notice that a rectangular planar array as shown in Figure 6.18b is a special case of a cylindrical array. Cylindrical conformal arrays are favourable for side- and downlooking air- and spaceborne MTI radar.

8.2.3.2 Forward looking conformal arrays

With conformal array technology forward looking radar antennas can be designed which have a sidelooking capability as well. They might even be able to look to the rear to a certain extent, which means an azimuthal coverage of more than $180°$. The MTI

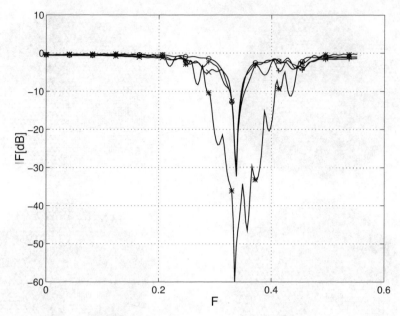

Figure 8.22: Number of subarrays (dartboard, FL, $\varphi_L = 45°$): ○ *32 subarrays;* ∗ *4 subarrays;* + *8 subarrays;* × *16 subarrays*

capability will be based on a similar principle as proposed for the crow's nest antenna. Instead of designing doublet or triplet sensors several identical tracks of sensors have to be arranged one below the other.

The principle can be seen in Figure 8.31. As seen from the top or from below there are three parabolic tracks which denote the sensor trajectories. Notice that they are displaced by a certain distance in the flight direction (Figure 8.31a). In practice these parabolas (or similar trajectories, depending on the aircraft shape) will be installed one below the other as can be seen in Figure 8.31b. All sensors belonging to one trajectory will be combined by a beamformer network. As a result we obtain in the given example three channels which all have the same beam pattern and, hence, the same clutter spectrum. This is, as carried out before, the prerequisite for successful clutter suppression at low computational expense.

8.3 Array concepts with omnidirectional coverage

In this section we discuss briefly a couple of array concepts which may be applied to a surveillance radar with 360° azimuthal coverage. Such antennas will be mounted on an aircraft and may in the future replace mechanically rotating devices like the AWACS antenna, rendering all the benefits of electronically steered array antennas including the potential of spatial and space-time signal processing. Some more details are given in KLEMM [350, Chapter 5],[338, 347, 348].

Array concepts with omnidirectional coverage

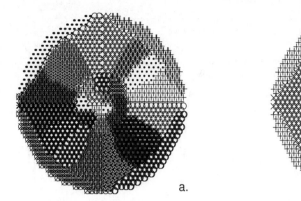

Figure 8.23: Subarray combination based on irregular structure: a. 32 subarrays; b. 7 subarrays

8.3.1 Four linear arrays

By using four linear or rectangular arrays as depicted in Figure 8.33 360° azimuthal coverage can be achieved. Since the linear array is best suited for use with STAP good moving target suppression performance can be expected. However, this antenna concept has the drawback that only one fourth of the implemented array is active while the remaining three fourths are idle.

The signal-to-noise ratio of an array radar is proportional to

- the transmitted power ($\propto N$)
- the transmit gain ($\propto N$)
- the receive gain ($\propto N$)

i.e., the SNR is $\propto N^3$. Therefore, if three fourths of the array elements are idle the SNR is about 18 dB smaller than in the case where all elements are active. This is certainly a significant disadvantage.

8.3.2 Circular ring arrays

The use of STAP with a circular ring array has been analysed by ZATMAN [741, 742], see Figure 8.32, BELL et al. [39], and FUHRMANN and RIEKEN [169]. FRIEDLANDER [166] proposes a subspace technique for the use with circular arrays. This antenna is a part of the UESA project (UHF Electronically Scanned Array). 60 antenna elements are distributed on a circle. 20 adjacent elements out of 60 will be excited so as to enable 360° azimuthal coverage. All the cited work is centred about the UESA program (UHF Electronically Scanned Array) sponsored by ONR, USA.

ZATMAN has demonstrated that optimum STAP works well with a circular array. Some problems which we identified earlier already in the context of forward looking arrays can degrade the performance of the circular array as well:

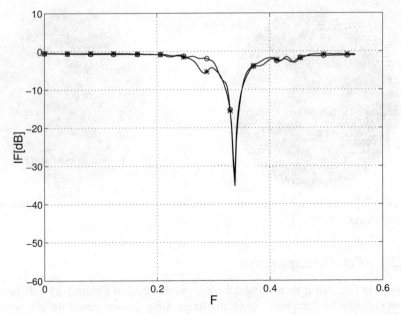

Figure 8.24: Number of subarrays (irregular subarrays, FL, $\varphi_L = 45°$): ○ *32 subarrays;* ∗ *7 subarrays*

- The clutter Doppler is range dependent as for forward looking arrays. This may lead to shortage of training data at short range and may require range dependent clutter suppression.

- The system bandwidth leads to broadening of the clutter notch which causes degradation in slow target detection.

- The clutter rank is slightly higher than for a comparable linear array which leads to a slight widening of the clutter notch and a slight SNIR loss outside the clutter Doppler area.

- Post-Doppler STAP techniques have proven to operate sufficiently well even for the circular antenna.

- According to [741] pre-Doppler techniques show significant degradation.

The last statement on pre-Doppler processing is questionable. We achieved good results for the circular planar forward looking array which also has a geometry quite different from a uniform linear array.

Since about two thirds of the array are idle the SNR (compared to a fully active array) is reduced by about 14 dB.

NGUYEN *et al.* [478] analyse the multistage Wiener filter for use with circular ring arrays.

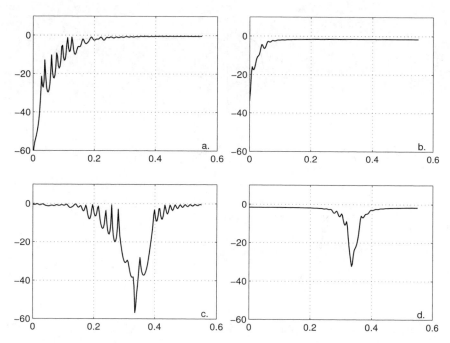

Figure 8.25: Σ-Δ processing for circular antenna (SL, $L = 2$, CNR = 20 dB): a. $\varphi_L = 90°$, no tapering; b. $\varphi_L = 90°$, tapering; c. $\varphi_L = 45°$, no tapering; d. $\varphi_L = 45°$, tapering;

8.3.3 Horizontal planar arrays

In the following we discuss briefly a few array concepts with special aptitude for 360° azimuthal coverage under the condition that all array elements are active. To meet this goal, radiating elements have to be used which have an omnidirectional radiation pattern in the horizontal dimension. For example, a horizontal ring dipole as have been designed for the Crow's Nest Antenna (WILDEN & ENDER[711]), see Figure 8.34.

8.3.3.1 Rectangular arrays

The properties of rectangular horizontal planar arrays operating in a down-look mode have been discussed by KLEMM [305]. Such an array may be mounted under the fuselage of an air vehicle in the xy-plane (for definition see Figure 2.1) and may be scanned over 360° in azimuth. Because of the vertical sensor directivity pattern this array will, however, operate properly only at depression angles larger than about 30°.

It has been stated in Chapter 3, p. 117, that the DPCA effect[8] makes use of the platform motion in that spatial decorrelation due to travel delays of incoming waves

---

[8]DPCA (displaced phase center antenna, see Section 3.2.2) means that each of the sensors of a sidelooking line array assumes the position (or, more correctly the clutter phase) of its predecessor after one PRI.

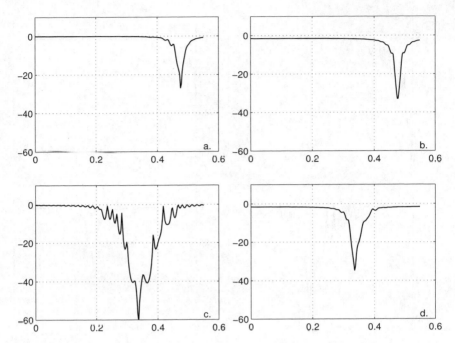

Figure 8.26: Σ-Δ processing for a circular antenna (FL, $L = 2$, CNR = 20 dB): a. $\varphi_L = 0°$, *no tapering;* b. $\varphi_L = 0°$, *tapering;* c. $\varphi_L = 45°$, *no tapering;* d. $\varphi_L = 45°$, *tapering;*

can be compensated for. This decorrelation effect appears only in wideband radar systems. Since this effect is based on the platform velocity it works normally only for sidelooking linear or planar arrays.

If the array has degrees of freedom (e.g., subarrays) in the y-direction the DPCA effect can be exploited to compensate for *lateral* velocity components as may occur due to wind drift, in particular with weather clutter. In addition, for broadband radar the lateral motion enables the processor to compensate for bandwidth-induced spatial decorrelation effects.

Rectangular horizontal arrays are not the best choice for a radar with 360° coverage because the beamshape varies with angle.

8.3.3.2 Displaced ring arrays

One of the findings of Chapter 6 was that the performance of a subarray based STAP architecture is optimal if the subarrays of the array have uniform size and shape. If this condition is fulfilled the clutter Doppler spectra received by the various subarrays are equal except for a complex factor which takes the displacement of the subarrays into account. Then clutter suppression can easily be performed by a simple weighted subtraction of the spectra. This is the key to extremely efficient STAP architectures.

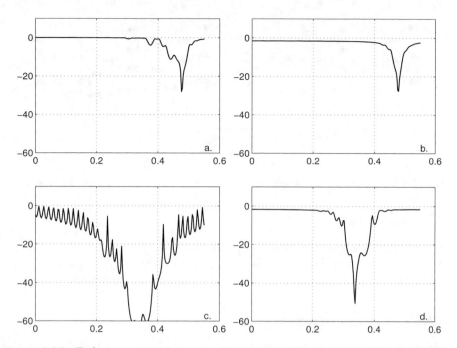

Figure 8.27: Σ-Δ *processing for a circular antenna (FL, L = 2, CNR = 40 dB):
a.* $\varphi_L = 0°$, *no tapering; b.* $\varphi_L = 0°$, *tapering; c.* $\varphi_L = 45°$, *no tapering;
d.* $\varphi_L = 45°$, *tapering;*

This feature occurs only in uniformly spaced linear arrays (ULA). The problem we are facing is to design an array which combines the STAP properties of a linear array with the capability of 360° azimuthal coverage.

Let us start from the UESA circular ring array. Let us assume a linear array composed of uniform subarrays which have a ring shape. All the elements of each individual ring are combined by a subarray beamformer. If such rings are displaced by a constant amount we come up with a linear array which has almost circular shape. This concept is shown in Figure 8.35. If the array is composed of omnidirectional elements like the dipole shown in Figure 8.34 we get approximately an omnidirectional array with displaced subarrays.

8.3.3.3 Randomly thinned arrays using element groups

The concept of displacing circular planar arrays via groups of radiating elements can also be applied to randomly thinned arrays as have been used in the ELRA phased array radar (WIRTH[716], GRÖGER et al. [216]). For illustration consider Figure 8.36. As can be seen there are groups of $K = 4$ radiating elements. All the elements in the first, second, third and fourth position are combined by beamformers steered networks forming $K = 4$ subarray beams directed in the same look direction. The

270   *Antenna related aspects*

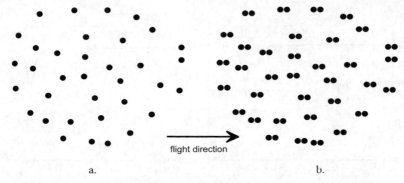

Figure 8.28: Randomly distributed arrays: a. single sensors; b. sensor doublets

$K = 4$ subarray outputs represent the spatial dimension of a space-time processor. If the geometrical orientation of the elements coincides with the flight path we have a sidelooking antenna with randomly thinned (approximately) circular subarrays.

8.3.3.4   Octagonal arrays

The octagonal array has the same properties as rectangular (or linear) arrays, that means, it allows for the formation of shift invariant subarrays. It means also that the element spacing should not be larger than half the wavelenght so as to avoid grating lobes or grating nulls.

Figure 8.37 illustrates the way uniform subarrays are formed in an octagonal array. As in the above examples a sidelooking array with overlapping octagonal subarrays can be obtained if the orientation of the subarrays coincides with the flight path.

8.3.3.5   *Some properties of* $360°$ *coverage arrays*

When used with STAP the three aforementioned array concepts have different properties (KLEMM [350, Chapter 5],[338, 347, 348, 351]) which can be summarised as follows

- The width of the clutter notch (and, hence, the minimum detectable velocity) depends on the number of subarrays K for the displaced rings as well as for the randomly thinned array. The larger the number of subarrays is the narrower the clutter notch becomes.

- The clutter notch produced by the octagonal array is narrower than for the displaced rings and randomly thinned arrays.

- For the same number of radiating elements the displaced rings and the randomly thinned arrays are much larger than the octagonal array. Therefore, the beam produced by the octagonal array is much wider than for the other two concepts.

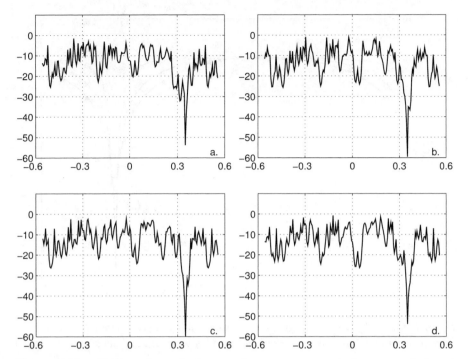

*Figure 8.29: Performance of the crow's nest antenna (subarray techniques):
a. 8 octants; b. 5 slices (cut in the cross-flight plane) with equal numbers of sensors;
c. 5 slices in the cross-flight plane with equal extension in the flight direction; d. 3-D
overlapping subarrays in the flight direction*

- As long as the arrays are operated in a sidelooking manner there is no range dependence of the clutter Doppler. Therefore, ambiguous clutter returns fall altogether on the same Doppler frequency so that no additional clutter notches are produced by the STAP processor (see Chapter 10).

- In the case of range ambiguous operation the clutter rejection performance is strongly degraded by ambiguous clutter notches if the array is rotated by a certain angle which may occur due to winddrift (crab angle). The octagonal array is less sensitive than the others because it has the smallest aperture (about $5°$ crab angle is tolerable).

- Horizontal planar arrays with two-dimensional subarray structure as discussed in Section 8.3.3 shown for instance in Figures 8.1, 8.2, 8.3 are insensitive to ambiguous clutter arrivals for any array rotation.

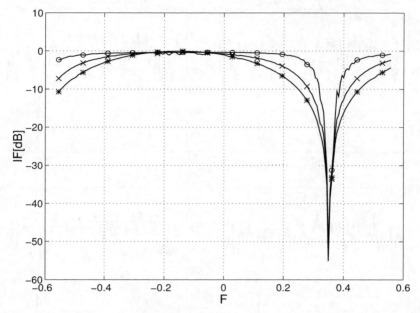

Figure 8.30: MTI performance of the crow's nest antenna ($N = 480$; $L = 2$; CNR = 20 dB): ○ optimum processing; ∗ doublet sensor approach; × triplet sensor approach

8.4 STAP and conventional MTI processing

8.4.1 Introduction

The worldwide enthusiasm about STAP should not make us forget that under certain conditions conventional MTI processing (beamforming and *temporal* adaptive clutter rejection) may be sufficient, especially in conjunction with tapering of the antenna aperture. One prominent example is the high PRF mode of an airborne radar where most targets of interest are faster than the maximum clutter speed. In this case a simple high-pass filter can be sufficient for clutter suppression.

It should noted that in most of the literature on space-time adaptive processing known to the author a comparison between space-time and simple temporal adaptive processing (beamforming, tapering, adaptive temporal filtering) is missing. In many papers different space-time architectures are compared among each other, but the comparison with conventional one-dimensional processing is not done. In some cases the numerical results achieved with space-time processing are quite poor so that simple temporal processing may compete with space-time processing. This may come true particularly for large antennas with narrow beam and low sidelobes. The enthusiasm about STAP should not make us overlook simple standard solutions.

Figure 8.38 shows such a conventional *temporal* processor. In the sequel all results obtained with the space-time processor will be compared with this conventional one-dimensional processor.

In the following numerical evaluation several problems are to be discussed:

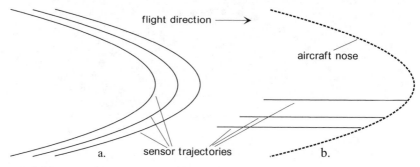

Figure 8.31: Parabolic trajectories of a forward looking array: a. top view, b. side view

- What is the effect of aperture tapering on the clutter rejection performance?
- Do we need space-time adaptive processing when tapering is applied?
- What is the dependence on the look direction?

In the following we consider mainly forward looking arrays (aperture in the cross-flight direction). Similar results can be obtained for sidelooking arrays.

8.4.2 Linear arrays

Let us first compare the optimum processor (Chapter 4), the space-time FIR filter processor (7.12), Figure 8.4, and also the conventional MTI according to Figure 8.38, for the case of a linear array. Figures 8.39 – 8.41 show the IF for a linear array.

In Figure 8.39 the upper curve has been calculated for the optimum processor. No tapering was applied. The look direction is $45°$. The clutter notch appears at the clutter Doppler frequency in the look direction. It should be noted that the area around the clutter notch relates to slow targets and is therefore of special importance.

It can be recognised that the optimum curve is quite well approximated by the space-time adaptive processor according to Figure 8.4. Beamforming plus temporal adaptive filtering leads to severe losses in the neighbourhood of the Doppler frequency in the look direction. That means a severe degradation of slow target detection.

Figure 8.40 shows the same constellation as before, however with look angle $0°$ (looking straight forward). Now we notice that the suboptimum space-time processor after Figure 8.4 yields an IF curve that coincides perfectly with the optimum. Furthermore, the conventional processor after Figure 8.38 shows a considerable improvement in performance. In view of the fact that this processor requires very low processing capacity this might be a good compromise between cost and performance.

In order to explain this behaviour let us have a look at the isodop map in Figure 3.1. Forward looking ($0°$) means looking from the center of the plot to the right in the x-direction. Notice that here each of the hyperbolas runs perpendicular to the look direction so that in a certain range-azimuth cell almost no Doppler variation occurs. The processor encounters more or less one single clutter Doppler frequency only! This

274 *Antenna related aspects*

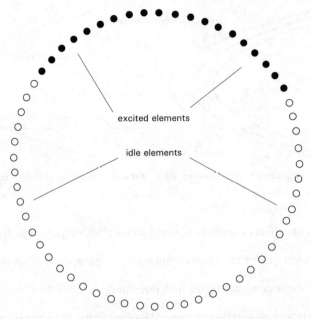

Figure 8.32: Scheme of 60-element circular ring array, with 20 elements excited

is the reason why the conventional temporal processor performs quite well. In the forward look direction clutter echoes are *Doppler narrowband*.

It should be noted that this narrowband effect is typical for a forward looking array. It does not occur for sidelooking arrays which normally do not look in the direction $\varphi = 0$ (endfire).

Looking under 45° (up or down in Figure 3.1) the isodops run almost in parallel with the constant azimuth lines. Therefore, each angle-range cell contains a lot of different Doppler frequencies (high Doppler gradient). Here we encounter *Doppler broadband* clutter. This is the reason why the conventional processor curve in Figure 8.39 deviates so strongly from the optimum. The response of the temporal filter is dictated by the clutter bandwidth while the space-time filter makes use of the fact that the clutter spectrum is just a narrow ridge in the azimuth-Doppler plane so that a narrow clutter notch is obtained regardless of the clutter bandwidth.

In Figure 8.41 Hamming tapering was applied to the array aperture. Again we consider the case of 45° look direction. Now some tapering loss can be noticed in the pass band (the curves do not reach the theoretical optimum). Secondly, one can notice that the stop band of the conventional processor is still considerably broadened. Tapering does not seem to give any advantage for clutter rejection.

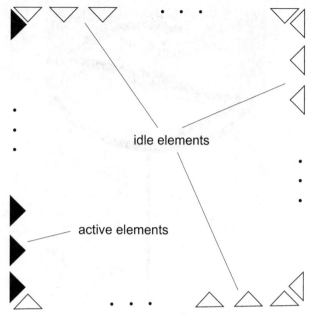

Figure 8.33: Four linear (or rectangular vertical) arrays

8.4.3 Circular planar array

The following considerations are focused on a circular planar antenna array with operational dimensions. The number of sensors is $N = 1204$, the number of echoes was chosen to be $M = 64$, the number of subarrays is $K = 32$, the temporal filter length is $L = 5$. Therefore, the dimension of the clutter covariance matrix at subarray level is $32 \cdot 5 \times 32 \cdot 5$ instead of $1204 \cdot 5 \times 1204 \cdot 5$ at the sensor level. The irregular subarray configuration shown in Figure 8.1 is assumed. This antenna configuration has already been analysed in Section 8.2 with respect to directivity patterns. We compare again the space-time FIR filter processor with conventional processing (beamformer + adaptive temporal filter). Calculation of the IF of the optimum processor for such a large antenna exceeds by far the capacity of available computers.

In the subsequent calculations we concentrate on the subarray structure shown in Figure 8.1. This irregular subarray structure has been optimised so as to yield optimum sidelobe control for the difference patterns, see NICKEL [489].

8.4.3.1 *No tapering*

Figure 8.42 shows the improvement factor versus the normalised target Doppler frequency for space-time processing after Figure 8.4. The four curves have been plotted for different azimuth angles ($\varphi_\mathrm{L} = 0°, 20°, 40°, 60°$). Figure 8.43 has been plotted for the conventional MTI processor (beamformer + adaptive temporal filter) after Figure 8.38. The parameter constellation is the same as before.

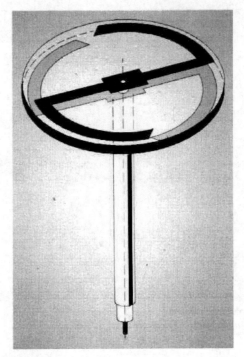

Figure 8.34: Circular dipole

The advantage of space-time processing over time processing only is obvious. Space-time processing achieves a narrow clutter notch at the Doppler frequency associated with the look direction. The temporal adaptive filter shows broad areas with low signal-to-clutter ratio which reflect the Doppler response of the antenna beam. The width of the clutter notches depends very much on the width of the clutter spectrum which in turn depends on the look angle (see the above remarks). It is a maximum for $\varphi_L = 60°$ and a minimum for $\varphi_L = 0°$. We see from Figure 8.43 that for $\varphi_L = 0°$ the performance is comparable to space-time processing (compare the clutter notches). For space-time processing the width of the clutter notches is nearly independent of the look direction.

While the IF curves for the linear array run quite smoothly we find some slight ripple for the circular array in Figure 8.42, which can be explained as follows.

The best subarray structure is given by uniform overlapping subarrays for linear arrays ([306] and Chapter 6). Only in this case do all subarrays have identical directivity patterns, and the subarray displacement is $\lambda/2$ which means that the Nyquist condition for the spatial dimension is perfectly fulfilled. In this sense the irregular subarrays of the circular antenna depicted in Figure 8.42 provide irregular spatial sampling where each 'spatial sample' has its individual clutter-to-noise ratio and individual clutter spectrum. Therefore, clutter output signals at different subarrays have mutual correlation losses which limits the potential of clutter rejection.

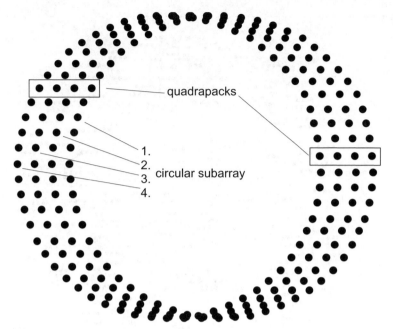

Figure 8.35: Displaced circular ring arrays

8.4.3.2 Influence of tapering

In Figures 8.44 and 8.45, a 40 dB Taylor weighting has been applied to the circular antenna (see Figure 8.1). In Figures 8.44 and 8.45 we find the IF curves achieved by the space-time processor and the temporal filter again for the look directions $\varphi_L = 0°, 20°, 40°, 60°$. We assumed CNR = 20 dB in both plots. It can be seen that tapering leads to a considerable improvement for the temporal filter. Even at the 20° look direction (Figure 8.45) we obtain a relatively narrow clutter notch. In the forward looking case ($\varphi_L = 0°$, narrowband clutter, Figure 8.45) the temporal filter performs as well as the space-time filter. Since there is only one clutter frequency present it can be cancelled well by the temporal filter. It should be noted that in the pass band about 2 dB tapering losses have to be taken into account.

For higher clutter-to-noise ratio (40 dB in Figure 8.46) the losses encountered by conventional processing are significant. Except for $\varphi_L = 0°$ space-time processing is superior to temporal processing. In the forward look direction the temporal filter again performs perfectly.

It should be noted that all space-time curves with tapered aperture run quite smoothly, which is an indication that tapering cuts down the number of degrees of freedom (or eigenvalues of **Q**) by reducing the sidelobe level.

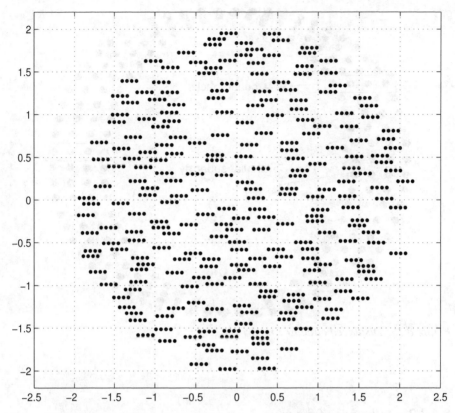

Figure 8.36: Randomly thinned array using sensor quadruplets (axes in [m])

8.4.3.3 *Effect of temporal filter length*

Earlier results with linear arrays (Figure 7.10) have shown that the clutter suppression performance is almost independent of the temporal filter length L. The reason for this behaviour is that with a linear array all subarrays can be made identical so that the directivity patterns of all subarrays are identical. Under such conditions the clutter Doppler spectra of the subarray outputs are identical so that clutter cancellation simply be done by mutual delay and subtraction.

If the subarrays are different the directivity patterns and, hence, the clutter spectra at the subarray outputs are different. Then clutter cancellation requires additional spectral shaping before delay and subtraction. This spectral shaping can be done by the space-time filter if a sufficient number of temporal degrees of freedom L is available.

Compare Figure 8.47 with Figure 8.48. Figure 8.47 has been calculated for $L = 2$ taps while in Figure 8.47 $L = 6$ was assumed. We notice that the adaptive space-time FIR filter shows some improvement due to increased filter length while for conventional processing (beamformer + adaptive temporal filter) no advantage is achieved by increasing the filter length.

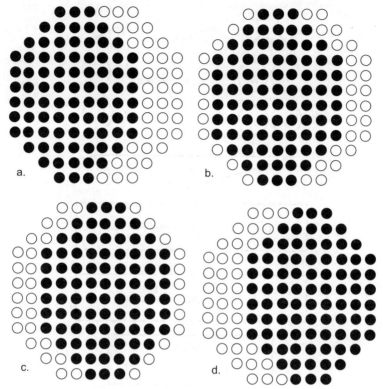

Figure 8.37: Formation of overlapping subarrays in an octagonal array

8.4.3.4 Effect of PRF

In most of the numerical calculations we assumed that the PRF was chosen to be the Nyquist frequency of the clutter bandwidth. Therefore, the Doppler response was unambiguous between $F = -0.5 \ldots 0.5$. In low PRF and medium PRF modes the PRF may be chosen so that the clutter spectrum is undersampled in the time dimension. This causes ambiguities within the clutter bandwidth, see for example Figures 3.20 and 3.21. The following results are based on the irregular subarray structure shown in Figure 8.1.

In Figure 8.50 the effect of the choice of the PRF is demonstrated. The upper curves show the space-time FIR filter results, while the lower curves have been calculated for the beamformer + adaptive *temporal* FIR filter. It can be seen that for Nyquist sampling (subplot a) we find only one clutter notch within the clutter bandwidth while for decreasing PRF (subplots b–d) the number of ambiguous clutter notches increases.

The clutter notches cause 'blind' velocities as usual in MTI systems. These ambiguous blind velocities result in degradation in target detection within the clutter band. However, as can be seen particularly from Figure 8.50d the 'clutter resolution' (width of the clutter notch) of the space-time filter is much higher than that of the

280 Antenna related aspects

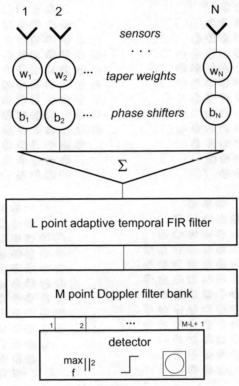

Figure 8.38: Beamformer and temporal clutter filter cascaded

conventional beamformer. Therefore, regardless of the losses at the blind velocities, the space-time filter is clearly superior to one-dimensional processing. The advantage of space-time processing shown by Figure 8.50 is impressive.

8.4.4 Volume array

For completeness we compare the space-time and the conventional temporal processor also for the case of the crow's nest antenna. As can be seen from Figure 8.49, the conventional MTI processor fails completely. It behaves similar to the subarray techniques which were briefly addressed in Figure 8.29.

8.5 Other antenna related aspects

8.5.1 Sparse arrays for spacebased radar

In this section we consider antenna related aspects of space-based GMTI radar. Some issues arising in the practical implementation of space-based GMTI radar have been

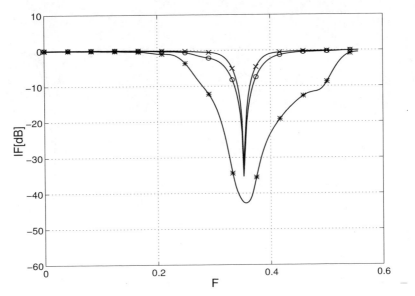

Figure 8.39: Performance of a linear array without tapering ($\varphi_L = 45°$, CNR = 20 dB, FL): × *optimum processing;* ○ *space-time adaptive FIR filter;* ∗ *non-adaptive beamformer + adaptive temporal filter*

discussed by DAVIS [111].

The question is, can a space-based radar be used to detect slow vehicles on the ground? The main difference compared with airborne radar lies in the platform velocity. For space-based radar it is typically in the region of 7000 m/s whereas the speed of an airplane is around 300 m/s. This means that the minimum detectable velocity (MDV) produced by the STAP receiver has to be an order of magnitude lower than for airborne applications. The minimum detectable velocity (i.e., the width of the clutter notch) depends on both the spatial (array) and the temporal (coherent processing interval) apertures. This will be illustrated by a couple of numerical examples which are based on the parameter set given in Table 8.1. This section contains material published in KLEMM [345, 344].

Increasing the physical aperture of the array antenna for space-based applications may be limited by constraints such as size, weight, and power consumption. It is much easier to increase the CPI by either transmitting a sufficient number of pulses or by temporal undersampling, that means, transmitting at lower PRF. In the example in Figure 8.51 an array with 768 elements with 0.5λ spacing was assumed. The PRF varies between the different curves. With lower PRF the temporal aperture increases and, hence, the clutter notch becomes narrower. As can be seen the width of the clutter notch is only a few m/s wide if PRF=f_{Ny}/16. If, however, the temporal sampling determined by the PRF is of the order of magnitude of the spatial sampling determined by the element spacing (two lower curves) the clutter notch widens up considerably. Discrimination of slow targets from clutter might be impossible.

Let us now replace the $\lambda/2$ spaced array assumed in Figure 8.51 by a sparse array.

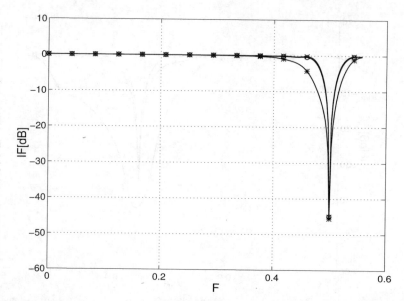

Figure 8.40: Performance of a linear array without tapering ($\varphi_L = 0°$, CNR = 20 dB, FL): × *optimum processing;* ○ *space-time adaptive FIR filter;* ∗ *non-adaptive beamformer + adaptive temporal filter*

In Figure 8.52 the PRFs were chosen as in Figure 8.51. In addition an array with 24 elements was used, and the spacing was varied in the same way as the PRF. Except for the lowest curve we have undersampling in both space and time while preserving the DPCA condition (spatial sampling equal to temporal sampling). As can be seen we obtain a performance similar to the fully filled array assumed in Figure 8.51.

In Figure 8.53 we deviate from the DPCA condition in that the array is rotated in the horizontal dimension. Clearly, for a rotated array there is no coincidence of array phase centres between subsequent PRIs. In the case of Nyquist sampling in both space and time we get a normal clutter notch, but very poor clutter resolution, because the spatial and temporal apertures are very small (lower curve). For a large array with $\lambda/2$ spacing but undersampled PRF we obtain the desired performance (upper curve). Finally, if both the PRF and the array spacing is undersampled array rotation causes dramatic losses.

In fact, such a, strongly undersampled system is extremely sensitive to any kind of deviation from the DPCA condition (coincidence of spatial and temporal samples). Even a very little rotation of the array by 1° causes dramatic losses (Figure 8.54). Another example showing the sensitivity of undersampled space-time processing is given in Figure 8.55. Here we assumed a sidelooking configuration without rotation, however, the spatial sampling interval (element displacement) was chosen to be 1% smaller than the temporal sampling interval. Again we notice dramatic losses in performance.

We can conclude that the minimum detectable velocity (MDV) can be reduced by increasing the spatial and/or temporal apertures (array aperture, CPI). However,

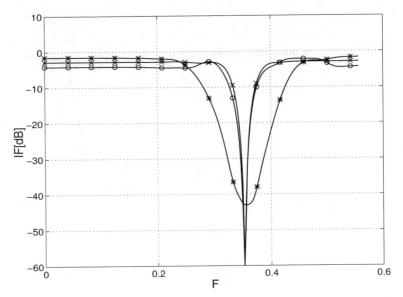

Figure 8.41: Effect of tapering ($\varphi_L = 45°$, CNR = 20 dB, linear array, FL): × *optimum processing;* ○ *space-time adaptive FIR filter;* ∗ *non-adaptive beamformer* + *adaptive temporal filter*

when undersampling is applied in both space and time the DPCA condition has to be met precisely. Any small deviations (tolerance, array orientation errors etc.) from the DPCA condition cause dramatic losses. Because of this sensitivity undersampling in space and time is not useful for practical applications.

However, the performance is robust against deviations from the DPCA condition if Nyquist sampling is used in *one* of the two dimensions. This can be explained as follows: If a waveform is sampled at Nyquist rate each point of the waveform can be recovered by interpolation techniques. Therefore, even if the DPCA condition is not met, that is, the temporal samples do not precisely coincide with the spatial samples, there is always sufficient correlation between temporal and spatial samples as long as one of the dimensions is sampled at Nyquist rate.

Since the STAP problem is symmetric in space and time one can decide which of both dimensions should be sampled at Nyquist rate. As stated earlier, an undersampled (thinned) array leads to a reduction in size, weight and power consumption. This would however, require that the PRF is chosen to be equal to the clutter Doppler bandwidth. In the numerical example at hand this is about 100 Khz. Such a high PRF would cause a very narrow unambiguous range increment and cause a large number of ambiguous clutter arrivals. Therefore, the better choice is an array sampled at half wavelength in conjunction with undersampling in the time domain.

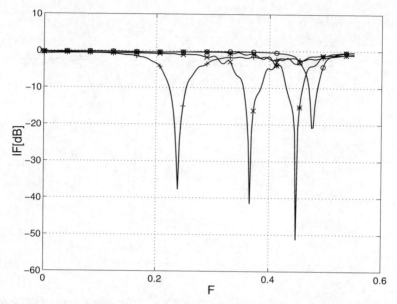

Figure 8.42: Space-time processing, no tapering (circular array, FL): ○ $\varphi_L = 0°$; * $\varphi_L = 20°$; × $\varphi_L = 40°$; + $\varphi_L = 60°$

8.5.2 Polarisation-space-time processing

8.5.2.1 *Overview of available literature*

The concept of space-time processing for clutter rejection in moving radar system can be extended to one more dimension by including the polarisation. This requires a polarimetric phased array antenna on receive.[9] Different polarisation properties of targets and clutter have been used for polarimetric clutter cancellation. The concept of polarimetric clutter cancellation can be included in the coherent processing by forming a *joint polarisation-space-time* processor which has been described in detail by PARK H.-R. *et al.* [522]. Short versions of this paper can be found in PARK H.-R. and WANG [521]. Joint polarisation-space-time processing promises improvement in target detection especially when clutter and target are closely spaced in both the angle and Doppler domains.

The fully adaptive processor (referred to by the authors as the 'specified polarisation-space-time generalised likelihood ratio' (SPST-GLR)) requires a filter bank for all possible target polarisations which makes this processor unattractive for practical use.

A suboptimum solution (referred to as PST-GLR) uses a steering vector in the space-Doppler domain, however, not in the polarisation dimension. It is shown by the authors that this techniques approximates well the performance of the SPST-GLR. Considerable improvement compared with space-time only processing can be obtained

---

[9]The full polarimetric information (HH, VV, HV, VH) is obtained by using polarisation diversity on receive *and transmit*. No processing technique for exploiting the full polarisation is known to the author.

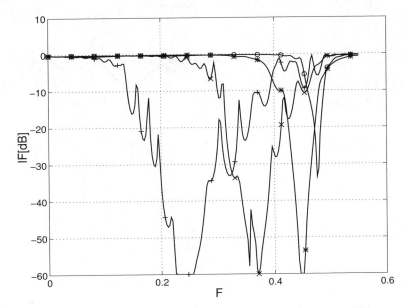

Figure 8.43: BF + adaptive temporal filter, no tapering (circular array, FL): ○ $\varphi_L = 0°$; * $\varphi_L = 20°$; × $\varphi_L = 40°$; + $\varphi_L = 60°$

by including the polarisation in the adaptive processor. The improvement depends on the degree of polarisation of the clutter.

A so called 'polarimetric discontinuity detector' has been described by PARK H.-R. *et al.* [523]. The PDD statistic involves the estimated polarimetric clutter covariance matrix and the primary data in the cell under test, but no signal replica is used. This avoids the construction of a filter bank for all possible target polarisations.

The problem of polarimetric discrimination of targets from clutter by a phased array radar has been discussed by HANLE [249, 250, 251]. The author has pointed out that polarimetric target discrimination is degraded at all look directions other than broadside. An error compensation technique is proposed.

DESMÉZIÈRES *et al.* [114] use polarisation space-time processing for channel identification in a multipath environment.

Some results on polarimetric STAP processing have been presented by SHOWMAN *et al.* [608]. The authors compare two processing schemes:

- the polarimetric matched filter which optimises the signal to clutter output. This concept leads to the eigenvector corresponding to the maximum eigenvalue of the generalised eigenvalue problem stated by the two covariance matrices of signal and clutter.

- the polarimetric whitening filter which is identical to the optimum adaptive processor (OAP).

For the polarimetric matched filter the target covariance matrix and, hence, the polarimetric properties of the target have to be known a priori. The results achieved

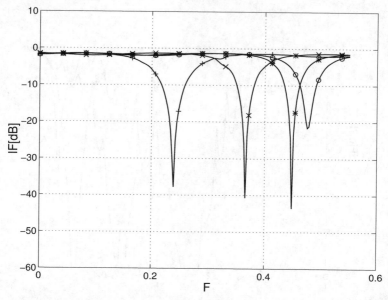

Figure 8.44: Space-time processing, 40 dB Taylor weighting (circular array, FL): ○ $\varphi_L = 0°$; * $\varphi_L = 20°$; × $\varphi_L = 40°$; + $\varphi_L = 60°$

by the authors have demonstrated that including polarisation into STAP may offer advantages if the polarisation properties of target and clutter are different enough. In most considered cases the polarimetric whitening filter has proven to be superior to the polarimetric matched filter.

8.5.2.2 Arrays with non-uniformly polarised elements

The geometry of a conformal may cause a variation of the polarisation of the individual radiating elements. such an example is described by WORMS [726]. In this example several Vivaldi antennas are mounted around a circular core (for example a seekerhead) in such a way that the polarisation varies from one to the next element by a certain angle increment.

In such case all spatial signal processing (beamforming, jammer nulling, STAP) has to be extended by the polarisation dimension in order not to lose any signal energy.

If, however, polarimetric space-time processing is available anyway, it may bring advantages in discrimination of slow moving targets from clutter (SHOWMAN *et al.* [608]). For more details see Section 8.5.2.

8.5.3 Radome effects

Normally the radar is covered by a radome to protect it against environmental effects, in particular weather. In some applications the radar has a dome or cone shape as in the case of a fighter aircraft nose radar. Typical features of such radomes are the so-called *flash lobes*. By flash lobes secondary arrivals (jamming or clutter) in the antenna

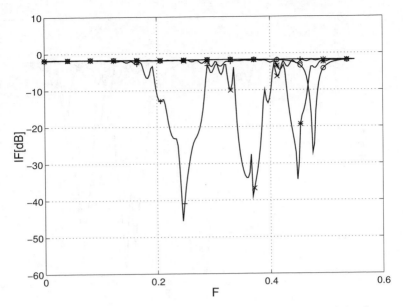

Figure 8.45: BF + adaptive temporal filter, 40 dB Taylor weighting (circular array, FL): ○ $\varphi_L = 0°$; * $\varphi_L = 20°$; × $\varphi_L = 40°$; + $\varphi_L = 60°$

mainbeam due to reflections at the inner side of the radome are understood. Dealing with a moving radar this means that the main beam receives a superposition of two clutter arrivals. In general these arrivals come from different directions and, hence, have different Doppler frequencies. The following aspects are relevant for judging the impact of flash lobes:

- The radome material is chosen so that the major part of the energy penetrates the radome. A small portion is absorbed and another small portion is reflected. Therefore the clutter energy reflected at the inner of the radome is supposed to be small compared with the clutter in the main beam direction. This is valid for transmit and receive.

- A part of the 'flash' clutter arrival (direct transmit, reflected receive) is attenuated by the sidelobe pattern of the transmit beam.

- The second part of the flash clutter arrival (reflected transmit, reflected receive) is attenuated twice by the reflection process at the inner radome.

- For forward looking radar the clutter power is distributed on a semi-circular trajectory as shown in Figure 3.19. The reflected clutter spectrum looks similar to Figure 3.32, but shifted by the reflection angle along the azimuth axis and with reduced power.

- Since the direct and reflected paths can be assumed to be uncorrelated the additional power spectrum may need additional degrees of freedom for clutter filtering.

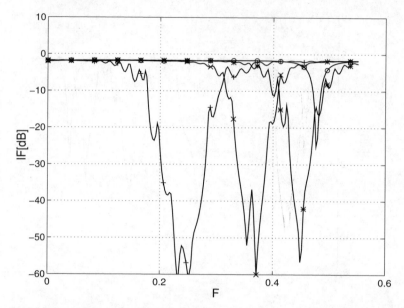

Figure 8.46: BF + *adaptive temporal filter, 40 dB Taylor weighting (CNR=40 dB; circular array, FL):* ∘ $\varphi_L = 0°$; ∗ $\varphi_L = 20°$; × $\varphi_L = 40°$; + $\varphi_L = 60°$

No theoretical or experimental results on clutter flash lobes are available in the open literature. From the above considerations, however, the effect of clutter flash lobes is expected to be small.

8.5.4 Alternating transmit approach

LOMBARDO and COLONE [417] propose a technique to increase the virtual aperture of a small antenna array. Instead of transmitting with the full array the aperture is subdivided in a leading and a trailing subarray which both tranmit in an alternating fashion. As a consequence, the number of two-way phase centres is doubled. Using this concept for STAP leads to an improvement in slow target detection. Moreover, the accuracy of bearing estimation is enhanced.

8.6 Summary

While all the results obtained in the previous chapters were based on linear arrays, in this chapter we focus on alternative antenna array configurations which are more adapted to practical requirements. The findings of this chapter can be summarised as follows

1. **Circular planar** antenna arrays will play a major role in future aircraft nose radars. It turns out that circular antennas exhibit favourable properties with

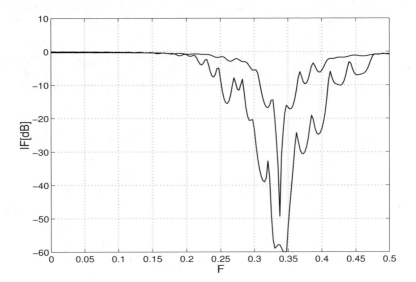

Figure 8.47: Effect of temporal filter length ($L = 2$, FL, $\varphi = 45°$). Upper curve: space-time FIR filter; lower curve: BF + adaptive temporal filter

respect to space-time adaptive clutter cancellation. In detail the following results were obtained:

- Space-time adaptive processing can be used with sideways or forward looking circular planar antennas.
- For large antennas the optimum processor is of no practical use due to computational complexity.
- The suboptimum STAP processor based on a subarray structure and a space-time FIR filter reaches almost optimum clutter rejection performance. In this context it should be noted that the circular shape of the antenna provides some horizontal tapering.
- Tapering results in slight tapering losses but may improve the performance of the conventional beamformer/temporal adaptive filter processor, especially in the vicinity of the clutter notch.
- In the broadside direction of the array (flight axis) clutter echoes are narrowband so that the conventional processor may be satisfactory.
- The clutter bandwidth increases with the off-broadside look direction so that the performance of the temporal filter is considerably degraded while the space-time processor exhibits the same performance as in the broadside direction. However, tapering mitigates the effect of the clutter bandwidth at moderate CNR values.
- In the case of high CNR values (40 dB), only the space-time processor yields sufficient clutter rejection.

290 Antenna related aspects

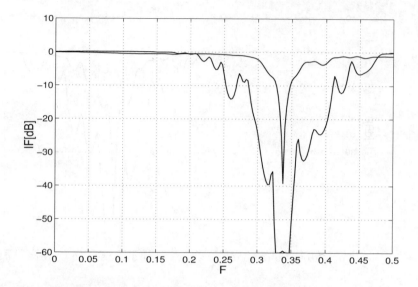

Figure 8.48: Effect of temporal filter length ($L = 6$, FL, $\varphi = 45°$). Upper curve: space-time FIR filter; lower curve: BF + adaptive temporal filter

- If additional broadband jammers are present the conventional processor fails because there is no temporal correlation that could be exploited.

2. Space-time adaptive clutter rejection can be performed with **conformal** arrays as long as they are cylindral and equispaced in the flight direction.

3. Other conformal array configurations are possible, for instance ellipse-like arrays at the nose of an aircraft (see Fig. 8.31).

4. A **volume array** like the crow's nest antenna cannot be subdivided into subarrays so that satisfactory space-time clutter rejection is achieved. One way of suboptimum processing may be a configuration with sensor doublets or triplets instead of single sensors. However, this will probably involve problems of practical implementation.

5. If the PRF is chosen below the Nyquist rate of the clutter bandwidth ambiguous clutter notches (blind velocities) occur. However, space-time processing is always superior to conventional beamforming + adaptive temporal clutter filtering.

6. For circular arrays (in contrast to linear or rectangular arrays) the performance of the space-time FIR filter can be improved by increasing the temporal filter length L. This has to do with the fact that for circular arrays it is impossible to generate equal subarrays. Increasing the number of temporal degrees of freedom has to be paid for in terms of computational complexity.

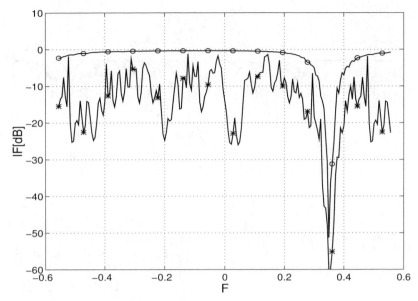

Figure 8.49: The crow's nest antenna: ○ *optimum space-time processing;* ∗ *beamformer + optimum temporal filter*

7. By using an omnidirectional radiating element horizontal planar arrays with 360° azimuthal coverage can be designed. Several of such arrays may be arranged at constant distances so that a uniform linear array with directive subarrays is formed. Such configuration can be applied to airborne surveillance GMTI radar.

8. In spaceborne radar systems undersampling can be a way of reducing the minimum detectable velocity. However, a system that is undersampled in both the spatial and the temporal dimension is extremely sensitive agains any deviation from the DPCA condition. Undersampling should be applied only in one dimension, preferably in the temporal dimension (reduced PRF).

9. Arrays with non-uniform polarisation need space-time-polarisation processing.

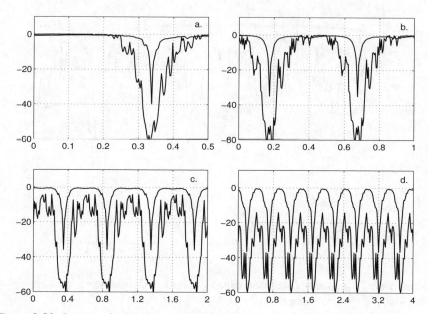

Figure 8.50: Impact of PRF (forward looking, $\varphi = 45°$; upper curves: space-time FIR filter, lower curves: BF + adaptive temporal filter): a. PRF = f_{Ny}; b. PRF = $f_{\mathrm{Ny}}/2$; c. PRF = $f_{\mathrm{Ny}}/4$; d. PRF = $f_{\mathrm{Ny}}/8$

Table 8.1 Radar parameters

platform height	$H = 400$ km
target slant range	$R = 800$ km
array type	linear, equispaced
array orientation	sidelooking
number of array elements	N=768, 24
number of spatial channels	K=3; 5
number of processed echoes	M=24
number of temporal channels	L=3; 5
clutter-to-noise ratio (CNR)	20 dB
wavelength	λ=0.03 m
PRF (Nyquist)	1.02 MHz
PRF (useful)	4 KHz

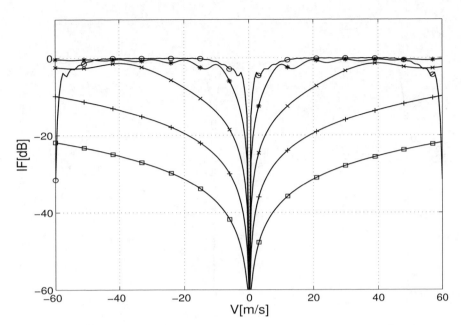

Figure 8.51: Dependence of the clutter notch on the PRF. □ $PRF = f_{Ny}$ *(Nyquist of clutter band width, DPCA case);* + $PRF = f_{Ny}/4$; × $PRF = f_{Ny}/16$; ∗ $PRF = f_{Ny}/64$; ○ $PRF = f_{Ny}/256$

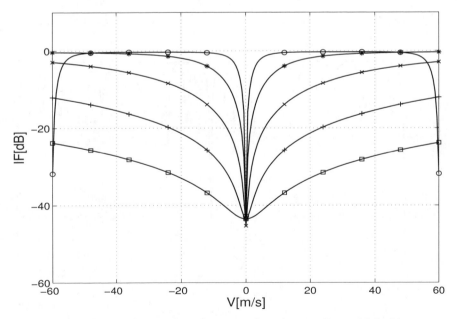

Figure 8.52: Effect of spatial and temporal undersampling, sidelooking array. Undersampling: □ *no undersampling: Nyquist of the clutter band, $\lambda/2$ spacing;* + *1/4;* × *1/16;* ∗ *1/64;* ○ *1/256*

294 Antenna related aspects

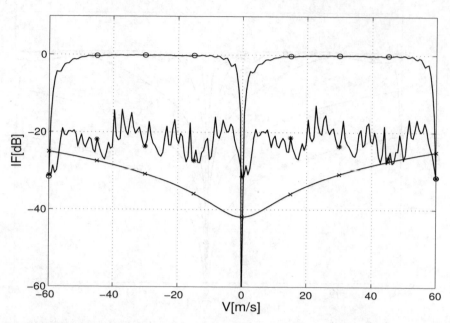

Figure 8.53: Effect of spatial and temporal undersampling with rotated array. ○ $\lambda/2$ *spacing, PRF undersampled by a factor of 256;* ∗ *both array and PRF undersampled by 256;* × *Nyquist sampling in space and time. Rotation angle:* $45°$

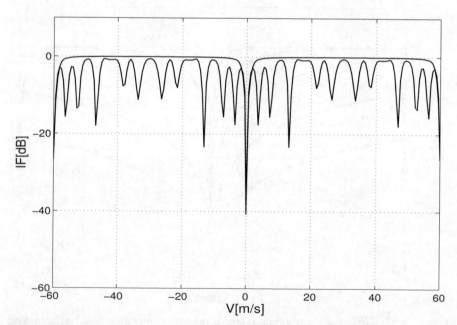

Figure 8.54: Effect of $1°$ *array rotation (lower curve), no rotation (upper curve),* $N = M = 24$, *PRF=Nyquist/256*

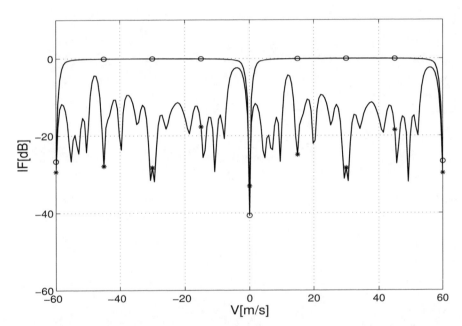

Figure 8.55: Deviation by 1% from the DPCA condition (lower curve), no deviation (upper curve), $N = M = 24$, *PRF=Nyquist/256*

Chapter 9

Space-frequency processing

9.1 Introduction

In the previous chapters several *space-time* processing algorithms were presented which are capable of adaptive real-time processing. In this chapter some consideration on *space-frequency* domain processing[1] are made. We discuss a total of five different processing schemes. Some of them are variants of the space-time processors described in Chapters 5, 6, and 7. All processors presented will be compared in terms of performance. A comparison in terms of computational complexity will be presented in Chapter 16. WRIGHT and WELLS [727] compare pre-Doppler and post-Doppler architectures for application with space-based radar. KEALEY and FINLEY [285] as well as BICKERT [47] compared pre- and post-Doppler STAP on the basis of airborne clutter data. They found both approaches equivalent.

The near-optimum techniques described in the previous chapters have been derived from the optimum receiver (1.3), which in essence means *pre-whitening* of the clutter component in the received echo signals and *matching* to the desired target signal. It has been shown in Section 1.2.6 that the optimum test can be formulated in the frequency domain as well as in the time domain. Now we have to extend this concept to space-time vector quantities.

The discrete Fourier transform can be written as a unitary matrix $\mathbf{F}$. Since we want to transform from time to frequency while not changing the space dimension the space-time Fourier transform becomes

$$\mathbf{F} = \begin{pmatrix} w_{11}\mathbf{I} & w_{12}\mathbf{I} & \cdots & w_{1M}\mathbf{I} \\ w_{21}\mathbf{I} & w_{22}\mathbf{I} & \cdots & w_{2M}\mathbf{I} \\ \vdots & \vdots & \ddots & \vdots \\ w_{M1}\mathbf{I} & w_{M2}\mathbf{I} & \cdots & w_{MM}\mathbf{I} \end{pmatrix} \quad (9.1)$$

---
[1] Frequently referred to as *post-Doppler* processing.

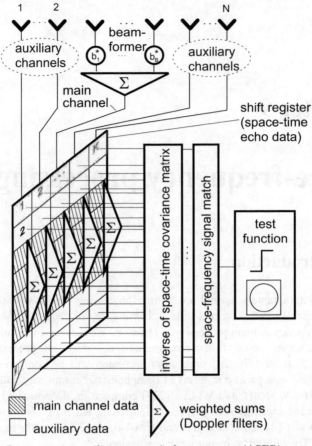

Figure 9.1: 2-D symmetric auxiliary sensor/echo processor (ASEP)

where **I** is the spatial $N \times N$ unit matrix.[2] The coefficients w_{nm} are given by (1.94)

$$w_{nm} = \frac{1}{\sqrt{M}} \exp(-j2\pi(n-1)(m-1)/M) \quad n, m = 1, \ldots, M \quad (9.2)$$

By similarity transform the clutter covariance matrix **Q** becomes a power spectral matrix

$$\mathbf{Q}_F = \mathbf{F Q F}^* \quad (9.3)$$

and the space-time vectors of the desired target signal and received echo signal become

$$\begin{aligned} \mathbf{s}_F &= \mathbf{F s} \\ \mathbf{x}_F &= \mathbf{F x} \\ \mathbf{c}_F &= \mathbf{F c} \end{aligned} \quad (9.4)$$

---

[2] Or $K \times K$ according to the processors in Chapter 6.

Introduction 299

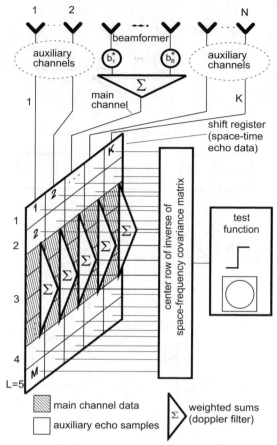

Figure 9.2: 2-D symmetric auxiliary sensor/echo processor, with signal contributions by auxiliary sensors/data omitted (ASEP)

$$n_F = Fn$$

Then the optimum space-time processing becomes in the frequency domain

$$y_F = x_F^* Q_F^{-1} s_F \qquad (9.5)$$

which is obviously identical with $y = x^* Q^{-1} s$ since $F^* F = FF^* = I$.

This well-known fact encourages us to look in more detail into Doppler frequency domain approaches for real-time adaptive air- and spaceborne clutter suppression. While we are dealing with space-time vectors and matrices it should be noted that some of the techniques work in the space-frequency domain, i.e., a one-dimensional Fourier transform will be applied in the time dimension of the space-time filter. Two other techniques work in the *angular-Doppler domain*, which means the space-time data will be transformed by a two-dimensional Fourier transform before adaptive processing. Five different techniques are described, two of which are new variants of the auxiliary sensor FIR filter (ASFF) processor described in Chapter 7.

300 Space-frequency processing

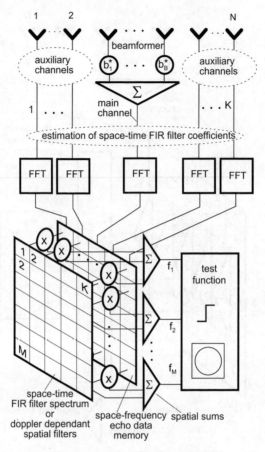

Figure 9.3: Frequency domain space-time FIR filter (FDFF)

9.2 The auxiliary space-time channel processor (ACP)

This processor has already been discussed in Chapter 5, Section 5.3, and Figures 5.6 and 5.7. Recall that a bunch of beams is distributed over the entire angular clutter domain so that all clutter is received by full gain beams.

To match the clutter in angle and frequency each of the beams is cascaded with a doppler filter matched to the respective clutter doppler frequency. A search channel (in practice a beamformer cascaded with a doppler filter bank) is added for matching the desired target.

Estimation of the clutter covariance matrix and clutter filtering takes place in the beam/doppler channel domain. Therefore, this technique belongs to the family of space-frequency processors and has been mentioned here for the sake of completeness.

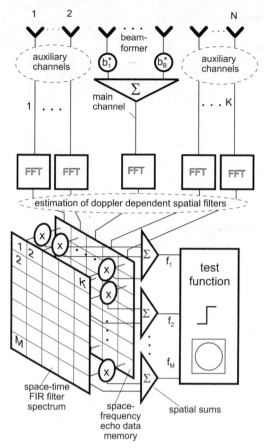

Figure 9.4: Frequency-dependent spatial filter (FDSP)

9.3 The symmetric auxiliary sensor/echo processor

The processor described below is based on auxiliary sensors and auxiliary echo samples. It is referred to as auxiliary sensor/echo processor (ASEP). Let us recall the considerations on the *symmetric auxiliary sensor processor* made in Section 6.2.1. The results given there have shown that this structure is quasi-optimum in terms of slow target detection. In this section we try to extend this concept to the time dimension.

In Figure 7.5, the symmetric auxiliary sensor/space-time FIR filter processor was shown. Let us now modify this scheme by replacing the FIR filter by an *symmetric auxiliary data* scheme as used in the spatial domain in the upper part of Figure 7.5. Recall that in (6.17) the symmetric auxiliary sensor transform was described by a spatial transform matrix of the form

$$\mathbf{T}_s = \begin{pmatrix} 1 & & & & & & & & 0 \\ & \ddots & & & & & & & \\ & & 1 & & & & & & \\ & & & b_1 & & & & & \\ & & & \vdots & & & & & \\ & & & b_B & & & & & \\ & & & & 1 & & & & \\ & & & & & \ddots & & & \\ 0 & & & & & & & & 1 \end{pmatrix} \qquad (9.6)$$

where the b_i are beamformer weights. Extending this scheme to the time dimension leads to a transform matrix of the form

$$\mathbf{T} = \begin{pmatrix} \mathbf{T}_s & & & & & & & 0 \\ & \ddots & & & & & & \\ & & \mathbf{T}_s & & & & & \\ & & & w_{f1}\mathbf{T}_s & & & & \\ & & & \vdots & & & & \\ & & & w_{fm}\mathbf{T}_s & & & & \\ & & & \vdots & & & & \\ & & & w_{f,M-L+1}\mathbf{T}_s & & & & \\ & & & & \mathbf{T}_s & & & \\ & & & & & \ddots & & \\ 0 & & & & & & & \mathbf{T}_s \end{pmatrix} \qquad (9.7)$$

L is the number of temporal channels after the transform, w_{fm} is the Fourier phase coefficient according to (9.2), with f denoting the Doppler frequency and m the time (index of successive PRIs). The processor and the improvement factor are given by (5.3), (5.4) and (5.5), respectively.

Figure 9.1 shows a block diagram of this processor. On the top we again have the symmetric auxiliary sensor configuration which was discussed in some detail in Section 6.2.1. $K - 1$ auxilary sensors are arranged in a symmetric fashion on both sides of a beamformer with beamformer weights b_i. This step is described by the spatial transform (9.6).

Recall that the symmetric auxiliary channel structure as defined by (9.6) is equivalent to overlapping subarrays (6.8). The *PRI-staggered* post-Doppler processor by WARD and STEINHARDT [692] generates additional temporal degrees of freedom by forming several Doppler filters for each Doppler frequency, all being based on L different data windows. These windows are used to form overlapping data segments which is a *time domain* equivalent to overlapping subarrays in the spatial domain. The processor is equivalent to that in Figures 9.1 and 9.2. According to the authors this technique is robust against tolerances of the element patterns because it belongs to the class of post-Doppler processors. The PRI staggered post-Doppler processor has proven to perform well for short CPI (PARKER [530]).

The output data are stored in a $K \times M$ data memory. The data in the shadowed part of the memory are fed into Doppler filters with equal Doppler frequency while the white data serve as *temporal* auxiliary channels. The sums in each of the columns indicate the Doppler filters, that is, all the shadowed data in each of the columns are weighted with Doppler filter coefficients before summation. Notice that we arranged here in the time domain the same auxiliary channel configuration as in the spatial dimension on top of the drawing. The Doppler filters assume the role of the beamformer while echo data before and after the Doppler filter serve as 'auxiliary channels'.

The number of output channels is now reduced in both the spatial and temporal dimension. At this level the adaptation (e.g. estimation of the space-frequency covariance matrix) has to be carried out. The size of the space-frequency covariance matrix is $KL \times KL$ (in our example 25×25).

In practice we want to have the numbers of auxiliary sensors and auxiliary echo data small compared with the number of beamformer elements, or the length of the echo data record, respectively. Therefore, according to Section 6.2.1.3, the target signal contributions of the auxiliary sensors and echo samples may be neglected. Then the transformed (or secondary) signal matching vector

$$\mathbf{s}_T = \mathbf{T}\mathbf{s} \tag{9.8}$$

reduces to

$$\mathbf{s}_T \approx \begin{pmatrix} 0 \\ \vdots \\ 0 \\ 1 \\ 0 \\ \vdots \\ 0 \end{pmatrix} \tag{9.9}$$

The adaptive processor

$$\mathbf{w}_T = \mathbf{Q}_T^{-1}\mathbf{s}_T \tag{9.10}$$

reduces to the centre column of $\mathbf{Q}_T^{-1}$. This simplified processor is shown in Figure 9.2. Since beamforming and Doppler filtering has been included in the pre-transform no further Doppler filter bank is required.

9.3.1 Computing the inverses of the spectral covariance matrices

The adaptive processing, that is, calculating the centre row of $\mathbf{Q}_T^{-1}$, requires the estimation and inversion of the space-time covariance matrices for all Doppler frequencies of interest.

Notice that only the main channel (shadowed) depends on the Doppler frequency. The auxiliary samples on the top and on the bottom of the memory (white) are the

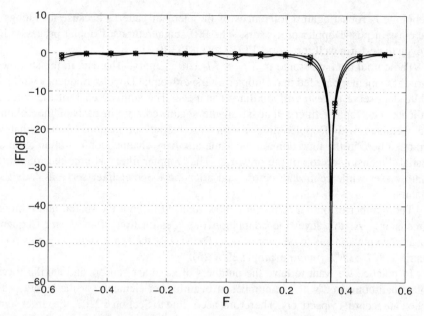

Figure 9.5: Comparison of space-frequency processors (sidelooking array, $\varphi_L = 45°$): ○ *OAP;* ∗ *ASEP;* × *ACP;* + *ASFF;* $N = M = 24$, $K = L = 5$

same for all Doppler frequencies. Therefore, the space-time clutter covariance matrices associated with different Doppler frequencies differ only in the centre column and row. This property can be exploited by a dedicated algorithm for efficient calculation of the matrix inverses. This algorithm is based on matrix partitioning and has been described in Chapter 5, Section 5.5.

9.4 Frequency domain FIR filter (FDFF)

Let us again come back to the auxiliary sensors FIR filter (ASFF) described in (7.12), (7.15) and Figure 7.5. Notice that the FIR filter operation involves a convolution in the time dimension. The temporal convolution of the space-time least squares FIR filter with the space-time data sequence can basically be carried out in the frequency domain by multiplying the frequency response of the FIR filter with the spectra of the radar data (fast convolution).

Such a processor is depicted in Figure 9.3. The coefficients of the space-time FIR filter are estimated in the time-domain, and a set of spectral filter coefficients is calculated by applying a multichannel DFT (space-time filter spectrum). For each frequency the contributions of the K antenna channels are then summed up, and the resulting filtered spectrum is fed into the usual detection device.

On the one hand this approach needs a multichannel FFT instead of a single channel FFT; on the other hand the stationarity of the received echo sequences is exploited in that only *one* set of filter coefficients must be estimated for all frequencies. This

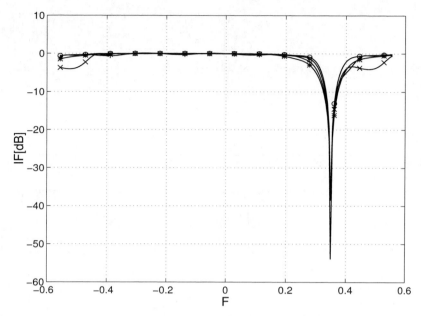

Figure 9.6: Comparison of space-frequency processors (forward looking array, $\varphi_L = 45°$): ○ OAP; ∗ ASEP; × ACP; + ASFF; $N = M = 24$, $K = L = 5$

frequency domain processor is equivalent to the space-time FIR filter shown in Figure 7.5, and described in (7.12) or (7.15).

It has been found in Chapter 7 that the clutter suppression performance is nearly independent of the temporal dimension of the space-time FIR filter. Under certain conditions additional degrees of freedom may be desirable (e.g. to compensate for tolerances of the sensor positions and for other array errors). The number of degrees of freedom can be increased by choosing a larger temporal filter length L. Increasing the temporal filter dimension requires additional operations if the filtering is carried out in the time domain. For a frequency domain FIR filter the computational expense is independent of the filter length as long as the clutter filter is shorter than the number of Doppler cells.

A specific problem associated with frequency-domain FIR filtering should be mentioned. The FIR filter is effective only as long as the temporal extension of the FIR filter lies inside the data sequence to be filtered. If the temporal filter dimension slides over the edges of the data record (that means, there is only a partial overlap between the temporal filter length and the data record) clutter suppression will not work any longer in this part of the convolution.

The Fast Convolution algorithm (replacing the convolution in the time domain by a multiplication in the frequency domain while exploiting the computational efficiency of the FFT) provides a complete convolution including the edge data. If the data record has the temporal length M and the temporal filter dimension is L, then the resulting data sequence has the length $M + (L - 1)$. In order to obtain the clutter-free data sequence one has to truncate the resulting sequence by $L - 1$ values on both sides.

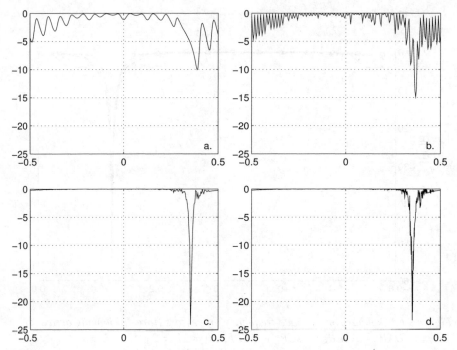

Figure 9.7: *Frequency-dependent spatial filtering (IF vs normalised Doppler; SL, CNR = 20 dB)*: a. $M = 16$; b. $M = 64$; c. $M = 256$; d. $M = 1048$

Then the resulting data sequence has the length $M + 1 - L$. Compared with the time domain implementation, instead of one FFT three of them have to be calculated.

9.5 Frequency-dependent spatial processing (FDSP)

A similar processor can be designed when the estimation of the filter coefficients from clutter data is carried out in the frequency domain, that means, *after* the FFT, see Figure 9.4. This filter is based on the frequency-dependent space-frequency covariance matrices which are based on adjacent Doppler bins and array elements. These matrices have the dimension $KL \times KL$ where L means here a certain number of FFT channels. This kind of processing is similar to the JDL-GLR by WANG H. and CAI [668] which will be addressed further below.

Such post-Doppler processors tend to be more robust against errors of the sensor patterns (see WARD and STEINHARDT [692]). The individual Doppler filters provide narrowband processing which is equivalent to using only a small sector of the element pattern. Therefore, the Doppler spectral effects caused by different sensor patterns is mitigated.

An attractive processor is found for $L = 1$ which means that *spatial* clutter cancellation filters are cascaded with the individual Doppler channels. The spectral cross-terms of the space-frequency clutter covariance matrix are not taken into account

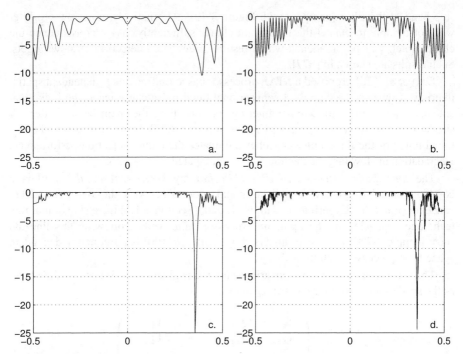

Figure 9.8: Frequency-dependent spatial filtering (IF vs normalised Doppler; SL, CNR = 40 dB): a. $M = 16$; b. $M = 64$; c. $M = 256$; d. $M = 1048$

for calculating the space-frequency clutter filter. This is a significant simplification of the processing based on the fact that signals at the outputs of different Fourier channels tend to become mutually uncorrelated as the observation time MT approaches infinity.

This frequency-dependent spatial clutter cancellation technique has been applied for instance in the MCARM program by SURESH BABU et al. [626] and FENNER et al.[3] [162], GOLDSTEIN et al. [197] using MOUNTAINTOP data TITI [646], and in the AER II program by ENDER [127]. Several techniques based on a Doppler filter bank are presented by BAO et al. [29]. A comparison of frequency-dependent spatial clutter cancellation ('factored STAP') and DPCA for use on a spaceborne platform is given by NOHARA [501]. A statistical analysis of this 'partially adaptive STAP detector' for Doppler target detection and azimuth estimation is given by REED et al. [558]. BERGIN et al. [44] consider frequency dependent spatial processing in the context of multi-resolution SAR/GMTI radar.

Some authors (WANG X. [676], WANG Y. et al. [679], WANG Y. and PENG [677], WU R. et al. [729], XIONG et al. [733]) combine frequency-dependent spatial filtering with a temporal MTI pre-filter in all antenna channels. This pre-filter is to reduce mainbeam clutter. It can be anticipated that such a conventional temporal MTI filter will not yield sufficient improvement in slow target detection. If the pre-filter is matched to the clutter bandwidth determined by the main beam this leads to a broad

---

[3]The authors refer to this as Doppler-factored STAP.

stop band of the filter (compare with Figure 3.40). Alternatively, a narrow pre-filter will not be able to cancel the mainbeam clutter sufficiently. As a consequence the clutter notch will be broadened which results in degraded detection of slow targets. Such results are shown in [677].

BAO et al. [28] proposed a STAP processor based on frequency-dependent spatial filtering. However, they apply a conventional three-pulse canceller to the individual antenna channel prior to Doppler filtering for cancelling the main beam clutter. A similar receiver structure is proposed by WANG et al. [677]. It can be expected that the broad notch of the three-pulse canceller dominates the clutter rejection performance. Degradation in slow target detection can be anticipated.

The *time-space cascaded STAP architecture* by BRENNAN et al. [66] is a combination of frequency-dependent spatial processing with the two-pulse delay canceller principle. Experiments with NRL data (Section 1.1.6) have shown that this technique approximates the optimum processor quite well. Following this line the 'post-Doppler STAP processor' analysed by COOPER [100] is a frequency-dependent space-time processor with three delays.[4]

The space-frequency clutter covariance matrix as given by (9.3) has the same form as the space-time covariance matrix $\mathbf{Q}$ in (3.22)

$$\mathbf{Q}_F = \begin{pmatrix} \mathbf{Q}_{11} & \mathbf{Q}_{12} & \cdots & \mathbf{Q}_{1M} \\ \mathbf{Q}_{21} & \mathbf{Q}_{22} & \cdots & \mathbf{Q}_{2M} \\ \vdots & \vdots & & \vdots \\ \mathbf{Q}_{M1} & \mathbf{Q}_{M2} & \cdots & \mathbf{Q}_{MM} \end{pmatrix} \tag{9.11}$$

where now the indices of the spatial submatrices $\mathbf{Q}_{nm}$ denote the Doppler frequency. A frequency-dependent spatial filter is obtained by omitting the cross-variance matrices in (9.11) and taking the inverse

$$\mathbf{H} = \begin{pmatrix} \mathbf{Q}_{11} & 0 & \cdots & 0 \\ 0 & \mathbf{Q}_{22} & \cdots & 0 \\ \vdots & \vdots & & \vdots \\ 0 & 0 & \cdots & \mathbf{Q}_{MM} \end{pmatrix}^{-1} = \begin{pmatrix} \mathbf{Q}_{11}^{-1} & 0 & \cdots & 0 \\ 0 & \mathbf{Q}_{22}^{-1} & \cdots & 0 \\ \vdots & \vdots & & \vdots \\ 0 & 0 & \cdots & \mathbf{Q}_{MM}^{-1} \end{pmatrix} \tag{9.12}$$

so that the processor becomes

$$\mathbf{w}_F = \mathbf{H}\mathbf{s}_F \tag{9.13}$$

where

$$\mathbf{s}_F = \mathbf{F}\mathbf{s} \tag{9.14}$$

is the Fourier spectrum of the target signal replica.

---

[4]In this application the discrete Fourier transform might be efficiently computed by use of a recursive algorithm (DILLARD [115]).

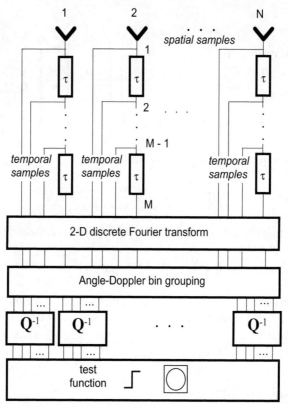

Figure 9.9: Angle-Doppler adaptive processor

If the target signal is a sine wave as assumed in (2.31), the transformed signal vector has the form

$$\mathbf{s}_F = \begin{pmatrix} 0 \\ \vdots \\ 0 \\ \mathbf{s}_m \\ 0 \\ \vdots \\ 0 \end{pmatrix} \qquad (9.15)$$

where we assumed that the target signal frequency denoted by the index m coincides with the frequency of the m-th channel of the discrete Fourier transform matrix. The processor then reduces to

$$\mathbf{w}_F(m) = \begin{pmatrix} \mathbf{Q}_{11}^{-1} & 0 & \cdots & 0 \\ 0 & \mathbf{Q}_{22}^{-1} & \cdots & 0 \\ \cdot & \cdot & \cdots & \cdot \\ \cdot & \cdot & \cdots & \cdot \\ 0 & 0 & \cdots & \mathbf{Q}_{MM}^{-1} \end{pmatrix} \begin{pmatrix} 0 \\ \cdots \\ 0 \\ \mathbf{s}_m \\ 0 \\ \cdots \\ 0 \end{pmatrix} = \begin{pmatrix} 0 \\ \cdots \\ 0 \\ \mathbf{Q}_{mm}^{-1} \mathbf{s}_m \\ 0 \\ \cdots \\ 0 \end{pmatrix} \quad (9.16)$$

According to (5.5) the improvement factor is

$$\mathrm{IF} = \frac{\mathbf{w}_F^* \mathbf{s}_F \mathbf{s}_F^* \mathbf{w}_F \cdot \mathrm{tr}(\mathbf{Q})}{\mathbf{w}_F^* \mathbf{Q}_F \mathbf{w}_F \cdot \mathbf{s}^* \mathbf{s}} \quad (9.17)$$

Inserting (9.16) into (9.17) leads to the following expression for the improvement factor

$$\mathrm{IF}(m) = \mathbf{s}^*_m \mathbf{Q}_{mm}^{-1} \mathbf{s}_m \cdot \frac{\mathrm{tr}(\mathbf{Q})}{\mathbf{s}^* \mathbf{s}} \quad (9.18)$$

Since $\mathbf{s}_m$ is a spatial signal vector except for a factor of $\sqrt{M}$ according to (1.94), and a complex phase factor it can be replaced by a beamformer vector

$$\mathrm{IF}(m) = \mathbf{b}^* \mathbf{Q}_{mm}^{-1} \mathbf{b} \cdot M \cdot \frac{\mathrm{tr}(\mathbf{Q})}{\mathbf{s}^* \mathbf{s}} \quad (9.19)$$

9.5.1 Spatial blocking matrices

Since the space-time clutter filtering has been reduced to a sequence of *spatial* filters associated with different Doppler frequencies *spatial* blocking matrices according to Section 6.3.1 may be applied to each Doppler channel in order to mitigate the effect of target signal cancellation if the target signal is included in the adaptation.

9.6 Comparison of processors

In Figures 9.5 and 9.6 the IF has been plotted versus the normalised target doppler frequency as in the previous chapters. The individual curves are associated with different frequency domain processors. The curves in Figure 9.5 have been plotted for a sidelooking antenna configuration, while Figure 9.6 shows the forward looking case.

As can be seen the performance of all processors is very similar. The usual clutter notch occurs at the clutter Doppler frequency in the look direction (clutter match of the processor). All four curves approximate the optimum IF very well.

Figures 9.7 and 9.8 show the performance of the frequency-dependent spatial filtering technique (FDSF) in its dependence on the number of echo samples ($M = 16, 64, 256, 1048$). Figure 9.7 has been calculated for CNR = 20 dB, Figure 9.8 for CNR = 40 dB. A sidelooking array was assumed. Similar results can be obtained for forward looking radar.

Recall that the theoretical maximum of the improvement factor is $\text{IF}_{\max} = \text{CNR} \cdot NM$, i.e., it is proportional to the number of echoes. The four curves in Figure 9.7 have been normalised to their individual $\text{IF}_{\max}$.

As can be seen from Figures 9.7 and 9.8 the optimum IF is quite well approximated for $M \geq 256$ (curves c). For shorter pulse sequences some losses can be observed. These losses are the penalty for assuming that the clutter power spectral matrix (9.3) of finite order is block diagonal. The block diagonal form is perfectly reached when $M \to \infty$.

9.7 Angle-Doppler subgroups

9.7.1 General description

There is one more option of designing space-time adaptive clutter cancellers. The principle is illustrated in Figure 9.9. The space-time echo samples are transformed by a two-dimensional Fourier transform into the angle-Doppler domain. The total number of angle-Doppler cells is subdivided into two-dimensional groups of angle-Doppler cells. Clutter suppression will be carried out for each of the subgroups separately.

The individual way of grouping is at the choice of the designer. One straightforward choice is to use *adjacent* cells in both the angular and Doppler dimensions. In the spatial dimension subarray beamforming may be carried out in the RF domain. If a fully digitised array is available subarray beamforming can be done digitally.

The principle of adaptive subgroup processing in the Doppler domain has been discussed for one-dimensional adaptive clutter suppression by KLEMM [292]. The doppler domain is subdivided into small subgroups of adjacent Doppler channels. The adaptive processing is applied in parallel or successively to these subgroups. It was found that the achievable performance depends on the subgroup size. Groups of two Doppler bins only are particularly attractive because the spectral covariance matrices have size 2×2 so that matrix inversion does not require any arithmetic operations. The *real-time STAP* technique by MEYER-HILBERG [461] also belongs to the class of frequency domain approaches.

9.7.1.1 The JDL-GLR

The *Joint Domain Localised Generalised Likelihood Ratio Detector* (JDL-GLR) by WANG H. and CAI [668] is a prominent example for an extension of this kind of processing to space-time signals received by an airborne radar. The authors demonstrate that joint space-time processing is superior to cascaded space-time or time-space processing. In comparison with the optimum processor, some losses are encountered.

The JDL-GLR uses auxiliary space-time vectors in a similar way as the processors described in Chapter 5. In this sense it also may belong to the class of space-time transforms as treated in Chapter 5.

However, when operating on measured data (MCARM) the JDL-GLR showed favourable properties in terms of clutter suppression performance and required

Space-frequency processing

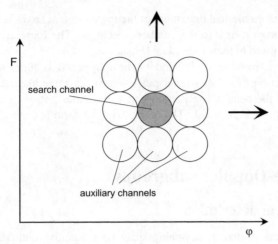

Figure 9.10: Principle of the JDL-GLR

secondary data support (MELVIN and HIMED [449]). A similar multi-beam approach is proposed by WANG and CHEN [684].

Some improvement can be obtained if the pre-calculated steering vectors are replaced by measured sets of coefficients so as to compensate for sensor directivity patterns and mutual coupling between sensors (ADVE and WICKS [4, 7]).

In Figure 9.10 the principle of the JDL-GLR is illustrated. The circles denote positions of space-time search and auxiliary channels in the $F - \varphi$ plane. The centre channel is the search channel which is surrounded by auxiliary channels. The total of auxiliary and search channels span the signal vector subspace for further processing, such as clutter rejection via the inverse of the beam-Doppler subspace covariance matrix. If only the signal energy in the search beam is to be taken into account the JDL-GLR clutter filter is just a vector given by the row of the beam-Doppler covariance matrix associated with the search channel.

The distance between the channels can be chosen arbitrarily. It turns out, however, that close spacing (about half beamwidth, half Doppler channel width) is optimum for slow target detection. Arranging the channels even closer leads to numerical problems.

In Figure 9.11 the normalised improvement factor is shown to illustrate the effect of auxiliary channel spacing on the detection performance of the JDL-GLR. As can be seen best performance is achieved for the closest spacing (3 deg, 3 m/s) which corresponds roughly to half the beamwidth/Doppler channel width.

The JDL space-time adaptive processor can be optimised so as to achieve good low Doppler target detection performance with minimum support of secondary training data (PADOS et al. [519]). Two variants are shown. The first is suitable when the eigenspectrum is spread out whereas the second one is more adapted to cases where the eigenspectrum is concentrated on a few dominant eigenvalues. WANG et al. [688] propose different spacings of Doppler and angular channels to improve the performance of the JDL.

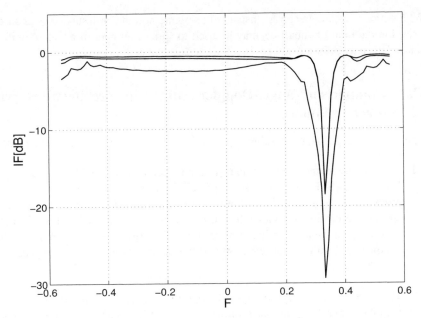

Figure 9.11: Performance of the JDL-GLR. Impact of the distance between auxiliary channels: 3 deg / 3 m/s (upper curve); 6 deg / 6 m/s (middle curve); 12 deg / 12 m/s (lowest curve)

ZHANG et al. [749] modify the classical JDL processor by replacing pairs of auxiliary beams by difference patterns, thus achieving an even more economic processor architecture.

A similar processing scheme has also been used by the authors for the evaluation of the Σ-Δ processor [671]. WICKS et al.[706] compare the performance of the JDL-GLR with a DPCA based radar. It is shown that the JDL-GLR outperforms the DPCA processor, especially when the PRI is not matched to the platform velocity[5] (see Chapter 4, Figure 4.17).

In the paper by COOPER [100] an 'adjacent Doppler bin processor' is discussed. This processor has an architecture similar to the JDL-GLR by H. WANG. The 'M-CAP' processor by Y. WANG et al. [682] uses all array channels but only a few adjacent Doppler bins, which is a special case of the JDL-GLR. The authors demonstrate robustness against channel errors and good performance for various antenna orientations.

The JDL-GLR may have a beneficial property. In many applications the array consists of sensors having different directivity characteristics. Let us consider, for instance, an array subdivided into non-uniform subarrays as shown in Figure 11.13. The associated subbeams have different shapes which receive different clutter spectra. It was stated earlier that uniform subbeams are the key to extremely cost

---

[5]Like other adaptive processors the JDL-GLR will work also in the forward looking arrangement. This is not the case for a non-adaptive DPCA.

efficient processor architectures. Instead of using the non-uniform subarrays as spatial dimension the total of subarrays may be used to form a number of adjacent uniform beams. This may offer advantages in STAP performance.

9.7.2 Comparison angle-Doppler subgroup architectures with other techniques

Comparing the angle-Doppler subgroup techniques a few observations can be made:

1. Let us compare the angle-Doppler subgroup techniques with the auxiliary sensor FIR filter (ASFF) approach, see Figure 7.5. It should be noted that the ASFF requires fewer arithmetic operations for adaptation because only *one* small space-time covariance matrix has to be updated and inverted. This advantage follows from the assumption of temporal stationarity which is given if the PRF is constant during the observation and the radar moves at constant speed on a straight course.

2. The ASFF seems to approximate the optimum processor better than the angle-Doppler subgroup technique (compare the results in Chapter 7 with those given in [668]).[6] On the other hand, as the angle-Doppler subgroup techniques do not assume temporal stationarity they might work also with staggered PRF. Of course, in this case the Fourier transform cannot make use of the FFT algorithm.

3. The number of different spectral processors offers more degrees of freedom than the ASFF does. This may be an advantage over the ASFF if errors in the antenna channel have to be compensated for.

4. If the number of elements of each angle-Doppler subgroup is set equal to 1 (one beamformer, one Doppler bin) then the angle-Doppler subgroup technique becomes identical to the *frequency-dependent spatial filtering* technique described in Section 9.5.

5. If the number of elements of each angle-Doppler subgroup is set equal to NM (full gain beamformers, full gain Doppler filters) then the angle-Doppler subgroup technique becomes identical to the *space-time auxiliary channel technique* discussed in Chapter 5, Section 5.5.3.

6. We can conclude from 4 and 5 that the performance of the angle-Doppler subgroup technique will be somewhere between the *auxiliary space-time channel processor* and the *frequency-dependent spatial filtering* technique. The first technique is sensitive to additional degrees of freedom through bandwidth effects or antenna channel errors. The second technique is close to optimum only for large data sequences ($M \geq 256$).

---

[6]Actually this is difficult to compare. The authors calculate P_D versus Doppler while we use the IF versus Doppler.

9.7.3 Other post-Doppler techniques

Post-Doppler STAP processors beyond those discussed in this chapter have been described by several authors, for instance SHAW and MCAULEY [602], WARD [690, pp. 95 – 153] and DAY [112]. SU and ZHOU [620] describe a post-Doppler generalised sidelobe canceller using eigenvectors of Doppler dependent *spatial* covariance matrices as auxiliary channels.

The processor suggested by SUN *et al.* [622, 623] includes a Doppler filter bank for each antenna element and several Doppler filters cascaded with the main beam. The covariance matrix used for clutter cancellation includes the antenna elements at a certain target Doppler frequency plus the main beam Doppler filter outputs. The technique is supposed to be more robust against array errors than the ACP (Section 5.3).

9.8 Summary

In the previous chapters various techniques for *space-time* clutter rejection were analysed. Chapter 9 deals with some possible *space-frequency* approaches. Among a large variety of possible receiver structures (there are no limits to the fantasy of the system designer) the following techniques have been identified:

1. The **auxiliary channel processor** described in Chapter 5 is in essence a space-frequency domain approach.

2. The concepts of **overlapping subarrays** or **symmetric auxiliary sensors** for reducing the signal vector space (number of antenna channels) in the spatial dimension treated in Chapter 6 can be applied in the time dimension as well. The clutter rejection performance of such processor is quasi-optimum.

3. **Space-time FIR filtering** after Chapter 7 can be implemented by using the properties of the discrete Fourier transform. The Fourier transform is taken for all channels along the time axis and so for the FIR filter coefficients. Then the convolution involved in the FIR filter operation is done by multiplication of frequency responses.

4. **Frequency-dependent spatial filtering** is a technique which is based on the statistical independence of Fourier channels. This is satisfied for long data sequences (typical $M > 256$, increasing with the CNR). Applications are predominantly in multichannel SAR systems. *Spatial* blocking matrices (Section 6.3.1) can be applied for each Doppler frequency to avoid target signal cancellation by the clutter filters.

5. The **JDL-GLR** processor uses adjacent Doppler bins and beams as clutter reference.

A comparison of all techniques in terms of computational complexity is presented in Chapter 16.

Chapter 10

Radar ambiguities

A pulse Doppler radar can be ambiguous in either range or Doppler frequency. The ambiguity of a radar depends on the selected PRF. Three modes of airborne radar operation are well known: the high PRF (HPRF), medium PRF (MPRF), and low PRF (LPRF) mode.

In the HPRF mode the PRF is chosen so that the Doppler response is unambiguous for all possible target velocities. As a consequence of the high PRF this mode is usually range ambiguous, that is, the range of a detected target may come either from the indicated range or any range determined by multiples of the PRI. This mode is useful whenever the radial target velocity has to be known unambiguously, for instance in a fighter nose radar in a long range look down situation. In such applications the radial target velocity can be much higher than any relative clutter velocity. In this case the target response appears outside the clutter Doppler band of the radar so that space-time processing is not necessary for clutter rejection.

For the LPRF mode the PRF is chosen so low that the received echoes are unambiguous in range. As a penalty, the Doppler response is ambiguous. The LPRF mode is particularly useful for target search. The MPRF mode is a compromise of the LPRF and HPRF modes and is usually ambiguous in both range and Doppler. For details on the three PRF modes see SCHLEHER [589, pp. 59-73], LACOMME *et al.* [373]. The effect of radar clutter ambiguities on ground target tracking by STAP radar has been discussed in KLEMM [337, 349]. KOGON and ZATMAN [363] propose techniques for the mitigation of ambiguous clutter returns in space-based GMTI radar.

In the first section of this chapter we focus on the effect of range ambiguities on the performance of space-time adaptive clutter cancellation. The effect of range ambiguity causes additional ambiguous clutter returns (multiple-time-around clutter). Such clutter echoes may have a different Doppler frequency than the primary clutter if the clutter Doppler is range dependent (all antenna configurations except sidelooking) which may cause additional clutter notches in the Doppler response of the space-time filter. These effects will be demonstrated. Moreover, we show how these ambiguous clutter returns can be compensated for.

The second section deals with Doppler ambiguities. Doppler ambiguities influence the radar performance in two ways. First, the velocity of Doppler targets can be estimated only modulo PRF; secondly, the clutter notch on an MTI or STAP filter is repeated every PRF (blind velocities). We demonstrate that by using staggered PRI the effect of blind velocities can largely be suppressed in an appropriate space-time processor.

10.1 Range ambiguities

In this section some aspects of range dependent clutter rejection are summarised. As previously considered in Section 3.1.2, the clutter Doppler frequency is range independent only for sidelooking arrays, that is, linear or planar arrays, whose aperture extends in the flight direction. For other configurations, particular forward looking, the clutter Doppler frequency is range dependent, especially at near range. Near range means about $5 \times H$ which can be seen from Figures 3.4 and 3.6.

10.1.1 Multiple-time-around clutter, linear arrays

Multiple clutter returns occur if the radar operation is range ambiguous. This means that the PRF is chosen such that the PRI is smaller than the maximum echo delay. This happens usually in the medium and high PRF modes (MPRF, HPRF). In this case the clutter echo of a certain range increment includes clutter contributions from other range increments[1] which are separated by

$$\Delta R = \frac{c}{2f_{\mathrm{PR}}} \tag{10.1}$$

Then the clutter echo from one range increment R becomes according to (2.38) and (2.36)

$$c_{i,m}(R, \varphi, v_p) = \sum_{r=0}^{r_{max}} \int_{\varphi=0}^{2\pi} AD(\varphi) L(R_r, \phi) G(\phi, m) \Phi_m^t(R_r, \varphi, v_p) \Phi_i^s(R_r, \varphi) \mathrm{d}\varphi \tag{10.2}$$

with

$$\Phi_m^t(R_r, \varphi, v_p) = \exp\left[\mathrm{j}\frac{2\pi}{\lambda} 2 v_p mT \cos\varphi \cos\theta_r\right] \tag{10.3}$$

$$\Phi_i^s(R_r, \varphi) = \exp\left[\mathrm{j}\frac{2\pi}{\lambda}(x_i \cos\varphi + y_i \sin\varphi)\cos\theta_r - z_i \sin\theta_r\right]$$

where

$$\theta_r = \arcsin\frac{H}{R_r} \tag{10.4}$$

and

$$R_r = R + r\Delta R \tag{10.5}$$

where θ_r and R_r are the elevation angles and ranges of the ambiguous echoes.

---

[1] These ambiguous clutter contributions have been neglected in all previous considerations.

We assume that clutter contributions from different range increments are independent. Then the clutter covariance matrix becomes

$$\mathbf{Q}^{(c)} = \sum_{r=0}^{r_{max}} \mathbf{Q}_r^{(c)} \tag{10.6}$$

so that the elements of $\mathbf{Q}^{(c)}$ become, by using (3.28),

$$\begin{aligned}
q_{ln}^{(c)} &= E\{c_{i,m} c_{k,p}^*\} \\
&= \sum_{r=0}^{r_{max}} E\Big\{ \int_{\varphi=0}^{2\pi} A(\varphi) D(\varphi) L(R_r, \varphi) G(\varphi, m) \Phi_m^t(R_r, \varphi, v_p) \Phi_i^s(R_r, \varphi) \mathrm{d}\varphi \\
&\quad \times \int_{\phi=0}^{2\pi} [A(\phi) D(\phi) L(R_r, \phi) G(\phi, m) \Phi_p^t(R_r, \phi, v_p) \Phi_k^s(R_r, \phi)]^* \mathrm{d}\phi \Big\}
\end{aligned} \tag{10.7}$$

where the indices of the covariance matrix l, n are related to the sensor and echo indices i, k, m, p through (3.23) and (3.24).

10.1.1.1 Forward looking array

For a forward looking array the Doppler frequency of clutter returns is range dependent as has been shown by (3.20) and Figure 3.6. This range dependence can also be recognised by comparing isodops and beam traces as shown in Figure 3.4. Notice that there are a lot of crossings between isodops and beam traces, especially at near range. The range dependence of clutter Doppler decreases at far range.

Multiple clutter returns come from different ranges. Therefore they are expected to have different clutter Doppler frequencies. The variation in Doppler will be larger at near range than at far range.

The effect of multiple clutter returns on adaptive clutter rejection is illustrated by Figure 10.1. A forward looking *linear* array was assumed. The curves are based on clutter rejection by means of the fully adaptive FIR filter processor after Figure 7.1. As in the previous chapters the subplots show the IF as a function of the normalised Doppler frequency.

Case a represents the near range situation ($R/H = 1.66$). Instead of one clutter notch we recognise two of them. The first notch is the one due to primary clutter at near range. The second notch is due to multiple returns which come from farther distances.

Since the clutter Doppler frequency is proportional to the cosine of the depression angle θ (see for example (3.25)) the multiple clutter arrivals from larger ranges show higher Doppler frequency. Most of the multiple arrivals come from large ranges. Therefore they altogether have about the same Doppler frequency so that their influence sums up in *one* additional clutter notch. In the subplots b – d the range of the primary clutter increment is increased so that the primary clutter notch approaches the secondary one. In cases c and d the processor is not able to resolve the primary and secondary clutter responses.

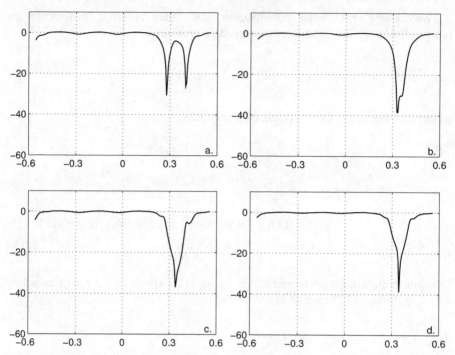

Figure 10.1: Influence of multiple-time-around clutter (IF vs F; fully adaptive FIR filter processor, linear array, FL): a. $R_0/H = 1.66$; b. $R_0/H = 3$; c. $R_0/H = 4.33$; d. $R_0/H = 5.66$

In Figure 10.2 we consider the same situation as before. However, instead of a linear array a *planar* array has been used (six horizontal rows with 24 elements each). Again the fully adaptive FIR filter has been used so that the processor is adaptive in both the horizontal and the vertical dimension. As can be noticed there is only one clutter notch in each of the plots. Obviously the primary clutter notch is preserved while the notches due to ambiguous clutter echoes are suppressed.

In Figure 10.2 the processor[2] was focused on the *nearest* of a total of ambiguous range intervals. Let us discuss the case where the processor is focused on a range increment at far range. We selected the tenth out of a total of 15 ambiguous range intervals. The width of each range interval is

$$\Delta R = \frac{c}{2f_{\text{PR}}} \tag{10.8}$$

For $f_{\text{PR}} = 12000$ Hz one obtains a range interval of 12.5 km. That means, the beamformer has been focused on $R = R_0 + 9 \times 12.5 = R_0 + 112.5$ km where R_0 is the range of the first range increment used in Figures 10.1 and 10.2. Then the ratio R/H is almost independent of R. In Figure 10.3 this is reflected by the fact that the clutter notch appears at the same Doppler frequency for all values of R. Notice

---

[2]That is, the beamformer included in the space-time FIR filter according to (7.12).

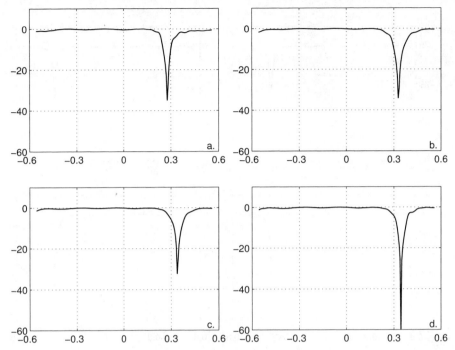

Figure 10.2: Influence of multiple-time-around clutter (IF vs F; fully adaptive FIR filter processor, planar array with six rows of 24 elements each, FL, focused on the first range increment): a. $R_0/H = 1.66$; b. $R_0/H = 3$; c. $R_0/H = 4.33$; d. $R_0/H = 5.66$.

that no multiple clutter notches appear at *lower* Doppler frequencies. Multiple clutter returns have again been cancelled by the vertical adaptivity of the planar array. The use of vertical adaptivity to cancel ambiguous clutter from short range in the context of an element digitised array has been proposed by BRIDGEWATER [49].

We can conclude that multiple-time-around clutter in a forward looking array radar can be suppressed by space-time processing provided the antenna array is adaptive in both the horizontal and the vertical dimensions. The multiple clutter returns appear under different depression angles and, therefore, can be cancelled via *vertical nulling*.

The effect of range dependence of the clutter Doppler has been mentioned in reference [49] in the context of space-based radar. In addition to the effect of radar platform motion, the effect of earth rotation is considered. The range ambiguous arrivals lead to an increase in the number of eigenvalues of the clutter covariance matrix. AYOUB et al. [24, 25] analyse the effect of range ambiguities on the performance of different clutter cancellers. It is shown that ambiguous clutter can basically be cancelled, provided that the number of degrees of freedom is sufficient. WANG, Y.-L. et al. [683] have discussed the use of space-time adaptive processing for non-sidelooking array geometries. Staggered PRIs are suggested as a remedy against range ambiguities. A strategy for choosing the PRIs is presented.

For example, let us consider a spaceborne radar geometry where the radar is at

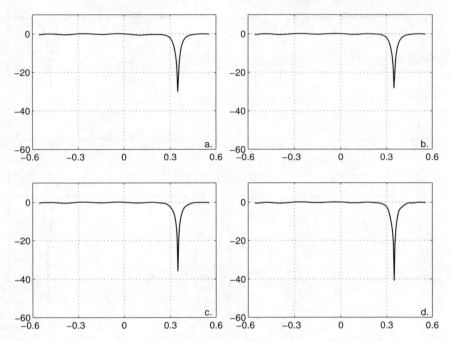

Figure 10.3: Influence of multiple-time-around clutter (IF vs F; fully adaptive FIR filter processor, planar array, FL, focused on tenth ambiguous range increment): a. $R_0/H = 1.66$; b. $R_0/H = 3$; c. $R_0/H = 4.33$; d. $R_0/H = 5.66$.

altitude H and the visible range extends from ground range R_{g1} to R_{g2}. H is typically of the order of magnitude of $500\ldots 1000$ km. We notice some details:

- The high velocity of the radar satellite (several km/s) requires high PRFs for unambiguous clutter suppression operation. This in turn results in a large number of range ambiguous clutter echoes.

- The depression angle can be quite large. Therefore, the Doppler dependence with range is significant over all the visible range interval. This is different from airborne radar where a clutter Doppler gradient is encountered only at near range.

- The two slant ranges R_{s1} and R_{s2} have the same order of magnitude. Therefore, the power levels of the multiple echo returns are almost equal for the whole visible range.

It follows from this consideration that the effect of multiple-time-around clutter is more severe for spaceborne than for airborne radar.

10.1.1.2 Sidelooking array

It has been shown by (3.19) that the clutter Doppler frequency for a sidelooking array is range independent. This can also be concluded from the fact that beam traces on

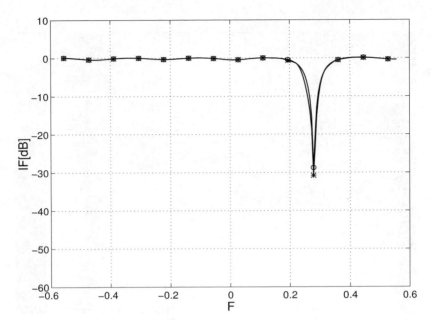

Figure 10.4: Influence of multiple-time-around clutter (fully adaptive FIR filter processor, SL, $R/H = 1.66$): ○ *linear array;* ∗ *planar array*

the ground and isodops coincide for sidelooking arrays, see Figure 3.3. Therefore, all multiple clutter returns coincide on the same Doppler frequency.

Figure 10.4 shows a numerical example. It can be noticed that both the linear and the planar array produce only one clutter notch. Such results have been presented by KLEMM in [313, 314, 315]. Further results on STAP with range ambiguous radar have been presented by AYOUB et al. [24].

10.1.1.3 Multiple clutter in the range-Doppler IF matrix

In Figures 10.5–10.7 we find the multiple-time-around clutter in a range-Doppler representation. The three plots have been produced by varying the PRF. The plots show the IF in signal to clutter + noise ratio as achieved by the optimum processor, see Chapter 4. It was assumed that the clutter-to-noise ratio is 20 dB at $R = 15000$ m. The clutter power varies with range according to the radar range equation, i.e., the power due to a single point scatterer on the ground is $P_c \propto \frac{1}{R^4}$. Accordingly the CNR and, hence, the obtainable IF is larger at near range than at far range. This is reflected in all three figures by the grey level which increases with range. Recall that the target velocity ranges from $-100\ldots 100$ m/s, see Table 2.5.2.1.

In Figure 10.5 the PRF has been chosen so that the clutter Doppler bandwidth is sampled at the Nyquist frequency (12 000 Hz which corresponds to an unambiguous range of 12 500 m). [3] On the right we notice mainly two black lines. The curved

---

[3]This PRF has been assumed for most of the calculations presented in this book, see Table 2.5.2.1.

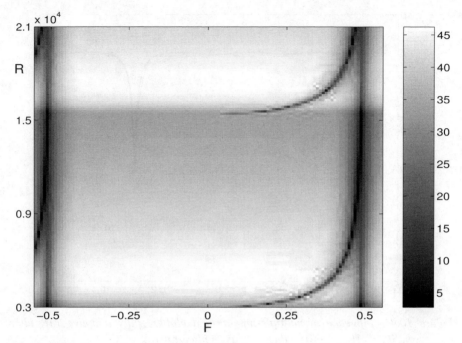

Figure 10.5: Range-Doppler matrix (IF[dB] vs R/m and F; FL, $\varphi_L = 0°$, PRF = Nyquist of clutter band)

line which starts at F $= 0$ is the notch due to primary clutter. This notch runs along the range-Doppler clutter trajectory for a forward looking array as given by (3.20). It shows clearly the range dependence of the clutter Doppler frequency at near range. The curve approximates F $= 0.5$ for larger range which is the Doppler frequency determined by the platform velocity. The curve starts at F $= 0$ which means a look direction perpendicular on the ground.[4] In this direction there is no radial velocity component between clutter and radar.

The vertical line on the far right of Figure 10.5 denotes the multiple-time-around clutter. The unambiguous range interval is 12.5 km in this example. As one can see at $R = 12.5$ km the primary clutter trajectory has reached the value F $= 0.5$. Therefore, secondary and even later clutter arrivals will show up close to F $= 0.5$.

At a range of 15 500 m (platform height + 12 500 unambiguous range) the clutter pattern showing up at 3000 m is repeated. Now the role of primary clutter and multiple clutter is interchanged. The straight line on the right is mainly the primary clutter whereas the non-linear part[5] originates from multiple clutter at short range. At 3000 m the IF decreases, which is caused by the vertical sensor pattern according to (2.21).

Figure 10.6 shows the same situation, however with PRF = 48 000 Hz, i.e., four times the Nyquist frequency of the clutter bandwidth. Now the unambiguous range interval is only 3.125 km. The primary clutter runs as before, the clutter trajectory,

---

[4]The platform altitude was assumed to be $H = 3000$ m, see Table 2.1.
[5]Called 'J-hook' by some authors, AYOUB *et al.* [24, 25].

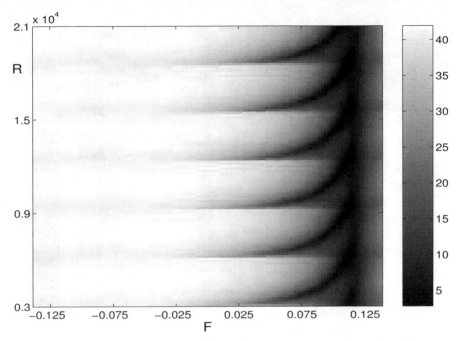

Figure 10.6: Range-Doppler matrix (IF[dB] vs R/m and $\mathbf{F}$; FL, $\varphi_L = 0°$, PRF = 4 × Nyquist of clutter band)

however, approximates only a quarter of the PRF at larger range. The trajectory of the multiple-time-around clutter response is also slightly curved. This happens because at 6000 m range the secondary clutter comes from an area where even the primary clutter has not yet reached the maximum Doppler frequency. We also notice that the ambiguous response is somewhat spread in Doppler. This happens because multiple arrivals with different Doppler frequencies are superimposed in each of the range cells. The range ambiguous intervals can be clearly seen.

For low PRF we get a situation as shown in Figure 10.7 (PRF = 0.25 · Nyquist). Now we have some Doppler ambiguity inside the Doppler interval determined by $v_{\text{rad}} = -100, \ldots, 100$ m/s. It can also be noted that the clutter notch according to multiple-time-around clutter is less deep than before (compare with the grey levels of the multiple-time-around clutter trajectories in the previous figures). This happens because the unambiguous clutter interval is now 50 km so that multiple arrivals are strongly attenuated. No ambiguities from short range show up because the unambiguous range interval is larger than the maximum range in the plot.

10.1.1.4 Aspects of implementation

For radar/ground geometries large depression angles occur at short range, i.e., at ranges in the order of magnitude of the radar height. For large two-dimensional antennas with narrow vertical beamwidth the beam might have to be steered in elevation so as to cover

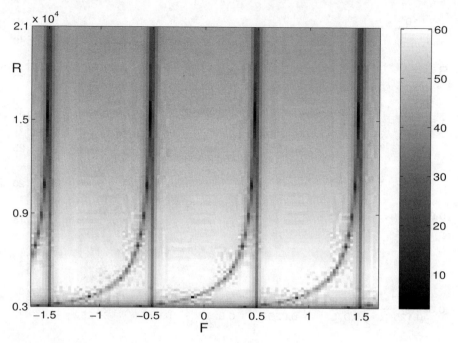

Figure 10.7: Range-Doppler matrix (IF[dB] vs R/m and F; FL, $\varphi_L = 0°$, PRF = 0.25 × Nyquist of clutter band)

the whole visible ground range.

For a space-time clutter filter which operates in two spatial dimensions basically the filter should be re-adapted with every range.

10.1.2 Multiple-time-around clutter, circular planar arrays

We found in the previous section that an array with vertical degrees of freedom can, in conjunction with a space-time processor, cancel multiple-time-around clutter returns via spatial nulling in the vertical dimension. In the following we want to discuss how far circular planar antennas are capable of cancelling multiple-time-around clutter in a forward looking array geometry.

For the following numerical examples we use the checkerboard subarray structure shown in Figure 8.19. Figure 8.19a shows the circular aperture subdivided into 32 square subarrays. By grouping these subarrays one can obtain different subarray patterns as can be seen in Figure 8.19b – d. The number of resulting subarrays varies accordingly (6, 12, 18). It should be noticed that Figure 8.19b has no vertical degrees of freedom, Figure 8.19c has one, Figure 8.19d has two, and Figure 8.19a has five vertical degrees of freedom.

The question to be answered is how many degrees of freedom are required for efficient cancellation of multiple clutter returns. A look at Figure 10.5 tells us that there is only one additional clutter arrival in each range cell (right hand part of the

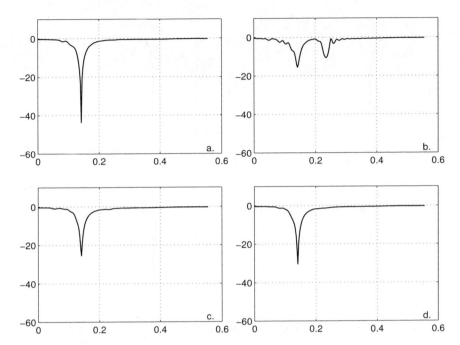

Figure 10.8: Influence of multiple-time-around clutter (IF vs F; fully adaptive FIR filter processor, circular planar array, omnidirectional transmission, FL, CNR = 20 dB). Checkerboard subarrays after Figure 8.19: a. 32 square subarrays; b. 6×1 subarrays; c. 6×2 subarrays; d. 6×3 subarrays

clutter response) if the PRF is low enough. In this case we can expect that only one vertical degree of freedom is required for cancellation of ambiguous clutter. For higher PRF the clutter response gets broadened as can be seen in Figure 10.6 which might require more degrees of freedom for spatial cancellation.

Figures 10.8 and 10.9 show numerical examples calculated for the four array structures given in Figure 8.19. In Figure 10.8 we eliminated the influence of the vertical transmit pattern by assuming an omnidirectional transmit antenna, while in Figure 10.9 a transmit array of the same size as the receive array was assumed.

In Figure 10.8b the effect of multiple clutter arrivals of the IF curve can be seen clearly as a second clutter notch because this subarray configuration has no vertical degree of freedom. All other subarray configurations can cancel the ambiguous clutter by means of vertical nulling.

Figure 10.9 shows the same situation as before, however with a transmit antenna of the same size as the receive antenna. Now the second clutter notch due to ambiguous clutter returns has almost disappeared in Figure 10.9b, which indicates that the multiple clutter returns have been attenuated by the transmit beam pattern.

Figures 10.10 and 10.11 show the same scenarios as before, however with 40 dB CNR. Now the second clutter notch cannot be cancelled without vertical nulling

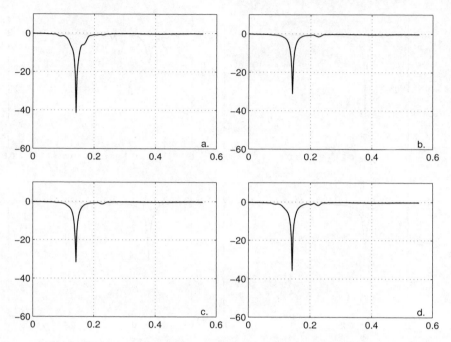

Figure 10.9: Influence of multiple-time-around clutter (IF vs F; fully adaptive FIR filter processor, circular planar array, transmit array = receive array, FL, CNR = 20 dB). Checkerboard subarrays after Figure 8.19: a. 32 square subarrays; b. 6×1 subarrays; c. 6×2 subarrays; d. 6×3 subarrays

(compare Figure 10.11b with 10.9b).

REES and SKIDMORE [561] and MATHER et al. [441] describe a spatial only technique for cancellation of ambiguous clutter returns in HPRF mode. For each individual range bin under test echoes from associated ambiguous range gates are cancelled by spatial nulling. In this way the temporal dimension inherent in space-time processing is not required; however, adaptation has to be carried out for all individual range gates or, at least, for sufficiently small range segments. The simulations shown are based on a large fully digitised array antenna decomposed into 'domains'. The first domain is a fully populated small subarray. Higher domains include phase centre locations for small subarrays of first order. The full weight solution is a convolution of weights from lower domains. Details on adaptive processing for this type of antenna can be found in PAINE [516, 517, 515].

HALE et al. [243, 245] analysed a hybrid STAP architecture consisting of an adaptive vertical beamformer cascaded with a partial adaptive STAP processor. In contrast to a 2D STAP architecture the adaptive vertical beamformer cancels clutter from a full range ring. Such technique can be applied to cancellation of ambiguous clutter returns. In another paper by the same authors [244] the cascaded and the 3D joint-domain approach in presence of range ambiguous clutter (space-time processing

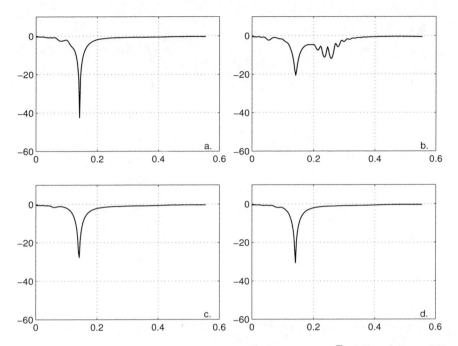

Figure 10.10: Influence of multiple-time-around clutter (IF vs F; fully adaptive FIR filter processor, circular planar array, omnidirectional transmission, FL, CNR = 40 dB). Checkerboard subarrays after Figure 8.19: a. 32 square subarrays; b. 6×1 subarrays; c. 6×2 subarrays; d. 6×3 subarrays

with 2D array) are compared. The 3D STAP approach is slightly superior to the cascaded architecture, however, shows reasonable performance. In KLEMM [335] the effect of range ambiguous clutter on the performance of a sidelooking array in a crabbing platform has been investigated.

10.2 Doppler ambiguities

As stated above Doppler ambiguities occur mostly in low PRF radar operation (LPRF). In this section we investigate the effect of Doppler ambiguities and ways to compensate for them.

10.2.1 Preliminaries

Clutter echoes received by a moving radar are composed of arrivals with Doppler frequencies

$$f_{\rm D} = \frac{2v_{\rm p}}{\lambda} \cos\alpha \qquad (10.9)$$

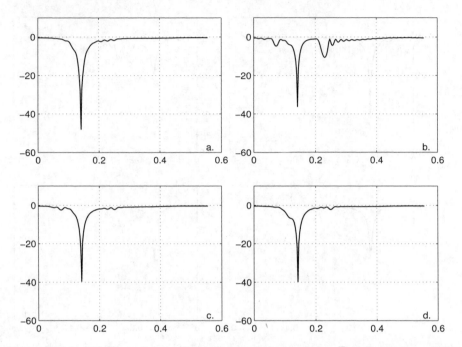

Figure 10.11: Influence of multiple-time-around clutter (IF vs F; fully adaptive FIR filter processor, circular planar array, transmit array = receive array, FL, CNR = 40 dB). Checkerboard subarrays after Figure 8.19: a. 32 square subarrays; b. 6×1 subarrays; c. 6×2 subarrays; d. 6×3 subarrays

where v_p is the platform velocity and α the angle of clutter arrival (see Figure 2.1). Therefore, ground clutter returns received by a moving radar are spread over a Doppler band whose limits are determined by the platform velocity: $[-2v_p/\lambda, 2v_p/\lambda]$. The capability of detecting slowly moving targets by conventional MTI techniques is greatly degraded if they appear within the clutter bandwidth.

In this section we analyse the effect of pulse-to-pulse staggered PRI on the performance of space-time adaptive filters for suppression of air-/spaceborne clutter. It is well known that a staggered PRI offers a number of attractive features:

- The radial target velocity can be estimated unambiguously.

- Blind velocity zones (ambiguous clutter notches) are suppressed.

- If a pseudorandom stagger code is chosen which varies from burst to burst staggering may be a countermeasure against coherent spot jammers.

Some disadvantages associated with a staggered PRI should be mentioned:

- The FFT algorithm cannot be used for designing a Doppler filter bank.

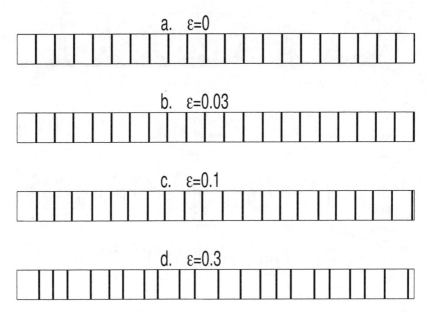

Figure 10.12: Pseudorandom stagger patterns (©1999 SEE)

- The coefficients of an adaptive MTI FIR filter vary with time during the filtering operation (see further below). They have to be updated at every PRI which requires more computational effort than a filter with constant coefficients.

- The sidelobes of the Doppler filter bank are raised, which leads to an increased false alarm rate.

If conventional MTI weightings are used with staggered PRI some losses in SCNR will occur due to distortions of the Doppler response of the filter (SKOLNIK [610, pp. 15.35]). Such losses can be avoided in an adaptive system by adapting the filter coefficients at every PRI.

The numerical results shown below have been calculated for a sidelooking linear array according to Figure 2.1.

10.2.2 Clutter and target models

Consider a radar geometry as shown in Figure 2.1. It shows two antenna array configurations (sidelooking, forward looking) moving at constant speed v_p parallel to the plane ground. Let the radar transmit coherent pulses of the form

$$s_\mathrm{t}(t) = \Re[A_\mathrm{t} E(t) \exp(\mathrm{j}\omega_\mathrm{c}(t+mT))]$$
$$m = 1,\ldots,M \quad (10.10)$$

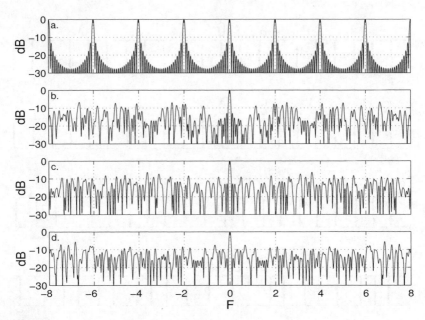

Figure 10.13: Doppler channel response for pseudorandom staggering: a. $\varepsilon = 0$*; b.* $\varepsilon = 0.03$*; c.* $\varepsilon = 0.1$*; d.* $\varepsilon = 0.3$ *(©1999 SEE)*

where A_t is the power of the transmitted waveform, $E(t)$ the signal envelope, ω_c the circular carrier frequency, and T the pulse interval. Then the i-th sensor of an array receives at time mT a partial clutter echo from a fixed scatterer at angle (φ, θ) which has been given by (2.29).

The total clutter echo is obtained by integrating over all signals arriving from all possible angles (2.34). In addition to clutter returns we assume receiver noise according to (2.49), (2.50) which is assumed to be uncorrelated in space and time. Target signals are given by the expression (2.31).

10.2.3 Pseudorandom staggering

Now let us introduce a pseudorandom stagger offset $\varepsilon r(m)$ where the random part $r(m)$ is uniformly distributed in the interval $[-1, 1]$ and varies with the temporal index m. Then the individual clutter echo coming from a certain reflector on the ground becomes

$$c_{im}(\varphi) = A(\varphi) \exp\left[j\frac{2\pi}{\lambda}\{(x_i + 2v_p mT(1 + \varepsilon r(m))) \right.$$
$$\left. \times \cos\varphi + y_i \sin\varphi\} \cos\theta - z_i \sin\theta)\right] \quad (10.11)$$
$$m = 1,\ldots,M \quad i = 1,\ldots,N$$

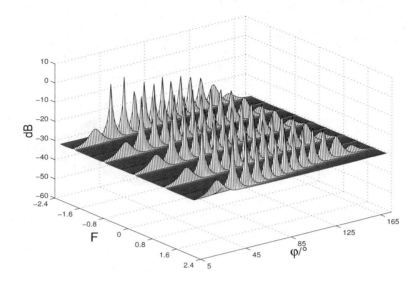

Figure 10.14: Azimuth-Doppler clutter spectrum without staggering (©1999 SEE)

and the target signal is

$$s_r(\varphi) = A \exp\left[j\frac{2\pi}{\lambda}(2v_{\text{rad}}mT(1+\varepsilon r(m)) + (x_i \cos\varphi_t \right.$$
$$\left. + y_i \sin\varphi_t)\cos\theta - z_i \sin\theta)\right] \quad (10.12)$$
$$m = 1,\ldots,M \quad i = 1,\ldots,N$$

Figure 10.12 shows stagger patterns for different values of ε. Figure 10.13 shows responses of staggered Doppler channels for different values of ε. For $\varepsilon = 0$ we get the well-known periodic frequency response. With increasing ε the mainlobe becomes unambiguous, however at the price of higher sidelobes.

In Figures 10.14 and 10.15 minimum variance azimuth-Doppler clutter spectra after (1.104) are shown.[6]

In these examples the PRF has been chosen to be 1/4 of the considered Doppler range [-400 m/s,400 m/s] so that four ambiguous clutter ridges appear in the case of constant PRF (Figure 10.14). With pseudorandom staggering (Figure 10.15) the ambiguous Doppler responses are cancelled.

In Figure 10.16 the improvement in signal-to-clutter + noise ratio achieved by

---

[6]The spiky nature of these plots are a consequence of 'graphical undersampling' which may occur if high-resolution estimators are used with an insufficient number of points. Using more points would lead to a totally black image. In reality there are continuous clutter ridges.

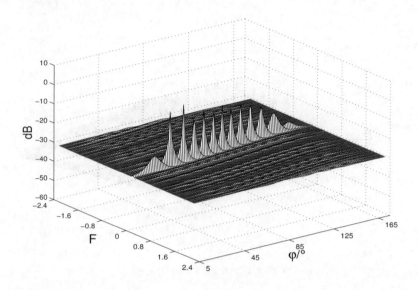

Figure 10.15: Azimuth-Doppler clutter spectrum with pseudorandom staggering (©1999 SEE)

optimum processing is shown. Constant PRF was assumed. The optimum processor was given by (1.3) and the optimum improvement factor by (1.19).

In Figure 10.16 the optimum IF has been plotted for different PRFs versus the normalised Doppler frequency F. f_{Ny} is the Nyquist Doppler frequency of the clutter band in the case of $\varepsilon = 0$. As can be seen there are many ambiguous clutter notches. The associated velocities are called 'blind velocities'. In Figure 10.17 a pseudorandom stagger code was applied. As one can see the ambiguities are removed at the expense of a little ripple in the pass band of the STAP filter.

10.2.4 Quadratic staggering

Another way of staggering is to increase (or decrease) the PRI T in certain steps, for instance in a quadratic fashion

$$\begin{aligned} c_{im}(\varphi) & = A(\varphi) \exp\left[j\frac{2\pi}{\lambda} \left\{ \left(x_i + 2v_{\text{p}} mT\left(1 + \varepsilon\frac{m}{M}\right) \right) \right.\right. \\ & \left.\left. \times \cos\varphi + y_i \sin\varphi \right\} \cos\theta - z_i \sin\theta) \right] \\ & m = 1,\ldots,M, \quad i = 1,\ldots,N \end{aligned} \qquad (10.13)$$

The corresponding stagger patterns are shown in Figure 10.18, and the frequency responses of the associated Doppler filters are shown in Figure 10.19.

Figure 10.20 shows an example for a Doppler-azimuth clutter spectrum. In comparison with Figure 10.14 we notice that the ambiguities caused by a constant PRF

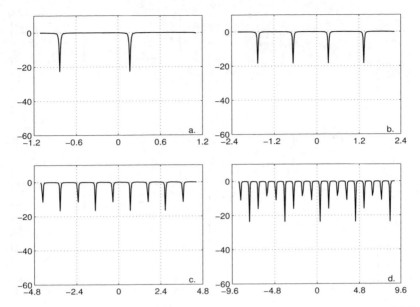

Figure 10.16: Improvement factor vs normalised target Doppler: constant PRF. a. PRF = $2f_{Ny}$; b. PRF = f_{Ny}; c. PRF = $\frac{1}{2} \times f_{Ny}$; d. $PRF = \frac{1}{4}f_{Ny}$ (©1999 SEE)

are almost perfectly suppressed. Figure 10.20 is very similar to Figure 10.15. This gives the impression that the individual stagger technique does not have significant influence on the capability of supression of ambiguity.

The optimum improvement factor after (1.19) has been shown for quadratic staggering in Figure 10.21. The result is similar to Figure 10.17. The ambiguous clutter notches are removed. A slight ripple in the pass band is the penalty for obtaining an unambiguous clutter response.

ADVE et al. [11] propose logarithmic array spacing in conjunction with orthogonal waveforms transmitted by the individual radiators of the array to reduce grating lobes in highly thinned (distributed) arrays. The concept of orthogonal transmit waveforms and aperiodic array spacing to mitigate Doppler blind zones is also discussed by LEATHERWOOD et al. [379]. Transmission of orthogonal waveforms by a transmit array may be used for adaptive beamforming with a single receive antenna (CALVARY and JANER et al. [82]).

10.2.5 Impact of platform acceleration

It should be noticed that the term $2v_{\mathrm{p}}mT(1 + \varepsilon\frac{m}{M})$ is ambiguous in T and v_{p}, that is, the quadratic factor applies to either of them. Quadratic staggering as perceived by the radar is, therefore, equivalent to a constant acceleration of the platform.

In practice the perturbations of the radar platform motion are not known. They may be measured by an inertial navigation system (INS) and then incorporated in the processing.

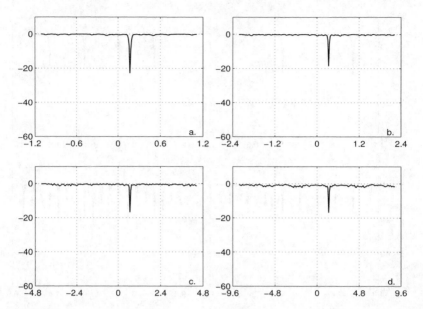

Figure 10.17: Improvement factor vs normalised target Doppler (pseudorandom staggering, $\varepsilon = 0.3$): a. PRF $= 2f_{Ny}$; b. PRF $= f_{Ny}$; c. PRF $= \frac{1}{2}f_{Ny}$; d. PRF $= \frac{1}{4}f_{Ny}$ (©1999 SEE)

Let us assume that the perturbations of the platform motion are unknown. Therefore, for the steering vector we assume constant speed, that is, the equidistant stagger pattern shown in Figure 10.18a. For simplicity we further assume that the platform moves at constant acceleration in such a way that the echo sequences are based on the stagger patterns shown in Figure 10.18b–e. It would be interesting to know how far mismatch between deviations from a constant speed of the radar has an impact on the performance of a space-time MTI filter.

Figure 10.22 shows the improvement factor of a space-time adaptive filter versus the normalised target Doppler frequency for the four acceleration patterns (Figure 10.18a–d). Comparing the curve $\varepsilon = 0$ (Figure 10.18a) with those plotted for $\varepsilon > 0$ shows the expected sensitivity to mismatch. In conclusion, the motion perturbations of the radar platform should be known as accurately as possible and incorporated in the temporal part of the steering vector whose elements are given by (10.12). KRIKORIAN [371] discusses the principle of matching a filter to a pulse sequence received from an accelerating air target.

10.2.6 Space-time FIR filter processing

All the numerical examples given above are based on the optimum processor as given by (1.3). This processor exploits the information of all sensors and all echo samples. Due to its computational complexity it is not useful for practical applications, in particular for real-time application.

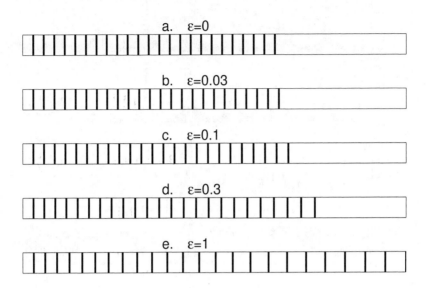

Figure 10.18: Quadratic stagger pattern (©1999 SEE)

A considerable amount of work has been devoted to the design of subspace techniques for suboptimum clutter suppression (see for instance Chapters 5, 6, 7, 9, WARD [690], and many others). An extremely efficient processor is given by the space-time least squares FIR filter. For details on FIR filter based STAP processors see Chapter 7.

Notice that these processors operate as FIR filter (convolution) only in the time dimension. For a certain class of antenna arrays these processors approximate the clutter rejection performance of the optimum detector (1.3) within one dB or less at extremely low expense. About 4–9 space-time filter coefficients are sufficient to achieve quasi-optimum clutter rejection. This low number of coefficients is achieved only for the class of cylindrical arrays whose array axis is arranged parallel to the ground and whose sensor elements are equally spaced. Special cases of cylindrical arrays are: linear, rectangular planar, and certain conformal arrays (for instance, integrated in the hull of an aircraft).

A FIR filter processor is based on the assumption of temporal stationarity of clutter echoes which requires in particular that the PRF is constant. If the PRF is not constant because of staggering the FIR filter becomes time dependent, that is, the filter coefficients have to be adapted with each (irregular) pulse repetition interval (KLEMM [334]). The space-time submatrix chosen for calculating the FIR filter coeffients has to be selected for each PRI anew. Figure 10.23 illustrates how the sliding covariance submatrices relate to the total $N \times M$ (or $K \times M$) covariance matrix. The fat lines indicate which submatrix is associated with a certain step of the filter operation.

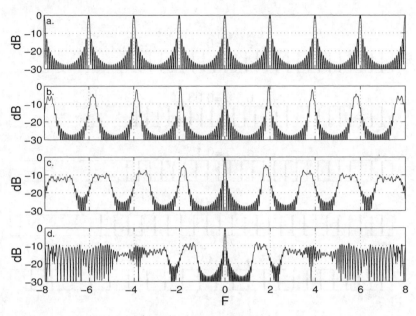

Figure 10.19: Doppler channel response for quadratic staggering: a. $\varepsilon = 0$; b. $\varepsilon = 0.03$; c. $\varepsilon = 0.1$; d. $\varepsilon = 0.3$ (©1999 SEE)

A numerical example can be seen in Figure 10.24. Notice that the abscissa extends from $-2B_c$ to $2B_c$ (four times the clutter bandwidth). Figure 10.24a shows the Doppler response of a space-time FIR filter with constant coefficients using an echo sequence with constant PRI. Consequently, ambiguous clutter notches show up. In Figure 10.24b we assumed a pseudorandom stagger code. The plot shows the IF achieved by the optimum processor when the PRI is staggered. The ambiguities are removed at the expense of a slight decrease in IF and a little ripple close to the clutter notch. Using a constant FIR filter with staggered PRI leads to heavy losses in IF (Figure 10.24c). Matching the FIR filter coefficients to the stagger code by re-adapting (KLEMM [336]) at each PRI gives reasonable clutter suppression performance as shown in Figure 10.24d. The loss compared with the optimum processor (Figure 10.24b) is about 3 dB. The width of the clutter notch is slightly broadened.

10.3 Summary

The findings of this chapter on radar ambiguities can be summarised as follows:

1. **Multiple-time-around** clutter is no problem for sidelooking array radar since the clutter Doppler is range independent. All contributions from various ranges coincide in the same clutter frequency, and hence fall into the same clutter notch of the space-time filter.

2. For forward looking radar the multiple clutter contributions have different

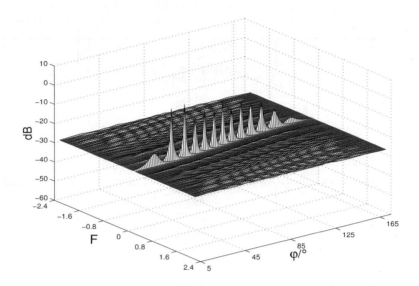

Figure 10.20: Azimuth-Doppler clutter spectrum with quadratic staggering (©1999 SEE)

Doppler frequencies which may lead to a second clutter notch. This second clutter notch can be compensated for by using an antenna with vertical *adaptivity*. For instance, circular planar arrays with two-dimensional subarray structure are capable of cancelling the multiple arrivals by vertical nulling.

3. For a practical implementation, it is necessary that the space-time clutter filter is updated at every range bin or at least at range segments which are so small that the ambiguous clutter returns fall into the vertical null of the antenna pattern.

4. The effect of multiple-time-around clutter is mitigated by the vertical directivity of the main beam of a planar antenna. For moderate clutter strength there is no need for vertical adaptivity for cancellation of multiple clutter returns.

5. If optimum processing is applied, the ambiguous clutter notches that occur in the case of constant PRF are removed when staggered PRIs are employed. The losses in SCNR in the filter pass band are negligible.

6. The optimum adaptive processor takes care of the stagger pattern. If it is to be replaced by a cost-efficient space-time FIR filter (Chapter 7) the filter coefficients have to be updated at every PRI (time-variable FIR filter).

7. Staggering the PRF in a quadratic fashion is equivalent to acceleration of the radar platform. It has been shown that in the case of errors of the platform velocity the temporal part of the steering vector (target replica) has to be matched to the motion perturbation with high accuracy.

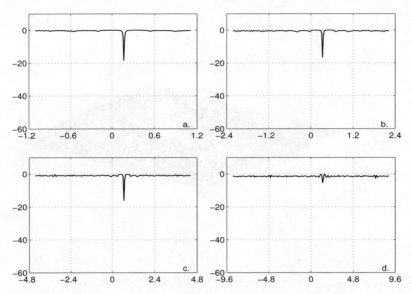

Figure 10.21: *Improvement factor vs normalised target Doppler (quadratic staggering, $\varepsilon = 1$): a. PRF = $2f_{\text{Ny}}$; b. PRF = f_{Ny}; c. PRF = $\frac{1}{2}f_{\text{Ny}}$; d. PRF = $\frac{1}{4}f_{\text{Ny}}$ (©1999 SEE)*

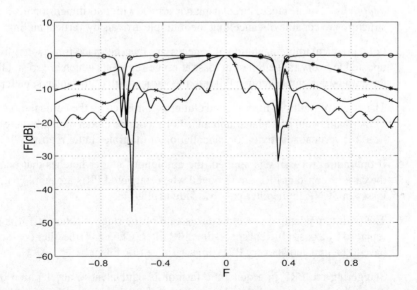

Figure 10.22: *Accelerating platform: Effect of Doppler filter mismatch. Processor: $\varepsilon = 0$. Platform: ○ $\varepsilon = 0$; ∗ $\varepsilon = 0.03$; × $\varepsilon = 0.1$; + $\varepsilon = 0.3$ (©1999 SEE)*

Summary

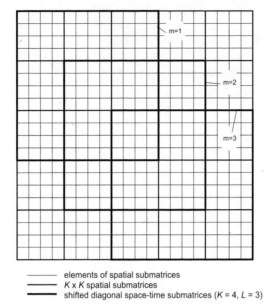

Figure 10.23: Matrix scheme for space-time FIR filtering, $K = 4$, $M = 5$

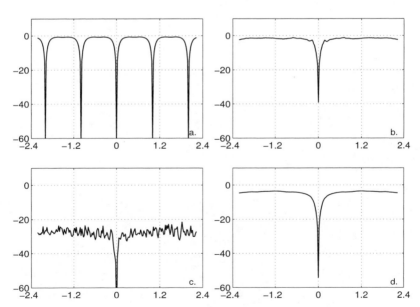

Figure 10.24: FIR filter with staggered PRI (pseudorandom stagger code, $\epsilon = 0.3$, PRF = 3000 Hz, overlapping subarrays). a. no staggering; b. optimum processor with staggered PRI; c. fixed FIR filter with staggered PRI; d. FIR filter with variable coefficients and staggered PRI

Chapter 11

STAP under jamming conditions

11.1 Introduction

In this chapter the effect of additional jamming on the performance of clutter filters will be analysed. We assume that the jammers transmit continuously (CW) and are broadband, that is, the jammer bandwidth is larger than the usable Doppler bandwidth. In the numerical analysis the jammer model given in Section 2.4.3 is used. The effect of MTI under jamming conditions has been discussed by the author [308, 317].

There are basically two ways of performing jammer and clutter rejection. One possibility is *simultaneous* jammer and clutter suppression. This means that during normal radar operation the *jamming + clutter* space-time covariance matrix is estimated and a space-time filter for rejection of jamming and clutter is designed on the basis of the covariance matrix. The number of degrees of freedom has to be chosen so that the jammer + clutter scenario is adequately covered.

The second possibility is to perform jammer and clutter rejection in two steps (KLEMM [308], FANTE [144]). The first step includes the estimation of a *spatial* jammer covariance matrix in a passive radar mode, that is, before transmit, to make sure that the covariance matrix estimate is free of clutter. In the second step the jammer + clutter covariance matrix is estimated after transmit and *after* the spatial anti-jamming filter. The resulting space-time clutter filter has to cope with clutter only if the jammer cancellation is perfect.

Both of these techniques will be analysed in the following sections. Optimum and suboptimum techniques discussed in previous chapters will be considered. Most of the numerical material is based on linear arrays. Circular planar arrays are addressed briefly.[1]

It should be noted that the effect of jamming will not be visible in most of the figures shown in this chapter, that is, we do not see the jammer itself but only the

---

[1] For clutter rejection and sidelobe control with circular arrays see Section 8.2.

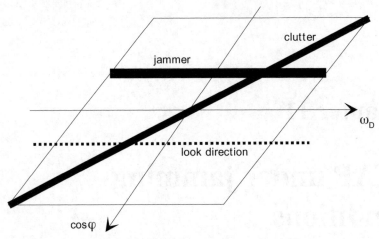

Figure 11.1: φ-ω_D *trajectories of clutter and jammer*

effect of jamming on clutter rejection performance. To understand this let us have a look at Figure 11.1. It shows the trajectories of a typical jammer-clutter scenario for a sidelooking array radar in the $\cos\varphi$-ω_D plane. While the space-time clutter spectrum extends along the diagonal of the plot the jammer is located at a certain position and extends over the whole Doppler domain. Forming a search beam is a *spatial* operation (like jamming) which can be illustrated by another straight line (dotted) parallel to the jammer trajectory. The IF plots used for judging the performance of clutter rejection techniques are cross-sections through 3-D IF plots in the look direction, i.e., along the dotted line in Figure 11.1.[2] Therefore, only the clutter notch which appears at the crossing between the look direction and the clutter trajectory will be visible in the subsequent figures.

11.2 Simultaneous jammer and clutter cancellation

In this section we want to find out what the impact of additional jamming on clutter rejection is. We discuss this problem for several processors which have already been analysed in earlier chapters with respect to clutter suppression. The interference covariance matrix now becomes

$$\mathbf{Q} = E\{\mathbf{q}\mathbf{q}^*\} \qquad (11.1)$$

where the interference vector

$$\mathbf{q} = \mathbf{c} + \sum_{j=1}^{J} \mathbf{j}_j + \mathbf{n} \qquad (11.2)$$

---

[2] Such 3-D plots have already been shown in Figures 4.2 and 4.3.

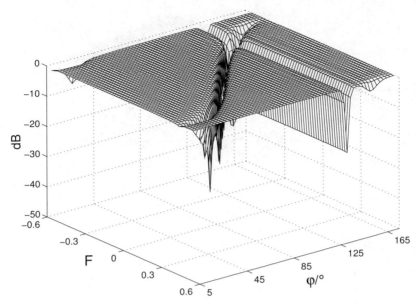

Figure 11.2: Improvement factor for a jammer + clutter scenario (SL array, $J = 1$)

includes clutter, jamming and noise. J is the number of jammers. Since we assumed clutter, jamming and noise to be statistically independent the covariance matrix becomes a sum of clutter, jammer and noise covariance matrices

$$\mathbf{Q} = \mathbf{Q}_c + \mathbf{Q}_j + \mathbf{Q}_n \qquad (11.3)$$

11.2.1 Optimum adaptive processing (OAP)

First we consider the performance of the optimum processor after (4.1) and Figure 4.1 for jammer and clutter rejection. The improvement factor is given by (4.5). It is emphasised again that the optimum processor is not useful for practical applications but will serve as a reference when discussing suboptimum processors.

Spectra of clutter + jammer scenarios have already been shown in Figures 3.38 and 3.39. In Figures 11.2 and 11.3 we now find the associated IF plots (Figure 11.2: sidelooking, one jammer; Figure 11.3: forward looking, three jammers). The trench-like stopbands of clutter and jammers can be seen clearly. Everywhere else the optimum improvement factor is reached.

In Figure 11.4 the impact of jamming on clutter rejection is illustrated in more detail. A sidelooking array configuration with $N = 12$ sensors and $M = 12$ echo pulses has been assumed.

The uppermost curve (o) shows the IF for the case that $J = 5$ jammers of equal strength radiate into the sidelobe region of the antenna pattern. The IF curve shows optimum behaviour (like the optimum curves shown before). No influence of the

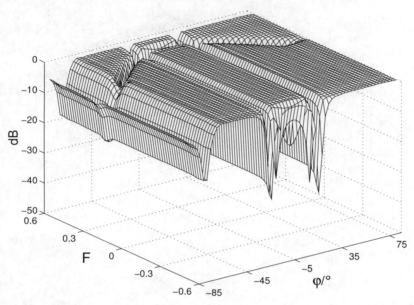

Figure 11.3: *Improvement factor for a jammer + clutter scenario (FL array, $J = 3$)*

jammers can be noticed. In fact, if the look direction is different from any jammer direction the beam looks into the 'optimum IF' plane (see Figures 11.2 and 11.3) so that no degradation in clutter suppression occurs.

The second curve ($\times$) deals with the same jammer scenario as before, however, with one out of five jammers placed in the look direction. As can be seen the target detection performance is considerably degraded. Even in the pass band (i.e., far away from the clutter notch) some 4 dB are lost. This degradation becomes more dramatic if an even higher percentage of the jammer energy were focused on the look direction (e.g., one out of two jammers), or if the jammer-to-clutter power is higher (here: 0 dB). Notice that the clutter notch is seriously broadened so that detection of slow targets is questionable.

For the third curve ($*$) 11 jammers of equal strength have been uniformly distributed in azimuth (no jammer in the look direction). Each of the jammers produces M eigenvalues of the space-time covariance matrix (recall Figure 3.36). The total number of eigenvalues is then $N + M - 1$ (for clutter) $+ J(M - 1) = 23 + 132 = 144$ (see (3.71)) which is equal to the maximum possible number $NM = 144$. In other words, with this jammer + clutter scenario the optimum processor is saturated in terms of degrees of freedom. Therefore, the improvement factor is reduced dramatically.

As long as the number of jammers is such that the number of eigenvalues of $\mathbf{Q}$ is smaller than NM perfect clutter and jammer rejection can be expected from the optimum processor. This is shown in the two examples Figure 11.5 (sidelooking) and Figure 11.6 (forward looking), where the four curves have been calculated for $J = 0, 1, 2, 3$ sidelobe jammers in addition to clutter.

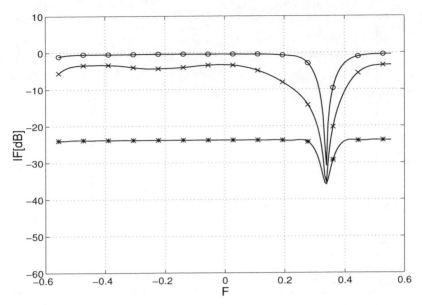

Figure 11.4: IF for jamming + clutter (SL, $N = 12$, $M = 12$): ○ $J = 5$, *jammers in sidelobe area;* × $J = 5$, *one jammer in look direction;* * *11 jammers in sidelobe area (all jammers have equal power, the total jammer power was assumed to be 20 dB)*

11.2.2 Coherent repeater jammers

The effect of coherent repeater jammers on the detection performance of the optimum processor has been analysed by WANG H. *et al.* [669, 670]. A coherent repeater jammer generates artificial target signal sequences at arbitrary ranges and Doppler frequencies. Thus the jammer may contaminate the training data while the cell under test is jammer-free. This causes a mismatch between the interference in the cell under test and the adapted anti-jamming filter which leads to losses in target detection.

11.2.3 Space-time auxiliary channel processors

This class of processors was described in Chapter 5. Linear *space-time* transforms containing space-time vectors according to (5.1) can be used to reduce the signal vector space from NM down to $C = N + M$ if only clutter and noise are present. This means that the computational load required by this class of processors increases with the antenna dimensions and the length of the pulse burst. The main problem is that the number of auxiliary channels required for interference suppression is strictly connected with the number of interference eigenvalues of the interference + noise covariance matrix. If the number of degrees of freedom is chosen smaller than the number of eigenvalues losses in SCNR will occur.

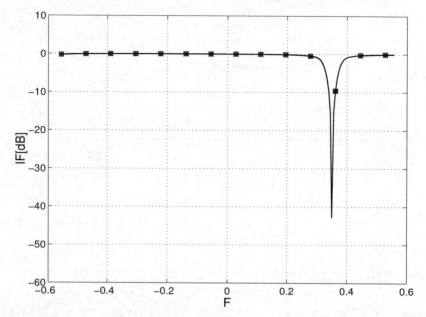

Figure 11.5: Effect of jamming on clutter rejection (optimum processor, SL): ○ $J = 0$; * $J = 1$; × $J = 2$; + $J = 3$

11.2.3.1 Auxiliary eigenvector processor (AEP)

The AEP was introduced in Section 5.2. It uses $N + M - 1$ clutter eigenvectors of the space-time clutter + noise covariance matrix as auxiliary channels. An additional signal matched search channel is provided so that the total number of channels amounts to $C = N + M$.

It has been shown in Chapter 3 (Figures 3.36 and 3.37) that each broadband jammer[3] produces $M - 1$ additional eigenvalues. That means, for each additional jammer, $M - 1$ additional auxiliary channels (eigenvectors) in the space-time transform **T** in (5.1) are required. This leads to an unbearable additional burden for the space-time processor.

As found in Figure 5.3, the auxiliary eigenvector processor is relatively tolerant against reduction of the number of auxiliary channels because some of the clutter eigenvectors are associated with small eigenvalues and, hence, can be omitted without significant losses in SCNR. The decay of the eigenvalues is a property of the clutter spectrum. The shape of the clutter spectrum is determined by the transmit beam, tapering, etc. As can be seen from Figure 3.36, there is no smooth decay in the eigenvalues of jammers. Therefore, additional jamming must be paid for by increasing the number of channels (M channels for each jammer).

---

[3] Bandwidth larger than unambiguous Doppler domain.

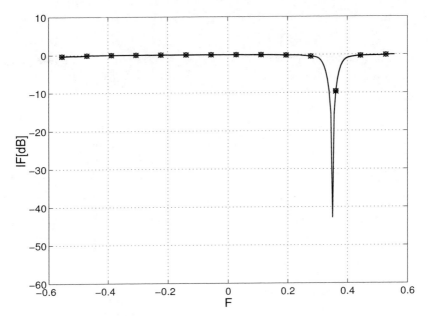

Figure 11.6: Effect of jamming on clutter rejection (optimum processor, FL): ○ $J = 0$; ∗ $J = 1$; × $J = 2$; + $J = 3$

11.2.3.2 Auxiliary channel processor (ACP)

The auxiliary channel processor was described in Section 5.3. It involves a number of beamformers so as to cover the whole azimuthal range. Each of the beamformers is cascaded with a Doppler filter matched to the clutter Doppler frequency in the look direction of the individual beam, see Figure 5.7. Each of these space-time channels is matched in angle and Doppler to a certain clutter patch on the ground.

These clutter channels have the property that only $N + M - 1$ vectors are linearly independent. This means that the transform **T** in (5.1) cannot have more than $N+M-1$ linearly independent columns of this kind of beamformer/Doppler filter structure[4] so that the covariance matrix becomes singular. Therefore, additional jamming cannot be rejected.

The problem of a singular covariance matrix may basically be circumvented by diagonal loading, i.e., adding 'artificial noise'. However, since there are much more promising solutions for space-time filters (see the next section) we do not pursue the work on the ACP any further.

11.2.4 Spatial auxiliary channel processors

In the next four examples the jammer and clutter rejection performance of the auxiliary sensor FIR filter processor (ASFF) after (7.12) and Figure 7.5 is illustrated. The dependence of the number of channels on the number of jammers is of primary interest.

---
[4]For an explanation see Section 5.5.3.

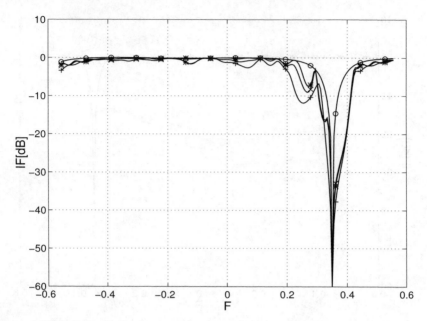

Figure 11.7: Effect of jamming (symmetric auxiliary sensor processor, $K = 3$, SL): ○ $J = 0$; ∗ $J = 1$; × $J = 2$; + $J = 3$

11.2.4.1 Sidelooking linear array

Figure 11.7 shows IF curves for different numbers of jammers ($J = 0, 1, 2, 3$). The total jammer-to-noise ratio and the clutter-to-noise ratio were both assumed to be 20 dB. The number of channels is $K = 3$ which is the absolute minimum for this type of processor.

If no jamming is present (○) a quasi-optimum IF curve is obtained. For $J > 0$ some losses close to the clutter notch can be noticed. This happens because the number of spatial channels is smaller than the number of interfering sources. It should be noted that the assumed jamming scenario is relatively weak (three jammers whose total power is equal to the clutter power). In a stronger jamming scenario these losses will increase considerably.

If the number of channels is increased so that the number of channels is sufficient to cope with the jammer + clutter scenario ($K = 7$ in Figure 11.8) the effect of jamming is suppressed.

11.2.4.2 Forward looking array

Similar results are obtained for a forward looking linear array (Figures 11.9 and 11.10). Notice that even for $K = 7$ channels (Figure 11.10) some slight losses can be observed. Even the (○) curve (no jamming) in Figure 11.9 does not run as smoothly as the corresponding curve for the sidelooking array (Figure 11.7).

This behaviour was expected. The forward looking array receives two clutter

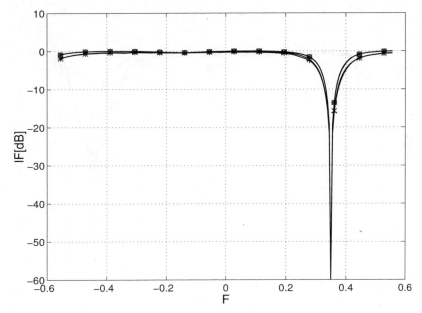

Figure 11.8: Effect of jamming (symmetric auxiliary sensor processor, $K = 7$, SL): ○ $J = 0$; ∗ $J = 1$; × $J = 2$; + $J = 3$

arrivals for each Doppler frequency (one from the left and one from the right) and, therefore, requires more *spatial* degrees of freedom for the clutter rejection filter. This effect has already been addressed in Sections 3.2.3 and 6.2.1. Further increase of the number of channels reduces the losses.

11.2.4.3 Comparison of spatial and space-time auxiliary channel processing

There are fundamental differences between spatial and space-time subspace transforms. A space-time transform $\mathbf{T}$ contains space-time vectors as described by (5.1). Provided that the covariance matrix $\mathbf{Q}$ has full rank NM (which is fulfilled when white noise is included) the transformed covariance matrix $\mathbf{Q}_T = \mathbf{T}^*\mathbf{Q}\mathbf{T}$ has rank L which is the number of space-time channels. In other words, each space-time channel produces *one* eigenvalue of $\mathbf{Q}_T$.

On the other hand in general a broadband jammer causes M eigenvalues (for a sidelooking array and DPCA case only $M - 1$) so that for each jammer M additional space-time channels have to be added to the transform matrix in order to provide a sufficient number of degrees of freedom.[5]

The spatial pre-transform was defined by (6.1), where the spatial submatrices $\mathbf{T}_s$ may assume different forms, for example overlapping subarrays (6.2), disjoint

---

[5]Remember that for the ACP (Section 5.3) no channels beyond the number of clutter channels can be added – the transformed matrix becomes singular.

352 STAP under jamming conditions

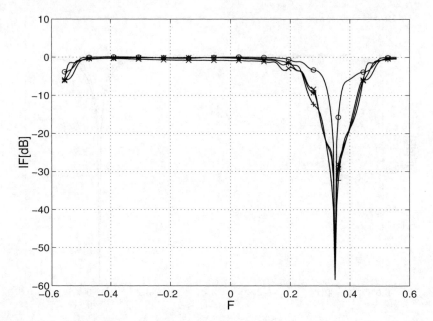

Figure 11.9: Effect of jamming (symmetric auxiliary sensor processor, $K=3$, FL): ○ $J=0$; ∗ $J=1$; × $J=2$; + $J=3$

subarrays (6.10), and beamformer/symmetric auxiliary sensors (6.17). Each of these matrices has KM columns (K is the number of spatial channels) so that the rank (or the number of eigenvalues) of the transformed covariance matrix $\mathbf{Q}_T = \mathbf{T}^*\mathbf{QT}$ is KM. That means, adding *one* spatial channel the transform $\mathbf{T}$ results in M additional eigenvalues of $\mathbf{Q}_T$.

Therefore, for each jammer only one additional spatial channel is required. The above consideration can be summarised as follows: Jamming is a *spatial* problem and has to be countered efficiently by modifications in the *spatial* dimension of the space-time processor.

11.3 Circular arrays with subarray processing

In this section we consider again circular planar arrays as have been discussed already in Chapter 8.

11.3.1 Adaptive space-time processing versus temporal clutter filtering

For the sake of completeness we present here two examples for clutter and jammer suppression for the circular planar array with irregular subarray structure as shown in Figure 8.1. The upper curve in Figure 11.11 shows the performance of a tapered space-time processor after Figure 8.4. We notice the usual IF curve with clutter notch.

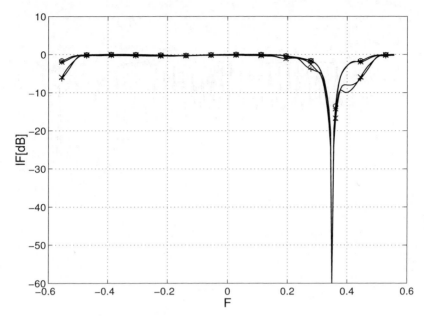

Figure 11.10: Effect of jamming (symmetric auxiliary sensor processor, $K = 7$, FL): ○ $J = 0$; ∗ $J = 1$; × $J = 2$; + $J = 3$

Except for the usual 2 dB tapering loss compared with the optimum, no losses in SNIR occur. We considered here a case with abundant degrees of freedom (32 channels, three jammers).

The lower curve shows the result for beamforming and temporal clutter filtering according to Figure 8.38. Notice that in the presence of a broadband jammer no clutter suppression is possible with a temporal filter. A forward looking antenna geometry was chosen. The lower curve in Figure 11.11 shows the clutter Doppler response for this case. Since in the forward look direction the clutter is narrowband we encounter the inverse of the Doppler filter response. At $\varphi_L = 45°$ clutter returns are Doppler broadband. Therefore the main lobe response is much broader (Figure 11.12) and the nulls of the clutter response are flattened out (compare with Figure 11.11).

11.3.2 Two-dimensional arrays in multi-jammer scenarios

The capability of an antenna array of cancelling jammers depends on the number of degrees of freedom of the adaptive array as well as on the number and geometry of the jammer scenario. In the following we assume an antenna array as shown in Figure 11.13. Such an array can be obtained by combining every two subarrays in the dartboard array, see Figure 8.3. Two jammer scenarios are assumed, scenario 1 with all jammers on the ground, and scenario 2 with a 2-dimensional jammer distribution, that means, a mixture of ground and airborne jammers. For a more detailed description the reader is referred to KLEMM [346].

We want to find out how far the geometry of the jammer scenario has influence on

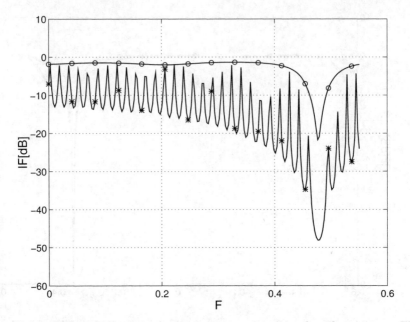

*Figure 11.11: Effect of jamming on clutter suppression (circular planar array, FL, $\varphi_L = 0°$): ○ space-time processing; * beamformer + adaptive temporal filter*

the jammer suppression capability. Let us consider a trivial example. Let all the signals received by the upper and lower halves of the array be combined by two beamformers. Subtracting the upper beamformer output from the lower one (as is done in monopulse antennas) results in a radiation pattern with a horizontal trench. All jammers falling in this trench will be suppressed. Notice that in this example the number of jammers exceeds the number of degrees of freedom (which is only one). Therefore, it can be expected that the jammer geometry has some influence on the capability of jammer suppression. The given results are based on the optimum processor at subarray level, see Chapter 6.

11.3.2.1 *Jamming only*

In in the bar plots in Figures 11.14 to 11.16 some results are given for scenario 1 (ground jammers). The improvement factor has been plotted versus the number of jammers. Notice that the lower ends of the bars denote the loss in improvement factor. The total jammer-to-noise ratio (JNR) is varied between 20 dB and 60 dB.

First of all it can be noticed that no significant jamming effect takes place if the total JNR is 20 dB (Figures 11.14). In this case the sidelobe level of the array is obviously sufficiently low to suppress the jammers. If the jammer power gets larger (Figures 11.15, 11.16) we recognise that about 4 jammers can be suppressed without significant loss in improvement factor.

A look at the results obtained for scenario 2 (2-dimensional jammer distribution, Figures 11.17 to 11.19) shows that about 7 jammers can be suppressed without

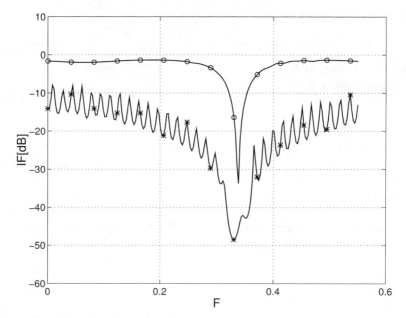

Figure 11.12: Effect of jamming on clutter suppression (circular planar array, FL, $\varphi_L = 45°$): ○ *space-time processing;* * *beamformer + adaptive temporal filter*

significant loss.

The different responses of the array to the two jammer scenarios can be interpreted as follows. If the jammers are more or less located in the horizontal as in the case of ground jammers (scenario 1) only the horizontal degrees of freedom are effective. A look at the array shown in Figure 11.13 tells us that there are only four subarrays in the horizontal dimension which is equivalent to 3 degrees of freedom. Obviously one more degree of freedom is generated by the horizontal asymmetry of the upper and lower subarrays. If, however, the jammers are distributed in both the horizontal and vertical dimensions (scenario 2) the full number of degrees of freedom of the array are exploited (8 subarrays equivalent to 7 degrees of freedom). Therefore suppression of seven jammers can be expected. This is clearly reflected in Figures 11.17, 11.18, and 11.19.

11.3.2.2 Jammer plus clutter

In a clutter + jammer scenario the number of jammers that can be cancelled is reduced because a few degrees of freedom are absorbed for clutter suppression. We know from the results in Chapter 6 (Figures 6.2, 6.10, 6.11) that for clutter suppression in a forward looking array radar the number of *spatial* degrees of freedom should be larger than 1, that means, the number of antenna channels K should be larger than 2.

Some numerical results are shown in Figures 11.20 and 11.21. It can be seen in Figure 11.20 (all jammers on the ground) that in presence of clutter only 1 jammer can be cancelled without significant loss. As remarked above the array in Figure 11.13 has

356 STAP under jamming conditions

Table 11.1 Parameters for multi-jammer scenario

PRF	2 KHz
number of elements N	902
number of subarrays K	8
platform height H	1000 m
range R	35 km
platform velocity	200 m/s
look direction	33°
antenna orientation	forward looking
JNR	20, 40, 60 dB
jammer azimuth angles [°]	0; -12, 10; 43; -50; 61 -60; 74; -71; 82; 86
jammer depression angles, scenario 1	all jammers on ground at target range
jammer depression angles [°], scenario 2	airborne and groundbased jammers -32; -25; -21; -15; -10; -5 0; 16; 24; 30; 44
jammer power	same power for all jammers
CNR	4.6 dB

three to four horizontal degrees if freedom. At least two spatial DoF are absorbed by clutter rejection, only one DoF remains for jammer cancellation. Figure 11.21 is based on scenario 2 (2D jammer distribution). In this case up to 5 jammers can be suppression without large losses. For seven jammers considerable losses arc encountered the total number of degrees of freedom of the interference (7 jammers, 2 for clutter) exceeds the number of degrees of freedom of the array (7).

11.4 Separate jammer and clutter cancellation

Since broadband jammers are *spatial* interference by nature and clutter returns are *space-time* the question arises how far it would be possible to reject the jammers by a *spatial* filter prior to clutter suppression by *space-time* processing techniques. The anti-jamming filter has to be adapted in a passive mode, that means, before transmit. The space-time clutter filter is then adapted in the active mode, in the presence of jamming and clutter, however after jammer suppression. In this way the space-time anti-clutter filter might be made small because it has to cope with clutter only. This principle was proposed by KLEMM [308] and has been used with MOUNTAINTOP data (TITI [646]) by MARSHALL [437]. HALE et al. [243, 245] propose a hybrid STAP architecture consisting of an adaptive *vertical* beamformer cascaded with a partial adaptive STAP processor. Such architecture may offer advantages in the case of jammer/clutter scenarios or range ambiguous clutter returns. In [244] the same authors compare such a cascaded approach with 3D STAP (horizontal, vertical, time).

For a sidelooking array RICHARDSON [566] has derived the number of degrees of

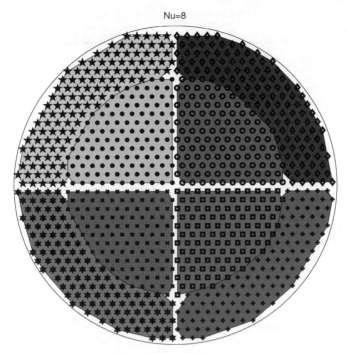

Figure 11.13: Circular array with 8 subarrays

eigenvalues of the clutter + jammer space-time covariance matrix to be

$$N_e \leq N + (\beta + J)(M - 1) \tag{11.4}$$

where β is the ratio of Doppler ambiguities and spatial ambiguities.

In Figure 11.22 the principle of cascaded anti-jamming and anti-clutter filtering is illustrated by means of a schematic jammer-clutter spectrum. The jammer appears as a thin wall along the Doppler axis and has no extension in the azimuth axis. The jammer can first be cancelled by a *spatial* anti-jamming filter which can be seen on the left. The remaining clutter will be suppressed by a space-time clutter filter which has to cope with clutter only.

11.4.1 Optimum jammer cancellation and auxiliary channel clutter filter

As a first step we consider the case that the jammer suppression operation is fully adaptive according to the optimum processor (4.1). After jammer cancellation the data are transformed by a suitable *spatial* transform so that clutter rejection takes place in a reduced subspace.

358 STAP under jamming conditions

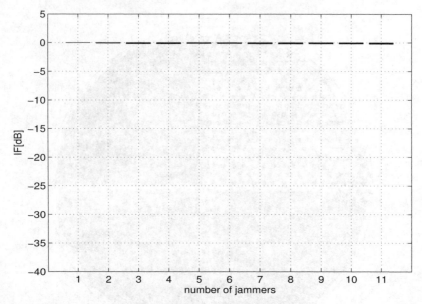

Figure 11.14: Detection performance in presence of jammers, scenario 1: ground jammers; total JNR: 20 dB

11.4.1.1 The principle

Figure 11.23 shows a block diagram of such a cascaded processor. The output signals of the array are multiplied with the inverse of the $N \times N$ spatial jammer + noise covariance matrix. At the output the jammer-free signals are given by the transformed vector

$$\mathbf{x}_1 = \mathbf{Q}_j^{-1} \mathbf{x} \tag{11.5}$$

where the matrix $\mathbf{Q}_j^{-1}$ becomes in space-time notation

$$\mathbf{Q}_j^{-1} = \begin{pmatrix} \mathbf{Q}_s^{-1} & & & \mathbf{0} \\ & \mathbf{Q}_s^{-1} & & \\ & & \ddots & \\ \mathbf{0} & & & \mathbf{Q}_s^{-1} \end{pmatrix} \tag{11.6}$$

and $\mathbf{Q}_s$ is the $N \times N$ spatial covariance matrix of jammers and noise.[6]

The jammer suppression stage is followed by a spatial subspace transform

$$\mathbf{x}_2 = \mathbf{T}^* \mathbf{x}_1 = \mathbf{T}^* \mathbf{Q}_j^{-1} \mathbf{x} \tag{11.7}$$

---

[6]$\mathbf{Q}_s$ is constant along the diagonal only as long as the echo sequence is stationary. In the case of manoeuvres of the radar platform this is not fulfilled (RICHARDSON [564]). For details see Chapter 16.

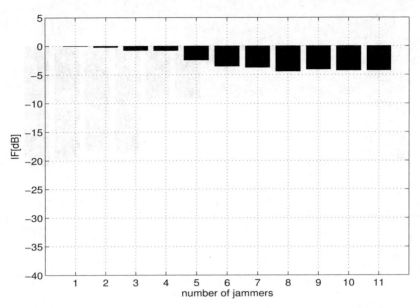

Figure 11.15: Detection performance in presence of jammers, scenario 1: ground jammers; total JNR: 40 dB

where

$$\mathbf{T} = \begin{pmatrix} \mathbf{T}_s & & & \\ & \mathbf{T}_s & & \mathbf{0} \\ & & \ddots & \\ \mathbf{0} & & & \mathbf{T}_s \end{pmatrix} \quad (11.8)$$

and $\mathbf{T}_s$ is a spatial subspace transform. For example, for beamforming with symmetric auxiliary sensors (see (6.17) and Figure 6.12) one gets

$$\mathbf{T}_s = \begin{pmatrix} 1 & & & & & & & \\ & \ddots & & & & & \mathbf{0} & \\ & & 1 & & & & & \\ & & & b_1 & & & & \\ & & & \vdots & & & & \\ & & & b_B & & & & \\ & & & & 1 & & & \\ & \mathbf{0} & & & & \ddots & & \\ & & & & & & 1 \end{pmatrix} \quad (11.9)$$

Below the dashed line in Figure 11.23 we find the usual space-time adaptive clutter suppression filter which is based on the space-time clutter + jammer + noise covariance matrix

$$\mathbf{w}_T = \mathbf{Q}_T^{-1}\mathbf{s}_T \quad (11.10)$$

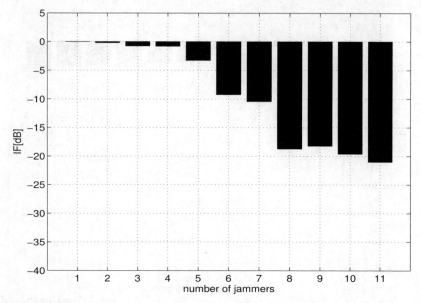

Figure 11.16: Detection performance in presence of jammers, scenario 1: ground jammers; total JNR: 60 dB

where
$$\mathbf{s}_T = \mathbf{T}^* \mathbf{Q}_j^{-1} \mathbf{s} \tag{11.11}$$
and
$$\mathbf{Q}_T = \mathbf{T}^* \mathbf{Q}_j^{-1} \mathbf{Q} \mathbf{Q}_j^{-1} \mathbf{T} \tag{11.12}$$

The improvement factor achieved by the processor (11.10) is given by (5.5).

11.4.1.2 A note on symmetry

The symmetric auxiliary beam/sensor configuration was shown to be equivalent to forming overlapping identical subarrays so that the clutter Doppler spectra at each of the subarray outputs are identical (Figure 6.8). This is a prerequisite for achieving a low-order subspace (small number of antenna channels). This principle works only for *equidistant* linear, rectangular planar or cylindrical arrays. For this array type the spatial submatrices are Toeplitz and the block structure of the space-time clutter covariance matrix is Toeplitz as well. Therefore, this kind of matrix is persymmetric, i.e., symmetric with respect to both matrix diagonals. As a consequence all columns (or rows) of the covariance matrix are symmetric so that the symmetric structure of the auxiliary sensor processor is maintained.

11.4.1.3 Results

Figure 11.25 shows the performance of the first stage of the cascaded processor depicted in Figure 11.23. The improvement in signal-to-interference ratio is plotted

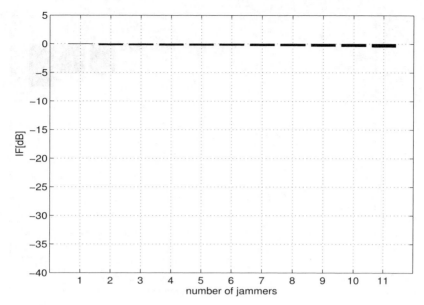

Figure 11.17: Detection performance in presence of jammers, scenario 2: ground jammers; total JNR: 20 dB

versus the azimuth angle φ. The three notches indicate clearly the positions of the jammers used in this analysis. For the sake of brevity we use a sidelooking array only. The results obtained for the forward looking array are similar.

In Figure 11.26 we see IF curves for $J = 0, 1, 2, 3$ jammers. The number of channels was chosen to be $K = 3$. Notice that this is the absolute minimum number of channels for the symmetric auxiliary channel processor. It would not be sufficient to suppress clutter and jammers simultaneously (compare with Figure 11.7). Obviously the jammer suppression filter (the inverse of the jammer covariance matrix) cancels the jammers quite well.

Even better performance is obtained for $K = 5$ (Figure 11.27) or $K = 7$ (Figure 11.28) channels. It should be noted, however, that with $K = 7$ channels the number of spatial degrees of freedom of the subspace is sufficient to suppress three jammers and clutter *simultaneously*, see Figure 11.8.

11.4.2 Jammer and clutter auxiliary channel filters cascaded

In this approach the auxiliary sensor transform which has proven to be very useful when being used with linear arrays is applied twice, firstly before jammer suppression, secondly, before clutter cancellation.

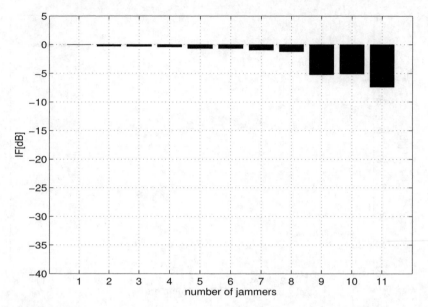

Figure 11.18: Detection performance in presence of jammers, scenario 2: ground jammers; total JNR: 40 dB

11.4.2.1 The principle

The cascaded processor structure of Figure 11.23 is modified in such a way that an auxiliary sensor transform $\mathbf{T}^{(j)}$ is applied to the received data before jammer suppression. $\mathbf{T}^{(j)}$ is block diagonal like (11.4), with spatial submatrices of the form

$$\mathbf{T}_s^{(j)} = \begin{pmatrix} 1 & & & & & & & \\ & \ddots & & & & & 0 & \\ & & 1 & & & & & \\ & & & b_1 & & & & \\ & & & & \ddots & & & \\ & & & & & b_{B_j} & & \\ & & & & & & 1 & \\ & 0 & & & & & & \ddots \\ & & & & & & & & 1 \end{pmatrix} \qquad (11.13)$$

Jammer suppression is carried out in a K_j dimensional subspace where K_j denotes the number of columns of $\mathbf{T}^{(j)}$ (technically, the number of antenna channels). K_j has to be chosen so that it provides a suffient number of degrees of freedom to cope with the jammer + clutter scenario.

A secondary transform $\mathbf{T}^{(c)}$ is introduced to reduce the vector space even further down to K_c channels. If the jammer suppression operation works perfectly we have only clutter and noise in this vector space. The secondary transform has again the form

Separate jammer and clutter cancellation

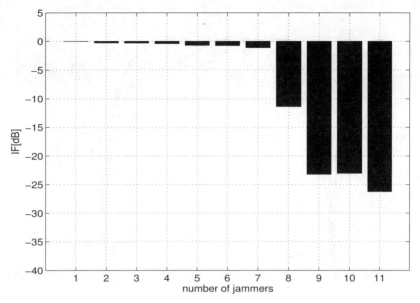

Figure 11.19: Detection performance in presence of jammers, scenario 2: ground jammers; total JNR: 60 dB

of (11.8), with spatial submatrices

$$\mathbf{T}_s^{(c)} = \begin{pmatrix} 1 & & & & & & & \\ & \ddots & & & & & 0 & \\ & & 1 & & & & & \\ & & & b_1 & & & & \\ & & & & \cdot & & & \\ & & & & & \cdot & & \\ & & & & & b_{B_c} & & \\ & & & & & & 1 & \\ & 0 & & & & & & \ddots \\ & & & & & & & & 1 \end{pmatrix} \qquad (11.14)$$

Notice that the beamformer weights in (11.14) are different from those in (11.13). Both sets of beamformers are related to each other through the transform (11.13).

The number K_c denotes the number of columns of the spatial transform $\mathbf{T}_c$. It should be chosen as small as possible because at this level space-time processing is carried out.

Now the adaptive processor in the $K_c \times M$ dimensional clutter suppression subspace is

$$\mathbf{w}_T = \mathbf{Q}_T^{-1} \mathbf{s}_T \qquad (11.15)$$

where

$$\mathbf{s}_T = \mathbf{T}_c^* (\mathbf{T}_j^* \mathbf{Q}_j \mathbf{T}_j)^{-1} \mathbf{T}_j^* \mathbf{s} \qquad (11.16)$$

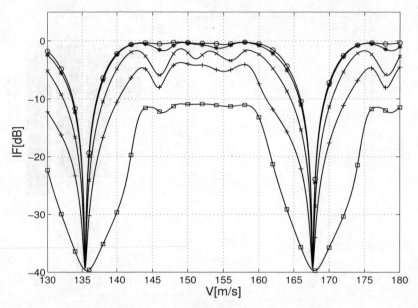

*Figure 11.20: Impact of jamming on clutter suppression Nr. of jammers: ○ 0; * 1; × 3; + 5; □ 7; scenario 1 (ground jammers); total JNR: 60 dB*

and
$$\mathbf{Q}_T = \mathbf{T}_c^*(\mathbf{T}_j^*\mathbf{Q}_j\mathbf{T}_j)^{-1}\mathbf{T}_j^*\mathbf{Q}\mathbf{T}_j(\mathbf{T}_j^*\mathbf{Q}_j\mathbf{T}_j)^{-1}\mathbf{T}_c \tag{11.17}$$

The transform at the jammer suppression level $\mathbf{T}_j$ is block diagonal according to (11.8), with submatrices given by (11.14). Likewise, the transform at the clutter suppression level $\mathbf{T}_c$ is block diagonal according to (11.8) with submatrices given by (11.15). As for all transform techniques the improvement factor is given by (5.5).

Figure 11.24 shows a block diagram of the two-stage auxiliary sensor processor. In practice a space-time FIR filter according to (7.12) may be used for clutter rejection.

The spatial submatrices of the total transform then take the form

$$\mathbf{T}_s = \mathbf{T}_s^{(j)}(\mathbf{T}_s^{(j)*}\mathbf{Q}_s^{(j)}\mathbf{T}_s^{(j)})^{-1}\mathbf{T}_s^{(c)} = \begin{pmatrix} \mathbf{q}_1 & \cdots & \mathbf{q}_{(K-1)/2} & \mathbf{v} & \mathbf{q}_{(K+1)/2} & \cdots & \mathbf{q}_K \end{pmatrix} \tag{11.18}$$

where the subscript 's' means 'spatial submatrix' according to (11.6) and (11.8). The vectors $\mathbf{q}_i$ denote those columns of $(\mathbf{T}_s^{(j)*}\mathbf{Q}_s^{(j)}\mathbf{T}_s^{(j)})^{-1}$ associated with the auxiliary channels, and

$$\mathbf{v} = \begin{pmatrix} 0 \\ \vdots \\ 0 \\ \mathbf{K}_c\mathbf{b}_c \\ 0 \\ \vdots \\ 0 \end{pmatrix} \tag{11.19}$$

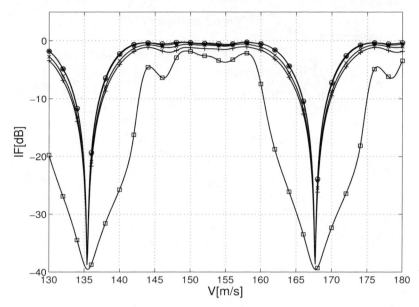

*Figure 11.21: Impact of jamming on clutter suppression Nr. of jammers: ○ 0; * 1; × 3; + 5; □ 7; scenario 2; total JNR: 60 dB*

$\mathbf{K}_c$ is the $B_c \times B_c$ submatrix of $(\mathbf{T}_s^{(j)*}\mathbf{Q}_s^{(j)}\mathbf{T}_s^{(j)})^{-1}$ associated with those central channels which are fed into the secondary beamformer $\mathbf{b}_c = \mathbf{T}_s^{(c)*}(\mathbf{T}_s^{(j)*}\mathbf{Q}_s^{(j)}\mathbf{T}_s^{(j)})^{-1}\mathbf{T}_s^{(j)*}\mathbf{b}$.

11.4.2.2 Results

Some numerical results obtained with the cascaded jammer and clutter auxiliary channel processor after Figure 11.24 are shown in Figures 11.29 and 11.30. The number of jammers was assumed to be $J = 3$, the number of channels in the anti-clutter stage is $K_c = 3$ in Figure 11.29, and $K_c = 5$ in Figure 11.30. The four curves in each of the plots have been calculated for different numbers of channels of the first stage ($K_j = 5, 7, 9, 15$).

Remember that $K_c = 3$ is the minimum number of channels for this symmetric auxiliary channel processor (one auxiliary channel left of the beamformer, one on the right). Therefore, three additional jammers will require at least $K_j = 7$ channels in the first stage. As can be seen in Figure 11.29 even for $K_j = 7$ some losses, especially in the clutter notch area, have to be tolerated.

The inverse of the jammer covariance matrix obviously causes some distortion of the clutter echoes in such a way that the anti-clutter filter with $K_c = 3$ antenna channels does not work perfectly because of a lack of degrees of freedom.

If the number of channels becomes $K_c = 5$ (Figure 11.30) the curves for $K_j \geq 7$ are almost optimum.

366 STAP under jamming conditions

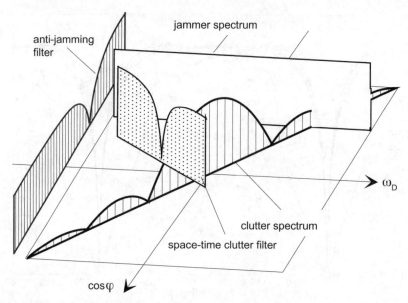

Figure 11.22: *Cascading jammer and clutter filters*

11.4.2.3 *Computational expense*

Both processors (simultaneous jammer and clutter rejection/cascaded auxiliary channel anti-jamming and anti-clutter processor) are compared in terms of arithmetic operations in Tables 11.2 and 11.3.

Table 11.2 shows the number of operations required for the inversion of the covariance matrices of jammers and clutter. The third column contains the number of operations for simultaneous jammer and clutter suppression while column 4 shows the operations for jammer and clutter filters cascaded. We assumed that the covariance matrix is estimated and inverted.

For the inversion of the space-time covariance matrices we assumed $8K^3M$ operations which is the number of operations required for inverting a block Toeplitz matrix (see AKAIKE [12], and GOVER and BARNETT [209]). For the inversion of the spatial covariance matrix we assumed K^3 operations which is normal for a non-Toeplitz matrix.

It can be seen that for larger numbers of antenna channels K_j the cascaded solution needs considerably less operations than is required for simultaneous jammer and clutter suppression.

For the filtering we assumed that just a scalar vector product (see (11.19) and (7.12)) is needed. In the case of the cascaded processing scheme we added another column in Table 11.3 which contains the number of operations for the second auxiliary channel transform (before clutter rejection). Remember that the first auxiliary sensor transform (before jammer cancellation) can be implemented in the RF domain.

The numbers for the matrix inverse in Table 11.4.2.2 denote only those operations

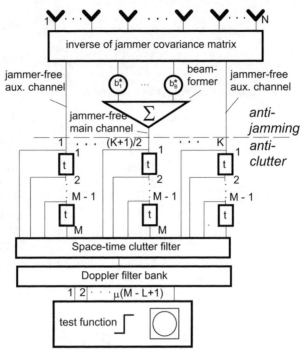

Figure 11.23: *Cascading fully adaptive anti-jamming and auxiliary sensor clutter filters*

which are required for the inversion operation. In addition we have the number of operations required for estimating the covariance matrices. This requires about $2(.)^3$ operations (REED *et al.* [557]). Of course, due to the fact that the space-time covariance matrices are block Toeplitz only one block row has to be estimated. As can be seen cascading jammer and clutter filters can achieve considerable savings of operations.

The numbers of operations for space-time FIR filtering per range gate and echo pulse are relatively small for both techniques. FIR filtering through all range gates in real-time should not be a problem with modern digital technology.

11.4.2.4 *Sideband jammer cancellation*

The concept of cascading a spatial adaptive anti-jamming filter with a space-time adaptive clutter filter has been refined by RIVKIN *et al.* [570]. A sideband is generated outside the radar band followed by a sideband filter so that no radar returns will enter this channel. Given the case that a white noise jammer covers *both* the radar band as well as the sideband, data from the latter can be used to estimate a clutter-free spatial jammer covariance matrix (or to adapt a suppression filter from jammer only data). The filter weights have to be matched to the radar frequency band by a linear transform. A similar concept (cascading a spatial sidelobe canceller with Σ-Δ space-time clutter cancellation) has been presented by ZHANG *et al.* [748].

Table 11.2 Operations required for matrix inverse update

$M=5$		simultaneous	cascaded
K_j	K_c	$8MK_j^3$	$8MK_c^3 + K_j^3$
5	3	5000	1205
7	3	13720	1423
9	3	29160	1809
5	5	5000	5125
7	5	13720	5343
9	5	29160	5729
11	5	53240	6331
13	5	87880	7197
15	5	135000	8375

Table 11.3 Operations required for filtering (per range gate)

$M=5$		simultaneous	cascaded	transform
K_j	K_c	MK_j	$MK_c + K_j$	$(K_c - 1)K_j$
5	3	25	20	10
7	3	35	22	14
9	3	45	24	18
5	5	25	30	20
7	5	35	32	28
9	5	45	34	36
11	5	55	36	44
13	5	65	38	52
15	5	75	40	60

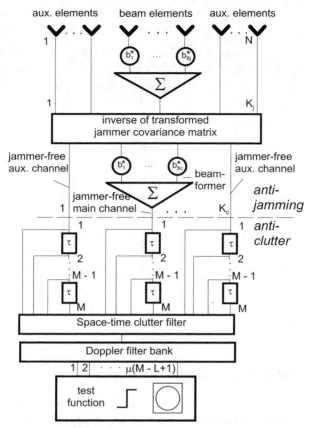

Figure 11.24: Cascading auxiliary sensor anti-jamming and clutter filters

11.5 Jamming in the range-Doppler IF matrix

Figures 11.31 and 11.32 show the improvement factor achieved by the optimum processor (see Chapter 4) for a jammer-clutter scenario with CNR = 20 dB and JNR = 60 dB. The IF plots are quite similar to those obtained for clutter only (see 4.31 and 4.32). The clutter notches can be recognised clearly while the jammer power is uniformly distributed over the whole range-Doppler matrix. Notice that for sidelooking radar the clutter Doppler frequency is constant with range while for forward looking radar we see the Doppler dependence with range according to (3.20).

Comparing the range-Doppler matrices for clutter only with clutter + jamming we recognise a basic difference. For clutter only (Figures 4.31 and 4.32) the radar range equation causes the clutter returns to be range dependent while the jammer power is range independent. If the jammer is strong compared with the clutter returns the IF plots are dominated by the jammer, that is, the improvement factor becomes range independent (Figures 11.31 and 11.32).

370 STAP under jamming conditions

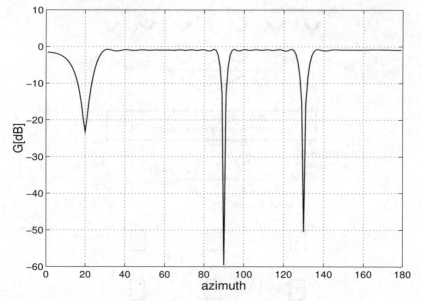

Figure 11.25: Performance of the anti-jamming processor

11.6 Terrain scattered jamming

If a jammer radiates on to a radar antenna, part of the energy will reach the radar antenna via reflection from the ground. Since the ground is in general a diffuse reflector the following effects can be noticed:

- Multipath arrivals may reach the radar antenna under different angles.
- Multipath arrivals are in general a superposition of reflections from various scatterers at different ranges. Therefore, each of the individual reflections are time-shifted versions of the direct arrival.
- The radar bandwidth may require broadband processing.[7]

11.6.1 Transmit waveform

DOHERTY [118] proposes a technique for pre-filtering of received echoes in the TIME domain (fast time) for a pulse compression radar using polyphase codes. Polyphase codes are uncorrelated for all delays except for the PRI. Clutter (reflections of radar pulses from the ground) will, therefore, be uncorrelated in time except for the PRI delay. This property is exploited to separate clutter from terrain scattered jamming. If the received echo contains correlated portions they are due to terrain scattered jamming. A predictor based FIR filter is used to suppress the terrain scattered jamming before

---

[7]Broadband jammer suppression via space-time processing is discussed in Chapter 13.

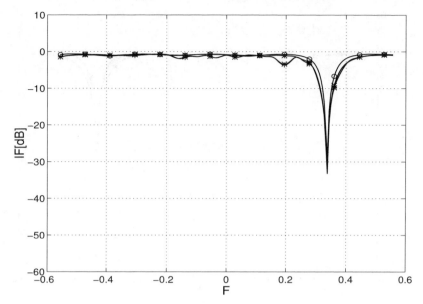

Figure 11.26: Optimum jammer suppression and auxiliary channel clutter filter cascaded ($N = 25$, $K = 3$, SL): ○ $J = 0$; * $J = 1$; × $J = 2$; + $J = 3$

pulse compression. According to the author some significant improvement in SNIR is achieved.[8]

11.6.2 Adaptive multipath cancellation

FANTE [140] analyses the behaviour of a broadband two-antenna canceller for suppression of terrain scattered jamming. Two cases are considered: 1. One specular arrival; 2. reflection on a diffuse ground. The technique proposed is in essence a space-TIME (TIME being the 'fast time') processor including a spectral analysis for each antenna and a jammer canceller for each FREQUENCY (frequency associated with fast time).

It should be noted that the cancellation of two arrivals is accomplished with a two-element antenna which has one spatial degree of freedom only. Additional degrees of freedom are taken from the TIME dimension. On the other hand, it is interesting to find out how far the temporal degrees of freedom can be reduced when the number of antenna elements is increased.

An extension of this technique to *airborne* radar has been given by FANTE and TORRES [142]. The problem of cancellation of diffuse terrain scattered jamming as well as clutter is discussed. While for cancellation of diffuse jamming a space-TIME processor architecture is required, airborne clutter suppression needs a space-time ('slow' echo-to-echo time) processor. In order to handle both problems simultaneously

---
[8]It should be noted that this technique of terrain scattered jammer cancellation does not require a multi-channel antenna. Bandwidth effects have not been considered in this book.

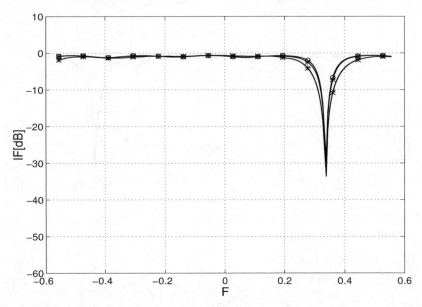

Figure 11.27: Optimum jammer suppression and auxiliary channel clutter filter cascaded ($N = 25$, $K = 5$, SL): ○ $J = 0$; ∗ $J = 1$; × $J = 2$; + $J = 3$

a space-time-TIME processor architecture has to be designed.

JOUNY and CULPEPPER [283] propose an adaptive space-TIME FIR filter technique for use with a sensor array to cope with different arrivals as may occur in the case of diffuse reflection. Simulations indicate that significant interference cancellation ratios can be achieved. Similar results using a space-TIME FIR filter have been obtained with MOUNTAINTOP data by WICKS et al. [707].

The technique proposed by KOGON et al. [359] uses beamspace-TIME processing for adaptive suppression of terrain scattered jamming. A bunch of orthogonal auxiliary beams points at the directions of arrival of the terrain scattered jamming. There are two motivations for using the beamspace instead of the element space. First, the auxiliary beams form a 'blocking matrix' which may mitigate the effect of signal inclusion in the main beam.[9] Secondly, the auxiliary beams form a subspace where the individual scattered jammer arrivals can be used mutually as reference signals while the main channel receives mainly the direct path. As the delay between scattered arrivals is much smaller than between scattered arrivals and the direct path, the beamspace technique results in a reduction of the temporal filter length.

In another paper by KOGON et al. [360] it is demonstrated that space-TIME adaptive processing can be used to cancel main beam jamming by exploiting coherent multipath.

GRIFFITHS [212] uses a constrained adaptive processor with MOUNTAINTOP data (TITI [646]) for hot clutter cancellation. A multiple constraint is used to preserve the characteristics of the desired signal.

---

[9]This concept has also been used by NORDEBO et al. [505] for broadband adaptive beamforming.

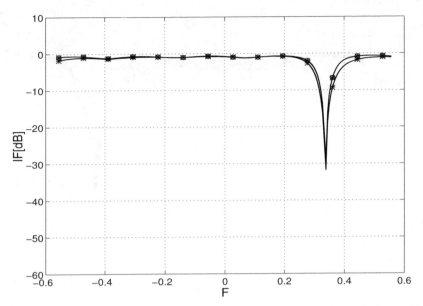

Figure 11.28: Optimum jammer suppression and auxiliary channel clutter filter cascaded ($N = 25$, $K = 7$, SL): $\circ$ $J = 0$; $*$ $J = 1$; $\times$ $J = 2$; $+$ $J = 3$

11.7 Summary

Clutter suppression in the presence of noise jamming has been discussed in this chapter. Noise jammers are broadband compared with the clutter Doppler bandwidth but discrete in space. Because of the large bandwidth each jammer causes M additional eigenvalues of the space-time covariance matrix.

Jammer suppression is basically a *spatial* problem while clutter suppression is a space-time problem. From this consideration the question arises of whether clutter and jammers should be cancelled *simultaneously* by one space-time filter, or a spatial anti-jamming filter should be cascaded with a space-time anti-clutter filter.

The following insights have been obtained from the model study:

1. The **optimum processor** suppresses clutter and jammers optimally as long as the number of jammers does not cause a saturation of degrees of freedom of the clutter + jammer space-time covariance matrix $NM - 1$. For clutter rejection about $N + M - 1$ degrees of freedom are required. Since each jammer requires $M - 1$ degrees of freedom (assumption: sidelooking array, DPCA conditions, see (3.71)) the maximum number of jammers that can be suppressed is $J_{\max} = (NM - 1 - N - M + 1)/(M - 1)=\text{INT}(N - M/(M - 1))$, $M > 1$.

2. **Space-time auxiliary channel processors** after Chapter 4 have to match the number of channels to the number of clutter plus jammer eigenvalues. This increases the complexity considerably because with every additional jammer $M - 1$ additional channels are required.

374 STAP under jamming conditions

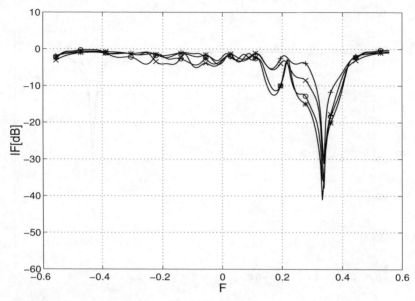

Figure 11.29: Auxiliary channel jamming and clutter filters cascaded ($N = 25$, $K_c = 3$, $J = 3$, SL): ○ $K_j = 5$; * $K_j = 7$; × $K_j = 9$; + $K_j = 15$

3. **Spatial transform/FIR filter** based processors (Chapters 6 and 7) have to be matched to the number of jammers only in the spatial dimension since jammers are *spatial* by nature. For each jammer only *one* additional antenna channel has to be provided. Of course, the number of jammers must not exceed the capacity of the array: $J_{max} = \text{INT}(N - M/(M-1))$. For linear arrays the clutter rejection performance is very close to optimum.

4. **Simultaneous clutter and jammer** cancellation with a subarray/FIR filter processor functions well also with circular planar antenna arrays.

5. Jammer cancellation by means of a **fully adaptive spatial** filter cascaded with a space-time adaptive clutter filter (symmetric auxiliary sensor type) works optimally for a linear array.

6. **Cascading** a spatial symmetric auxiliary sensor filter for anti-jamming with a space-time symmetric auxiliary sensor FIR filter may suppress both jammer and clutter sufficiently well with a linear array. The use of other array configurations (for example, circular planar) is an open issue.

7. For **two-dimensional arrays** the clutter+jammer suppression performance depends very much on the jammer scenario. If the jammers are located mainly in the horizontal dimension as in the case of ground based jammers only the horizontal degrees of freedom of the array are effective. For a two-dimensional jammer geometry the full number of degrees of freedom of the array is effective; in this case more jammers can be cancelled.

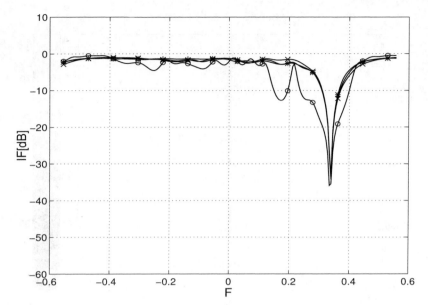

Figure 11.30: Auxiliary channel jamming and clutter filters cascaded ($N = 25$, $K_c = 5$, $J = 3$, *SL*): ○ $K_j = 5$; ∗ $K_j = 7$; × $K_j = 9$; + $K_j = 15$

8. In **jammer+clutter scenarios** a few spatial degrees of freedom (2-3) are absorbed for clutter rejection. The number of jammers which can be cancelled is reduced accordingly.

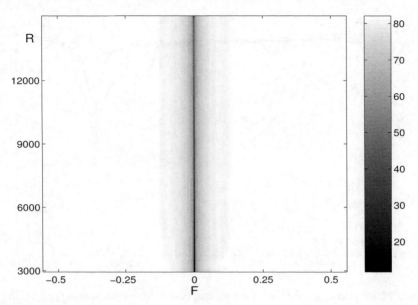

Figure 11.31: Range-Doppler matrix (IF/dB, R/m, SL, $\varphi_L = 90°$; one jammer with 60 dB JNR)

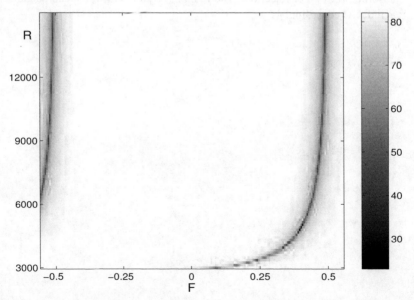

Figure 11.32: Range-Doppler matrix (IF/dB, R/m, FL, $\varphi_L = 0°$, one jammer with 60 dB JNR)

Chapter 12

Space-time processing for bistatic radar

All the previous considerations were related to the usual monostatic radar operation. In this chapter the impact of bistatic radar operation on the performance of air-/spaceborne MTI radar based on space-time adaptive processing is discussed. Bistatic radar makes covert observation possible which will play an increasing role in future reconnaissance systems (LE CHEVALIER [381]). Summaries of bistatic radar have been given by HANLE [248, 247], SKOLNIK [610, Chapter 25], and DUNSMORE in GALATI [175]. A bistatic SAR experiment has been described by MARTINSEK and GOLDSTEIN [438]. CHEN and BEARD [91] propose optimum flight paths for minimising the clutter Doppler spread. Space-TIME processing for suppression of jammers in a broadband bistatic radar configuration has been discussed by SEDYSHEV et al. [596]. Application of space-time adaptive processing to bistatic spaceborne geometries has been analysed by FANTE [145, 146] and KLEMM [330]. A bistatic experiment has been conducted using the MCARM array and a stationary balloon-borne transmitter (SANYAL et al. [584], BROWN et al. [71]).

Localisation of emitting sources (for instance, jammers) by use of out-of-plane multipath is in essence a passive bistatic localisation technique. COUTTS [102] estimates the target parameters range, heading, velocity, and altitude by evaluating attenuated, delayed, and Doppler shifted replicas of the emitter waveform. The technique has been verified successfully by using MOUNTAINTOP data (TITI [646]). HARTNETT and DAVIS [252] present design parameters for a bistatic airborne adjunct to a space based radar. GMTI performance plays a predominant role.

It has been shown in the previous chapters that monostatic STAP radar has the following properties:

- For a horizontal flight path and a planar earth the curves of constant clutter Doppler (isodops) are hyperbolas.

- For a sidelooking linear antenna geometry the clutter Doppler is range independent.

- Clutter trajectories in the $\cos\varphi$-F plane (F = normalised Doppler) are in general ellipses (or straight lines for a sidelooking array).

- For linear arrays the beam traces on the earth are hyperbolas symmetrical about the array axis.

- For sidelooking antenna geometry the clutter Doppler is range independent.

- Owing to the equivalence of azimuth and Doppler, clutter spectra are by nature narrow ridges in the φ-F plane which leave sufficient space for the detection of moving targets.

- The number of clutter eigenvalues is under certain conditions (Nyquist sampling in space and time, no bandwidth effects) $N_\mathrm{e} = N + M - 1$.

In this chapter we demonstrate that some of these well-known properties are destroyed by the displacement between transmitter and receiver in a bistatic configuration. It is shown that even for a sidelooking array geometry the clutter Doppler is range-dependent which requires adaptation of the STAP processor for each individual range gate. Conclusions for the design of STAP processors are drawn. Some related results can be found in HIMED et al. [263] and KLEMM [330, 331, 333].

Since the angle of arrival changes with range in bistatic radar configurations the Doppler of bistatic clutter arrivals is not only range dependent but depends also on the angle of arrival. HIMED [266] proposes a technique for joint angle and Doppler compensation for bistatic clutter data. Further properties of bistatic clutter have been identified by HIMED and HERBERT and RICHARDSON [258] identified several properties of bistatic clutter

- Interactions of the mainlobe of one antenna with the sidelobes of the other.

- Clutter regions where large changes of Doppler frequency correspond to small changes of angular location.

- Variation of the clutter angle-Doppler relationship due to topographic features (e.g., variations in ground height).

In the thesis by ONG [511] various aspects of signal processing for bistatic airborne radar such as Doppler and power compensation, are discussed.

12.1 Effect of bistatic radar on STAP processing

12.1.1 Discussion of the bistatic clutter Doppler

In the following we will discuss the effect of bistatic transmit-receive configurations on the performance of the STAP processor. Typical properties of bistatic radar are: Transmitter and receiver are displaced by a certain distance, transmitter and receiver are moving on different flight paths and at different velocities.

The geometry of bistatic radar is shown in Figure 12.1. The receiver is at point R at height H_R above the xy ground plane, and the transmitter at point T at height

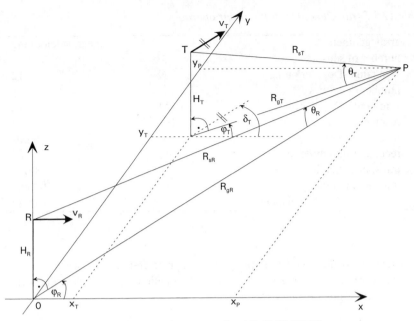

Figure 12.1: Geometry of airborne bistatic radar (©1998 IEEE)

H_T. The receiver moves in the x-direction at speed v_R while the transmitter moves in the direction δ_T at speed v_T. A transmit pulse hits the ground at point P after passing the transmit slant range R_{sT}, and the reflected pulse is received by the receiver after passing the receiver slant range R_{sR}.

The total Doppler frequency of the received clutter echo is then

$$f_D = \frac{v_T}{\lambda}\cos(\varphi_T - \delta_T)\cos\theta_T + \frac{v_R}{\lambda}\cos(\varphi_R - \delta_R)\cos\theta_R \qquad (12.1)$$

which is a 3-D version of the Doppler formula given in SKOLNIK [610, p. 25.14]. δ_T and δ_R are the directions of the flight paths of transmitter and receiver. In the geometry shown in Figure 12.1 and following the convention used throughout this book we have $\delta_R = 0$, i.e., the receiver flies in the x-direction. Notice that the first part of (12.1) describes the one-way transmit Doppler while the second part stands for the one-way receive Doppler.

By inspection of (12.1) we notice a number of interesting properties of the bistatic clutter Doppler.

1. **Stationary transmitter**. Existing sources of opportunity such as radio or TV stations may be used as transmitters. The clutter Doppler reduces to

$$f_D = \frac{v_R}{\lambda}\cos\varphi_R \cos\theta_R \qquad (12.2)$$

Except for a factor of 2 we recognise the Doppler relation for monostatic radar, see (3.2). The term $\cos\varphi_R$ causes the clutter Doppler to be symmetric with

Table 12.1 *Parameters for bistatic radar operation*

array geometry		linear, sidelooking
number of sensors in range-Doppler plots	N	12
number of echoes in range-Doppler plots	M	12
receiver height	H_R	3000 m
receiver velocity	v_R	90 m/s
receiver flight direction	δ_R	0°
transmitter height	H_T	3000 m
transmitter velocity	v_R	90 m/s
receiver/transmitter displacement		6000 m

respect to the receiver flight path. Doppler frequencies are positive in the forward and negative in the backward direction. For sidelooking arrays the clutter Doppler is range independent.

2. **Flight paths** of receiver and transmitter are **aligned**. This means, $\delta_T = \delta_R = 0$, the Doppler is symmetric to the flight path.

3. **Transmit velocity is great compared with receiver velocity**. This is a typical configuration where the transmitter is on a satellite while the receiver is airborne. The isodops are dominated by the flight path of the transmitter.

4. Except for a sidelooking receive array using a stationary transmitter (case 1) the clutter Doppler can be expected to be range dependent.

12.1.2 Numerical examples

The following numerical examples will be presented for further illustration of the statements made above. This example (and all subsequent numerical examples) are based on the parameters listed in Tables 12.1 and 2.1.

We assume a configuration where the flight paths of receiver and transmitter are crossing each other ($\delta_R = 0°, \delta_T = 90°$, see Figure 12.2). The transmitter-receiver geometry changes rapidly with time and, therefore, may not be useful for practical radar operation. It is useful, however, for explaining some properties of bistatic STAP radar.

Recall that in monostatic radar the sidelooking array receives clutter returns whose Doppler is range dependent. The following considerations are based on a sidelooking linear array.

Effect of bistatic radar on STAP processing 381

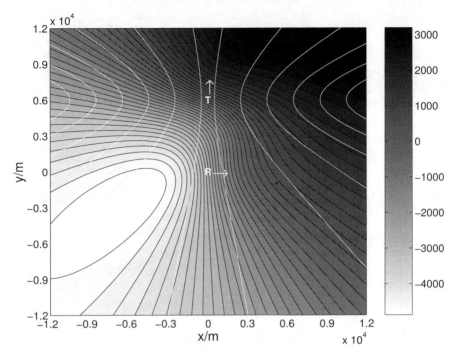

Figure 12.2: Bistatic isodops (black) and transmit beam traces (white), monostatic, SL

12.1.2.1 *Isodops*

Figure 12.2 shows the transmitter-receiver geometry and the associated isodops[1] (black lines). The grey levels denote the Doppler frequency. As can be seen the isodops are no longer hyperbolas as in the monostatic case (Figures 3.3, 3.4). The isodops exhibit significant distortions. It is obvious that there is no symmetry relative to the receiver flight path.

The white hyperbolas denote the footprints of the transmit beam on the ground. The transmit beam is more relevant in the context of Doppler dependency than the receive beam, because the transmit beam is fixed during a PRI while the receive beam (or at least, the receive direction) varies while the echo is coming in. Many intersections between transmit beams and isodops can be noticed which indicates that even for a sidelooking array the clutter Doppler is range dependent.

12.1.2.2 *Doppler-angle trajectories (Lissajou patterns)*

The Doppler-angle trajectories of clutter are obtained by plotting the Doppler frequency as given by (12.1) versus the directional cosine $\cos \varphi_R \cos \theta_R$. Such curves can be seen in Figure 12.3 (R_{2w} is the 2-way slant range). The fat part of the curves denotes

---

[1] As in the monostatic case it would be desirable to have a closed form solution for the isodops like the one for monostatic radar (3.1). However, such a closed form solution cannot easily be obtained. The isodop plots shown in the following are numerical solutions generated by use of MATLAB contourf.

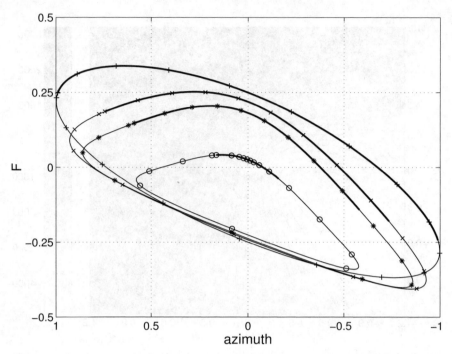

Figure 12.3: Bistatic Lissajou patterns, SL, R_{2w}/m = ○ 10 000; * 14 000; × 18 000; + 100 000

the front lobe of the antenna pattern while the thin part stands for the backlobe. The variation of the clutter Doppler with range is obvious (compare with the monostatic case in Figure 3.5a). As expected there is no symmetry with respect to azimuth or Doppler.

12.1.2.3 Power spectrum

Figure 12.4 shows an MV power spectrum of bistatic clutter with a sidelooking receive array (similar to those presented in Chapter 3). The backlobe has been suppressed by means of a sensor pattern according to (2.20). It can be noticed that the spectrum follows a Doppler-azimuth trajectory as shown in Figure 12.3.

Let us compare this power spectrum with a monostatic spectrum like that shown in Figure 3.18. We realise that on the one hand the Doppler-angle trajectory (the fooprint) of the bistatic spectrum has a different shape. For the effectiveness of STAP, however, we can expect that it will work in a similar efficient way as with monostatic radar. There will be a clutter notch at the Doppler frequency in the look direction and a flat SNIR (or IF) elsewhere.

By inspection of Figure 12.4 we may even assert that the number of clutter eigenvalues of the bistatic space-time clutter covariance matrix will be of the same order of magnitude as for the monostatic case. The rationale for this assertion is that

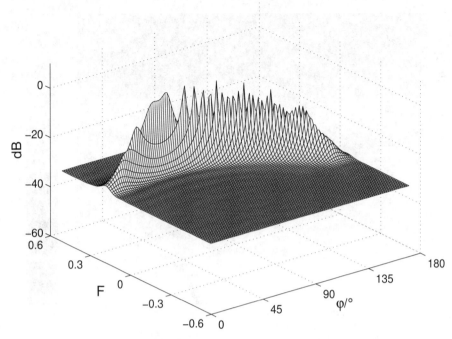

Figure 12.4: Bistatic clutter power spectrum, SL

the spectrum in Figure 12.4 occupies about the same amount of surface of the Doppler-angle plane as the monostatic spectrum in Figure 3.18, only the shape of the trajectory has changed.

For certain transmitter–receiver configurations the effect of the dependence of the clutter Doppler on range may be quite complicated and leads in general to losses in STAP performance (MELVIN *et al.* [455, 456]). KOGON and ZATMAN [362] have pointed out that the range dependence which causes a non-stationarity may lead to a shortage in training samples required for adapting the clutter filter. *Doppler warping* (BORSARI [58], adaptive Doppler compensation KREYENKAMP and KLEMM [370]) and *derivative based updating* (HAYWARD [256]) are ways of mitigating the effect of Doppler dependence with range in monostatic radar. In bistatic configurations the variation of the receive angle has to be taken into account (HIMED[266]).

12.1.2.4 *Range-Doppler matrix*

In Figure 12.5 we find the IF plotted versus Doppler and range. Notice that in all bistatic range Doppler plots the range scaling denotes half the 2-way distance. These range-Doppler plots illustrate best the range dependence of the clutter Doppler. Recall that for a monostatic sidelooking array the clutter Doppler was independent of range (Figure 4.31). Instead, the clutter trajectory is quite similar to Figure 4.32 which was

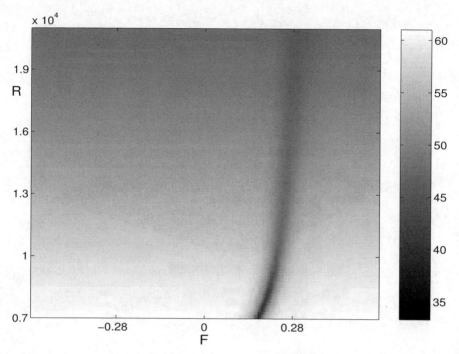

Figure 12.5: Range-Doppler matrix, SL, $\varphi_T = 45°$ (R = half two-way distance, grey levels = IF)

calculated for a forward looking array ($\varphi_L = 0°$).

It is noteworthy that both the clutter trajectories in Figures 4.32 and 12.5 have similar shapes. In the first case the array axis is rotated by $\varphi_L = 90°$ relative to the isodop field, in the second case a part of the isodop field (transmit) is rotated by $\varphi_L = 90°$ relative to the array axis.

12.1.2.5 Effect on eigenspectrum and IF plot

In Figure 12.6 the eigenspectrum for a bistatic radar configuration using a sidelooking receive array has been plotted. For the o curve we assumed an omnidirectional sensor pattern whereas for the ∗ curve arrivals of one halfplane of the ground has been suppressed by use of directive elements. As can be seen the sidelooking bistatic array behaves like the monostatic forward looking array (compare with Figure 3.13). The number of clutter eigenvalues is approximately doubled if the sensors and transmission is omnidirectional. Obviously the clutter Doppler field is non-symmetric with respect to the array axis which is a consequence of the bistatic operation. Recall that the clutter Doppler field of a monostatic sidelooking array is symmetric.

In Figure 12.7 the improvement factors of monostatic and bistatic, sidelooking and forward looking arrays are compared. The optimum processor (1.3) has been used in this calculation. As can be seen, the performance of all constellations is about the same.

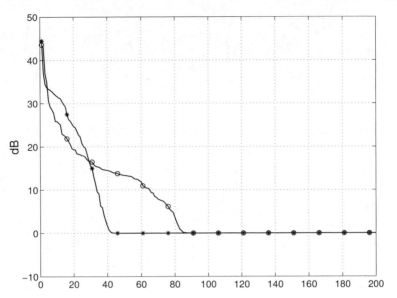

Figure 12.6: Clutter eigenspectrum (configuration as in Figure 1.3, SL), ○ omnidirectional sensor patterns; ∗ directive sensor patterns (©1998 IEEE)

In the bistatic case, the clutter notch is slightly shifted in Doppler.

12.2 Realistic bistatic geometries (tandem configuration)

In this section three different transmit–receive geometries are discussed which might be useful for ground observation by bistatic airborne radar. For the sake of brevity we consider only sidelooking arrays in order to demonstrate that the favourable property of range independence of sidelooking monostatic STAP gets lost when operating in a bistatic geometry. The transmit–receive geometry was chosen to be: receiver height = transmitter height = 3000 m; receiver–transmitter distance: 6000 m.

12.2.1 Two aircraft with aligned flight paths

Let us consider a transmit–receive arrangement as shown in Figure 12.8. This represents a two-aircraft configuration flying at the same height and speed so that the geometry does not change during the flight. The isodops still look like hyperbolas, however a closer look at the area between T and R reveals that we deal with functions of higher complexity. We notice a lot of intersections between the transmit beam traces and the isodops which indicates that there is some range dependency of the clutter Doppler frequency.

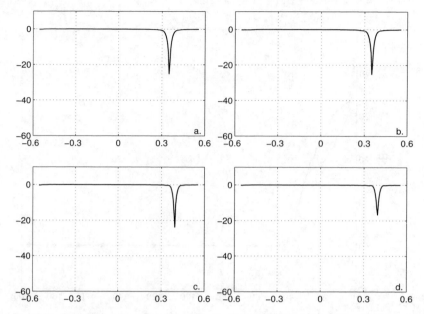

Figure 12.7: Normalised improvement factor vs normalised Doppler: a. monostatic, SL; b. monostatic, FL; c. bistatic (configuration as in Figure 12.2, SL); d. bistatic (configuration as in Figure 12.2, FL, (©1998 IEEE))

This impression is confirmed by the associated Lissajou patterns shown in Figure 12.9. The four curves stand for four different target ranges. As can be seen the shape and position of the Lissajou curves vary considerably with range. For the far range ($R_{2w} = 10000$ m) we obtain the monostatic case as shown in Figure 3.5a. The symmetry of the T-R configuration (Figure 12.8) results in a symmetric isodop field. This is reflected in Figure 12.9 by the fact that forward lobe response and backlobe response coincide (opposite to Figure 12.3).

The range dependence of the clutter Doppler can best be seen in the range-Doppler matrix as shown in Figure 12.10. Recall that for a monostatic sidelooking array radar the clutter Doppler is range independent (compare with Figure 4.31). Notice that the ordinate denotes half the two-way range.

12.2.2 Two aircraft with parallel flight paths (horizontal across-track)

Figure 12.11 shows another airborne configuration where two aircraft are flying on parallel paths at the same height. As one can see there is a certain displacement between the transmit beam traces and the bistatic isodops. At the far range, i.e., at ranges large compared to the T-R baseline the isodops approximate hyperbolas as in monostatic operation. In the vicinity of the T-R baseline, however, one can recognise that the isodops are no longer hyperbolas but higher order functions. Again the intersections

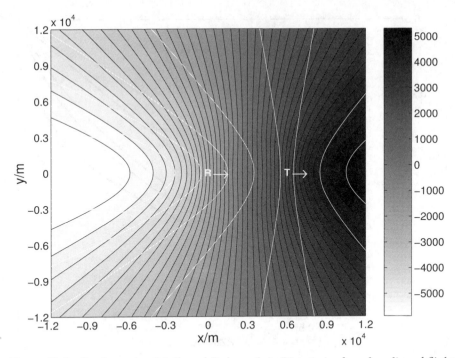

Figure 12.8: Configuration 1 (aligned flight paths): Bistatic isodops for aligned flight paths (black) and transmit beam traces (white), SL

between the transmit beam trace indicate that there is some range dependence of the clutter Doppler.

This is confirmed by the Lissajou patterns shown in Figure 12.11. We notice that for different bistatic range, different Doppler-azimuth trajectories are obtained. At zero azimuth (for a sidelooking array this corresponds to 90°, i.e. broadside) all curves coincide which means that in the broadside direction there is no range dependence. Moreover, one can notice that the Doppler-azimuth trajectories are different for the forward lobe and backlobe of the antenna pattern (fat and thin lines).

In the range-Doppler matrix in Figure 12.13 we chose the look direction to be $\varphi_L = 45°$ in order to demonstrate the range dependence. For $\varphi_L = 0°$ the range-Doppler matrix looks as shown in Figure 4.31.

12.2.3 Two aircraft, transmitter above receiver (vertical across-track)

In Figure 12.14 we see two aircraft flying on parallel paths, with the transmitter above the receiver. Now the isodops are again hyperbolas; however, their focus is different from the focus of the transmit beam hyperbolas. Therefore, again some range dependence of the clutter Doppler can be expected.

This is confirmed by the azimuth-Doppler trajectories depicted in Figure 12.15.

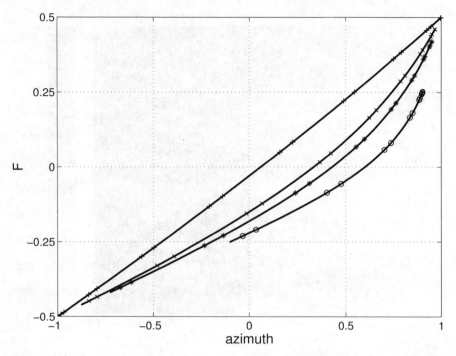

Figure 12.9: Bistatic Lissajou patterns for configuration 1, SL, R_{2w}/m = ∘ 10 000; ∗ 14 000; × 18 000; + 100 000

Notice that we get for this transmit–receive configuration straight lines as is the case of monostatic STAP radar, however with different slopes. Again all the curves coincide for the broadside direction. For all other look directions the clutter Doppler is range dependent as can also be recognised from Figure 12.16. It has been noted by MELVIN *et al.* [455] that the range dependence of the clutter Doppler may lead to significant losses in STAP performance if not compensated for.

In summary we conclude that even for a sidelooking bistatic radar in general the clutter Doppler is range dependent.

12.2.4 A note on range dependence

In the previous sections we showed that for bistatic radar configuration the clutter Doppler depends on range. However, the problem is more complex. Recall that bistatic operation runs in the following way. The transmitter transmits a pulse in a fixed direction. The bistatic clutter return coming from different ranges arrives at the receiver at different angles because of the bistatic geometry. In conclusion, in bistatic radar configurations not only the Doppler but also the receive direction changes with range. The effect of clutter dispersion due to the range dependence of Doppler and angle has been described by HIMED [266], HIMED *et al.* [265], and HIMED *et al.* [268]. LAPIERRE *et al.* [376, 378] propose a technique for aligning the complete DD-

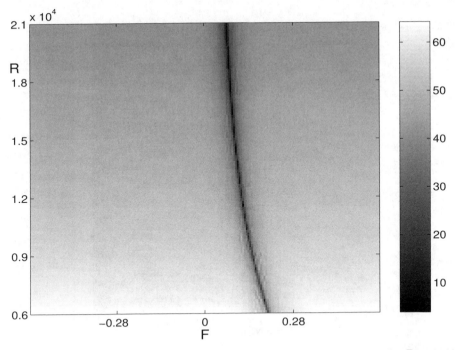

Figure 12.10: Range-Doppler matrix for configuration 1, SL, $\varphi_T = 90°$, (R = half two-way distance, grey levels = IF)

curves (Doppler-direction) to compensate for the range dependence of Doppler and angular direction in bistatic radar configurations.

12.3 Ambiguities in bistatic STAP radar

Since in most bistatic configurations the clutter Doppler is range dependent we have to expect some broadening of the width of the clutter notch or even additional clutter notches. For monostatic radar such effects have been discussed in Section 10.1. In both cases (broadening, multiple notches) losses in moving target detection have to be taken into account.

12.3.1 Range ambiguities

Two different examples of range ambiguity are shown in Figures 12.17 and 12.18. The first belongs to a sidelooking, the second to a forward looking array. The first is based on the bistatic configuration 1, the second configuration 3. These two figures are just examples chosen at random. There are many possibilities to create various bistatic configurations, and consequently, many different ambiguous clutter patterns will occur. We confine ourselves here to showing some basic properties of ambiguous bistatic STAP radar based on the two given examples.

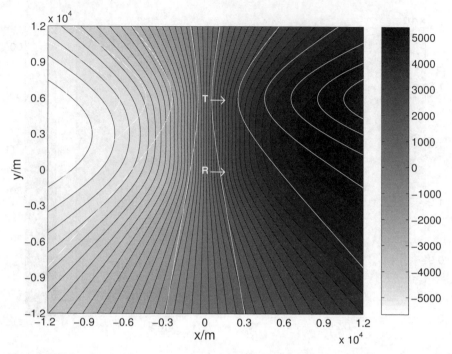

Figure 12.11: Configuration 2 (parallel flight paths): Bistatic isodops (black) and transmit beam traces (white), SL

Figure 12.17 is in essence the same range-Doppler plot as given by Figure 12.10, however with range ambiguous clutter returns. We notice the sharp dark line which denotes the primary clutter. In contrast to Figure 12.13 we notice now the ambiguous clutter returns as a shading left of the primary clutter. The first range ambiguity can be seen clearly at $R_{\min} + R_{\mathrm{amb}} = 4.85 + 8.33 = 13.18$ km. For $\varphi_T = 90°$ we get $R_{\min} = H + \sqrt{D^2 + H^2}$ as the minimum clutter range for this configuration, and R_{amb} is the unambiguous range interval as determined by the PRF. It can be seen that the primary clutter notch in the first unambiguous range interval (dark line in the lower part of the plot) is repeated at different Doppler frequencies (near range clutter in the second range interval). This reflects the fact that the receive look angle φ_L varies with range while the transmit angle is constant. The varying receive look angle leads to a different interpretation of the clutter Doppler. In the first range interval on the right we find the far range ambiguous returns.

The example shown in Figure 12.18 has been calculated for the bistatic configuration 3 (transmitter above receiver, same velocity and direction) and a forward looking array. Again we see the first ambiguity at $R_{\min} + R_{\mathrm{amb}} = 6 + 8.33 = 14.33$ km (in configuration 3 we have $R_{\min} = 6$ km). In the first unambiguous range interval we recognise the primary clutter on the left side and the additional ambiguous response on the right. At short range we notice Doppler frequencies in the ambiguous response which do not appear as primary clutter at longer range. The reason for this behaviour is

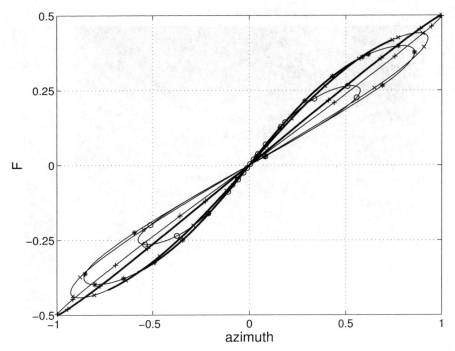

*Figure 12.12: Bistatic Lissajou patterns for configuration 2, SL, R_{2w}/m = ∘ 10 000; *
14 000; × 18 000; + 100 000*

that we calculated the individual lines of these range-Doppler plots for the actual target range, that is, the processor was focused on each range cell. More detailed explanation has been given in Chapter 10, Section 10.1.1.3.

In the second unambiguous range interval ($R > 14.33$ km) we find the primary and ambiguous clutter on the right and some ambiguous clutter from short ranges on the left. It can be noticed that at larger range the clutter notch is considerably broadened whereas at short range two separate clutter notches occur.

12.3.2 Range and Doppler ambiguities

In the following two examples we demonstrate the effect of range and Doppler ambiguities and ways of compensation.

Figure 12.19 shows IF curves versus target velocity for a sidelooking array in the bistatic configuration 1 (two aircraft with aligned flight paths). In Figure 12.19a the ideal case is shown. No range ambiguities have been assumed, and Doppler ambiguities do not occur inside the clutter band ($v_p = -90, \ldots, 90$ m/s) since the PRF has been chosen to be the Nyquist frequency of the clutter bandwidth (12 000 Hz). In Figure 12.19b the PRF was chosen to be 4 times the Nyquist rate so that many range ambiguous clutter returns can be expected. We notice a considerable broadening of the

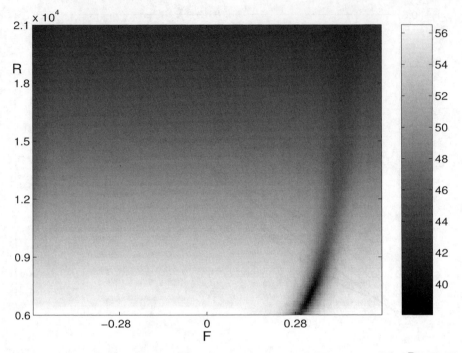

Figure 12.13: Range-Doppler matrix for configuration 2, SL, $\varphi_T = 45°$, (R = half two-way distance, grey levels = IF)

clutter notch.[2] Using a planar array with additional vertical adaptivity similar to Section 10.1.1.1 compensates to a certain extent for the effect of ambiguous clutter returns so that the resulting clutter notch is narrowed (Figure 12.19c). Additional PRI staggering according to Section 10.2 gives no difference since the clutter band is oversampled anyway with the chosen PRF.

Figure 12.20 shows IF curves versus target velocity for a forward looking array in the bistatic configuration 3 (two aircraft, parallel flight paths, transmitter on top of the receiver). In Figure 12.20a we see for comparison the optimal case (no range ambiguities included, PRF Nyquist rate). Including range ambiguities and reducing the PRF down to 0.25 Nyquist results in IF pattern as shown in Figure 12.20b. Without knowing Figure 12.20a it would be hard to decide which of the various notches is primary, or range or Doppler ambiguous. Using again an adaptive planar array removes the range ambiguities (Figure 12.20c). The remaining Doppler ambiguities can be removed by PRI staggering (Figure 12.20d).

---

[2]Which in part is a consequence of the short duration of the coherent processing interval owing to the chosen PRF.

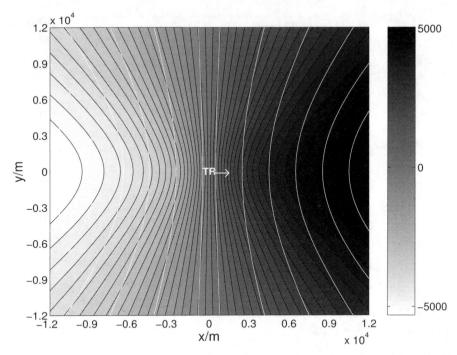

Figure 12.14: Configuration 3 (transmitter above receiver): Bistatic isodops (black) and transmit beam traces (white), SL

12.4 Use of sparse arrays in bistatic spaceborne GMTI radar

12.4.1 Introduction

It has been pointed out in Chapter 8 that the effectiveness of space-time adaptive processing is very sensitive to any deviation from the DPCA condition (coincidence of spatial and temporal samples, (3.39)) if the system is undersampled in both the temporal and the spatial dimension (PRF lower than the Nyquist frequency of the clutter band, array element spacing larger than half the wavelength). Basically optimum clutter suppression performance can be achieved with a system which is undersampled in both dimensions. Undersampling is a technique that can be used to enlarge the spatial (array) and temporal (CPI) apertures. Prerequisite for proper STAP operation with a system undersampled in both dimensions is that the DPCA condition is met precisely. In practice deviations from the DPCA condition will certainly occur, mainly caused by array and timing errors. It is in particular not maintained if the PRI is staggered (see Chapter 10).

In this section we discuss another aspect which prohibits enlarging the spatial and temporal apertures by undersampling, namely the space-time clutter Doppler distortion in bistatic radar configurations. Let us consider a space-based bistatic configuration

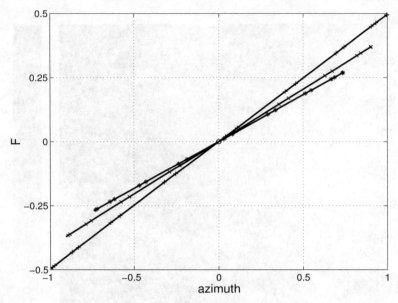

*Figure 12.15: Bistatic Lissajou patterns for configuration 3, SL, R_{2w}/m = ○ 10 000; * 14 000; × 18 000; + 100 000*

with parameters as given in Table 12.1. These results were taken from KLEMM [344].

12.4.2 DPCA in bistatic configurations

Recall that the phase term in the clutter covariance (3.38) becomes for a sidelooking equispaced array

$$\Phi_{m-p}(v_p, \varphi)\Psi_{i-k}(\varphi) \qquad (12.3)$$
$$= \exp\left[j\frac{2\pi}{\lambda}(2v_p(m-p)T + (i-k)d)\cos\varphi\cos\theta\right]$$
$$= \exp[j2\pi(f_D(m-p)T + (i-k)d\cos\varphi\cos\theta)]$$

with f_D denoting the Doppler frequency.

Replacing the Doppler frequency f_D by the bistatic Doppler (12.1) yields

$$\Phi_{m-p}(v_p, \varphi)\Psi_{i-k}(\varphi) \qquad (12.4)$$
$$= \exp\left[j2\pi(\frac{v_T}{\lambda}(m-p)T\cos(\varphi_T - \delta_T)\cos\theta_T)\right.$$
$$\left.+j2\pi((\frac{v_R}{\lambda}(m-p)T + (i-k)d)\cos\varphi_R\cos\theta_R)\right]$$

If the parameters are chosen so that the DPCA condition is met the receiver phase correlation term (lower line of (12.4)) becomes equal to 1 for *all* receive angles of arrivals. However, the total phase correlation is determined by both the receiver and

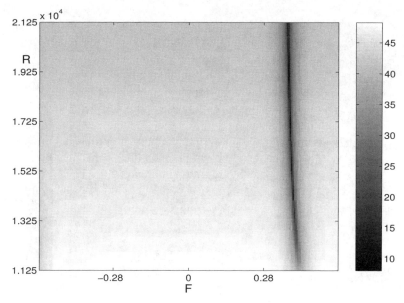

Figure 12.16: Range-Doppler matrix for configuration 3, SL, $\varphi_T = 45°$, (R = half two-way distance, grey levels = IF)

transmit Doppler. Therefore, meeting the DPCA condition on the receive side

$$(\frac{v_R}{\lambda}(m-p)T = -(i-k)d \tag{12.5}$$

does not in general maximise the space-time correlation. This is illustrated in the following numerical examples.

12.4.3 Some numerical examples

In this section we want to demonstrate that undersampling in both the spatial and the temporal dimension leads to losses in clutter suppression performance in bistatic space-based STAP radar. Notice that the subfigures in Figures 12.21, 12.22 and 12.23 have been produced for different baselines. In these plots the PRF was chosen to be 1/256 of the Nyquist frequency of the clutter band, and the element spacing of the array is 128λ.

Figure 12.21 has been calculated on the basis of the tandem configuration. In Figure 12.21a the baseline is 0.1 km which is almost a monostatic configuration. Notice that the performance shown is close to optimum. The maximum improvement factor is reached in the passband, and two narrow clutter notches can be seen. In Figure 12.21b (1 km) we see already a significant broadening of the clutter notches which indicates that the effect of the bistatic configuration becomes noticeable. For even larger baselines (Figures 12.21c and 12.21d) we notice a large number of additional notches.

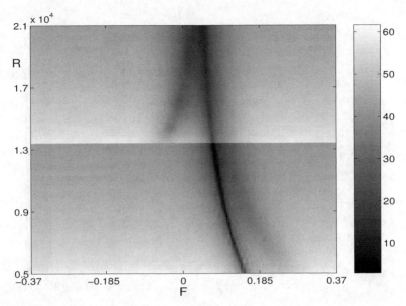

Figure 12.17: Range-Doppler matrix for configuration 1, SL, $\varphi_T = 90°$, PRF = 18 000 Hz (greytone denotes IF/dB, R = range/m)

Similar results have been obtained for the horizontal across-track configuration (Figure 12.22) and the vertical across-track configuration (Figure 12.23).

12.4.4 Comparison with fully filled array

If undersampling is done only in one dimension, that means, in the other dimension we have Nyquist sampling, then we can expect good STAP performance because the interpolation property of Nyquist sampling provides sufficient correlation between spatial and temporal samples. The fat curve has been computed for the full array (240 elements, $\lambda/2$ spacing), the thin one for a sparse array with the same aperture, but only 12 elements with 10λ spacing. As can be seen the full array achieves optimum performance whereas the sparse array fails completely.

12.5 Summary

Some considerations on the applicability of STAP to clutter suppression in bistatic radar configurations have been made. Some observations related to STAP processing are summarised in the following:

- In general, the clutter Doppler is range-dependent (even for a sidelooking array geometry) which requires adaptation of the STAP processor for each individual range gate. If the Doppler depends only weakly on range, small range segments

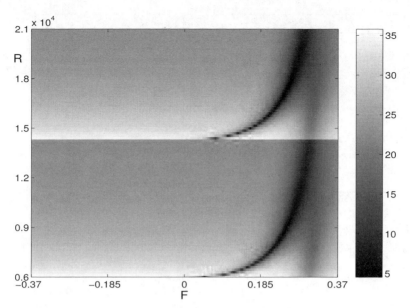

Figure 12.18: Range-Doppler matrix for configuration 3, FL, $\varphi_T = 0°$, PRF=18 000 Hz (greytone denotes IF/dB, R = range/m)

can be chosen so that no widening of the clutter notch due to range dependent Doppler spread can occur.

- The range dependence of the clutter Doppler may be mitigated by appropriate choice of the transmitter–receiver configuration, for instance, by chosing the transmitter height much larger than the receiver height above ground. Such configuration might be desirable anyway for operational reasons.

- For monostatic radar the clutter trajectories in the $\cos\varphi$-F plane are straight lines or ellipses. The displacement between transmitter and receiver in bistatic radar and different flight velocities and flight paths of transmitter and receiver cause distortions of the clutter trajectories.

- As a consequence of the above-mentioned, the clutter notch will be Doppler shifted. For certain receiver–transmitter configurations the isodops will be asymmetric relative to the array axis which causes a second clutter notch even for a sidelooking array if omnidirectional sensors are used.

- The shape of clutter trajectories in the φ-F plane depends very much on the individual bistatic configuration and the velocities of transmitter and receiver. Even for sidelooking arrays different clutter Doppler responses of the backlobe and forward lobe of the antenna pattern may occur.

- Owing to the equivalence of azimuth and Doppler, clutter spectra are by nature narrow ridges in the φ-F plane (as in the case of monostatic radar) which leave sufficient space for the detection of moving targets.

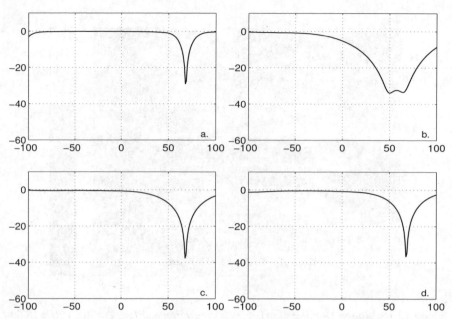

Figure 12.19: IF vs target velocity (m/s, sidelooking array, configuration 1, R = 19,5 km, N = M = 12. a. PRF = Nyquist of clutter bandwidth, no range ambiguities; b. PRF = 4× Nyquist, with range ambiguities; c. adaptive planar array; d. planar array + randomly staggered PRI

- In contrast to monostatic sidelooking radar precautions for cancellation of multiple-time-around clutter have to be taken. This is necessary also for monostatic forward looking radar.

- The number of clutter eigenvalues is (like in monostatic radar) under certain conditions (Nyquist sampling in space and time, no bandwidth effects) $N_e = N + M - 1$ if the Doppler distribution on the ground is the same in the forward and backward direction (left or right from the array axis), or if directive sensors are used, which cancel the backlobe area. In most applications directive sensors are used so that only one clutter notch occurs as in the case of monostatic radar.

- For linear arrays in either a sidelooking or forward looking configuration, ambiguous clutter responses cause a broadening of the clutter notch or create secondary clutter notches.

- The bistatic configuration is contradictory to the DPCA condition (phase coincidence of spatial and temporal clutter samples) because the clutter Doppler is composed of a transmit and a receive term. Therefore, Nyquist sampling is necessary, either by using a fully filled half weavelength spaced array or by choosing the PRF to be equal to the clutter bandwidth, or both.

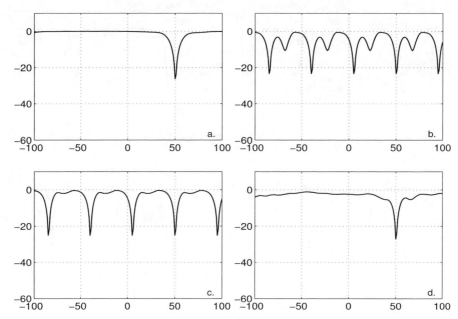

Figure 12.20: IF vs target velocity (m/s, forward looking array, configuration 3, R = 10 km, N = M = 12. a. PRF = Nyquist of clutter bandwidth, no range ambiguities; b. PRF = 0.25×Nyquist, with range ambiguities; c. adaptive planar array; d. planar array + randomly staggered PRI

Table 12.1 Radar parameters (R Receiver, T transmitter

platform height	$H = 400$ km
target slant range	$R = 800$ km
baseline	$0 \ldots 10$ km
array type	linear, equispaced
array orientation	sidelooking
array spacing	128λ
number of array elements	N=24, 32, 240
number of processed echoes	M=24, 32
clutter-to-noise ratio (CNR)	20 dB
wavelength	$\lambda = 0.03$ m
PRF (Nyquist)	1.02 MHz
PRF (used)	3.99 KHz
Bistatic configurations:	
Tandem	R and T aligned
Horizontal across-track	R and T on parallel paths
Vertical across-track	T above R

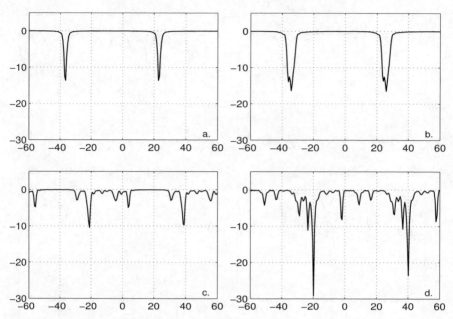

Figure 12.21: Sparse DPCA array (Nyquist/256): tandem configuration, baseline [km]: a. 0.1; b. 1; c. 10; d. 100

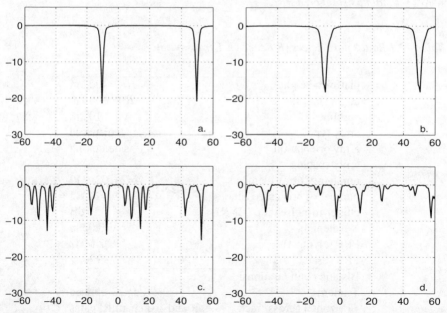

Figure 12.22: Sparse DPCA array (Nyquist/256): Horizontal across-track configuration, baseline [km]: a. 0.1; b. 1; c. 10; d. 100

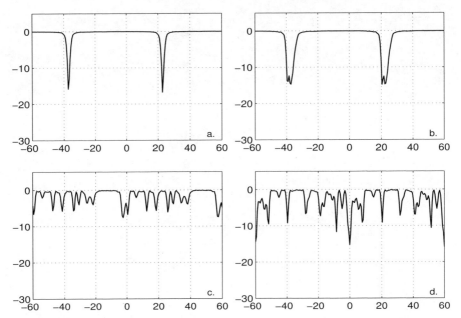

Figure 12.23: Sparse DPCA array (Nyquist/256): (T on top of R, parallel paths), baseline [km]: a. 0.1; b. 1; c. 10; d. 100

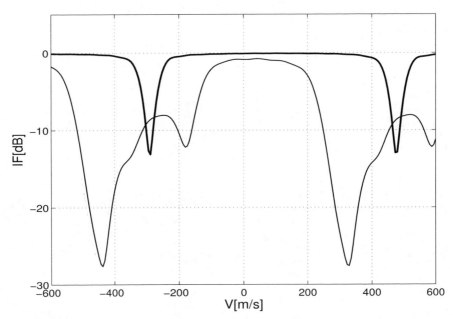

Figure 12.24: Sparse (thin) and full (fat) array in DPCA mode: PRF=Nyquist/20; N=240/12 (full/sparse); M=12; (tandem, baseline: 100 km)

Chapter 13

Interrelated problems in SAR and ISAR

In this chapter we discuss two specific problems of space-time adaptive processing for SAR and ISAR.

The first problem is strongly related to clutter suppression as has been treated in previous chapters. The question to be answered is how far inverse synthetic aperture (ISAR) imaging of moving targets by a moving phased array radar is degraded by space-time clutter filtering. ISAR techniques exploit the target motion rather than the radar platform motion for imaging of moving targets. Details of the principles of ISAR can be found in WEHNER [701, pp. 341].

Since space-time filtering techniques include the *time* dimension and ISAR is a *temporal* (pulse-to-pulse) compression technique some effect of clutter filtering on the ISAR resolution can be expected. We consider here space-*time* filtering. The effect of space-*frequency* clutter rejection techniques (see Section 9.5) will be similar. Since it is closer to our previous considerations on space-time processing we start with the ISAR problem. It should be noted that we will treat here the more complex problem of space-time clutter suppression for ISAR with a moving radar platform. Similar results can be obtained for a *stationary* radar with *temporal* clutter filtering only. Even conventional processing (applying a *temporal* clutter rejection filter to SAR data, see for instance MEDLIN [445, 446]) will cause some distortions of the ISAR chirp due to a moving target.

One of the earliest publications on ISAR was by WIRTH[1] [712]. The author used data received from a conventional surveillance radar with rotating antenna to image a Boeing 737 close to Cologne-Bonn airport (Germany). This early paper was written in German and, therefore, received little attention outside Germany. Later this paper was republished in English (WIRTH [712]).[2]

ISAR imaging of moving targets requires the knowledge of the target motion. In

---

[1] In the chapter on ISAR in the book by WEHNER [701, pp. 432] no earlier publications than 1980 have been quoted.

[2] Notice the editor's note by D. K. Barton.

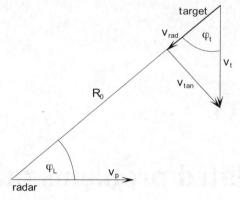

Figure 13.1: ISAR geometry

the case of air- or spacecraft mostly straight motion occurs while imaging of ground targets is based on curved motion paths. Eaton and Coe [121] have shown that military vehicles moving off-road may exhibit complex motion behaviour.

It should be noted that the considerations made in the sequel are based again on an array antenna (multichannel SAR) so that space-time processing can be used for clutter rejection with the capability of slow target detection. In other publications (for instance, D'ADDIO et al. [3]) system concepts for single channel SAR-MTI are presented.

The second problem addressed is space-time jammer suppression for broadband arrays as may be used specifically for SAR systems.[3] We assume that the radar transmits linear FM pulses which are compressed on receive by a matched filter. Such pulse compression techniques have been applied in SAR systems and other high-resolution radar as well as in real aperture solid state phased array radar (WIRTH [717]).

Jammer suppression in broadband requires space-TIME (COMPTON [96], LIN and YU [401], WHITE [705]) or space-FREQUENCY techniques (COMPTON [97], GODARA [194]). By TIME we denote the delay time of incoming echoes which is equivalent to range. If frequency domain techniques are applied we use the term FREQUENCY for the range equivalent frequency. We want to answer the question of how far the matched filter response[4] (range compression) which takes place in the TIME domain is degraded by space-TIME anti-jamming techniques. Again we prefer to consider space-TIME domain techniques. The achieved results hold for space-FREQUENCY domain techniques as well. It should be noted that these considerations apply to any kind of broadband radar, regardless of whether it is moving or not. It should further be noted that broadband anti-jammer filtering is not specifically related to MTI radar. However, there is quite an analogy between space-time and space-TIME processing which will contribute to a deeper understanding of the problems. The properties of space-time-TIME processing for clutter rejection will briefly be addressed. The content of this chapter has been published by KLEMM [322].

---

[3] For experimental and operational multichannel radars see Section 1.1.6.
[4] Also known as the cross-section of the point spreading function in the range dimension.

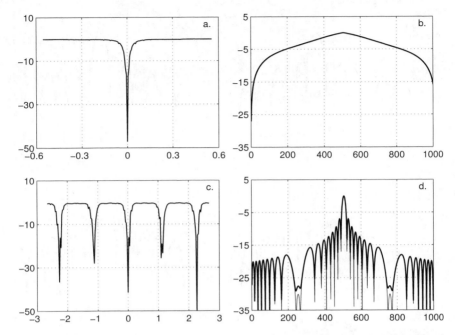

Figure 13.2: Influence of PRF ($M = 511$, $L = 5$): a. IF vs F (PRF = 12 000 Hz); b. PSF vs delay (PRF = 12 000 Hz); c. IF vs F (PRF = 2500 Hz); d. PSF vs delay (PRF = 2500 Hz)

It should be noted that these two examples treated below are not the only applications of space-time adaptive processing in SAR or ISAR. One other prominent example is the estimation of the correct position of a moving target in a SAR image. This technique and a variety of other signal processing methods for multichannel SAR have been presented by ENDER [124, 126].

Alternatively, processing can be carried out in the space-frequency-FREQUENCY domain. This may have advantages when using broadband array antennas for high range resolution (STEINHARDT and PULSONE [617] and HOFFMAN and KOGON [272]).

13.1 Clutter rejection for multichannel ISAR

Inverse synthetic aperture radar (ISAR) is a technique that exploits the target motion for target imaging. Basically a stationary radar can be used for this purpose. However, ISAR can also be applied with moving radar. Then we have a combination of SAR and ISAR where the difference between the velocities of radar platform and target are exploited for target imaging.

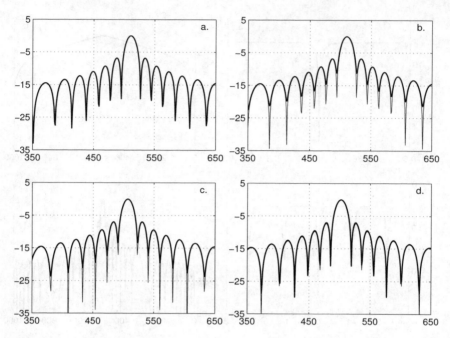

Figure 13.3: Dependence of matched filter response (azimuthal PSF vs delay) on the clutter filter length ($M = 511$): a. $L = 1$; b. $L = 2$; c. $L = 5$; d. $L = 10$.

13.1.1 Models

Consider the geometry shown in Figure 13.1. Let us define a radial target velocity v_t relative to the radar. Then the two-way phase progression with time becomes approximately (see HOVANESSIAN [274, p. 28])

$$\phi(t) = \phi(t_0) + \frac{4\pi}{\lambda} v_t \cos\varphi_t \, t + \frac{2\pi}{\lambda} \frac{(v_t \sin\varphi_t)^2}{R_0} t^2 \qquad (13.1)$$

In (13.1) we have a linear term which includes the radial velocity and a quadratic term which stands for the tangential motion. The velocity components are

$$\begin{aligned} v_{\mathrm{rad}} &= v_t \cos\varphi_t \\ v_{\mathrm{tan}} &= v_t \sin\varphi_t \end{aligned} \qquad (13.2)$$

13.1.1.1 Signal

Inserting the quadratic term in (13.1) into the target signal model (2.31) gives

$$s_{\mathrm{r}}(R_{\mathrm{s}},\varphi) = A \exp\left[j\frac{2\pi}{\lambda}\left\{2v_{\mathrm{rad}} mT + \frac{v_{\mathrm{tan}}^2}{R_0}(mT)^2\right.\right.$$

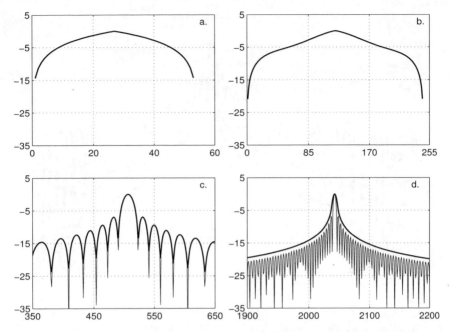

Figure 13.4: Dependence of matched filter response on matched filter length (PSF vs delay, solid: $L = 5$, thin: $L = 1$): a. $M = 31$; b. $M = 127$; c. $M = 511$; d. $M = 2047$

$$+(x_i \cos\varphi + y_i \sin\varphi)\cos\theta - z_i \sin\theta \Big\}\Big] \quad (13.3)$$
$$m = 1,\ldots,M, \quad i = 1,\ldots,N$$

The linear FM chirp as given by the quadratic phase term is of course a strong idealisation based on the assumption of a constant target velocity and constant direction. In practice other effects such as target motion and maneuvering have to be taken into account (HAYWARD [256]). For the purpose of analysing the effect of clutter filtering this simple target model is sufficient.

13.1.1.2 Clutter

Because in ISAR we deal with long data sequences a quadratic term has to be added to the clutter model. For the stationary clutter background the tangential motion of each clutter scatterer on the ground causes a quadratic phase term. By inserting the quadratic phase term into (2.34) we get

$$\begin{aligned} c_{im} &= \int_{\varphi=0}^{2\pi} s_\mathrm{r}(\varphi)\,\mathrm{d}\varphi \\ &= \int_{\varphi=0}^{2\pi} A\exp\left[\mathrm{j}\frac{2\pi}{\lambda}\Big\{(x_i + 2v_\mathrm{p}mT)\cos\varphi\right. \end{aligned}$$

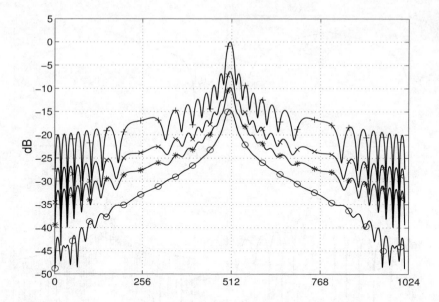

Figure 13.5: *Distortion of matched filter response close to clutter notch ($N = 24$, $L = 5$):* ○ $\varphi_t = 30°$; ∗ $\varphi_t = 30.2°$; × $\varphi_t = 30.4°$; + $\varphi_t = 30.6°$

$$+ \left(y_i + \frac{v_p^2 \sin \varphi}{R_0} (mT)^2 \right) \sin \varphi \cos \theta - z_i \sin \theta \right\} \bigg] \mathrm{d}\varphi \quad (13.4)$$

Let us consider a simple example. We assume

- A sidelooking linear equidistant array with spacing d. Then the sensor coordinates y_i and z_i become zero.

- That the motion of the transmit pattern during the coherent processing interval is negligible, i. e., $G(\varphi, m) = G(\varphi, p) = G(\varphi)$ which is satisfied if the temporal dimension of the clutter covariance matrix is small. This requirement will be fulfilled by using a short FIR filter according to Chapter 7.

- That the quadratic term of clutter echoes is negligible. This is also satisfied if the temporal dimension of the clutter covariance matrix is small.

- That no decorrelation due to internal clutter motions or system bandwidth takes place.

Then (13.4) becomes

$$c_{im} = \int_{\varphi=0}^{2\pi} AD(\varphi)L(\varphi)G(\varphi) \exp\left[\mathrm{j}\frac{2\pi}{\lambda}(di + 2v_p mT)\cos\varphi\cos\theta\right] \mathrm{d}\varphi \quad (13.5)$$

Assuming independent echoes from different clutter scatterers the elements of the space-time covariance matrix are then given, according to (3.31), by

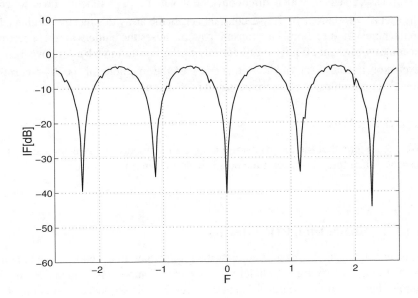

Figure 13.6: IF for small array ($N = 6$, PRF = 2500)

$$q_{ln}^{(c)} = E\{c_{im}c_{kp}^*\}$$
$$= P_c \int_{\varphi=0}^{2\pi} D^2(\varphi)L^2(\varphi) \mid G^2(\varphi) \mid$$
$$\times \exp\left[j\frac{2\pi}{\lambda}(d(i-k) + 2v_p(m-p)T)\cos\varphi\cos\theta\right]d\varphi \quad (13.6)$$

where $P_c = E\{AA^*\}$ is the clutter power. The indices l, n and i, k, m, p are related through (3.23). With the substitution $u = \cos\varphi$, $u = -1, \ldots, 1$, and integrating with respect to u gives

$$q_{ln}^{(c)} = 2P_c\tilde{D}(i) * \tilde{D}(i) * \tilde{G}(i) * \tilde{G}^*(i)$$
$$* \tilde{L}(i) * \tilde{L}(i)$$
$$* \frac{\sin[\frac{2\pi}{\lambda}(d(i-k) + 2v_p(m-p)T)\cos\theta]}{\frac{2\pi}{\lambda}(d_x(i-k) + 2v_p(m-p)T)\cos\theta} \quad (13.7)$$

The asterisks denote discrete convolutions with respect to the variable $i - k$. The quantities $\tilde{D}(i)$, $\tilde{G}(i)$, and $\tilde{L}(i)$ are the inverse Fourier transforms of the azimuth dependent variables $D(\varphi)$ (sensor pattern), $G(\varphi)$ (transmit pattern), and $L(\varphi)$ (reflectivity) at the sensor positions.

If for example $G(\varphi)$ is constant with φ we have the case of omnidirectional transmission. The inverse Fourier transform of a constant function is a single pulse (or, in this case, a sensor) which, after convolution with the $\frac{\sin x}{x}$ function does not change the spatial correlation. If $\tilde{G}(i)$ is constant across the array this means forming a beam in the boresight direction. The transmit pattern causes the clutter Doppler spectrum to become narrowband. Equivalently, convolving the $\frac{\sin x}{x}$ term with a window of constant excitation factors $\tilde{G}(i)$ yields a broadening of the mainlobe of the $\frac{\sin x}{x}$ correlation function and, hence, an increase in spatial correlation.

13.1.1.3 Noise

We can assume that the receiver noise is independent between sensors and from echo to echo. Therefore, the space-time noise covariance matrix becomes

$$\mathbf{N} = P_\mathrm{n}\mathbf{I} \tag{13.8}$$

13.1.2 Space-time FIR filtering

For imaging of moving targets close to the ground the target echoes have to compete with strong clutter returns. Therefore, some clutter suppression operation should be applied before azimuth compression of the received echo sequence. In order to obtain a reasonable target Doppler bandwidth out of the tangential target motion the observation time has to be sufficiently long. This means in turn that clutter suppression can either be done in the frequency domain (see Section 9.5) or by use of the FIR filter techniques described in Chapter 7. Recall that the temporal length of the FIR filter L can be chosen independently of the length of the total number of echoes M. We will use here the FIR filter technique described in Section 7.1.

The question to be answered is how far the chirp signal sequence generated through the tangential motion of the target will be distorted by the clutter filter operation. Since the clutter filter after (7.12) operates in time and space the temporal filter dimension will have some effect on the chirp signal. We want to find out how far clutter filtering may degrade the ISAR resolution.

In order to judge the resolution capability of ISAR processing after clutter filtering we calculate the azimuthal cross-section of the point spreading function. This is in essence the modulus of the output of the azimuth compression filter. The optimum compression filter in the SNR sense is given by the signal itself. Since the signal may be distorted by the clutter filter the *filtered* signal is used as a compression filter. In summary, the azimuthal point spreading function is given by the autocorrelation of the filtered signal sequence.

13.1.2.1 Choice of PRF

The proper choice of the PRF is determined by the clutter bandwidth (PRF = Nyquist rate). If the PRF is chosen smaller than the maximum Doppler frequency, ambiguities of the clutter filter response will occur. This effect has been demonstrated in Figures 4.12 and 4.13.

As shown before, the chirp signal generated by the tangential motion of the target has much lower frequency components than the clutter bandwidth. This signal is,

Table 13.1 Parameter set for ISAR examples

M	length of coherent pulse sequence	511
L	temporal filter length	5
N	Number of antenna elements	24
PRF	pulse repetition frequency	2500 Hz
φ_L	look direction	90°
φ_t	target flight direction	28°
v_p	radar platform velocity	90 m/s
v_t	target velocity	300 m/s
	array orientation	sidelooking
R_0	target range	6000 m
λ	wavelength	0.03 m

therefore, strongly oversampled when the PRF is matched to the clutter bandwidth. For the same reason the coherent processing interval (interval between first and last echo) is very short. Therefore, the achieved signal bandwidth is quite small, and the obtainable resolution is poor.

This dilemma is illustrated by a numerical example in Figure 13.2. Figures a. and b. show on the left the IF for PRF = 12 000 Hz which is the Nyquist frequency of the clutter band if there is only a radial target velocity (i.e., $\varphi_t = 0$). The look direction was chosen to be $\varphi_L = 90°$ (cross-flight). Therefore, the clutter notch appears at the Doppler frequency F = 0. As one can see the filter response (a.) is unambiguous.

Figure b. shows the azimuthal part of the point spreading function (PSF) which in essence is the autocorrelation of the chirp signal with and without filtering (both curves coincide). Notice that all point spreading functions are in a logarithmic scale. As can be noticed the PRF is so high that an echo sequence of 511 samples shows more or less a triangular autocorrelation function only (logarithmic scale) which indicates that the chirp component in the signal is negligible. Decreasing the PRF (c. and d.: PRF = 2500 Hz) leads to an expected chirp correlation, however at the expense of ambiguities in the clutter filter response (fat curve: with filtering; thin curve: without).

This is a general dilemma which cannot be circumvented. The results presented in the following section are based on the parameters used in Figure 13.2 c and d.

13.1.3 Effect of clutter cancellation on ISAR resolution

In the following numerical examples the effect of clutter filtering on ISAR resolution is analysed by calculation of the azimuthal point spreading function. The parameters listed in Table 13.1 have been kept constant if not denoted otherwise. For all other parameters see Table 2.1.

It should be pointed out that the Doppler frequency associated with $\varphi_t = 30°$ falls right into one of the multiple clutter notches. The choice $\varphi_t = 28°$ brings the target Doppler very close to the clutter notch.

13.1.3.1 Temporal filter length

In Figure 13.3 we see the azimuthal PSF for different temporal lengths of the space-time clutter FIR filter, plotted versus the sample number. Only the mainlobe region is shown. The four plots show the azimuthal point spreading function in a logarithmic scale after filtering (fat) and without filtering (thin). For $L = 1$ both curves coincide because there is no temporal filter dimension and hence no effect on the point spreading function. For $L > 1$ we notice that the width of the main beam is not affected by the clutter filter, but slightly shifted. Also, the sidelobe level is the same as without filtering. Slight differences can be noticed in the nulls of the point spreading function. The clutter filter tends to smear out the nulls.

13.1.3.2 Length of coherent echo burst

The four plots in Figure 13.4 show the azimuthal point spreading function for various numbers of echo samples ($M = 31, 127, 511, 2047$). It can be noticed that for short data sequences (here up to $M = 511$) there is no significant effect of filtering on the point spreading function.

However, for $M = 2047$ we notice that the sidelobe pattern of the point spreading function has been completely smoothed and the main beam has been slightly broadened. This effect can be explained as follows. For shorter echo sequences (in our example up to $M = 511$) the bandwith of the chirp signal due to the tangential target motion is sufficiently small. In an extreme case like $M = 31$ (Figure 13.4a) the Doppler frequency due to the radial velocity dominates entirely so that the point spreading function becomes a triangle (in linear scale). A single frequency, however, is *not distorted* by a linear filter but just attenuated or amplified. As the bandwith increases (Figure 13.4d), the distortion through the clutter filter becomes noticeable.

13.1.3.3 Behaviour close to clutter notch

Distortions of the point spreading function will occur mainly in areas where the filter transfer characteristics are not constant with the Doppler frequency. This happens in particular close to the clutter notches as can be seen in Figure 13.2.

The four curves shown in Figure 13.5, have been calculated for different target angles φ_t. The Doppler frequencies associated with the target angles have been chosen close to the clutter notch. For $\varphi_t = 30°$ the least signal power is received, and the distortion is maximum. Moving out of the clutter notch ($\varphi_t > 30°$) leads to increased signal power and decreasing distortion of the point spreading function.

It should be noted that regions with constant transfer characteristics as shown in Figure 13.2 occur only if the resolution capability and, hence, the aperture of the array is large enough. If the resolution capability is poor no flat pass band of the transmission characteristics is obtained (Figure 13.6) so that distortions of the PSF can be expected for large chirp bandwidths.

13.2 Jammer nulling for multichannel radar/SAR

Radar with a multichannel antenna has the potential of jammer rejection by forming nulls in the antenna directivity pattern. For broadband radar (imaging radar, particularly synthetic aperture radar) jammer cancellation requires broadband processing techniques. The use of such technique for mainbeam jammer cancellation has been analysed by FANTE [143]. BUCKLEY [74] presents a subspace technique for cancellation of broadband sources based on the Karhunen–Loève transform (eigenvector decomposition of the interference covariance matrix).

The theory of adaptive radar (or sonar) has been developed by BRENNAN and REED [62]. In their theory the authors take care of the *broadband* radar case by using space-TIME samples of signal and interference. Accordingly vectors and matrices have the same form as given by (3.21) and (3.22).

In the following discussion we make the following assumptions

- There is no impact of platform motion during one PRI.

- A sufficient amount of data is available in each PRI so that the space-TIME filter can be adapted during one PRI. This is verified particularly for SAR where the number of range increments is usually high due to the required range resolution. Therefore, the 'slow' time and platform velocity effects do not enter the discussion.

- Jammers are stationary during one PRI.

- The transmit waveform is a linear FM chirp. Basically the results hold for other waveforms as well; however this commonly used waveform corresponds nicely with the chirp occurring in ISAR imaging as shown in Section 11.1.

In the case where the slow (pulse-to-pulse) time has an influence on the adaption of the space-time filter, motions of the platform lead to significant losses in jammer suppression performance (HAYWARD [256]). This may occur especially in HPRF radar modes where the number of range gates can be very small.

13.2.1 Models

In the following we develop briefly the models used for the analysis of space-TIME filtering for broadband jammer cancellation in a phased array radar. We follow here closely the paper by COMPTON [96].

Suppose the radar transmits a signal which has a rectangular power spectral density of bandwidth $-B/2, \ldots, B/2$, with B being the bandwidth, centred at carrier frequency ω_0. Applying the inverse Fourier transform one obtains the correlation function

$$\rho(\tau) = P_s \frac{\sin B\tau/2}{B\tau/2} \exp(j\omega_0 \tau) \tag{13.9}$$

where P_s is the power of the radiating source. Let us specify the delay τ in terms of the parameters of a backscattered wave arriving at the sensors of a linear equispaced antenna array. Replacing τ by delays of a sampled waveform gives

$$\rho[(m-p)T_t + (i-k)T_s] = P_s \frac{\sin[\frac{B}{2}\{(m-p)T_t + (i-k)T_s\}]}{\frac{B}{2}\{(m-p)T_t + (i-k)T_s\}}$$
$$\times \exp[j\omega_0\{(m-p)T_t + (i-k)T_s\}] \quad (13.10)$$

where T_t and T_s are delay times between adjacent samples in the TIME and space dimension. The indices m and p are temporal (pulses) while i and k are spatial (sensors).

The product BT_s can be written as

$$BT_s = \frac{B}{\omega_0}(\omega_0 T_s) = B_r \phi \quad (13.11)$$

where

$$B_r = \frac{B}{\omega_0} \quad (13.12)$$

is the relative bandwidth and

$$\phi = \frac{2\pi}{\lambda_0} d \cos\varphi \quad (13.13)$$

is the spatial phase at the centre frequency due to delays across the array. d is the interelement spacing. φ is the angle between the direction of arrival and the array axis, λ_0 is the wavelength at the centre frequency. Following the notation used by COMPTON [96] we normalise T_t to the delay associated with a quarter wavelength delay at the carrier frequency. The time delay required to produce a 90° phase shift at frequency ω_0[5] is

$$T_{90} = \frac{\pi}{2\omega_0} \quad (13.14)$$

Then we can express T_t as

$$T_t = r_{90} T_{90} = \frac{\pi r_{90}}{2\omega_0} \quad (13.15)$$

where r_{90} is the number of quarter-wave delays in T_t at frequency ω_0. Furthermore we get

$$BT_t = B\frac{\pi r_{90}}{2\omega_0} = \frac{\pi}{2} B_r r_{90} \quad (13.16)$$

Inserting the normalised quantities in (13.10) gives for the correlation function

$$\rho[(m-p)T_t + (i-k)T_s] = P_s \frac{\sin[\frac{B_r}{2}\{\frac{\pi}{2}(m-p)r_{90} + (i-k)\phi\}]}{\frac{B_r}{2}\{\frac{\pi}{2}(m-p)r_{90} + (i-k)\phi\}} \quad (13.17)$$
$$\times \exp\left[j\frac{\pi}{2}(m-p)r_{90} + (i-k)\phi\right]$$

---

[5]We follow here the pass band notation of COMPTON [96]. If, however, the normalised sampling interval is chosen a quarter wave of $\omega_0 + B_r/2$ instead of ω_0 the baseband case ($\omega_0 = 0$) is included.

13.2.1.1 Jamming

Equation (13.17) describes the space-TIME correlation function due to a wave with relative bandwidth B_r arriving at angle φ. Using the two sets of temporal and spatial indices we can then formulate the space-TIME covariance matrix according to (3.22):

$$\mathbf{Q} = \begin{pmatrix} \mathbf{Q}_{11} & \mathbf{Q}_{12} & \cdots & \mathbf{Q}_{1M} \\ \mathbf{Q}_{21} & \mathbf{Q}_{22} & \cdots & \mathbf{Q}_{2M} \\ \vdots & \vdots & & \vdots \\ \mathbf{Q}_{M1} & \mathbf{Q}_{M2} & \cdots & \mathbf{Q}_{MM} \end{pmatrix} \quad (13.18)$$

where the temporal indices $m, p = 1, \ldots, M$ denote the spatial submatrices $\mathbf{Q}_{mp}$ while the spatial indices $i, k = 1, \ldots, N$ run inside the submatrices. In the case of multiple mutually uncorrelated jammers the total covariance matrix becomes

$$\mathbf{Q} = \sum_{j=1}^{J} \mathbf{Q}_j \quad (13.19)$$

where the elements of $\mathbf{Q}_j$ are given by (13.17) with the individual jammer powers $P_s = P_j$ and jammer directions $\varphi = \varphi_j$.

13.2.1.2 Noise

In all past chapters it was assumed that noise is independent from pulse to pulse and from sensor to sensor, see (2.49). In space-TIME processing noise is still independent between sensors, i.e., the noise is spatially white. In the TIME dimension, however, noise is correlated according to the band limitation. Therefore, the elements of the noise covariance matrix become

$$\rho_{m-p} = P_n \frac{\sin\{\frac{B_r}{4}\pi(m-p)r_{90}\}}{\frac{B_r}{4}\pi(m-p)r_{90}} \exp\left[j\frac{\pi}{2}(m-p)r_{90}\right] \quad (13.20)$$

with P_n being the noise power. For $N = 4$ and $M = 3$ the noise covariance matrix then looks as follows

$$\mathbf{N} = \begin{pmatrix}
\rho_0 & \cdot & \cdot & \cdot & \rho_1 & \cdot & \cdot & \cdot & \rho_2 & \cdot & \cdot & \cdot \\
\cdot & \rho_0 & \cdot & \cdot & \cdot & \rho_1 & \cdot & \cdot & \cdot & \rho_2 & \cdot & \cdot \\
\cdot & \cdot & \rho_0 & \cdot & \cdot & \cdot & \rho_1 & \cdot & \cdot & \cdot & \rho_2 & \cdot \\
\cdot & \cdot & \cdot & \rho_0 & \cdot & \cdot & \cdot & \rho_1 & \cdot & \cdot & \cdot & \rho_2 \\
\rho_1 & \cdot & \cdot & \cdot & \rho_0 & \cdot & \cdot & \cdot & \rho_1 & \cdot & \cdot & \cdot \\
\cdot & \rho_1 & \cdot & \cdot & \cdot & \rho_0 & \cdot & \cdot & \cdot & \rho_1 & \cdot & \cdot \\
\cdot & \cdot & \rho_1 & \cdot & \cdot & \cdot & \rho_0 & \cdot & \cdot & \cdot & \rho_1 & \cdot \\
\cdot & \cdot & \cdot & \rho_1 & \cdot & \cdot & \cdot & \rho_0 & \cdot & \cdot & \cdot & \rho_1 \\
\rho_2 & \cdot & \cdot & \cdot & \rho_1 & \cdot & \cdot & \cdot & \rho_0 & \cdot & \cdot & \cdot \\
\cdot & \rho_2 & \cdot & \cdot & \cdot & \rho_1 & \cdot & \cdot & \cdot & \rho_0 & \cdot & \cdot \\
\cdot & \cdot & \rho_2 & \cdot & \cdot & \cdot & \rho_1 & \cdot & \cdot & \cdot & \rho_0 & \cdot \\
\cdot & \cdot & \cdot & \rho_2 & \cdot & \cdot & \cdot & \rho_1 & \cdot & \cdot & \cdot & \rho_0
\end{pmatrix} \quad (13.21)$$

Table 13.2 Comparison of target models

phase component	space-time	space-TIME
linear temporal term	radial target velocity	carrier frequency
quadratic temporal term	tangential target velocity	chirp modulation
spatial term	target direction	target direction

where all the dots denote zeroes.

13.2.1.3 Signal

A commonly used signal waveform in broadband radar is the linear FM chirp which is given by

$$s(k,m) = A_s \exp\left[j\frac{\pi}{2}r_{90}m\left(1 + m\frac{B_r}{M}\right) + k\phi_s\right] \quad m = \left[-\frac{M}{2}, \frac{M}{2}\right] \quad (13.22)$$

where the linear m-term in the argument denotes the carrier frequency and the quadratic part the chirp modulation. According to (13.13) the spatial phase term

$$\phi_s = \frac{2\pi}{\lambda}d\left(1 + m\frac{B_r}{M}\right)\cos\varphi \quad (13.23)$$

includes the signal direction depending on time m. B_r is the relative signal bandwidth.

13.2.2 Comparison of models

Let us briefly compare the models for space-TIME jammer rejection with those for space-time clutter rejection presented in Section 13.1.

13.2.2.1 Signal

The target signals (compare (13.3) and (13.22)) both have the same form. They differ just in scaling according to the individual physical conditions. The phase is composed of three components which are summarised in Table 13.2.

13.2.2.2 Interference

The interference models are given by (13.7) for clutter, (13.17) and (13.19) for jamming. Both formulas have basically the same form in that the decorrelation effect due to the bandwidth effect is expressed by a $\frac{\sin x}{x}$ function. In (13.17) m and p denote 'slow' time while in (13.20) they denote 'fast' TIME.

In the case of clutter the *Doppler* bandwidth is generated by superposition of clutter arrivals with different Doppler frequency arriving from all possible directions. The $\frac{\sin x}{x}$ function stands for an omnidirectional echo field. This may be modified by

Table 13.3 Comparison of interference models

interference correlation	space-time (13.7)	space-TIME (13.17)
temporal term	clutter bandwidth $4v_p/\lambda$	system bandwidth B_r
spatial term	omnidirectional	jammer directions

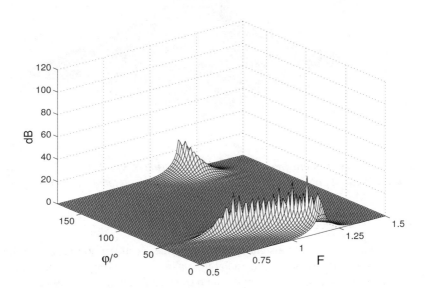

Figure 13.7: MV azimuth-frequency jammer spectrum ($J = 1$, $\varphi_1 = 30°$; $d/\lambda = 0.5$, beamformer calculated at carrier frequency)

additional directivity of the background reflectivity ($L(\varphi)$) or directivity of the sensor elements ($D(\varphi)$) and the transmit antenna ($G(\varphi)$).

In the case of broadband jammers the *system* bandwidth determines the space-time correlation. In Table 13.3 the roles of the correlation arguments for clutter and jamming are summarised.

13.2.2.3 Noise

As stated before noise is independent between sensors. It is also independent from pulse to pulse (space-time processing) while in space-TIME processing it may show some temporal correlation due to the system bandwidth.

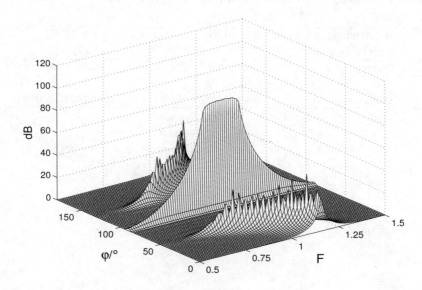

Figure 13.8: MV azimuth-frequency jammer spectrum: ($J = 3$, $\varphi_1 = 30°$; $\varphi_2 = 90°$; $\varphi_2 = 150°$; equal powers; $d/\lambda = 0.4$, beamformer calculated at carrier frequency)

13.2.3 MV spectra of jammers and noise

Let us now get some more insight by calculating the spectra of noise and jamming.

The bandwidth of a source radiating on a sensor array is interpreted by a beamformer as a spatial uncertainty. A source at frequency $\omega_0 + \omega$ arriving from angle φ_s at the sensors of a linear array with spacing d causes a signal at the k-th sensor which has the form

$$s(k) = A_s \exp\left(j\frac{\omega_0 + \omega}{c} dk \cos\varphi_s\right) \quad (13.24)$$

Matching this signal with a beamformer steered over the look angle φ_L with coefficients

$$s(k) = \exp\left(j\frac{\omega_0}{c} dk \cos\varphi_L\right) \quad (13.25)$$

shows that the array interprets the bandwidth as direction. When matching the beamformer (13.25) with the signal (13.24) we find the relation

$$\cos\varphi_L = \cos\varphi_s \left(1 + \frac{\omega}{\omega_0}\right) \quad (13.26)$$

Notice that this relation corresponds to (3.16)

$$f_r = \cos\beta \quad (13.27)$$

Equation 13.27 relates the relative clutter Doppler frequency to the angle of arrival of clutter echoes for the case of a sidelooking array. It appears for instance as a trajectory in clutter spectra as shown in Figure 3.18, or in improvement factor plots such as Figure 4.2.

Let us define a space-TIME steering vector with elements of the form

$$s_{km}(\varphi, \mathsf{F}) = \exp\left[j\frac{\pi}{2}r_{90}\mathsf{F}m + k\phi(\varphi)\right] \qquad (13.28)$$

where $\phi = \frac{2\pi}{\lambda}d\cos\varphi$ is the phase due to the steering angle φ and $\mathsf{F}(\omega) = 1 + \frac{\omega}{\omega_0}$ is the normalised frequency centred at the carrier. Inserting this steering vector and the covariance matrix of jamming and noise given by (13.17) and (13.20) into (3.67), we can obtain a $\varphi - \mathsf{F}$ minimum variance spectrum of broadband jammers.

Figure 13.7 shows the MV spectrum of a jammer with $B_r = 0.2$ located at $\varphi = 30°$. It is similar to the spectra of clutter received by a moving sidelooking array (see Figure 3.18) in that the jammer power is a narrow ridge in the φ-F plane.

As we assumed half wavelength spacing of the array ($\frac{d}{\lambda} = 0.5$ in (13.13)) the array is slightly undersampled because the maximum jammer frequency is $(1 + B_r)\omega_0 = 1.2\omega_0$. Therefore, we obtain some aliasing in the back of the plot. For the sake of clarity we will avoid this effect in all subsequent examples by assuming the spacing to be $d = 0.4\lambda$.

Looking into the plot from the right one is faced with the bandwidth of the jammer. Looking from the left shows the uncertainty of the jammer direction caused by the bandwidth.

Figure 13.8 shows the case of three jammers at $\varphi = 30°$, $\varphi = 90°$, and $\varphi = 150°$. As can be seen, at $90°$ there is no bandwidth effect on the spatial resolution of the jammer. The jammer appears as a point in space. Looking from the right shows clearly the approximately rectangular shape of the system transfer function.

13.2.4 Space-TIME FIR filter approach

As stated in the previous section the signal used to calculate the gain for the optimum processor was assumed to be narrowband, that is, the relative signal bandwidth B_r in (13.22) was set equal to zero. This enabled us to work with short data sequences (small M). For long data sequences as required to model a broadband chirp signal the optimum processor is not applicable due to excessive consumption of memory and computing time.

In order to handle long broadband signal sequences we come back to the concept of the space-time FIR filter described in Section 7.1.1. For a block diagram see Figures 7.1 and 7.2.

A major result of Chapter 7 was that such a space-time FIR filter achieves near-optimum clutter suppression performance with relatively short *temporal* filter length L. In contrast to the optimum processor the temporal filter length L is independent of the coherent processing interval (length of the echo data sequence). This makes this kind of processor an ideal tool for handling broadband signals. Of course, for long coherent processing intervals frequency domain techniques may be even more economic, see Chapter 9.

In contrast to Section 13.1 and Chapter 7 we will now use the FIR filter in the space-TIME domain, where TIME means the fast time (or range) domain. Figure 13.9

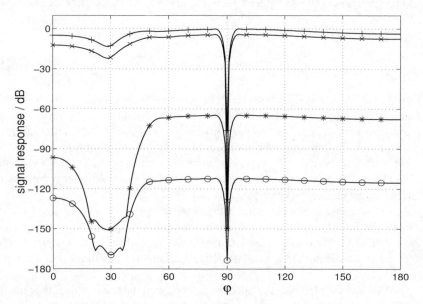

Figure 13.9: Signal response of space-TIME FIR filter, dependence on filter length ($M = 255$, $J = 2$, $r_{90} = 1$, $\varphi_1 = 90°$, $\varphi_2 = 30°$, equal power, $B_r = 0.2$; $d/\lambda = 0.4$): ○ $L = 1$; ∗ $L = 2$; × $L = 5$; + $L = 10$

shows the signal response versus the angle of arrival[6] of the space-TIME FIR filter for different temporal filter lengths[7] L.

In addition to the use of a space-TIME FIR filter instead of the space-TIME covariance matrix inverse we used a broadband signal according to (13.22). It can be noticed that the filter response to a bandpass signal with relative bandwidth $B_r = 0.2$ leads to a quite smooth jammer notch at $\varphi = 30°$.

13.2.5 Effect of broadband jammer cancellation on SAR resolution

Since the anti-jamming space-TIME FIR filter has a temporal dimension it is likely that the received radar echo is distorted in the TIME dimension by the filtering operation. In the following we analyse this effect by first filtering the desired target echo, secondly by calculating the matched filter response (MFR).[8] The matched filter used here takes care of the filter distortion, that is, we use the target signal replica *after* FIR filtering. This is in accordance with the whitening principle included in the optimum receiver (1.23), and also with the FIR filter receiver shown in Figure 1.7b. The matched filter output is given by the autocorrelation of the filtered signal replica.

---

[6]We show this plot only to verify that the FIR filter works properly. The gain calculation for this case is quite complex and has, therefore, been omitted.

[7]The differences in signal level between the four curves are due to the fact that the filter coefficients have not been normalised.

[8]The matched filter operation in SAR is called 'range compression'.

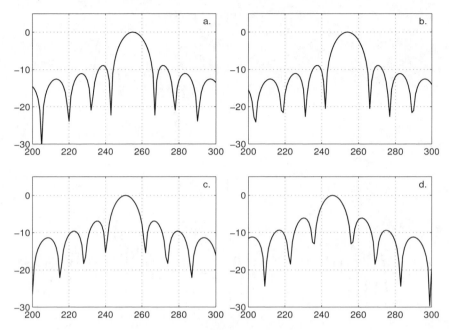

Figure 13.10: Dependence of matched filter response on filter length ($M = 511$, $J = 2$, $\varphi_1 = 90°$, $\varphi_2 = 30°$, $\varphi_L = 120°$, equal power): a. $L = 1$; b. $L = 2$; c. $L = 5$; d. $L = 10$

13.2.5.1 Behaviour in the filter passband

In Figure 13.10 the influence of the temporal filter length ($L = 1, 2, 5, 10$) is shown. Notice that only the centre part of the autocorrelation has been plotted. It can be seen that by increasing the temporal filter length (from a. to d.) the sidelobes are slightly raised.

Figure 13.11 shows the dependence of the matched filter response on the coherent processing interval M. The fat curves show the filtered MFR while the thin curve is the original autocorrelation function. There is obviously no significant dependence.

13.2.5.2 Behaviour close to the interference direction

There is obviously no significant effect of the space-TIME anti-jamming filter on the matched filter response of a broadband radar as long as the look direction is sufficiently different from the jammer directions. Now let us consider the angular areas in the vicinity of the jammers.

In Figure 13.12 the look angle was varied between $90°$ (jammer direction) and $105°$. Two interesting points can be noticed. Firstly, when moving out of the filter notch at $90°$ the level of the matched filter output increases. Secondly, the shape of the four curves is very similar. This is due to the fact that at $\varphi = 90°$ the filter is *spatial* only, which follows from Figure 13.8. It illustrates that at $\varphi = 90°$ the filter transfer

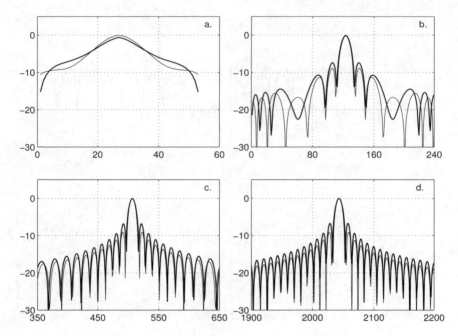

Figure 13.11: Dependence of matched filter response on coherent processing interval ($N = 24$, $L = 5$, $r_{90} = 1$, $J = 2$, $\varphi_1 = 90°$, $\varphi_2 = 30°$, $\varphi_L = 120°$, equal power): a. $M = 511$; b. $M = 2047$; c. $M = 8191$; d. $M = 32703$

characteristics are frequency independent. Therefore, no effect on the spectrum of the signal takes place; the matched filter response is about the same in the jammer direction as in the pass band (except for the signal strength).

It is obvious from Figure 13.9, that for a jammer direction of $30°$ there is a strong dependence between apparent jammer direction and frequency. Accordingly, we have to expect some distortion of the signal. This is illustrated in Figure 13.13. The upper curve is already outside the jammer notch and, hence, more or less undistorted. The other curves, however, show significant distortion. Especially right in the jammer direction (o) the first sidelobes are raised almost to the level of the main lobe which means a significant loss in resolution.

The effect of anti-jammer filtering on SAR resolution has been analysed by ENDER [130] using AER II data. Moreover, the author presents a technique to compensate in part for the distortion of the point spreading function in the range dimension.

13.3 Summary

Some conclusions can be summarised as follows

1. **Two related problems**. It was demonstrated that two different radar applications (space-time clutter suppression for ISAR with a moving radar, space-TIME

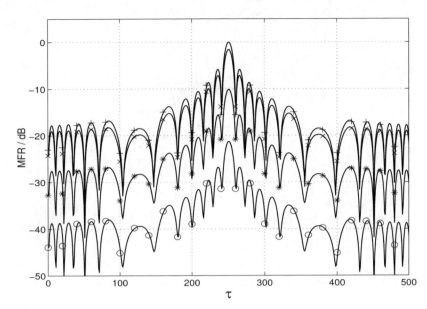

Figure 13.12: Distortion of matched filter response close to jammer location ($N = 24$, $L = 5$, $r_{90} = 1$, $J = 2$, $\varphi_1 = 90°$, $\varphi_2 = 30°$, equal power, L=5; $d/\lambda = 0.4$): ○ $\varphi_L = 90°$; ∗ $\varphi_L = 93°$; × $\varphi_L = 95°$; + $\varphi_L = 105°$

jammer suppression for SAR) have several aspects in common. While the received echo sequence of an ISAR target is approximately a linear FM chirp the same kind of waveform is commonly used to obtain range resolution in SAR systems. While clutter received by a moving radar has a certain bandwidth determined by the platform speed the jammer bandwidth in SAR is given by the system bandwidth. In both cases the space-time (space-TIME) covariance function matrix has elements that are based on a $\frac{\sin x}{x}$ function where x includes both the spatial and temporal dimensions.

2. **Receiver noise**. In the case of space-time processing (ISAR) the receiver noise is white in space and time. For jammer cancellation in SAR via space-TIME processing the noise is correlated by the system bandwidth. This, however, can not be exploited in practice because the signal has the same spectral shape as the noise.

3. **Effect of filtering**. In both applications the resolution of the ISAR (SAR) imaging system may be degraded by the space-time (space-TIME) filter for interference suppression. Broadening of the mainlobe of the point spreading function as well as a rise of sidelobes are typical effects. This happens particularly close to the filter notches in Doppler (azimuth) where the filter transfer characteristics are not constant with Doppler (azimuth). The influence of clutter (jammer) notches can be reduced by using a large array so as to keep clutter (jammer) notches narrow.

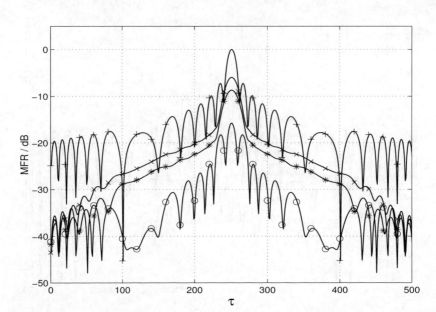

Figure 13.13: Distortion of matched filter response close to jammer location ($N = 24$, $L = 5$, $r_{90} = 1$, $J = 2$, $\varphi_1 = 90°$, $\varphi_2 = 30°$, equal power, $L = 5$; $d/\lambda = 0.4$): ○ $\varphi_L = 30°$; * $\varphi_L = 37°$; × $\varphi_L = 40°$; + $\varphi_L = 60°$

4. **Resolution**. Sensitivity to filtering increases with higher SAR (ISAR) resolution.

5. **Space-time-TIME processing**. ISAR imaging in both azimuth and range requires a broadband array. Then clutter suppression will have to be carried out by space-time-TIME processing. This may cause some distortions of the point spreading function in the azimuth and range dimension.

6. **Data tapering**. SAR and ISAR data are usually tapered to suppress the azimuth sidelobes of the real and synthetic antenna and the range sidelobes of the matched filter. By sidelobe rejection before generating the image the superposition of sidelobe based spurious image components is avoided. Tapering takes place in the spatial (array) and in both temporal (time, TIME) dimensions.

It has been shown by NICKEL [489] that adaptive processing tends to negate the effect of tapering on the sidelobes of an array antenna because the adaptive filter has the property of normalising the noise power across the array. This can be avoided by using a processor as shown in Figure 8.4. This processor includes a normalisation in such a way that all antenna channels have the same noise power. A comparison of beampatterns with and without normalisation was given in Figures 8.9 and 8.10.

Chapter 14

Target parameter estimation

The major part of this book is concerned with the *detection* of slow moving targets by a moving radar. Specifically, the suppression of clutter returns by means of space-time filter techniques is discussed in detail. A prerequisite for space-time techniques is a multichannel phased array antenna.

A major advantage of a phased array radar is the capability of multifunction operation. Multifunction operation means in essence that both search and tracking of targets can be fulfilled simultaneously.

The search function includes all kinds of signal processing tools, such as clutter or jammer reduction, target detection, and estimation of target angle and velocity. The accuracies of direction and velocity estimates are essential inputs for tracking algorithms. In this chapter we discuss the problem of estimating parameters of targets buried in heavy clutter. As before we assume that the target's radial velocity falls into the clutter bandwidth as determined by the radar platform velocity.

It is well known from spatial applications (parameter estimation by use of an array radar in the presence of interference) that efficient interference suppression is an essential part of the estimation procedure (NICKEL [488, 494, 491, 497]). In this section we investigate how far estimation of target velocity and azimuth is possible under heavy clutter conditions. Some results by WARD [693] indicate that good parameter estimates can be achieved by use of optimum space-time processing. In this section we continue this work by analysing the problems in some more detail. The impact of various radar parameters as well as the performance of suboptimum processors will be considered.

14.1 CRB for space-time ML estimation

In this section the Cramér–Rao bounds (CRB) for maximum likelihood (ML) estimation of the azimuth and radial velocity target parameters are briefly derived. For the optimum STAP processor, the CRB of target parameter estimates have been calculated by WARD [693]. The CRB has proven to be a powerful tool in comparative analyses of array architectures and STAP performance (SHOWMAN *et al.* [607]).

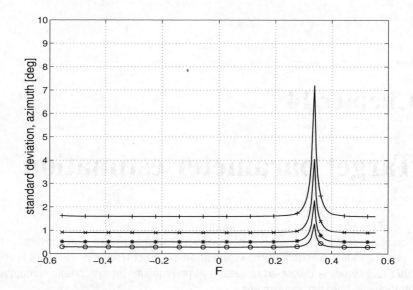

Figure 14.1: Effect of signal-to-noise-ratio on azimuth estimation: SNR = ○ 10 dB; ∗ 0 dB; × −10 dB; + −20 dB

Here we follow his derivation closely. Other publications on parameter estimation include routines for efficient azimuth and velocity estimation in clutter (WARD [694] and WARD and HATKE [696]. LEGG and GRAY [387] calculate the CRB for moving SAR targets in noise. SAR detection and imaging of moving targets has also been addressed by PERRY *et al.* [534]. DOGANDZIC and NEHORAI [117] derive the CRBs for estimation of target range, velocity and direction for active radar or sonar arrays.

14.1.1 The principle

We assume that the radar receives space-time data

$$\mathbf{x} = A\mathbf{s}(\varphi_t, v_{\mathrm{rad}}) + \mathbf{q} \tag{14.1}$$

where $\mathbf{q}$ is the interference component which may include clutter, jammers, and noise. $\mathbf{q}$ is characterised by the space-time covariance matrix $\mathbf{Q} = E\{\mathbf{q}\mathbf{q}^*\}$. We want to estimate the parameters φ_t and v_{rad} of a moving target whose radar response is given by the space-time vector $A\mathbf{s}(\varphi_t, v_{\mathrm{rad}})$.

The received data vector is assumed to obey a multivariate Gaussian p.d.f.

$$p(\mathbf{x} \mid \varphi, v, A) = \frac{1}{\pi^{NM} \det \mathbf{Q}} e^{-(\mathbf{x}-A\mathbf{s})\mathbf{Q}^{-1}(\mathbf{x}-A\mathbf{s})} \tag{14.2}$$

Let us define a vector of unknown real parameters

$$\theta = (\varphi \ v \ a \ b)^T \tag{14.3}$$

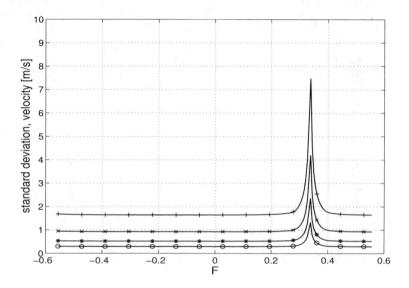

Figure 14.2: Effect of signal-to-noise-ratio on velocity estimation: SNR = ○ 10 dB; ∗ 0 dB; × −10 dB; + −20 dB

where a and b denote the amplitude and phase of the signal, i.e., $A = ae^{jb}$. Basically we are interested in direction and velocity estimates. Since, however, the amplitude and phase of the target echo are unknown they have to be included in the estimation process (so-called nuisance parameters).

For a given vector of received data $\mathbf{x}$, the maximum likelihood estimate of the parameters is given by

$$\hat{\theta} = \underset{\theta}{\operatorname{argmax}}\, p(\mathbf{x} \mid \theta) \qquad (14.4)$$

This is equivalent to maximising the log-likelihood function

$$l(\theta) = -(\mathbf{x} - A\mathbf{s}(\varphi, v))^* \mathbf{Q}^{-1}(\mathbf{x} - A\mathbf{s}(\varphi, v)) \qquad (14.5)$$

Maximising (14.5) with respect to the nuisance parameters a and b yields

$$(\hat{\varphi}\,\hat{v}) = \underset{(\varphi\, v)}{\operatorname{argmax}}\, \frac{\mathbf{x}^*\mathbf{Q}^{-1}\mathbf{s}\mathbf{s}^*\mathbf{Q}^{-1}\mathbf{x}}{\mathbf{s}^*\mathbf{Q}^{-1}\mathbf{s}} \qquad (14.6)$$

The ML estimate is given by the position of the maximum of the angle-Doppler scan pattern

$$(\hat{\varphi}\hat{v}) = \underset{(\varphi\, v)}{\operatorname{argmax}}\, |\mathbf{w}^*(\varphi, v)\mathbf{x}|^2 \qquad (14.7)$$

with an adaptive weight vector (see for instance NICKEL [488, 497])

$$\mathbf{w}(\varphi, v) = \frac{\mathbf{Q}^{-1}\mathbf{s}(\varphi, v)}{\sqrt{\mathbf{s}^*(\varphi, v)\mathbf{Q}^{-1}\mathbf{s}(\varphi, v)}} \qquad (14.8)$$

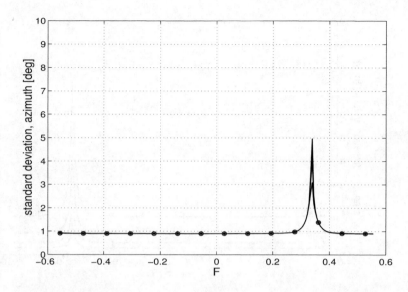

Figure 14.3: Effect of clutter-to-noise ratio: CNR = ○ 10 dB; ∗ 20 dB; × 30 dB; + 40 dB

14.1.2 The Cramér–Rao bound

The error covariance matrix of any unbiased estimator $\hat{\theta}$ is bound by the Cramér–Rao bound (CRB) which is given by the inverse of the Fisher information matrix (FIM) **J** (MARDIA *et al.* [436, p. 96])

$$E\{(\hat{\theta}-\theta)(\hat{\theta}-\theta)^*\} \geq \mathbf{J}^{-1} \qquad (14.9)$$

The elements of the Fisher information matrix can be obtained from the log-likelihood ratio by differentiation with respect to the individual parameters and taking the expectation

$$j_{ik} = -E\left\{\frac{\partial^2 l}{\partial \theta_i \partial \theta_k}\right\} \qquad (14.10)$$

Let us define the following vectors incuding the derivatives with respect to the individual parameters

$$\mathbf{s}_\varphi = \frac{\partial \mathbf{s}}{\partial \varphi}; \quad \mathbf{s}_v = \frac{\partial \mathbf{s}}{\partial v} \qquad (14.11)$$

Recall that the elements of the signal reference vector **s** are exponential terms as given by (2.31).

$$A\mathbf{s} = \begin{pmatrix} \vdots \\ A\exp[j\tfrac{2\pi}{\lambda}(2v_{\mathrm{rad}}mT + (x_i\cos\varphi + y_i\sin\varphi)\cos\theta - z_i\sin\theta)] \\ \vdots \end{pmatrix} \qquad (14.12)$$

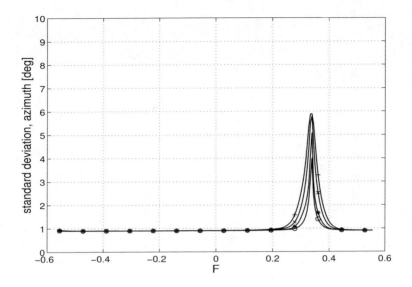

Figure 14.4: Effect of clutter fluctuations: $B_c =$ ∘ 0%; $*$ 1%; $\times$ 5%; $+$ 30%

It can easily be verified that for the special form of vectors given by (14.12) the elements of the Fisher information matrix (14.10) reduce to

$$j_{ik} = 2\Re \frac{\partial A\mathbf{s}^*}{\partial \theta_i} \mathbf{Q}^{-1} \frac{\partial A\mathbf{s}}{\partial \theta_k} \quad (14.13)$$

Then the Fisher information matrix has the form (WARD [693])

$$\mathbf{J} = 2 \begin{pmatrix} a^2\delta_\varphi & a^2\beta_x & a\beta_\varphi & a^2\gamma_\varphi \\ a^2\beta_x & a^2\delta_v & a\beta_v & a^2\gamma_v \\ a\beta_\varphi & a\beta_v & \xi & 0 \\ a^2\gamma_\varphi & a^2\gamma_v & 0 & a^2\xi \end{pmatrix} \quad (14.14)$$

with the definitions

$$\begin{aligned} \xi &= \mathbf{s}^*\mathbf{Q}^{-1}\mathbf{s} \\ \delta_\varphi &= \mathbf{s}_\varphi^*\mathbf{Q}^{-1}\mathbf{s}_\varphi \\ \delta_v &= \mathbf{s}_v^*\mathbf{Q}^{-1}\mathbf{s}_v \\ \beta_x + j\gamma_x &= \mathbf{s}_\varphi^*\mathbf{Q}^{-1}\mathbf{s}_v \\ \beta_\varphi + j\gamma_\varphi &= \mathbf{s}^*\mathbf{Q}^{-1}\mathbf{s}_\varphi \\ \beta_v + j\gamma_v &= \mathbf{s}^*\mathbf{Q}^{-1}\mathbf{s}_v \end{aligned} \quad (14.15)$$

The Cramér–Rao bounds for the individual parameters are given by the main diagonal elements of $\mathbf{J}^{-1}$ in (14.14).

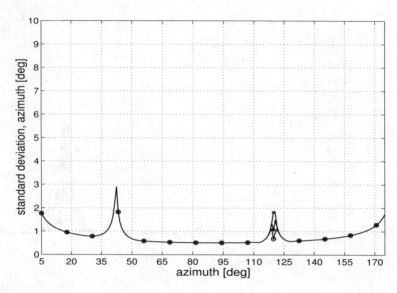

Figure 14.5: *Effect of noise jamming (SL, $v_t = 63.6$ m/s): JNR = ○ 10 dB; ∗ 20 dB; × 30 dB; + 40 dB*

14.1.3 Some properties of the Cramér–Rao bound

14.1.3.1 *Relation with the monopulse principle*

The spatial derivative of the steering vector becomes for a sidelooking linear array (for the geometry see Figure 2.1)

$$\mathbf{s}_\varphi = \begin{pmatrix} \alpha x_1 & & & & & & \\ & \ddots & & & & 0 & \\ & & \alpha x_N & & & & \\ & & & \alpha x_1 & & & \\ & & & & \ddots & & \\ & & & & & \alpha x_N & \\ & & & & & & \ddots \\ & 0 & & & & & \alpha x_1 \\ & & & & & & & \ddots \\ & & & & & & & & \alpha x_N \end{pmatrix} \mathbf{s} \quad (14.16)$$

where

$$\alpha = -\mathrm{j}\frac{2\pi}{\lambda}\sin\varphi\cos\theta \quad (14.17)$$

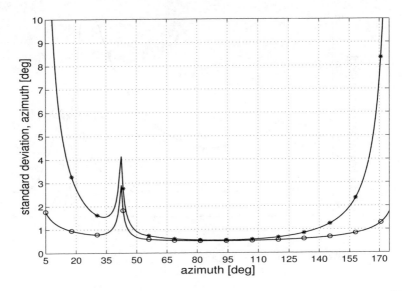

Figure 14.6: Effect of sensor directivity patterns (azimuth estimation, SL, $v_\mathrm{t} = 63.6$ m/s): ○ omnidirectional sensor patterns; ∗ directive sensor patterns

and x_n denote the positions of the array elements along the flight axis.

The derivative with respect to the target velocity becomes, for a sidelooking linear array,

$$\mathbf{s}_v = \begin{pmatrix} \beta & & & & & & 0 \\ & \ddots & & & & & \\ & & \beta & & & & \\ & & & 2\beta & & & \\ & & & & 2\beta & & \\ & & & & & \ddots & \\ & & & & & & \\ 0 & & & & & M\beta & \\ & & & & & & \ddots \\ & & & & & & M\beta \end{pmatrix} \mathbf{s} \qquad (14.18)$$

where

$$\beta = \mathrm{j}\frac{4\pi}{\lambda}T \qquad (14.19)$$

Replacing x by y in (14.16) and

$$\alpha = \mathrm{j}\frac{2\pi}{\lambda}\cos\varphi\cos\theta \qquad (14.20)$$

432 Target parameter estimation

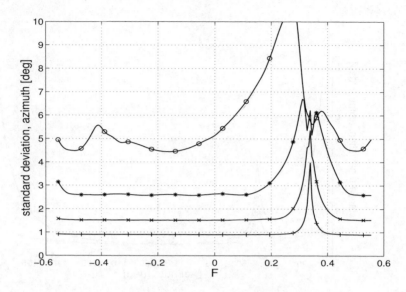

Figure 14.7: Impact of the array aperture size (azimuth estimation): $N = \circ\ 3;\ *\ 6;\ \times\ 12;\ +\ 24$

(14.16) becomes the spatial derivative for the forward looking array while the velocity derivative is the same as in (14.18).

As can be seen both of the derivative vectors include the original steering vector, multiplied by a diagonal space-time matrix which contains a linear weighting function in either space or time. As is well known such a linear weighting forms a difference pattern, in our application in the spatial as well as in the temporal dimension. Difference patterns are commonly used for high accuracy angle estimation in tracking radars. Some details on difference patterns in conjunction with space-time processing are given in Chapter 8.

The Hermitian or bilinear expressions occurring in the Fisher information matrix can be written as (WARD [693])

$$\mathbf{s}_\varphi^* \mathbf{Q}^{-1} \mathbf{s}_\varphi = \mathbf{s}_\varphi^* \mathbf{Q}^{-1} \mathbf{Q} \mathbf{Q}^{-1} \mathbf{s}_\varphi = \mathbf{g}_\varphi^* \mathbf{Q} \mathbf{g}_\varphi \qquad (14.21)$$

$$\mathbf{s}_v^* \mathbf{Q}^{-1} \mathbf{s}_v = \mathbf{s}_v^* \mathbf{Q}^{-1} \mathbf{Q} \mathbf{Q}^{-1} \mathbf{s}_v = \mathbf{g}_v^* \mathbf{Q} \mathbf{g}_v \qquad (14.22)$$

$$\mathbf{s}_v^* \mathbf{Q}^{-1} \mathbf{s}_\varphi = \mathbf{s}_v^* \mathbf{Q}^{-1} \mathbf{Q} \mathbf{Q}^{-1} \mathbf{s}_\varphi = \mathbf{g}_v^* \mathbf{Q} \mathbf{g}_\varphi \qquad (14.23)$$

In (14.21) we find the clutter power response of an adaptive beamformer (i.e., with clutter rejection) using a difference weighting as defined by (14.16). The same holds for the temporal dimension (14.22) where we find temporal difference weightings.

That means that the difference weighting vector $\mathbf{s}_\varphi$ is orthogonal to all clutter components contained in $\mathbf{Q}$. Comparing the expressions on the right side of (14.21), (14.22) with (1.16) we find that (14.21), (14.22) denote the SCNR at the output of the optimum processor, but with the steering vector being replaced by a space-time

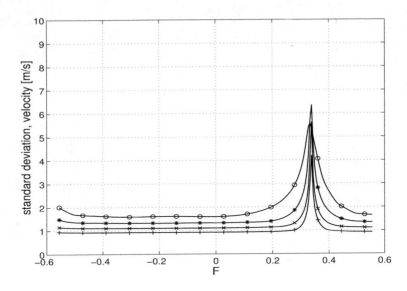

Figure 14.8: Impact of the array aperture size (velocity estimation): $N = \circ\ 3;\ *\ 6;\ \times\ 12;\ +\ 24$

difference weighting, i.e., a weighting which is orthogonal to the expected signal. Therefore, the product $\mathbf{g}_\varphi^* \mathbf{Q} \mathbf{g}_\varphi$ produces the maximum SNIR for a signal $\mathbf{s}_\varphi$.

Another view is as follows. Let us consider for simplicity the case of one parameter only, for example azimuth. Then the Fisher information matrix becomes a scalar

$$\mathbf{J} = 2A^2 \mathbf{s}_\varphi^* \mathbf{Q}^{-1} \mathbf{s}_\varphi \tag{14.24}$$

Let us consider the covariance matrix of target and clutter

$$\mathbf{R} = A^2 \mathbf{s}\mathbf{s}^* + \mathbf{Q} \tag{14.25}$$

Because of the orthogonality between sum and difference vector $\mathbf{s}^* \mathbf{s}_\varphi = 0$ one obtains

$$\mathbf{s}_\varphi^* \mathbf{R}^{-1} \mathbf{s}_\varphi = \mathbf{s}_\varphi^* \mathbf{Q}^{-1} \mathbf{s}_\varphi \tag{14.26}$$

Recall that the inverse of (14.24)

$$\mathbf{J}^{-1}(\varphi, v) = (\mathbf{s}_\varphi^*(\varphi, v) \mathbf{Q}^{-1} \mathbf{s}_\varphi(\varphi, v))^{-1} \tag{14.27}$$

has the form of the minimum variance (MV) spectral estimator given by (1.104), (3.67)[1] which minimises the output power under the constraint

$$\mathbf{w}^* \mathbf{s} = \text{constant} \tag{14.28}$$

---

[1] The MV estimator has been used to calculate clutter spectra in Chapter 3 (see Figures 3.18, 3.19, etc.).

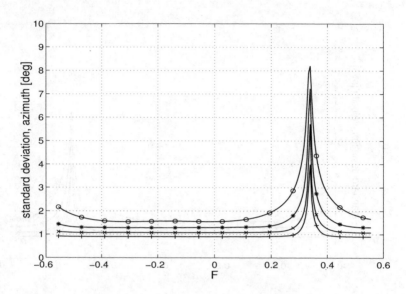

Figure 14.9: Impact of coherent processing interval (azimuth estimation): $M = \circ\ 3$; $*\ 6$; $\times\ 12$; $+\ 24$

given an input process characterised by the covariance matrix $\mathbf{Q}$ (HUDSON [275, p. 70]). The constraint can be satisfied by means of a difference pattern as used in monopulse angle estimation (see Chapter 8, Figures 8.14–8.18) because

$$\mathbf{s}_\varphi^* \mathbf{s} = 0 \tag{14.29}$$

The spatial difference vector $\mathbf{s}_\varphi$ was defined in (14.16). We conclude that the Cramér–Rao bound is given by the power output of a minimum variance power estimator using a difference weighting which is perfectly matched to the signal, divided by twice the signal power. The CRB can be reached by varying the parameter to be estimated until the output power of an optimally adapted processor with difference steering vector becomes a minimum.

The above considerations also hold for velocity estimation. Then the constraint becomes

$$\mathbf{s}_v^* \mathbf{s} = 0 \tag{14.30}$$

where $\mathbf{s}_v$ is a temporal difference vector as defined in (14.18).

14.1.3.2 Validity of the CRB

The Cramér–Rao bound is valid under the following conditions

1. The estimator must be unbiased (MARDIA *et al.* [436, p. 96]). This is fulfilled if the noise background is white (receiver noise only). In the case of correlated interference (jammer, clutter) the interference suppression may

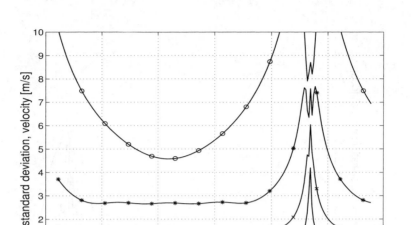

Figure 14.10: Impact of coherent processing interval (velocity estimation): $M = \circ\ 3;\ *\ 6;\ \times\ 12;\ +\ 24$

lead to deformations of the sum and difference patterns so that even the ML estimator will yield biased estimates. NICKEL [494, 497] has shown that a jammer radiating into one of the difference lobes of a monopulse antenna can cause severe off-sets of the angle estimates.

On the other hand, we have shown in Chapter 8 that space-time clutter suppression does not cause severe distortions of the difference pattern (see Figures 8.13–8.18). This has mainly to do with the fact that the clutter energy is spread out in space while a jammer is concentrated in one discrete direction. We can, therefore, assume that in our application the bias of parameter estimates is sufficiently small.

2. The estimator must be a linear function of the data from which the parameter is to be estimated (MARDIA *et al.* [436, pp. 99]). This is not fulfilled in the case of velocity and angle estimation. Both parameters are connected via non-linear functions (sine, cosine) with the phases of the measured complex signal samples. Therefore, the CRB is valid only for small variances so that the dependency on the parameter is approximately linear.

3. As a consequence of the conditions already detailed the signal-to-interference + noise ratio has to be sufficiently large. Then the covariance becomes so small that the estimator operates in an approximately linear environment. In the subsequent examples the SNRs at the single element and echo sample have been chosen so that the SNR at the beam output is sufficiently large (≥ 8 dB).

The requirement for high SNIR also implies that the interference suppression

436 Target parameter estimation

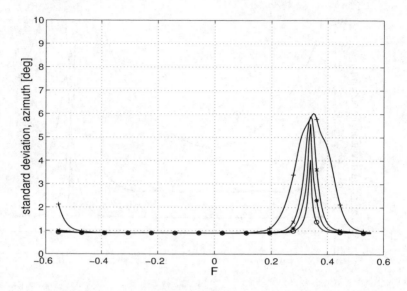

Figure 14.11: Effect of system bandwidth (FL): SNR = ○ $B_s = 0\%$; ∗ $B_s = 1\%$; × $B_s = 5\%$; + $B_s = 30\%$

inherent in $\mathbf{Q}^{-1}$ functions perfectly. This is guaranteed as long as the order of $\mathbf{Q}$ is larger than the number of clutter + jammer eigenvalues. If this condition is not satisfied the Cramér–Rao bound will produce meaningless results.

14.2 Impact of radar parameters on the CRB

The effect of various radar parameters on the accuracy of velocity and azimuth estimation is analysed in this section. This analysis will be carried out on the basis of the optimum processor (1.3). Some results on the Cramér–Rao bound for the optimum processor have been published by WARD [693].

By assuming the optimum processor we get some insight into the effect of the different radar parameters on the estimation accuracy, independently of the properties of the individual processor used. The optimum (fully adaptive) processor achieves the lowest Cramér–Rao bound possible. The transform techniques discussed in Section 14.3 are suboptimum. Therefore their performance is expected to be suboptimum.

The different parameters can be subdivided into those which have to do with the environment of the radar (Section 14.2.1) and others which are inherent in the radar system (Section 14.2.2). In the present analysis the Cramér–Rao bound curves for velocity estimation frequently look very similar to those for azimuth estimation. In these cases they will be omitted for the sake of brevity. A signal-to-noise ratio of -10 dB (single array element, single echo) was assumed for all numerical examples in this section. In most cases we assumed a sidelooking linear array.

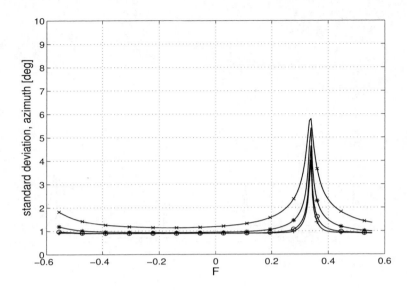

Figure 14.12: Influence of range walk: SNR = ∘ $\Delta_R = 100$ m; ∗ $\Delta_R = 10$ m; × $\Delta_R = 1$ m; + $\Delta_R = 0$ m

14.2.1 Environmental effects

14.2.1.1 Signal-to-noise ratio

Figure 14.1 shows the Cramér–Rao bound versus the normalised target Doppler frequency for azimuth estimation and Figure 14.2 for velocity estimation. The four curves have been calculated for different values of the SNR (signal-to-receiver noise ratio at the single antenna element and the single pulse). A comparison with Figure 14.24 confirms that the peak in standard deviation appears at the same Doppler frequency as the notch of the adaptive clutter filter. The flat parts of the curves reflect the Doppler sidelobe area. The higher the SNR is, the lower is the Cramér–Rao bound.

14.2.1.2 Clutter-to-noise ratio

In Figure 14.3 the clutter-to-noise ratio has been varied while the SNR is constant (SNR = -10 dB). We notice that the Cramér–Rao bound is almost independent of the clutter-to-noise ratio. There are only slight variations in the clutter peak region.

The optimum processor includes a clutter suppression filter ($\mathbf{Q}^{-\frac{1}{2}}$) which reduces the clutter portion down to the white noise level, regardless of the clutter-to-noise ratio. Recall that the clutter suppression filter ($\mathbf{Q}^{-\frac{1}{2}}$) forms a clutter trench in the Doppler-azimuth plane whose depth depends on the clutter-to-noise ratio. Examples for the signal response in azimuth and Doppler were given in Figures 4.2 and 4.3. This explains why the Cramér–Rao bound for the optimum processor is independent of the clutter-to-noise ratio.

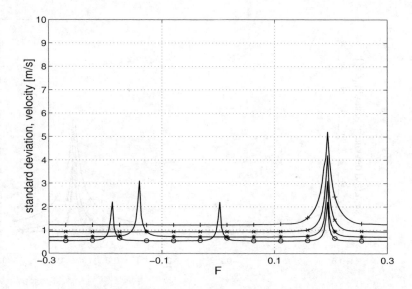

Figure 14.13: Influence of PRF: PRF = ∘ *4 KHz;* ∗ *7 KHz;* × *12 KHz;* + *21 KHz*

14.2.1.3 Clutter fluctuations

Clutter fluctuations due to internal motion of the backscattering scene (vegetation, weather or sea under windy conditions) cause a broadening of the clutter spectrum (Figure 3.32). This results in a broadening of the clutter notch (Figure 4.18) while outside the notch the optimum improvement factor is obtained.

This effect is reflected in Figure 14.4. Notice that the clutter peak is broadened while in the noise limited area the estimation error is independent of the amount of clutter fluctuation.

14.2.1.4 Additional jamming

Figure 14.5 shows the effect of one additional jammer for different jammer-to-noise ratios. In order to make the jammer visible (the jammer is discrete in azimuth, but white in Doppler, see the explanations in the introduction to Chapter 11) we plotted here the Cramér–Rao bound versus azimuth.

As in Figure 14.3 the jammer-to-noise ratio has almost no influence on the estimation performance of the optimum processor. There are some marginal differences in the clutter peak region.

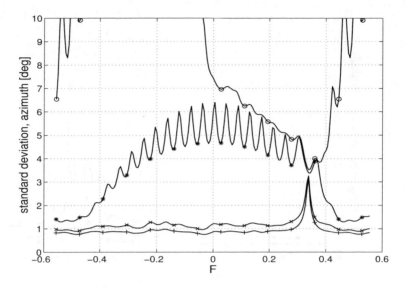

Figure 14.14: CRB for azimuth estimate (eigenchannel processor, SNR = −10 dB): C = ∘ 24; ∗ 48; × 96; + 144

14.2.2 Impact of system parameters

14.2.2.1 *Sensor directivity patterns*

The influence of the directivity patterns on the accuracy of azimuth estimation can be seen in Figure 14.6. In order to make the effect of sensor patterns visible we plotted once more the Cramér–Rao bound versus azimuth. The models defined by (2.20), (2.21) have been used.

We notice in both cases the clutter peak at about $47°$. It can be seen that the Cramér–Rao bound approaches infinity at $\varphi = 0°$ and $\varphi = 180°$. This happens because at these angles the sensor patterns have a null. Therefore, the signal amplitude a in (14.14) becomes zero in these directions so that the Fisher information matrix becomes singular. There is no such effect when using omnidirectional sensors. However, a slight increase in the azimuth standard deviation can be noticed on both sides. This reflects the well-known property of arrays that the angular estimation accuracy is better at the boresight than at the endfire direction. The CRB curves for velocity estimation look about the same, however, with the increase in both endfire directions. The velocity resolution of a Doppler filter bank (e.g., discrete Fourier transform) is normally constant over all the frequency range.

14.2.2.2 *Array aperture*

It can be expected that the capability of estimating target parameters increases with the antenna aperture size. In Figures 14.7 and 14.8 CRBs for various values of the

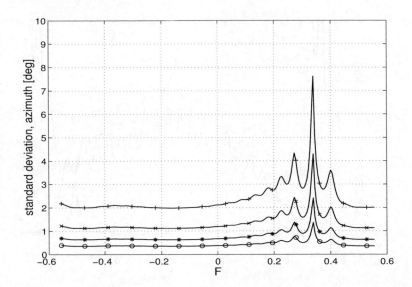

*Figure 14.15: CRB for azimuth estimate (symmetric auxiliary sensor processor, $K = 5$): SNR = ○ 10 dB; * 0 dB; × −10 dB; + −20 dB*

aperture size ($N = 3, 6, 12, 24$) are shown. As can be seen from Figure 14.7 the CRB for azimuth estimation depends heavily on the aperture size. For velocity estimation the influence is much smaller.

There are two effects responsible for the achievable estimation accuracy. On the one hand the accuracy of angle estimation depends on the resolution capability of the array, in other words, on the array size. On the other hand, decreasing the array size means also decreasing the output signal-to-noise ratio. This is the reason why we obtain different curves for velocity estimation.

14.2.2.3 Coherent processing interval

In Figures 14.9 and 14.10 we find CRB curves for different sizes of the coherent processing interval M ('temporal aperture'). Looking at Figures 14.9 and 14.10 we find that the curves are very similar to Figures 14.7 and 14.8, but with azimuth and velocity interchanged. Now we notice that the azimuth estimation accuracy depends on the size of the coherent processing interval.

14.2.2.4 System bandwidth

The impact of spatial decorrelation caused by travel delays of incoming waves across a broadband array has been discussed in Sections 3.4.3.3 and 4.4.5.3. A model was given in Section 2.5.2. It was shown that this spatial decorrelation effect does not occur for sidelooking arrays. However, for all other array configurations, especially

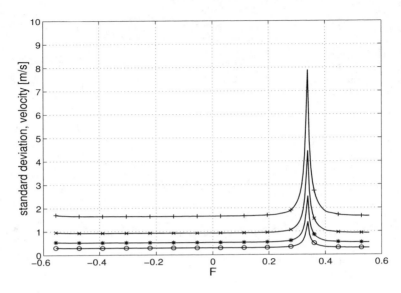

*Figure 14.16: CRB for velocity estimate (symmetric auxiliary sensor processor, $K = 5$): SNR = ○ 10 dB; * 0 dB; × −10 dB; + −20 dB*

forward looking arrays, the system bandwidth may cause a considerable broadening of the clutter notch and, hence, a degradation in the detection of slow moving targets.

In Figure 14.11 the influence of the system bandwidth on the CRB is illustrated. As can be seen the clutter peak is broadened which indicates a degraded estimation accuracy for azimuth estimation. For velocity estimation similar curves are obtained.

SORELIUS et al. [613] analyse the effect of system bandwidth on the estimation of the direction of arrival using the MUSIC, ESPRIT and WSF (weighted subspace fitting) methods. It was found that in difficult situations (closely spaced sources, large differences between source powers) the impact of small system bandwidth is significant.

14.2.2.5 Range walk

A model for the effect of range walk has been given in Section 2.5.1.2. Range walk is a temporal decorrelation effect which occurs particularly in broadband radar, i.e., radar with high range resolution. As has been shown range walk may cause a broadening of the clutter notch (Figures 4.19 and 4.20). It is shown in Figure 14.12 that the effect of range walk on angle or velocity estimation is similar to the effects of clutter motion or system bandwidth: the clutter notch is broadened and so is the clutter peak in the CRB plot.

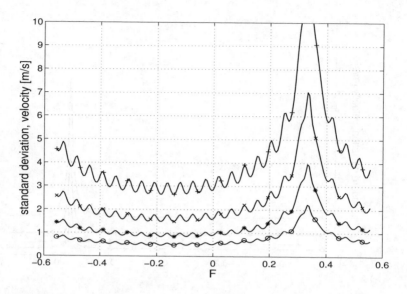

Figure 14.17: CRB for velocity estimate (beamformer and optimum temporal filter): SNR = ∘ 10 dB; ∗ 0 dB; × −10 dB; + −20 dB

14.2.2.6 Sampling effects

In Figure 14.13 the CRB for velocity estimation has been calculated for different pulse repetition frequencies (PRF). We can notice two different effects. First, with lower PRF the clutter response shows ambiguities which can be recognised as additional clutter peaks. Secondly, the CRB is different for different PRF outside the clutter region. This is due to the fact that with varying PRF the duration of the coherent processing interval is varied accordingly if the number of pulses is constant. It is obvious that the estimation accuracy depends on the duration of observation ('temporal aperture'). Under the same conditions the CRB curves for azimuth estimation coincide outside the clutter peak region.

Similar behaviour can be expected if the spacing between array elements is varied. In this case the same effects can be observed, but with azimuth and velocity interchanged. Variation of the element spacing leads to different CRB values in the clutter-free zone while the CRB curves for velocity estimation are independent of the array spacing.

14.3 Order reducing transform processors

It has been shown in Chapters 5, 6 and 9 that the dimension of the signal vector space can be reduced considerably by means of a suitable linear transform **T**. By 'suitable' it is understood that the clutter rejection performance in terms of the improvement factor (IF) of such a reduced order processor approximates the optimum processor (1.3) to a

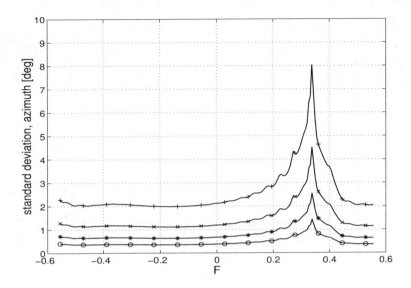

*Figure 14.18: CRB for azimuth estimate (ASEP, $K = 5$, $L = 5$): SNR = ○ 10 dB; * 0 dB; × −10 dB; + −20 dB*

large extent. The parameter estimation capability of various processor types has been discussed in KLEMM [332]. LOMBARDO and COLONE [416] propose a suboptimum STAP processor which includes both detection and angle estimation capability. The angle estimation part is based on a split array aperture.

As in Chapters 5, 6 and 9 the transformed data are defined by

$$\mathbf{s}_T = \mathbf{T}^*\mathbf{s}; \quad \mathbf{Q}_T = \mathbf{T}^*\mathbf{Q}\mathbf{T}; \quad \mathbf{x}_T = \mathbf{T}^*\mathbf{x}; \quad \mathbf{q}_T = \mathbf{T}^*\mathbf{q} \quad (14.31)$$

The derivatives of the steering vector become, after transform,

$$\mathbf{s}_{T\varphi} = \mathbf{T}^*\mathbf{s}_\varphi; \quad \mathbf{s}_{Tv} = \mathbf{T}^*\mathbf{s}_v \quad (14.32)$$

The Cramér–Rao bound can easily be formulated by substituting the above transformed quantities in (14.15) and (14.14). Notice that the subscript $_T$ denotes transformed quantities. With the above definitions the estimation performance of a wide class of STAP processors can be analysed.

In this section we present a number of numerical results which are to illustrate the angle and velocity estimation performance of a few transform type STAP processors. The numerical calculations are based on the parameters listed in Table 2.1.

14.3.1 Conventional processing

14.3.1.1 *Beamforming and adaptive temporal MTI*

This kind of adaptive MTI is used in stationary radar. Clutter cancellation takes place only in the temporal dimension. As a consequence, some velocity estimation is possible

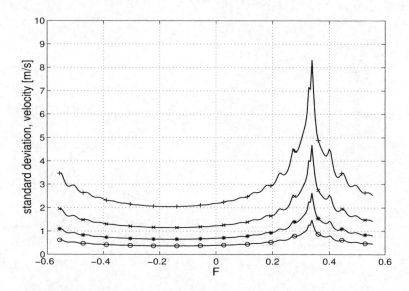

*Figure 14.19: CRB for velocity estimate (ASEP, $K = 5$, $L = 5$): SNR = ○ 10 dB; * 0 dB; × −10 dB; + −20 dB*

(Figure 14.17). For angle estimation the Cramér–Rao bound explodes.

14.3.1.2 Beamforming and Doppler filter cascaded

In this case no measures against clutter have been taken, and the detectability of targets is strongly degraded as can be concluded from the lowest curve in Figure 14.24. It turns out that the FIM becomes close to singular and the Cramér–Rao bound approaches infinity. No angle and velocity estimation can be done before rejecting the clutter and jamming.

14.3.2 Space-time transforms

Space-time transform processors have been treated in Chapter 5, Sections 5.2 and 5.2.3. The transform matrix includes a space-time search channel (space-time signal replica) **s** and several other space-time channels which serve as interference reference, see (5.2).

There are basically two kinds of space-time transforms. Both of them have to be tailored so as to match the clutter subspace of **Q**. This means that the number of space-time channels must not be less than the number of clutter eigenvalues which is under certain ideal conditions

$$N_e = N + M - 1 \tag{14.33}$$

see (3.59) and KLEMM [300].

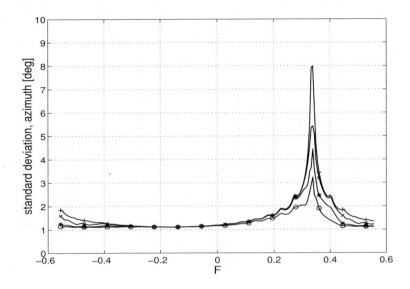

Figure 14.20: Effect of CNR (ASEP, K = 5, L = 5): CNR = ∘ 10 dB; ∗ 20 dB; × 30 dB; + 40 dB

14.3.2.1 Eigenvector transform

This technique is obtained if the submatrix $\mathbf{A}$ in (5.2) contains the first $N + M - 1$ clutter eigenvectors of $\mathbf{Q}$. As can be seen from Figure 14.24 (∗) the detection performance of the eigenvector transform processor is practically identical with the optimum processor (∘). Notice that the four curves in Figure 14.14 denote different numbers of channels. It can be noticed that for $C = 24$ channels according to (5.10) the calculation of the Cramér–Rao bound fails while for higher values of C reasonable Cramér–Rao bound curves are obtained.

For the chosen system dimensions $N = M = 24$ the minimum number of channels required for obtaining the optimum IF is $C = 48$. We notice, however, that in Figure 14.14 even for $C = 48$ poor performance is obtained. One can conclude that the minimum number of degrees of freedom according to (5.10) is sufficient for near-optimum detection. For angle or velocity estimation, however, a larger number is required.

This can be explained as follows. Suppose the number of channels of the eigenvector transform (5.10) is $N + M$ (in our example $C = 48$) and the eigenvectors associated with the $N + M - 1$ largest eigenvalues are used as auxiliary channels. That means that the number of auxiliary channels is just sufficient to cope with the degrees of freedom of the interference (DPCA conditions assumed). Under this assumption, the auxiliary eigenvectors form the interference subspace. For the noise subspace only one channel (the search channel) is left. A target signal can be detected only if it shows up in the noise subspace which is orthogonal to the interference subspace. The derivatives of a target signal vector, however, are orthogonal to the target signal vector. Therefore,

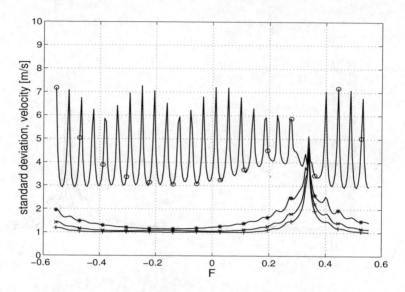

Figure 14.21: Effect of system dimensions on velocity estimation (ASEP): $K, L = \circ\ 3$; $*\ 5; \times\ 7; +\ 9$

they show up in the interference subspace and will be suppressed by the inverse of the covariance matrix. If the number of channels is increased beyond $N + M$ one obtains meaningful results for the CRB.

14.3.2.2 Auxiliary channel approach

In the case of the auxiliary channel processor the submatrix **A** in the space-time transform (5.2) includes a set of beamformers so as to cover the whole visible range with beams. Each of these beams is cascaded with a Doppler filter matched to the clutter Doppler in the beam direction. For the minimum number of degrees of freedom one obtains a meaningless Cramér–Rao bound curve like the ∘ curves in Figure 14.14. Reasonable performance may be achieved by increasing the number of channels which is not an attractive requirement.

We can conclude that space-time channel transform approaches require a considerably larger number of degrees of freedom for estimation than for detection, and hence, are not attractive solutions.

14.3.3 Spatial transforms

Spatial transforms have proven to be very effective for reducing the spatial dimension of a STAP system while preserving near-optimum clutter suppression performance. Typical transforms include subarray structures or sidelobe canceller type schemes (see Chapter 6). For equispaced linear (or more general: horizontally equispaced

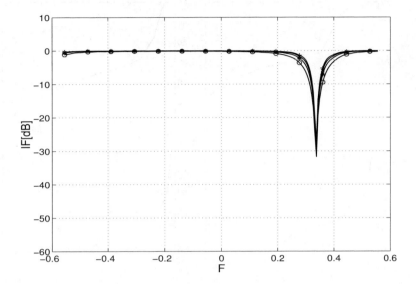

Figure 14.22: Effect of system dimensions on the improvement factor (ASEP): $K, L = \circ\ 3;\ *\ 5;\ \times\ 7;\ +\ 9$

cylindrical) arrays certain transforms yield near-optimum performance which is almost independent of the number of array channels K.

14.3.3.1 Symmetric auxiliary sensors

This concept (for details see Section 6.2.1) is equivalent to forming overlapping subarrays whose phase centres have the same spacing as the array elements (see Figure 6.8). Notice that the corresponding IF curve in Figure 14.24 ($\times$) coincides more or less with the optimum ($\circ$) curve.

The standard deviations of azimuth and velocity estimates are given in Figures 14.15 and 14.16. Comparing the results with those obtained with the optimum processor (Figures 14.1 and 14.2) we find that the standard deviations for velocity estimates are unchanged while a slight increase can be noticed for azimuth. This is the penalty for reducing the spatial dimension of the system.

14.3.4 Auxiliary sensor/echo processing (ASEP)

The concept of symmetric auxiliary elements may be applied also in the temporal dimension. This processor has been referred to as 'auxiliary sensor/echo processor' (ASEP), see Section 9.3. The space-time transform given by (9.7) reduces the dimensions of the signal vector space from NM down to KL with $K \ll N$ and $L \ll M$. The improvement factor (see + in Figure 14.24 and Figures 9.5, 9.6) is close to the optimum. The order of this processor has been reduced in both the spatial and the

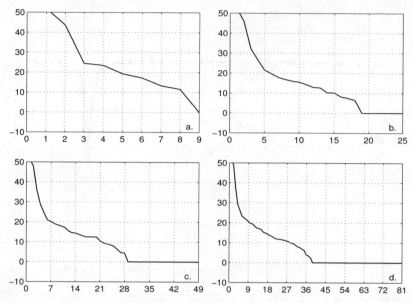

Figure 14.23: Effect of system dimensions on clutter eigenspectra (ASEP, eigenvalues in dB): $K, L =$ a. 3; b. 5; c. 7; d. 9

temporal dimension. Because of this reduction the ASEP is of particular importance for real-time on-board air- and spaceborne MTI systems. Therefore, we discuss its behaviour in more detail.[2]

The standard deviations for azimuth and velocity are shown in Figures 14.18 and 14.19, respectively. Once more we notice a certain degradation in estimation accuracy when comparing these results with the optimum (Figures 14.1 and 14.2). It should be noted, however, that the inversion of the 576×576 covariance matrix of the optimum processor requires about 12 000 times more operations than the inversion of the 25×25 matrix of the auxiliary sensor/echo processor.

In Figure 14.20 the response of the ASEP processor to varying clutter-to-noise ratio is shown. To some extent the curves are similar to those calculated for the optimum processor (Figure 14.3). There are some variations in the standard deviation in the clutter region (peak on the right side). In the clutter-free area the standard deviation is nearly independent of the CNR. Some small deviations are due to the fact that the ASEP is a suboptimum processor which does not provide optimum clutter rejection.

In Figure 14.21 the spatial and temporal dimensions K and L of the ASEP have been varied simultaneously. It can be noticed that from $K = L = 5$ upwards meaningful CRB curves are obtained. Notice that for $K = L = 3$ the CRB does not give a meaningful answer. Does this mean that the adaptive estimator has not sufficiently suppressed the clutter component? A look at the corresponding IF curves

---

[2] Recall from Chapter 7 that the space-time FIR filter in conjunction with the symmetric auxiliary sensor transform is another efficient technique. However, this technique cannot be formulated via a linear transform which means that no specific CRB can be calculated.

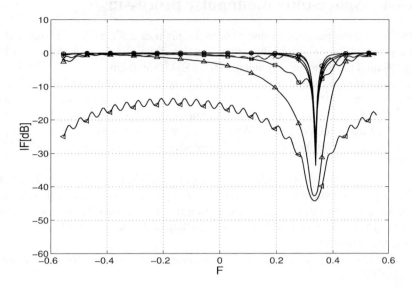

Figure 14.24: Improvement factor for various processors: ○ *optimum;* ∗ *eigenchannel;* × *symmetric auxiliary sensors;* + *auxiliary sensor/echo processing;* □ *asymmetric auxiliary sensors;* △ *beamformer and optimum temporal filter;* ◁ *beamformer and Doppler filter*

(14.22), however, reveals that this is not the case. Even for $K = L = 3$ almost optimum performance is obtained.

It is interesting to look at the associated eigenspectra (Figure 14.23). It can be noticed that the ASEP transform compresses the clutter part of the eigenspectrum. At the same time the ratio of the numbers of noise and clutter eigenvalues becomes smaller. Notice that for $K = L = 3$ there is only one noise eigenvalue left that assumes the original noise floor (0 dB). Obviously the Cramér–Rao bound is more sensitive to shortage in low noise eigenvalues than the improvement factor is. In conclusion, with $K = L = 3$ target detection appears to be feasible whereas for parameter estimation some more degrees of freedom ($K = L = 5$ or higher) are required.

Recall that similar effects have been observed in the performance of the eigenvector subspace processor (see the upper two curves in Figure 14.14). This processor uses space-time clutter reference channels whereas the ASEP is based on *separate* spatial and temporal auxiliary channels. Therefore, increasing the number of *spatial* degrees of freedom by one requires an additional M *space-time* channels (see Section 3.4.4). Therefore, the number of channels has to be significantly larger than $N + M$ (at least $N + 2M$ if $N > M$; $M + 2N$ if $M > N$) to provide angle and velocity estimation capability after clutter rejection.

14.4 Space-time monopulse processing

The monopulse parameter estimation technique uses a difference (Δ) and a sum (Σ) array pattern for estimating the target direction. Basically the look direction has to be varied until the monopulse quotient ($\Re\{\Delta/\Sigma\}$) is minimum.

Adaptive monopulse is a technique for parameter estimation in the presence of interference. NICKEL [488, 491, 494] describes adaptive monopulse techniques for two-dimensional angle estimation in the presence of jamming. He shows that target angle estimates tend to be biased particularly when the jammer is in the antenna main lobe. A detailed tutorial on the adaptive monopulse technique can be found in NICKEL [498].

The minimum variance adaptive monopulse technique by PAINE [513] is based on least squares estimation of the target's angular position. The same author has applied this technique to target Doppler estimation in clutter by combining monopulse with STAP ([514]). FANTE [147] describes a technique for achieving a linear slope of the monopulse quotient.

The results presented in this section are based on the adaptive monopulse algorithm by NICKEL.

14.4.1 Nickel's approach

In general the interference cancellation causes a distortion of the directivity and Doppler patterns. This leads to estimation errors which have to be compensated for. Nickel's adaptive monopulse algorithm is based on the assumption that the parameter estimates are a linear function of the monopulse quotient. Nickel considers the estimation of azimuth and elevation by means of a planar antenna array. The algorithms presented below is modified in that the parameter 'elevation' is replaced by 'radial target velocity'.

14.4.1.1 *The corrected adaptive monopulse algorithm*

In NICKEL [491, 494] the adaptive monopulse algorithm has been formulated at subarray level which can be expressed by a linear transform. We start here with transformed quantities in order to include suboptimum STAP techniques based on linear transform $\mathbf{T}$, see (5.3). The algorithm is presented without derivation and statistical background. Readers interested in more details are referred to Nickel's work.

In STAP applications the transform $\mathbf{T}$ may be spatial-temporal (Chapter 5), spatial (Chapter 6), or spatial and temporal cascaded (Chapter 9). The optimum processor in the transformed domain is

$$\mathbf{w}_T = \gamma \mathbf{Q}_T^{-1} \mathbf{s}_T \tag{14.34}$$

For parameter estimation we need some kind of difference patterns in space (array) and time (echo sequence). Such difference patterns can for instance be obtained by

- weighting one half of the aperture (or CPI) with 1 and the other half with -1;
- calculating the partial derivatives of the steering wector with respect to angle and velocity, see (14.43).

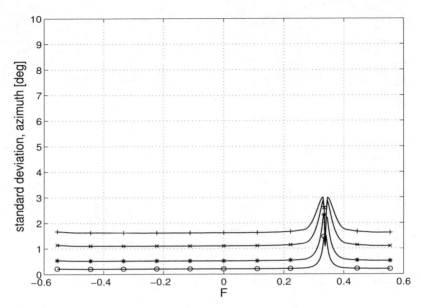

*Figure 14.25: Optimum adaptive processor (OAP): Azimuth estimation by adaptive monopulse; standard deviation; target in look direction (45°). SNR = ○ 10 dB; * 0 dB; × −10 dB; + −20 dB*

Let $\mathbf{d}_s$ and $\mathbf{d}_t$ be the weight vectors for generating spatial and temporal difference patterns. The transformed patterns (e.g., after subarray formation) are then

$$\mathbf{d}_{sT} = \mathbf{T}^* \mathbf{d}_s; \qquad \mathbf{d}_{tT} = \mathbf{T}^* \mathbf{d}_t \qquad (14.35)$$

T stands for "transformed" as in the previous chapters. For the difference patterns interference cancellation has to be done the same way as for the sum channels. The adapted difference patterns become

$$\mathbf{v}_{sT} = \gamma \mathbf{Q}_T^{-1} \mathbf{d}_{sT}; \qquad \mathbf{v}_{tT} = \gamma \mathbf{Q}_T^{-1} \mathbf{d}_{tT} \qquad (14.36)$$

It should be noted that $\mathbf{Q}_T^{-1}$ might be replaced by any other kind of clutter cancellation technique, for instance, the space-time adaptive FIR filter, Chapter 7.

The estimates for target azimuth and velocity are based on the monopulse quotients and correction values

$$\begin{pmatrix} \hat{\varphi} \\ \hat{v} \end{pmatrix} = \begin{pmatrix} \varphi_L \\ v_{DF} \end{pmatrix} + \mathbf{C} \begin{pmatrix} r_s - \mu_s \\ r_t - \mu_t \end{pmatrix} \qquad (14.37)$$

where the monopulse quotients for target azimuth and radial velocity are defined as

$$r_s = \Re\left\{\frac{\Delta_s(\varphi,v)}{\Sigma(\varphi,v)}\right\}; \qquad r_t = \Re\left\{\frac{\Delta_t(\varphi,v)}{\Sigma(\varphi,v)}\right\} \qquad (14.38)$$

and

$$\Delta_s = \mathbf{v}_{sT}^* \mathbf{x}_T; \qquad \Delta_t = \mathbf{v}_{tT}^* \mathbf{x}_T; \qquad \Sigma = \mathbf{w}_T^* \mathbf{x}_T \qquad (14.39)$$

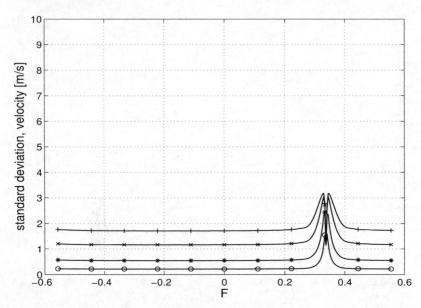

Figure 14.26: OAP: Velocity estimation; standard deviation versus normalised Doppler; target Doppler matched to Doppler filter. SNR = ○ 10 dB; ∗ 0 dB; × −10 dB; + −20 dB

In vector notation (14.37) can be written

$$\hat{\mathbf{u}} = \mathbf{u}_L + \mathbf{C}(\mathbf{r} - \mu) \tag{14.40}$$

where the vector $\mathbf{u}_L$ includes both the look angle and the velocity associated with the actual Doppler filter.

Δ_s and Δ_t are the output signals of the azimuth and velocity difference channels cascaded with the sum channels of velocity and azimuth, respectively, and $\Sigma(\varphi, v)$ is the output signal of the angle-velocity sum channel (beamformer and Doppler filter cascaded); φ_L is the look direction, v_{DF} the velocity associated with the actual Doppler filter; μ_s and μ_t are correction values which are to compensate for distortion caused by the interference. The subscripts s and t denote 'spatial' and 'temporal'.

The correction values can be determined as

$$\mu_s = \Re\left\{\frac{\mathbf{v}_{sT}^*\mathbf{s}_T}{\mathbf{w}_T^*\mathbf{s}_T}\right\}; \qquad \mu_t = \Re\left\{\frac{\mathbf{v}_{tT}^*\mathbf{s}_T}{\mathbf{w}_T^*\mathbf{s}_T}\right\}; \tag{14.41}$$

Recall that $\mathbf{v}_{sT}^*$ and $\mathbf{v}_{tT}^*$ are the *adapted* spatial and temporal difference weightings in the transformed domain. $\mathbf{s}_T$ is the transformed signal vector that would arrive from the look direction and would be matched to a certain Doppler filter. $\mathbf{w}_T$ is the *adapted* transformed weight vector according to (5.4), with the steering vector $\mathbf{s}_T$ being matched to the look direction and to the actual Doppler channel.

The elements of the inverse of the correction matrix are

$$c^{s,\varphi} = \frac{\Re\left\{\mathbf{v}_{sT}^*\mathbf{s}_{T\varphi}\mathbf{s}_T^*\mathbf{w}_T + \mathbf{v}_{sT}^*\mathbf{s}_T\mathbf{s}_{T\varphi}^*\mathbf{w}_T\right\}}{|\mathbf{s}_T^*\mathbf{w}_T|^2} - \mu_s 2\Re\left\{\frac{\mathbf{w}_T^*\mathbf{s}_{T\varphi}}{\mathbf{w}_T^*\mathbf{s}_T}\right\}$$

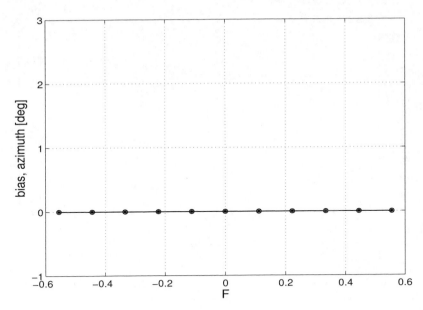

Figure 14.27: OAP: Bias of azmuth estimate; target in look direction. SNR = ∘ 10 dB; ∗ 0 dB; × −10 dB; + −20 dB

$$\begin{aligned}
c^{s,v} &= \frac{\Re\{\mathbf{v}_{sT}^*\mathbf{S}_{Tv}\mathbf{s}_T^*\mathbf{w}_T + \mathbf{v}_{sT}^*\mathbf{S}_T\mathbf{s}_{Tv}^*\mathbf{w}_T\}}{|\mathbf{s}_T^*\mathbf{w}_T|^2} - \mu_s 2\Re\left\{\frac{\mathbf{w}_T^*\mathbf{s}_{Tv}}{\mathbf{w}_T^*\mathbf{s}_T}\right\} \\
c^{t,\varphi} &= \frac{\Re\{\mathbf{v}_{tT}^*\mathbf{S}_{T\varphi}\mathbf{s}_T^*\mathbf{w}_T + \mathbf{v}_{tT}^*\mathbf{S}_T\mathbf{s}_{T\varphi}^*\mathbf{w}_T\}}{|\mathbf{s}_T^*\mathbf{w}_T|^2} - \mu_t 2\Re\left\{\frac{\mathbf{w}_T^*\mathbf{s}_{T\varphi}}{\mathbf{w}_T^*\mathbf{s}_T}\right\} \quad (14.42) \\
c^{t,v} &= \frac{\Re\{\mathbf{v}_{tT}^*\mathbf{S}_{Tv}\mathbf{s}_T^*\mathbf{w}_T + \mathbf{v}_{tT}^*\mathbf{S}_T\mathbf{s}_{Tv}^*\mathbf{w}_T\}}{|\mathbf{s}_T^*\mathbf{w}_T|^2} - \mu_t 2\Re\left\{\frac{\mathbf{w}_T^*\mathbf{s}_{Tv}}{\mathbf{w}_T^*\mathbf{s}_T}\right\}
\end{aligned}$$

where

$$\mathbf{s}_{T\varphi} = \mathbf{T}^*\frac{\partial \mathbf{s}}{\partial \varphi}; \qquad \mathbf{s}_{Tv} = \mathbf{T}^*\frac{\partial \mathbf{s}}{\partial v} \quad (14.43)$$

are the partial derivatives of a signal in look direction and for the actual Doppler filter with respect to azimuth and velocity.

14.4.1.2 Mean and variance for a Rayleigh fluctuating target

Let us now consider the space-time covariance matrix of target signal, clutter and noise in the transformed domain (e.g., subarray level)

$$\mathbf{R}_T = \mathbf{S}_T + \mathbf{C}_T + P_n\mathbf{T}^*\mathbf{T} \quad (14.44)$$

with $\mathbf{S}_T = \mathbf{s}_T\mathbf{s}_T^*$ being the signal dyadic. The monopulse operation can be described by another linear transform

$$\mathbf{R}_{MP} = \mathbf{T}_{MP}^*\mathbf{R}_T\mathbf{T}_{MP} = \begin{pmatrix} \mathbf{R}_D & \mathbf{R}_{DS}^* \\ \mathbf{R}_{DS} & \mathbf{R}_S \end{pmatrix} \quad (14.45)$$

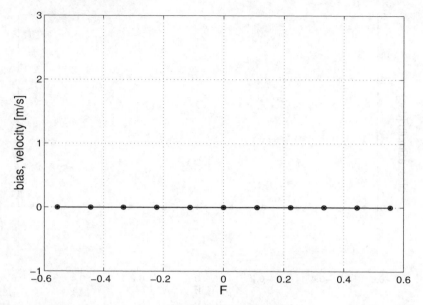

Figure 14.28: OAP: Bias of velocity estimate; target Doppler matched to Doppler filter; SNR = ○ 10 dB; ∗ 0 dB; × −10 dB; + −20 dB

with
$$\mathbf{T}_{\mathrm{MP}} = (\ \mathbf{v}_{\mathrm{sT}}\quad \mathbf{v}_{\mathrm{tT}}\quad \mathbf{w}_{\mathrm{T}}\ ) \tag{14.46}$$

Notice that the subscripts D and S refer to 'difference' and 'sum'. The mean of the monopulse quotient is
$$\mathbf{m}_{\mathrm{MP}} = \Re\{\mathbf{R}_{\mathrm{DS}}/R_{\mathrm{S}}\} \tag{14.47}$$

The covariance matrix of the monopulse quotient is
$$\mathrm{cov}(\mathbf{r}|P \geq \eta) = C_0 \cdot \mathbf{V} \tag{14.48}$$

where P is the sum channel output power and
$$\mathbf{V} = \Re\{\mathbf{R}_{\mathrm{D}|\mathrm{S}}/R_{\mathrm{S}}\} \quad \text{with} \quad \mathbf{R}_{\mathrm{D}|\mathrm{S}} = \mathbf{R}_{\mathrm{D}} - \mathbf{R}_{\mathrm{DS}} R_{\mathrm{S}} \mathbf{R}_{\mathrm{DS}}^* \tag{14.49}$$

The constant factor C_0 in (14.48) can be approximated by
$$C_0 = 0.5\left(\log(1 + \frac{\eta}{R_S})\right) \tag{14.50}$$

see NICKEL [498]. η is the detection threshold which should be chosen to be slightly higher than the white noise power at the sum channel output.

By means of (14.47) and (14.48) one obtains for the mean of azimuth and velocity
$$E(\hat{\mathbf{u}}|P \geq \eta) = \mathbf{u}_{\mathrm{L}} + \mathbf{C}[E(\mathbf{r}|P \geq \eta) - \mu] \tag{14.51}$$

and for the covariance matrix
$$\mathrm{cov}(\mathbf{r}|P \geq \eta) = \mathbf{C} \cdot \mathrm{cov}(\mathbf{r}|P \geq \eta)\mathbf{C}^{\mathrm{T}} \tag{14.52}$$

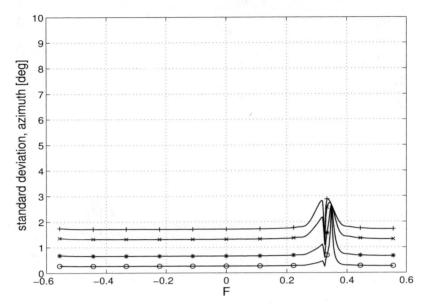

Figure 14.29: OAP: Standard deviation azimuth; $2°$ mismatch; SNR = ○ $10\ dB$; $*\ 0\ dB$; × $-10\ dB$; + $-20\ dB$

14.4.2 Numerical examples

In the following we show a couple of numerical examples in order to get some insight into the behaviour of azimuth and velocity estimation by space-time adaptive monopulse. First we consider the optimum adaptive processor (OAP) as was discussed in Chapter 4. Then we will have a look at our favourite suboptimum processors, the space-time FIR filter (Chapter 7) and the subarray based auxiliary echo processor (SAEP, see in Figure 14.35).

It should be noted that for the SAEP Cramér–Rao bounds can be calculated because the SAEP is obtained from the original signal space (sensor level, echo level) by a linear transform as shown in (9.7). The FIR filter, however, cannot be generated by a linear transform. Therefore, no Cramér–Rao bound exists.

14.4.2.1 *Optimum processing*

Figures 14.25 to 14.28 show the case of no mismatch between target azimuth and radar look direction. In some way the standard deviation curves are similar to the CRB curves in Figure 14.1 etc. In the passband of the STAP processor the curves are almost flat whereas in the direction of clutter match the standard deviation increases.

The similarity of the azimuth and velocity curves is remarkable. This can easily be explained. As in most parts of the book we consider a linear array with $N = 24$ elements and $M = 24$ pulses transmitted. In addition Nyquist sampling in space (array) and time (echo sequence) is assumed. That means, the spatial and temporal apertures are euqivalent. More over we can state that for a linear array azimuth ranges

456 Target parameter estimation

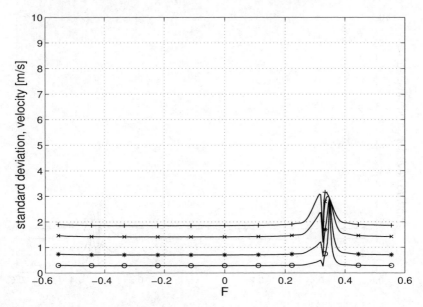

Figure 14.30: OAP: Standard deviation velocity; 2 m/s mismatch; SNR = ○ 10 dB; ∗ 0 dB; × −10 dB; + −20 dB

from 0 to $180°$ and the clutter velocity from -90 to 90 m/s (because we assumed 90 m/s platform velocity). That means, the range for velocity in m/s is the same as for azimuth in degrees. Notice the strong dependence of the standard deviation on the SNR.

In Figures 14.27 to 14.28 the bias values associated with the azimuth and velocity estimates are shown. These Figures tell us the well known fact that the bias is zero in the case of perfect match between look direction (Doppler) and target direction (Doppler filter).

The plots in Figures 14.29 to 14.32 display similar results, but now a mismatch of $2°$ in azimuth and 2 m/s in velocity was assumed. The standard deviation is increased only a little which suggests that the target is still in the linear part of the difference pattern so that the correction mechanism given by (14.32) functions properly. In contrast to perfect match in Figures 14.27 to 14.28 we have now a certain bias (Figures 14.31 to 14.32) which increases with decreasing SNR. It is remarkable that the bias exhibits negative values for large SNR and turns to positive values as the SNR decreases.

In Figures 14.33 and 14.34 the bias has been plotted versus the mismatch between look direction (Doppler) and target direction (Doppler filter). Again the mismatch was chosen $2°$ in azimuth, and 2 m/s in velocity, respectively. As long as the SNR is not below -10 dB reasonable bias is obtained up to a mismatch of about $3°$ (3 m/s).

Beyond $3°$ (2 m/s) the target is leaving the array beamwidth (Doppler channel) and, hence, the quasi-linear slope of the difference pattern. The 3 dB beamwidth of a linear

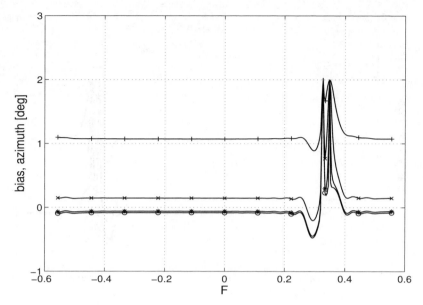

Figure 14.31: OAP: Bias of azimuth estimate; 2° mismatch; SNR = ∘ 10 dB; ∗ 0 dB; × −10 dB; + −20 dB

array is (SKOLNIK [609, p. 11-10])

$$\varphi_b = \frac{50.8}{\cos\varphi_L A/\lambda} \quad [\text{deg}] \tag{14.53}$$

which, for our parameters, gives $\varphi_b = 6°$. This indicates that beyond 3° (3 m/s) no reasonable parameter estimation can be expected. The curves in Figures 14.33 and 14.34 are in accordance with this consideration.

14.4.2.2 *Subarray-based auxiliary echo processing (SAEP)*

The subarray-based auxiliary echo processor processor (SAEP, see Figure 14.35) is a relative of the ASEP described in Chapter 9. The only difference is that the array has been subdivided into uniform subarrays while in the temporal dimension the sidelobe canceler structure is maintained.

For this processor we get a similar picture as for the space-time FIR filter (see Figures 14.36 to 14.39). Except for very strong targets (lowest curves) the standard deviation assumes very high values which are not tolerable for practical applications.

Comparing Figures 14.36 with 14.37 we find that the standard deviation is larger for the azimuth estimate than for the velocity. Obviously, the sidelobe canceler structure in the temporal part of the SAEP (lower part of Figure 14.35) is less suited for parameter estimation than the subarray structure used in the spatial dimension (upper part of Figure 14.35).

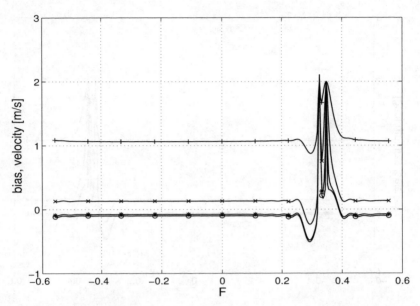

Figure 14.32: OAP: Bias of velocity estimate; 2 m/s mismatch; SNR = ○ 10 dB; ∗ 0 dB; × −10 dB; + −20 dB

14.4.2.3 The space-time FIR filter

It was found in Chapter 7 that the space-time FIR filter is an extremely efficient pre-Doppler STAP technique with the following properties

- except for very low losses associated with the shortening of the echo sequence by the FIR filter the performance is very close to optimum;

- the clutter suppression performance is almost independent of the filter dimensions K and L;

- As the filter dimensions can be made extremely small without losses in performance (e.g., $K = 3$, $L = 3$, resulting in a total of $3 \times 3 = 9$ coefficients) the FIR filter needs only small training support and can match to clutter inhomogeneities;

- The FIR-filter can be re-adapted at every PRI so as to match to non-stationary echo sequences (for instance, due to platform acceleration, staggered pulses, heterogeneous background).

Let us have now a look from the viewpoint of parameter estimation. Figures 14.40 to 14.43 show the standard deviation and bias of azimuth and velocity estimates. The mismatch between look direction (Doppler filter) and target direction (target Doppler) was chosen again to be 2° (2 m/s).

Compared with the optimum processor (Figures 14.25 to 14.26) a considerable increase in standard deviation can be noticed for both. It appears that reasonable

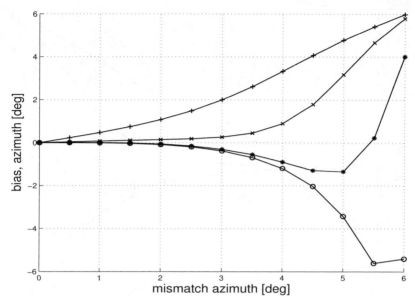

Figure 14.33: OAP: Bias of azimuth estimate vs mismatch; SNR = ○ *10 dB;* ∗ *0 dB;* × *−10 dB;* + *−20 dB*

estimates are obtained only for very strong targets (lowest curve, SNR=10 dB). The bias is in both cases less sensitive than the standard deviation.

It appears from these results that the space-time FIR filter, while being an ideal candidate for clutter suppression, does not achieve a satisfactory parameter estimation performance.

14.4.2.4 The JDL-GLR

As can be seen from Figures 14.44 to 14.47 the results achieved by the JDL-GLR (see subsection 9.7.1.1) are in principle similar to those achieved by the previous suboptimum processors. It is noticed, however, that both the standard deviations of the azimuth and velocity estimates as well as the bias are clearly lower than those found for the previous processors.

In the light of these results it appears that the JDL-GLR processor is the preferable choice among the available processors with real-time capability. This issue will be addressed in the next section in the context of ground target tracking.

14.4.3 Ground target tracking with monopulse radar

In the following we consider a radar-target scenario as depicted in Figure 14.48. This scenario has been used already in KLEMM [337, 349, 352], KOCH [358], and KOCH and KLEMM [357] to study aspects of target tracking. It is shown in [357] how an intelligent tracking algorithm can exploit the specific properties of the signal output of a STAP radar.

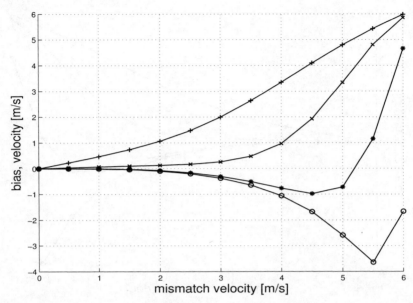

Figure 14.34: OAP: Bias of velocity estimate vs mismatch; SNR = ○ 10 dB; ∗ 0 dB; × −10 dB; + −20 dB

The results given in the quoted literature are based on the Cramér–Rao which is only a measure of theoretical interest. In the following we consider the application of the adaptive monopulse technique described above.

14.4.3.1 Modelling the tracking scenario

The target moves at low speed (10 m/s) on a straight path on the ground. The radar is on board an aircraft moving at 90 m/s on a straight path and at a height of z=3000 m (not visible in the top view in Figure 14.48) in the same direction. The distance between both paths was chosen to be 50 km. We assume further a linear sidelooking array. All those parameters not mentioned here can be found in the standard parameter set given in Table 2.1 in Chapter 2. The radar looks at the target every 20 s (revisit interval).

The modelling of the radar-target scenario is illustrated by Figures 14.49 and 14.50. In Figure 14.50a The slant range between radar and target is shown. Due to the higher speed of the aircraft the range decreases until it reaches a minimum determined by the distance between the two motion paths. Figure 14.49b shows the directivity pattern of the individual array element as seen from the target direction. For the chosen geometry the point of closest approach (which is the point where the target is overtaken by the aircraft) coincides with the maximum of the directivity pattern. Therefore, both the variable range range and the directivity pattern have influence on the CNR and the SNR observed by the radar. This is reflected in Figures 14.49c and 14.49d which show the CNR and SNR versus time. As expected both curves have a maximum at the point of closest approach.

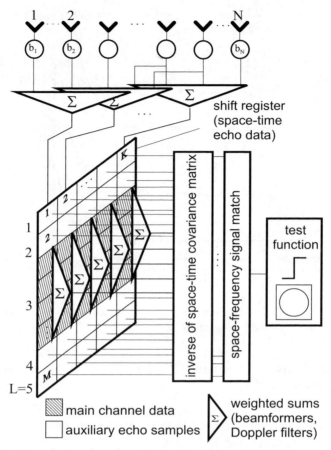

Figure 14.35: The subarray based auxiliary echo processor (SAEP)

Figure 14.50 shows the angle and Doppler relations associated with the given scenario. Figure 14.50a shows the azimuth angle versus time. At the point of closest approach the angle between target and radar becomes 90°. At this time the radar overtakes the target. The target motion is tangential to the radar so that the Doppler as measured by the radar is zero. This can be seen in Figure 14.50c showing the history radial target velocity. At the point of closest approach (slightly before the 50th revisit, see Figure 14.50a) the velocity curve crosses the zero line.

Two more effects should be noted. At about the 65th revisit a temporary stop of the target has been introduced (see Figure 14.50c). Furthermore, a random mismatch between

- the target direction and the look direction of the radar (uniformly distributed between $-3\ldots3$ degrees which corresponds roughly to the beamwidth)
- the target velocity and the velocity associated with the actual Doppler filter

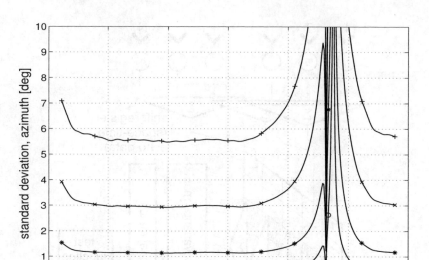

Figure 14.36: SAEP: Standard deviation of azimuth estimate; 2° mismatch. SNR = ∘ 10 dB; ∗ 0 dB; × −10 dB; + −20 dB

(uniformly distributed between $-3\ldots 3$ m/s which corresponds roughly to the Doppler filter width)

during the radar observation was introduced to model the difference between the actual target parameters and the look direction/velocity of the radar. The effect of randomized mismatch can be seen in Figure 14.50b and 14.50d.

In Figure 14.51 the SNIR after STAP processing versus the revisit number is shown for four different STAP processors:

- the optimum processor in Figure 14.51a (OAP, Chapter 4);
- the subarray based auxiliary echo processor in Figure 14.51b (SAEP, see Figure 14.35);
- the space-time FIR filter in Figure 14.51c (Chapter 7).
- the JDL-GLR in Figure 14.51d (Chapter 9).

First of all the impact of the range dependence of CNR and SNR can be noticed. The SNIR increases up to the point of closest distance between radar and target (about revisit 43). However, this is also the point where the radar overtakes the target. Therefore, the radial velocity is zero so that the STAP filter produces a clutter notch. At revisit 65 we assumed that the target stops for a while which causes a second clutter notch on the right. Then, with increasing range, the SNIR decreases again, both because of the range dependence and the sensor pattern (see Figure 14.49). The ripple in the curves is caused by the random mismatch between target direction and

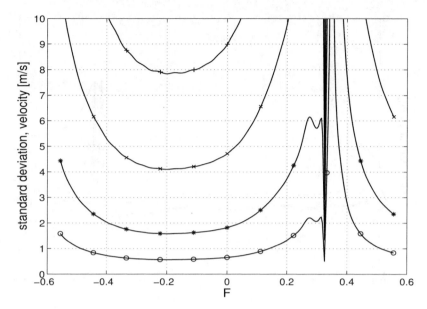

*Figure 14.37: SAEP: Standard deviation of velocity estimate; 2 m/s mismatch. SNR = ○ 10 dB; * 0 dB; × −10 dB; + −20 dB*

velocity and the look direction and search Doppler (centre of the actual Doppler filter) of the radar.

Comparing these four SNIR curves it appears that the clutter rejection performance is about the same. The similarity of the JDL-GLR curve (Figure 14.51d) with the optimum processor Figure 14.51a is remarkable.

For the results shown in the following two subsections it is assumed that the target echo does not fluctuate during the observation period.

14.4.3.2 Standard deviation of parameter estimates

In Figure 14.52 we find the standard deviation of the azimuth estimate in degrees versus the revisit time. The impact of the range dependence of the receive CNR and SNR is reflected by a decrease of the standard deviation when the radar approaches the target (left side), and increases when the radar has overtaken the target and the distance becomes larger again. In the locations of the clutter notches the standard deviation shows peaks. Once again it is remarkable that the optimum processer (Figure 14.52a) is clearly superior to all the suboptimum processor architectures. As above we find that many processor are capable of detecting a moving target in clutter, however, they differ considerably in their aptitude for parameter estimation.

In Figure 14.53 similar curves are shown for the standard deviation of velocity estimates[3]. It can be noticed that the performance of the FIR filter (Figure 14.53c)

---

[3] We present this example here for tutorial purpose. In practice the accuracy of the Doppler estimated by a convential Doppler filter bank is sufficient. Application of the monopulse principle in the time dimension

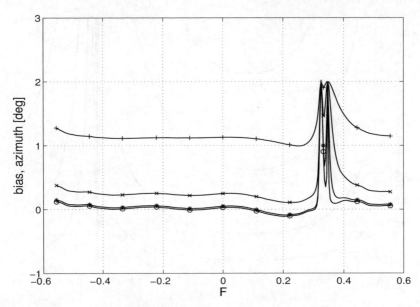

Figure 14.38: SAEP: Bias of azimuth estimate; $2°$ mismatch. SNR = ○ 10 dB; ∗ 0 dB; × −10 dB; + −20 dB

is very close to the optimum processor in Figure 14.53a. Except for the locations of the clutter notches the JDL-GLR produces also a near-optimum accuracy (see Figure 14.53d). The SAEP (Figure 14.53b) is not suited for velocity estimation. This has to do with the sidelobe canceller architecture in the temporal dimension (see Figure 14.35). Obviously uniform subarrays as used in the spatial dimension on top of Figure 14.35 are better suited for monopulse estimation, compare Figure 14.52b with Figure 14.53b.

The reason for the near-optimum performance for velocity estimation (Figure 14.53c) of the space-time FIR filter can be explained as follows. The FIR filter slides over the received echo signal sequence and shortens the echo length by $L - 1$, with L being the number of temporal filter coefficients. The filtered sequence of slightly reduced length $M - L + 1$ can still be used for proper Doppler estimation by temporal monopulse. In contrast, post-Doppler techniques such as ASEP or SAEP offer only a small number of temporal degrees of freedom which leads to an increase in standard deviation.

14.4.3.3 *Expectation of the bias of parameter estimates*

The azimuth bias for the four above mentioned processors in Figure 14.54a-d (and in Figure 14.55a-d for the velocity estimate). As one can see there is almost no bias. The FIR filter (Figure 14.54c) shows some bias at the locations of the clutter notches. It can be noticed that the best coincidence with the optimum processor is achieved by the JDL-GLR.

---

is unusual.

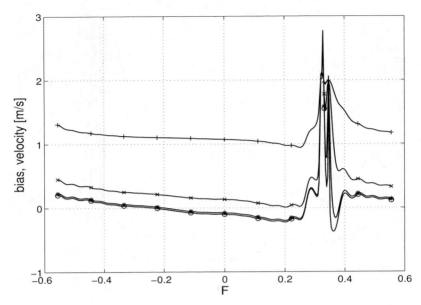

Figure 14.39: SAEP: Bias of velocity estimate; 2 m/s mismatch; SNR = ∘ 10 dB; ∗ 0 dB; × −10 dB; + −20 dB

14.4.3.4 Bias of parameter estimates

Figures 14.56 and 14.57 show the bias of the azimuth and velocity estimates as obtained by simulation. Such data are obtained by multiplying gaussian vectors with the lower left triangular matrix obtained by Cholesky factorisation of the Clutter + signal covariance marix as given by (14.44).

It can be seen that the optimum processor (Figures 14.56a and 14.57a) clearly outperforms the suboptimum techniques. The SAEP and the JDL-GLR show strong deviations from the true parameter value. It can be noticed that in particular for velocity estimation (Figures 14.57c) relatively good estimates are obtained by the space-time FIR filter. This can be explained as follows. The space-time FIR filter shortens the data sequence by $L - 1$, with L being the temporal filter length. If the filter length is short compared with the data sequence ($L \ll M$) then the shortening is negligible so that nearly the full signal length is preserved. This length is then available for forming a temporal difference pattern for Doppler estimation. Even azimuth estimation works better than with the other suboptimum techniques. Transform techniques such as SAEP or JDL-GLR operate in a strongly reduced subspace which is obviously not as suitable for parameter estimation.

14.5 Summary

The capability of angle and velocity estimation of various STAP (optimum and suboptimum) processors has been analysed. The suboptimum class of processors is

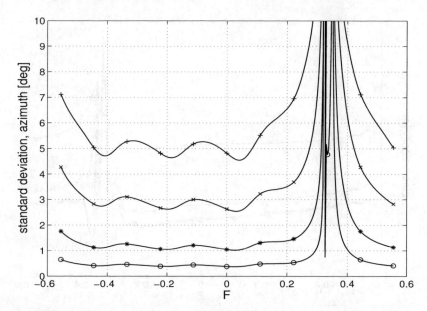

Figure 14.40: Space-time FIR filter: Standard deviation of azimuth estimate; 2° mismatch. SNR = ○ 10 dB; ∗ 0 dB; × −10 dB; + −20 dB

based on certain order reducing linear transforms. The analysis has been performed first by calculation of the Cramér–Rao bounds of angle and velocity estimates. In addition the adaptive monopulse principle has been discussed. The following observations have been made in the discussion of the CRB:

- The Cramér–Rao bound plotted versus either Doppler frequency or azimuth consists normally of two parts: the clutter response which is a peak centred about the clutter Doppler frequency in the look direction, and a clutter-free part which usually runs horizontally.

- The Cramér–Rao bound depends on various environmental parameters such as SNR, CNR, clutter fluctuations, and jamming.

- The dependence of the Cramér–Rao bound on different system parameters (sensor directivity patterns, aperture and CPI size, PRF, and bandwidth) has been demonstrated.

- The system bandwidth can degrade the detection and estimation performance in two ways: travel delay across the aperture and range walk. For sidelooking arrays only the range walk effect applies.

- Processors based on *space-time* transforms (eigenchannel, auxiliary space-time channel) can obtain near-optimum standard deviations of angle and velocity estimates if the numbers of channels C is chosen sufficiently high. C has to be chosen to be a higher number than required for detection.

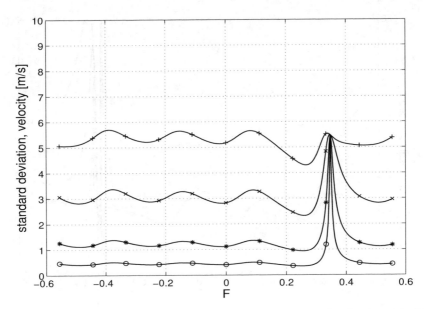

*Figure 14.41: FIR filter: Standard deviation of velocity estimate; 2 m/s mismatch. SNR = ∘ 10 dB; * 0 dB; × −10 dB; + −20 dB*

- There is no way of angle or velocity estimation if no clutter rejection measures are applied to the radar data.

- Conventional MTI processing used with a moving radar (beamformer and adaptive *temporal* clutter filter, as normally used in stationary radar) may give some poor estimates of the target velocity. Angle estimation is not possible.

- *Spatial* order reducing transforms lead to a slight degradation in angle estimation (symmetric auxiliary sensor structure). However, this degration appears to be tolerable, keeping the dramatic saving in computational load in mind.

- Subarray-based auxiliary echo processing (SAEP) leads to some tolerable increase of the standard deviations of both angular and velocity estimates. It can be observed once again that the number of degrees of freedom required for estimation is larger than for detection.

- It appears that uniform subarrays are better suited for parameter estimation than sidelobe canceler architectures.

- Order reduction in the time dimension can be achieved by replacing the inverse of the clutter covariance matrix in the optimum processor (1.3) by a space-time FIR filter (see Chapter 7). In conjunction with spatial transforms this concept leads to highly economic processor architectures with real-time operation capability. Since this processor cannot be easily formulated via a linear transform the CRB cannot be calculated.

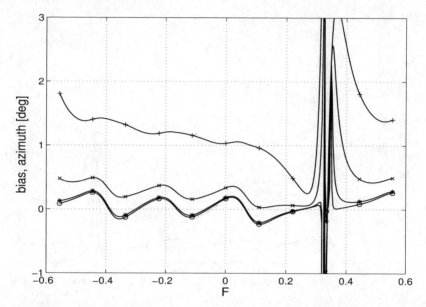

Figure 14.42: FIR filter: Bias of azimuth estimate; 2° mismatch. SNR = ∘ 10 dB; ∗ 0 dB; × −10 dB; + −20 dB

In contrast to the CRB which is only a theoretical lower bound of the variance of estimates the adaptive monopulse is a technique well suited for practical application. The following observations were made for target parameter estimation with adaptive monopulse in clutter:

- The optimum adaptive processor (sensor and echo pulse level) achieves estimation accuracies for both azimuth and target velocity which are in a useful order of magnitude even for weaker targets; the standard deviation increases with decreasing SNR;

- The bias of the azimuth an velocity estimates is zero for target match (look direction = target direction, Doppler filter = target Doppler);

- For mismatch between look and target direction (and similar for the velocity) the estimates are biased. The bias increases with decreasing SNR;

- Suboptimum STAP techniques (whose detection performance is close to optimum) exhibit strongly degraded estimation performance. Reasonable accuracy can obtained only for very strong targets.

- Probably the azimuth performance can be improved by using a larger number of antenna channels (in the examples we used $K = 5$). Velocity estimation is not as critical. In practice the accuracy given by a Doppler filter bank is sufficient. Increasing the spatial dimension of the STAP filter while keeping the temporal dimension low may be a reasonable trade-off between cost and performance.

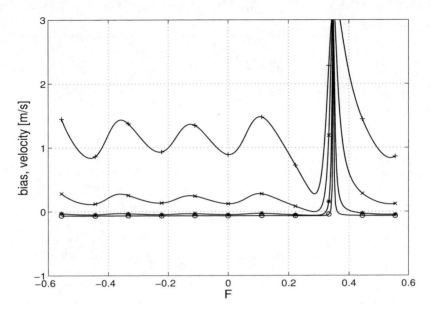

Figure 14.43: FIR filter: Bias of velocity estimate; 2 m/s mismatch; SNR = ∘ 10 dB; ∗ 0 dB; × −10 dB; + −20 dB

- The JDL-GRL produces azimuth and velocity estimates with standard deviations and expected biases close to the optimum.

- The standard deviation of the velocity estimate achieved by the space-time FIR filter is close to the optimum. Operating with simulated data the FIR filter is superior to all other techniques because almost the full data sequence can be used to form difference patterns.

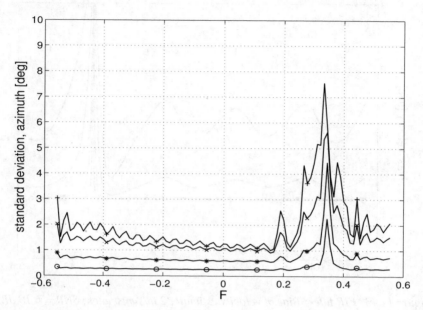

Figure 14.44: JDL-GLR: Standard deviation of azimuth estimate; 2° mismatch. SNR = ∘ 10 dB; ∗ 0 dB; × −10 dB; + −20 dB

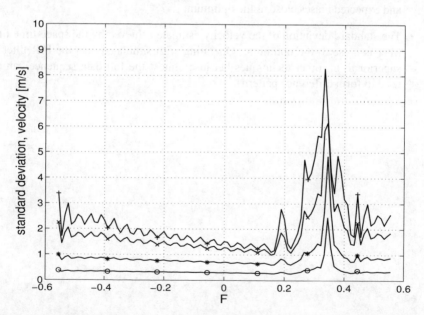

Figure 14.45: JDL-GLR: Standard deviation of velocity estimate; 2 m/s mismatch. SNR = ∘ 10 dB; ∗ 0 dB; × −10 dB; + −20 dB

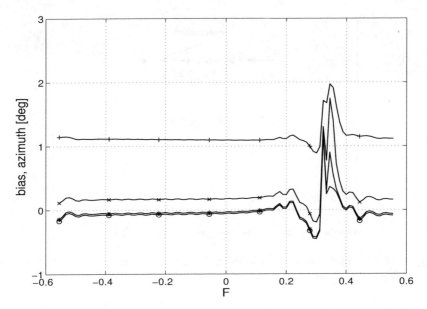

Figure 14.46: JDL-GLR: Bias of azimuth estimate; $2°$ mismatch. SNR = ○ 10 dB; ∗ 0 dB; × −10 dB; + −20 dB

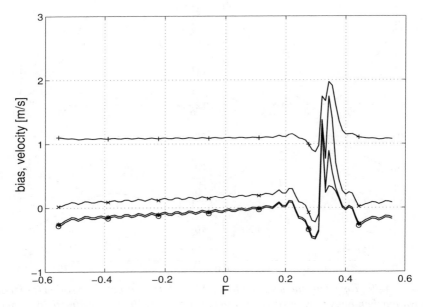

Figure 14.47: JDL-GLR: Bias of velocity estimate; 2 m/s mismatch; SNR = ○ 10 dB; ∗ 0 dB; × −10 dB; + −20 dB

472 Target parameter estimation

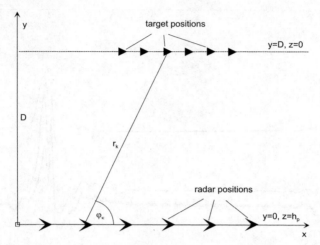

Figure 14.48: Track scenario: sidelooking linear array, ground target and airborne radar on parallel paths

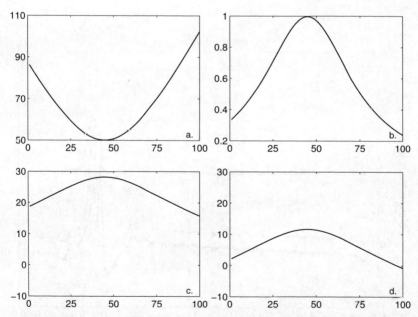

Figure 14.49: Time dependence of SNR and SNIR during target observation (abszissa: revisit number, revisit every 20 s): a. target range [km]; b. sensor directivity pattern; c. CNR [dB]; d. SNR [dB]

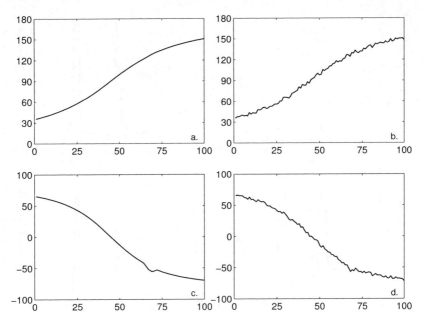

Figure 14.50: Modeling of the radar-target configuration (abszissa: revisit number). a. target azimuth [deg]; b. radar look direction [deg]; c. target velocity [m/s]; d. centre of Doppler filter [m/s]

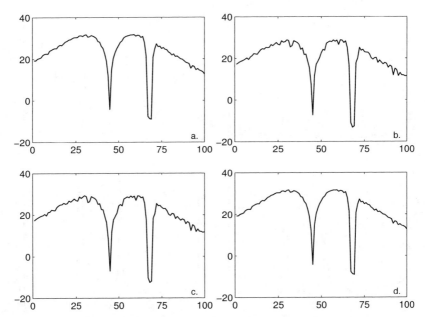

Figure 14.51: Detection performance of four STAP processors (SNIR versus revisit number): a. Optimum processor (OAP); b. Subarray based auxiliary echo processor (SAEP); c. space-time FIR filter; d. JDL-GLR

474 Target parameter estimation

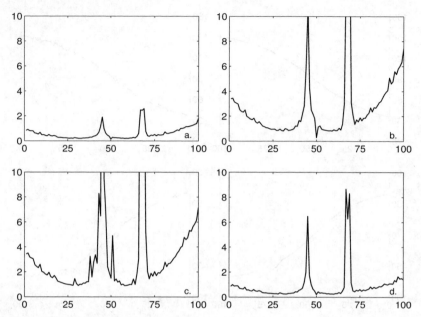

*Figure 14.52: Standard deviation of azimuth estimate [deg] (abszissa: revisit number):
a. Optimum processor (OAP); b. Subarray based auxiliary echo processor (SAEP); c. space-time FIR filter; d. JDL-GLR*

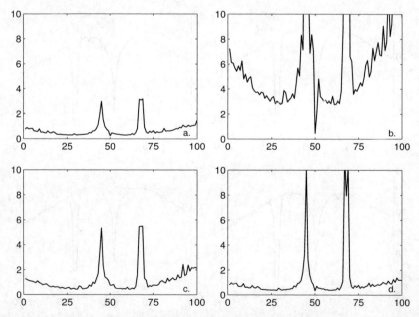

*Figure 14.53: Standard deviation of velocity estimate [deg] (abszissa: revisit number):
a. Optimum processor (OAP); b. Subarray based auxiliary echo processor (SAEP); c. space-time FIR filter; d. JDL-GLR*

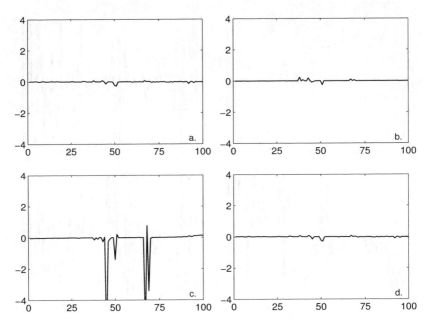

Figure 14.54: Expected bias of azimuth estimate [deg] (abszissa: revisit number): a. Optimum processor (OAP); b. Subarray based auxiliary echo processor (SAEP); c. space-time FIR filter; d. JDL-GLR

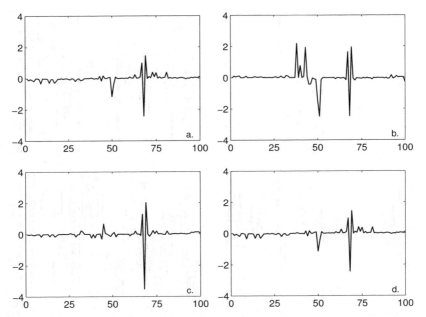

Figure 14.55: Expected bias of velocity estimate [deg] (abszissa: revisit number): a. Optimum processor (OAP); b. Subarray based auxiliary echo processor (SAEP); c. space-time FIR filter; d. JDL-GLR

476 Target parameter estimation

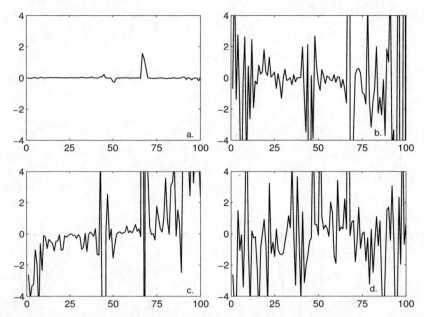

Figure 14.56: Bias of velocity estimate [deg] (abszissa: revisit number): a. Optimum processor (OAP); b. Subarray based auxiliary echo processor (SAEP); c. space-time FIR filter; d. JDL-GLR

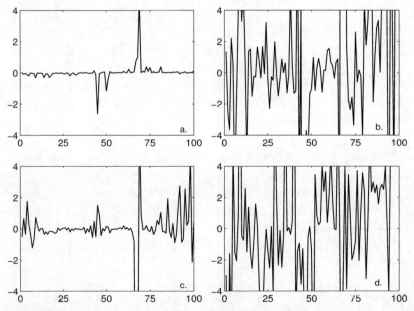

Figure 14.57: Bias of velocity estimate [deg] (abszissa: revisit number): a. Optimum processor (OAP); b. Subarray based auxiliary echo processor (SAEP); c. space-time FIR filter; d. JDL-GLR

Chapter 15

Influence of the radar equation

15.1 Fundamentals

15.1.1 From notional radar concepts to realistic operation

The design of near-optimum suboptimum space-time adaptive processors played a predominant role in the past chapters. It was pointed out that the optimum (fully adaptive) processor given by (4.1) is not applicable for practical radar operation because of various issues arising with large numbers of array elements N and processed echo data M.

A number of processor architectures was discussed in Chapters 5,6,7, and 9. It was found that several of them achieve near-optimum clutter suppression performance at extremely reduced computational burden. These architectures promise radar operation in real time. It was shown that the clutter notch achieved with suboptimum processors was very close the theoretical optimum so that the detection of slow targets appears to be possible.

The performance of the various processors was judged by calculating the improvement factor defined in (5.5), normalised to the theoretical optimum which is given by $CNR \cdot N \cdot M$. This normalised improvement factor is referred to by other authors as loss function (e.g., WARD [690]).

The normalised IF is independent of various radar and processing parameters such as range, the radar system parameters N and M, the clutter-to-noise-ratio (CNR) and the target signal-to-noise ratio (SNR). In fact, the CNR is part of the clutter covariance matrix and, hence, may have some influence on the IF curves, in particular the depth of the clutter notch. Because of these normalisations, the normalised IF is best suited to compare the clutter suppression performance of different processors. Calculating the normalised IF means ignoring the radar environment and the impact of the target strength.

In this chapter we discuss issues of STAP by taking the target properties and the clutter environment into consideration. This will be done in two ways, 1. by calculation SNR and CNR using the radar equation, 2. by calculating the probability of detection based on the SNIR achieved by STAP processing and the probability of false alarm.

15.1.2 The radar equation

There are various different forms of the radar equation in the literature. We use the form given in SKOLNIK [610]

$$\frac{P_{\rm r}}{P_{\rm t}} = \frac{G_{\rm r} G_{\rm t} \sigma \lambda^2 F_{\rm r}^2 F_{\rm t}^2}{(4\pi)^3 R^4 \cdot L} \tag{15.1}$$

where $P_{\rm r}$ and $P_{\rm t}$ denote receive and transmit power; $G_{\rm r}$ and $G_{\rm t}$ are antenna gains on receive and transmit; σ is the radar cross-section of the reflector; $F_{\rm r}$ and $F_{\rm t}$ are factors describing propagation effects; R denotes range. After amplification the signal-to-noise ratio is

$$SNR = \frac{P_{\rm r}}{k_0 T_0 B F_{\rm n}} \tag{15.2}$$

where k_0 is the Boltzmann constant, T_0 the absolute temperature ($k_0 T_0 = 10^{-23}$Ws), B the bandwidth and $F_{\rm n}$ the noise figure of the amplifier. Since we are interested in the SNR at the single array element we set $G_{\rm r} = 1$. Inserting (15.2) in (15.1) gives

$$SNR = \frac{P_{\rm signal}}{P_{\rm noise}} = \frac{P_{\rm t} G_{\rm t} \sigma \lambda^2 F_{\rm r}^2 F_{\rm t}^2}{(4\pi)^3 R^4 \cdot L \cdot k_0 T_0 B F_{\rm n}} \tag{15.3}$$

15.1.2.1 Target signal

Targets of interest are not in general point-shaped. Their radar cross-section $\sigma_{\rm tgt}$ may vary in the range of a few square meters to a couple of 1000 m^2. The actual value has to be inserted in (15.3).

15.1.2.2 Clutter

For clutter we use a constant-γ model. In this case the radar cross-section of clutter is modeled as

$$\sigma_c = \sigma_{0c} A_{\rm reflect} \tag{15.4}$$

with

$$A_{\rm reflect} = K \cdot R \cdot \frac{c}{2B} \tag{15.5}$$

being the reflecting clutter area. K is the transmit beamwidth. The scatter coefficient is

$$\sigma_{0c} = \gamma \sin \theta \tag{15.6}$$

where γ is the specific reflectivity of the ground. With $\sin \theta = H/R$ one obtains

$$\sigma_c = \gamma H K c / (2B) \tag{15.7}$$

15.1.3 SNIR and probability of detection

Let us consider the problem of detecting a deterministic target signal with a certain Doppler by means of a Doppler filter bank consisting of M Doppler channels. We assume that the output signals of the Doppler filter bank are squared, and the maximum is compared with a detection threshold. The signal is contained in one of the Doppler channels and has to compete with false alarms from all Doppler channels.

15.1.3.1 Ricean and Rayleigh distributions

The absolute value of the signal channel output follows a Ricean distribution, the outputs of the $M-1$ signal-free channels are Rayleigh distributed.

For a variable $X = |a+Z|$ with $a \in C$ and Z being $N_C(0,1)$ distributed the Ricean distribution becomes

$$\begin{aligned} F_a^{\text{Rice}}(x) &= \frac{1}{\pi} \int_{\{|z| \leq x\}} e^{-|z-a|^2} dz \\ &= \frac{1}{\pi} \int_0^x \int_0^{2\pi} e^{-r^2|a|^2 + 2|a|r\cos\varphi} r d\varphi dr \\ &= 2e^{-|a|^2} \int_0^x I_0(2|a|r) r dr \end{aligned} \qquad (15.8)$$

with the density

$$p^X(x) = 2xe^{-x^2-|a|^2} I_0(2|a|x) \qquad (15.9)$$

The probability that $X = |a+Z|$ exceeds a threshold η is then

$$P_D = 2e^{-|a|^2} \int_\eta^\infty e^{-r^2} I_0(2|a|r) r dr \qquad (15.10)$$

This value is obtained by calling the below listed procedure $perice(\eta^2, |a|^2)$.

For $a = 0$ (no signal) the Ricean distribution becomes the Rayleigh distribution which gives the probability that a noise sample exceeds the threshold

$$P_F = P(|Z| > \eta) = e^{-\eta^2} \qquad (15.11)$$

For a given P_f the threshold can be calculated as

$$\eta = \sqrt{-\ln(P_F)} \qquad (15.12)$$

15.1.3.2 The Doppler filter bank

Let P_F be the given false alarm probability. Then the false alarm probability in one out of M Doppler channels becomes

$$P_F^c = 1 - \sqrt[M]{1 - P_F} \qquad (15.13)$$

and the squared threshold for the single channel is

$$\eta^2 = -\ln P_F^c \tag{15.14}$$

Finally the probability that the output signal in one of the M channels exceeds the threshold according to (15.14) is

$$P_D = 1 - (1 - P_F^c)^{M-1}(1 - P_{SNIR}^{Rice}(\eta^2, SNIR)) \tag{15.15}$$

where P_{SNIR}^{Rice} is obtained from (15.10).

15.1.4 Mapping SNIR onto probability of detection

In the following a small MATLAB program is given which calculates the probability of detection for the parameters SNIR, the probability of false alarm and the number of Doppler filters used. This program was communicated to the author by J. ENDER. By using this program time consuming simulation runs can be avoided.

```
%Program snir2pd
%Pd for unknown target Doppler frequency,
%          FFT, uniformly distributed phase
%Pfa false alarm probability
%SNIRlog SNIR in dB
%NDoppF number of Doppler filters


p=1-exp(log(1-Pfa)/NDoppF);\\
eta=-log(p);\\
a=exp(log(10)*SNIRlog/10);\\
pp=(1-perice(eta,a));\\
if pp$\geq$1 ; Pd=0;\\
    else\\
       if pp$\geq 0$ , Pd=1;\\
           else\\
              Pd=1-exp((NDoppF-1)*log(1-p)+log(pp));\\
       end;\\
 end;\\\\

function y=perice(eta,SNIR)\\
%SNIR linear; eta = -log(Pfa) \\
  x=eta;y=SNIR;\\
  sum1=0;k=1;dx=exp(-x);dy=exp(-y);sum2=dy;\\
  for endlos=1:inf\\
    dx=x*dx/k;\\
    sum1=sum1+dx*sum2;\\
```

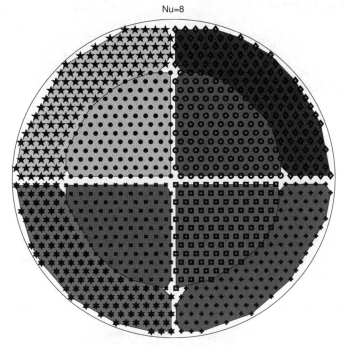

Figure 15.1: Circular array with 8 subarrays

```
dy=y*dy/k;\\
sum2=sum2+dy;\\
k=k+1;\\
   if ((k > x) \& (dx < 0.00000001));  y=1-sum1;return; end;\
end;\\\\
```

15.2 Numerical examples

In the following we illustrate the effect of the radar range equation on the capability of a realistic GMTI radar to detect slow moving targets.

15.2.1 Optimum space-time processing at subarray level

Optimum space-time processing has been treated to some extent in Chapter 6. The following analysis is based on a circular forward looking array[1] as shown in Figure 15.1, and some parameters summarised in Table 15.1 (all other parameter are given

---
[1] This array has been used already in Chapter 11

Table 15.1 Radar parameters

σ_{tgt} 10, 300 m^2	
γ	0.15
P_F	10^{-5}
transmit power	10 KW
number of array elements	N=902
number of subarrays	8
number of temporal channels	L=5
target range	10, 60 km
look angle	0°, 45°
PRF	2 KHz

in Table 2.1). Such an antenna will typically be installed in the nose of future fighter radars for operation in a forward looking mode. Therefore, the forward looking (FL) application is considered in the following.

We use the optimum processor at subarray level as defined by (5.4) and calculate the signal-to-noise+interference ratio at the processor output according to

$$\text{SNIR}_{\text{out}} = \frac{\mathbf{w}_T^* \mathbf{s}_T \mathbf{s}_T^* \mathbf{w}_T}{\mathbf{w}_T^* \mathbf{Q}_T \mathbf{w}_T} \tag{15.16}$$

Recall that the subscript $_T$ denotes transformed quantities according to (5.3). In this section we consider a spatial transform of the form (6.1) and (6.10) describing disjoint subarrays. Then the dimension of the signal vector space is $K \times M = 8 \times 24 = 192$.

The impact of the radar equation on the slow target detection capability with an antenna as shown in Figure 15.1 is illustrated by Figures 15.2 to 15.5.

In Figure 15.2 the SNIR is plotted versus the radial target velocity. The look direction is $\varphi_L = 0°$. Between left and right figures the target cross section is varied (10 and 300 m^2). Upper and lower figures differ in range (10 and 60 km).

Several observations can be made

- **Signal strength**. The right curves are shifted versions of the left curves. The shift is caused by the difference in target cross-section ($10 \log 30 = 14,77$ dB).

- **Range law**. The lower curves show the SNIR for 6 × the range of the upper subfigures. The decrease in SNIR is about 30 dB.

- **Clutter notch**. At short range (upper subfigures) the clutter notch appears at a lower target velocity than a larger range (lower curves). This reflects the impact of the depression angle which is much larger in the upper figures than below.

The associated P_D curves are shown in Figure 15.3. It is remarkable in which way the SNIR curves are mapped on P_D curves. Although the clutter notches in Figures 15.2a,b seem to be broader than those in Figures 15.2c,d the P_D curves show just the opposite behaviour.

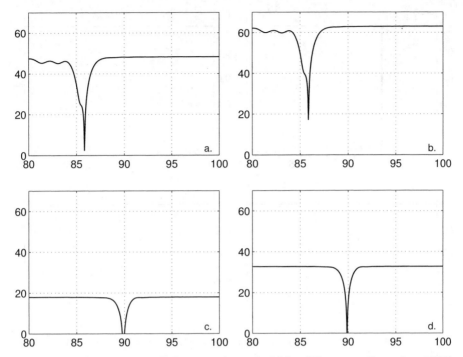

Figure 15.2: SNIR versus radial target velocity [m/s] for different configurations (OAP; FL, $\varphi_L = 0°$). a. $\sigma_{tgt} = 10$ m^2, $R = 10$ km; b. $\sigma_{tgt} = 300$ m^2, $R = 10$ km; c. $\sigma_{tgt} = 10$ m^2, $R = 60$ km; d. $\sigma_{tgt} = 300$ m^2, $R = 60$ km

The detection threshold was given by (15.12). For $P_F = 10^{-6}$ the squared threshold becomes $\eta^2 = 13.8$ corresponding to 11.4 dB. Notice that at 11.4 dB one encounters a very narrow clutter notch in Figure 15.2a. There is no clutter notch at 11.4 dB in Figure 15.2b. The clutter notches of the lower curves are clearly broader at 13.6 dB than those above. This illustrates nicely the impact of the signal strength on the detection behaviour. Notice that normalised IF curves as used in the previous chapters are useful to judge the effectiveness of STAP architectures, but the detection probability gives a clearer image of the detection performance under realistic operational conditions.

Recall that the improvement factor (5.5) is a quantity normalised to the target signal power and, hence, is independent of the signal. This is illustrated by Figure 15.6. Notice that the curves shown in Figures 15.2c. and d. differ only in the assumed signal strength. Therefore, the IF curve shown in Figure 15.6a applies to both of them. In contrast, the two P_D curves in Figure 15.3 c. and d. show distinct differences in the area of the clutter notch caused by a variation of the signal power.

Figures 15.4 and 15.5 show again the performance of the optimum processor at subarray lebel, however for a different look direction ($\varphi_L = 45°$). As can be seen the width of the clutter notch is broadened.

This aspect is illustrated in more detail in Figure 15.6. The four IF curves have

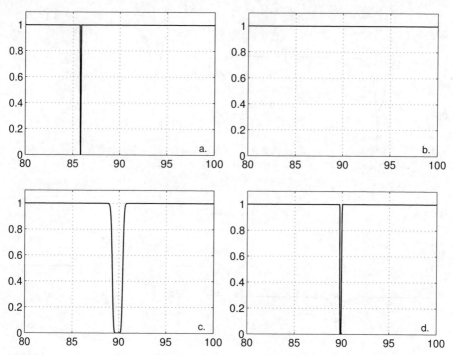

Figure 15.3: P_D versus radial target velocity for the configurations in Figure 15.2. a. $\sigma_{tgt} = 10$ m^2, $R = 10$ km; b. $\sigma_{tgt} = 300$ m^2, $R = 10$ km; c $\sigma_{tgt} = 10$ m^2, $R = 60$ km; d. $\sigma_{tgt} = 300$ m^2, $R = 60$ km

been calculated for different look angles. We can make the following observations

- The position of the clutter notch changes with the look direction. This is a natural consequence of the dependence of the Doppler on angular direction, see (3.2).

- With increasing look direction the clutter notches become broader.

- With increasing look direction the IF decreases even in the passband. This results from properties of the chosen antenna (Figure 15.1). Since clutter rejection is mainly a horizontal problem only the horizontal degrees of freedom of the antenna count. As we have four subarrays in the horizontal the number of spatial degrees of freedom is 3 which is about the minimum required for a forward looking array (recall that for each clutter Doppler frequency two arrivals from two different directions appear). For a directions other than broadside (0°) only the projection of the aperture in this direction is effective. Furthermore, the STAP performance of this array is degraded by the fact that the subarrays have quite different shapes.

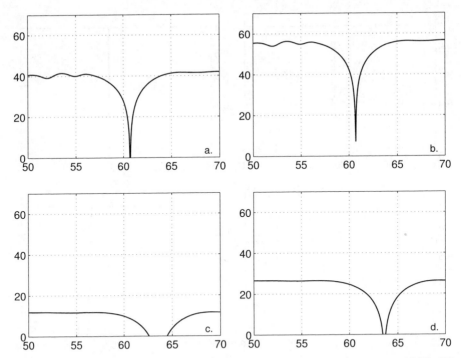

Figure 15.4: SNIR versus radial target velocity for different configurations (OAP; FL, $\varphi_L = 45°$). a. $\sigma_{tgt} = 10$ m^2, $R = 10$ km; b. $\sigma_{tgt} = 300$ m^2, $R = 10$ km; c $\sigma_{tgt} = 10$ m^2, $R = 60$ km; d. $\sigma_{tgt} = 300$ m^2, $R = 60$ km

15.2.2 Suboptimum space-time processors

In this subsection we discuss the behaviour of suboptimum processors with the capability of adaptive clutter suppression in real-time.

15.2.2.1 *Subarray based auxiliary echo processor (SAEP)*

The SAEP was discussed in Chapter 14, (see in Figure 14.35). This processor is closely related to the ASEP described in Chapter 9. It showed near-optimum clutter suppression performance. Figures 15.7 and 15.8 show IF and P_D for the SAEP. Comparing the results in Figure 15.8a. and b. with those in Figure 15.5c. and d. confirms once again that the SAEP is an excellent approximation of the optimum processor.

15.2.2.2 *Frequency dependent spatial processing*

Frequency dependent spatial processing[2] (see Chapter 9) is a very cost efficient adaptive clutter suppression technique which approximates, however, the optimum

---

[2]in the literature referred to as 'partial adaptive STAP' or 'post-Doppler STAP'.

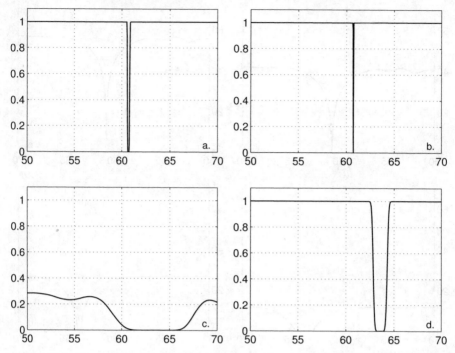

Figure 15.5: P_D *versus radial target velocity for the configurations in Figure 15.4.* a. $\sigma_{tgt} = 10 \text{ m}^2$, $R = 10$ km; b. $\sigma_{tgt} = 300 \text{ m}^2$, $R = 10$ km; c $\sigma_{tgt} = 10 \text{ m}^2$, $R = 60$ km; d. $\sigma_{tgt} = 300 \text{ m}^2$, $R = 60$ km

processor only if the data record is long, i.e., M is of the order of magnitude of several hundreds to thousands. For the the chosen number of echoes $M = 24$ we notice a considerable degradation in moving target detection performance (compare Figure 15.10a. and b. with Figure 15.8a. and b.).

15.2.3 One-dimensional processing

By one-dimensional processing we understand conventional techniques such as beamforming plus adaptive temporal clutter filter as can be used in stationary radar for adaptive clutter suppression of weather, sea and other kind of moving clutter (see the example in Section 1.2.5.2). This kind of processing does not take the spatial dimension of clutter and target returns into account and, therefore, gives in general poor performance for clutter rejection in moving radar systems. Beamforming plus Doppler filtering, that means, no specific clutter suppression, is even worse. A comparison of one-dimensional techniques with STAP was given in Figure 4.4.

15.3 Summary

The findings of this chapter can be summarised as follows:

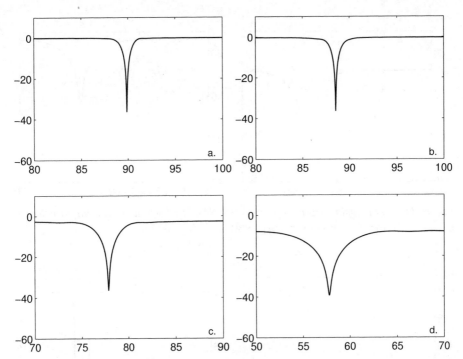

Figure 15.6: IF versus radial target velocity for look directions (OAP; $R = 60$, $\sigma_{tgt} = 10\,\text{m}^2$): a. $\varphi_L = 0°$; b. $\varphi_L = 10°$; c. $\varphi_L = 30°$; d. $\varphi_L = 50°$

1. The improvement factor (IF) is a measure suitable for judging the clutter suppression performance of space-time adaptive clutter filters. The IF is not sufficient to analyse the over-all target detection performance of a radar because it does not include the target signal power.

2. In order to analyse the detection performance the quantities CNR and SNR for the single antenna element and the single echo have to be calculated on the basis of the radar equation.

3. The target detection performance of a STAP radar can be determined by calculating the probability of detection which is a monotonic function of the signal-to-noise + interference ratio (SNIR), and, hence, includes the target radar cross-section.

488 Influence of the radar equation

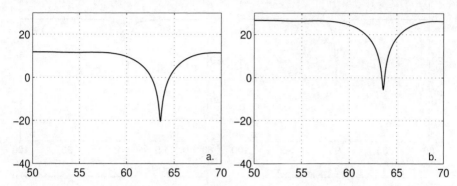

Figure 15.7: Dimension reduced processing (SAEP; IF versus radial target velocity).
a. $\sigma_{\text{tgt}} = 10 \text{ m}^2$, $R = 60 \text{ km}$; b. $\sigma_{\text{tgt}} = 300 \text{ m}^2$, $R = 60 \text{ km}$

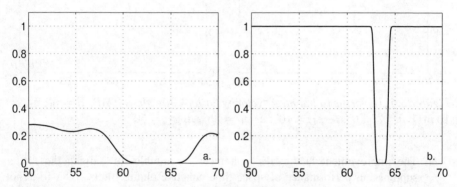

Figure 15.8: Dimension reduced processing (SAEP; P_D versus radial target velocity).
a. $\sigma_{\text{tgt}} = 10 \text{ m}^2$, $R = 60 \text{ km}$; b. $\sigma_{\text{tgt}} = 300 \text{ m}^2$, $R = 60 \text{ km}$

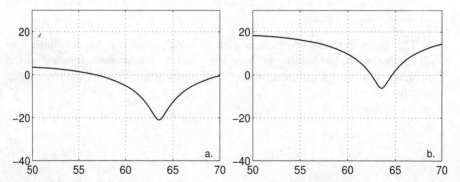

Figure 15.9: Frequency dependent spatial processing (IF versus radial target velocity).
a. $\sigma_{\text{tgt}} = 10 \text{ m}^2$, $R = 60 \text{ km}$; b. $\sigma_{\text{tgt}} = 300 \text{ m}^2$, $R = 60 \text{ km}$

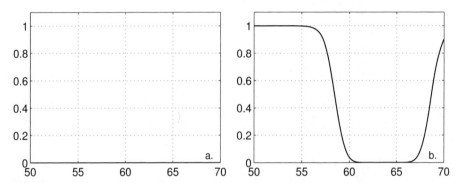

Figure 15.10: Frequency dependent spatial processing (P_D versus radial target velocity). a. $\sigma_{\text{tgt}} = 10 \text{ m}^2$, $R = 60$ km; b. $\sigma_{\text{tgt}} = 300 \text{ m}^2$, $R = 60$ km

Chapter 16

Special aspects of airborne MTI radar

16.1 Antenna array errors

In this section we discuss briefly the effect of antenna array errors on the clutter rejection performance. There are several kinds of error sources in an array antenna, such as tolerances of the sensor positions, amplification errors which result in amplitude and phase jitter as well as IQ and delay errors.

The impact of array errors on the beam patterns has been treated extensively by MAILLOUX in Chapter 7 of [428]. The effect of random amplitude and phase errors on the directivity, the beam pointing error and the peak sidelobe level is analysed. Furthermore the effect of quantisation errors due to amplitude, delay and phase quantisation on the directivity pattern is treated. In another paper by MAILLOUX [429] the effect of element failure on the performance of an array antenna is discussed. GERLACH and STEINER [181] present a technique for designing an adaptive matched filter so that the effect of IQ errors is a minimum. In the ELRA system (SANDER [583]) an automatic focusing routine is implemented which takes care of element failure during operation. WAX and ANU [699] discuss the behaviour of the minimum variance beamformer in the presence of steering vector errors. WU R. and BAO [732] show that channel errors lead to severe distortions of the beam pattern. An optimisation procedure is proposed which suppresses clutter in the presence of channel errors while minimising the antenna sidelobe level. HIMED et al. [267, 269] have demonstrated that the performance of the JDL-GLR (see Chapter 9) in presence of array element errors can be increased by suitable design of auxiliary beams.

VARADARAJAN and KROLIK [660] propose an interpolation technique for distorted linear arrays as may occur in sonar applications. This interpolation results in a virtual undistorted linear array. The interpolation leads to a reduction of the clutter rank of the clutter covariance matrix and significantly higher SNIR compared with the distorted array.

It is, however, not easy to exploit these results for the purpose of clutter or jammer rejection. Adaptive clutter rejection is, like superresolution, a high-resolution technique which is reflected by the narrow clutter or jammer notches in the Doppler or azimuth response of the processor. high-resolution means that the antenna with its associated processor gives strong reactions to small variations of the involved parameters. As long as these variations pertain to target parameters only, such as location and velocity, high-resolution offers advantages over any processing based on conventional resolution of a beamformer (or a Doppler filter) in terms of position and velocity accuracy. Errors in the array antenna, however, will have more impact in the case of a high-resolution processor than just for conventional beamforming.

The effect of errors on the clutter or jammer rejection performance is manyfold. If the tolerances are known a priori then the errors can be incorporated in the beamformer weights so that no mismatch between beamformer and received wavefronts will occur. Only those processing schemes which rely on a certain sensor configuration will exhibit degraded performance. For example, the optimum processor (see Chapter 4) is expected to cope with such tolerances. The effect of mismatch between the steering vector and arriving wavefronts as well as mismatch between the adaptive clutter filter and the clutter covariance matrix has been investigated theoretically by BLUM and McDONALD [51].

If the errors are unknown as in the case of amplitude and phase jitter due to the electronic chains in each of the array channels then the beamformer is mismatched in any case to the desired signal as well as to clutter echoes. Therefore, some degradation of the performance will be encountered even with the optimum processor.

16.1.1 Tolerances of sensor positions

The effect of tolerances of the sensor positions in the horizontal dimension is that arrays with intended uniform spacing are no longer uniformly spaced. Clutter rejection techniques that rely on uniform spacing are expected to be sensitive with respect to such tolerances. Some numerical results have been reported by KLEMM [312].

16.1.1.1 *The optimum adaptive processor (OAP)*

Let us first consider the optimum processor (see Chapter 4). Figure 16.1 shows four curves featuring the optimum improvement factor for different levels of array position tolerances. The position errors have been assumed to be uniformly distributed within an interval $-\varepsilon, \ldots, \varepsilon$. The four curves have been calculated for $\varepsilon = 0, 1, 5, 20\%$ of the nominal element spacing. It was assumed that the errors are known and have been incorporated into the angle-Doppler steering vector.

As can be seen sensor position tolerances have no effect on the performance of the optimum processor. This reflects in essence that the optimum processor is basically independent of any special sensor configuration. It works perfectly even with random sensor distribution as in the case of the crow's nest antenna, see Chapter 8. The curves in Figure 16.1 have been calculated for a forward looking array. Similar results are obtained for sidelooking arrays.

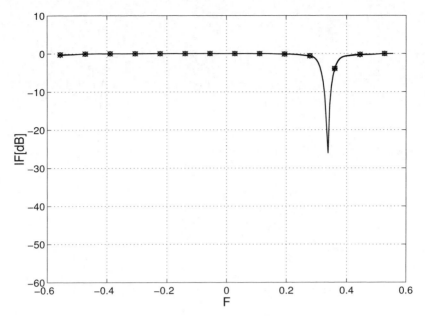

Figure 16.1: Influence of sensor position tolerances (optimum processor, FL): ○ $\varepsilon = 0\%$; ∗ $\varepsilon = 1\%$; × $\varepsilon = 5\%$; + $\varepsilon = 20\%$

16.1.1.2 *The auxiliary channel processor (ACP)*

Like the optimum processor the auxiliary channel processor (see Chapter 5) is not based on a specific configuration of the array sensors. However, it was found in Section 5.3 that the auxiliary channel processor is sensitive against any increase in the number of eigenvalues of the clutter covariance matrix. In this case the processor runs out of degrees of freedom which leads to losses in clutter rejection performance.

Recall that clutter echoes are characterised by a two-dimensional random process, which is band-limited in both space and time, see Section 3.5.2.2. The effect of sampling of a band-limited process can be illustrated by the following consideration. The correlation function of a band-limited process is obtained by inverse Fourier transform of a rectangular power spectrum which results in

$$\rho(k) = \frac{\sin(\frac{B}{2}kT)}{\frac{B}{2}kT} \exp(j\omega kT) \tag{16.1}$$

where ω is the carrier frequency and B the bandwidth.

For high sampling rates ($T \to 0$) one gets $\rho(k) \to 1$. The associated correlation matrix consists of unit elements only which means that there is one eigenvalue only.

For low sampling rates ($T \to \infty$) on gets $\rho(k) \to 0$, that is, the correlation matrix approaches the unit matrix which has N eigenvalues, N being the dimension of the correlation matrix.

Sensor position tolerances basically mean a change of the spatial sampling rate which has an effect on the eigenspectrum of the space-time clutter covariance matrix.

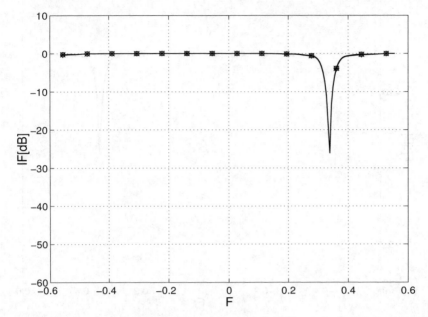

Figure 16.2: Influence of sensor position tolerances (auxiliary channel processor, FL):
$\circ \; \varepsilon = 0\%$; $* \; \varepsilon = 1\%$; $\times \; \varepsilon = 5\%$; $+ \; \varepsilon = 20\%$

If, however, the errors are small and have zero mean the effect on the clutter eigenspectrum can be expected to be small. This applies to the optimum and the auxiliary channel processor as well.

Figure 16.2 shows the IF curves for the auxiliary channel processor. No significant influence of the position tolerances can be noticed.

16.1.1.3 *The symmetric auxiliary sensor FIR filter (ASFF)*

In contrast to the two processors mentioned above the ASFF involves a reduction of the signal vector space by exploiting the properties of a linear equidistant antenna array. With a linear, uniformly spaced array overlapping subarrays of the same size can be formed. The concept of overlapping subarrays leads directly to the symmetric auxiliary sensor processor (see Figure 6.8). It was found in Chapters 6 and 7 that the symmetric auxiliary sensor processor in conjunction with a space-time FIR filter concept achieved near-optimum clutter rejection at very low computational expense.

The effect of sensor position tolerances on an overlapping subarray structure as shown in Figure 6.8 is that the subarray directivity patterns are all different. This means that the clutter spectra generated by these non-uniform subarrays are different. If follows that the subarray transform produces a transformed array whose outputs have different spectra. This results in spatial (sensor-to-sensor) decorrelation which finally causes losses in clutter rejection performance.

Figures 16.3 and 16.4 give an impression of the effect of sensor position tolerances on the performance of the ASFF. Figure 16.3 has been plotted for the forward looking

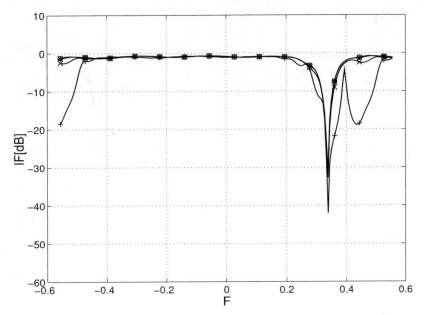

Figure 16.3: Influence of sensor position tolerances (auxiliary sensor FIR filter, FL): ○ $\varepsilon = 0\%$; * $\varepsilon = 1\%$; × $\varepsilon = 5\%$; + $\varepsilon = 20\%$

array while Figure 16.4 is for the sidelooking array. It can be noticed that only for large errors ($\varepsilon = 20\%$ of the error-free sensor spacing) do significant losses occur. In both cases the losses occur mainly at Doppler frequencies which correspond to the look direction of the sensor directivity pattern. For $\varepsilon = 5\%$ the losses are tolerable.

In Figure 16.5 the sensor position errors were chosen to be $\varepsilon = 20\%$. The four curves have been calculated for different lengths of the FIR filter L while keeping the data lengths M constant. It can be noticed that for $L = M = 24$ an almost optimum IF curve is obtained as far as clutter rejection is concerned. However, the curve runs about 14 dB below the optimum for the following reason. Remember that the number of filter output samples is $M - L + 1$. Therefore, as L approaches M there are no data left for subsequent Doppler filtering. Consequently the loss compared with the optimum is 10 log 24 = 13.8 dB.

The FIR filter technique is efficient only if the filter length L is small compared with the length of the data record. Therefore, if a FIR approach is under discussion, one should try to build the antenna as precisely as possible instead of using error compensation through long filters.

16.1.2 Array channel errors

The effect of amplification errors (amplitude and phase or IQ errors), delay errors has been analysed by NICKEL [486] for jammer cancellation applications. Some examples are also given by WIRTH and NICKEL [724]. In this work the influence of errors in conjunction with the system bandwidth on the jammer covariance matrix is presented.

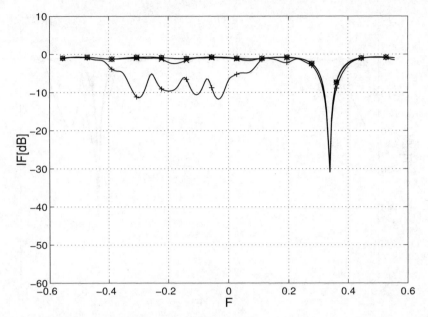

Figure 16.4: Influence of sensor position tolerances (auxiliary sensor FIR filter, SL): ○ $\varepsilon = 0\%$; $*\ \varepsilon = 1\%$; $\times\ \varepsilon = 5\%$; $+\ \varepsilon = 20\%$

Sensor position tolerances cause phase errors only. There is a fundamental difference between sensor position errors and channel phase errors. In the case of sensor position errors the response of the k-th sensor of a linear array to a wave arriving from angle φ is

$$s_k(t) = A \exp\left(j\omega t + \frac{2\pi}{\lambda}(x_k + \Delta x_k)\sin\varphi\right) \quad (16.2)$$

where Δx is the position error. In the case of channel errors one gets instead

$$s_k(t) = (A + \Delta A_k) \exp\left(j\omega t + \frac{2\pi}{\lambda}(x_k)\sin\varphi + \Delta\phi_k\right) \quad (16.3)$$

where ΔA and $\Delta \phi$ are amplitude and phase errors.

As one can see from these equations the effect of sensor position tolerances depends on the source direction. Channel errors, however, lead to a corruption of the received wavefront independently of the direction of arrival.

It can be expected that space-time clutter covariance matrices are affected by array errors in a similar way as spatial jammer + noise matrices. Therefore, a detailed analysis for space-time covariance matrices has been omitted here. Instead, we present just two typical numerical examples which show the effect of array errors and their relation to the system bandwidth clearly.

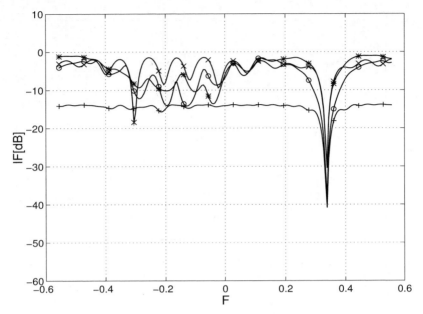

Figure 16.5: Dependence on temporal filter length (auxiliary sensor FIR filter, SL, $\varepsilon = 20\%$): ○ $L = 3$; ∗ $L = 6$; × $L = 12$; + $L = 24$

16.1.2.1 IQ errors

Figure 16.6 shows eigenspectra for the *spatial* covariance matrix for the case of four jammers radiating on a linear array. The ○ curve shows the ideal case (no bandwidth, no errors). The jammer positions have been chosen so that they can be resolved by the array. Therefore, they are represented in the eigenspectrum as four dominant eigenvalues.

The dashed curve shows additional eigenvalues caused by 2% system bandwidth. Similar eigenspectra have been shown for space-time clutter covariance matrices in Figures 3.29 and 3.31. The three solid curves show results of different computer runs with gaussian IQ errors ($\sigma = 0.1$) added.[1] An obvious increase of the number of eigenvalues can be noticed.

In Figure 16.7 the corresponding SNIR curves are shown. All curves are based on the optimum spatial processor. The vertical dashed lines denote the jammer positions. The ○ curve shows the optimum gain for the ideal case (no bandwidth, no errors).

The dashed curve shows the effect of system bandwidth and the three solid lines again show different samples when gaussian IQ errors are added. It can be seen that the bandwidth effect results mainly in a broadening of the jammer notches so as to limit the resolution capability of the array. However, like in Figures 4.22 and 4.23 outside the broadened interference notch (jammer or clutter) the optimum IF is obtained. In the case of additional IQ errors (solid curves) even in the pass band (on the far right and far left in Figure 16.7) the maximum SNIR is not reached. Also between the jammers

---

[1] The errors vary statistically across the bandwidth for each sensor.

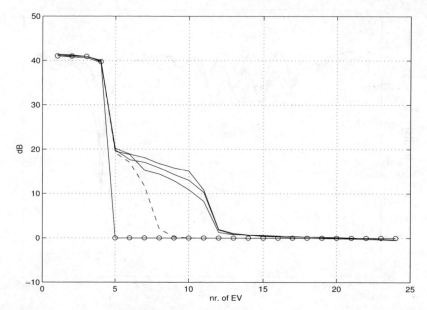

*Figure 16.6: Impact of IQ errors on the eigenspectrum (optimum jammer suppression):
○ no errors, zero bandwidth;* dashed line: *2% bandwidth;* solid lines: *IQ errors with
$\sigma = 0.1$ (By courtesy of U. Nickel)*

the erroneous curves run below the dashed line.

The impact of array errors may also influence the processor architecture. SUN et al. [622, 623] propose a post-Doppler processor scheme which is relatively robust against array errors. Post-Doppler processors tend to be less sensitive against errors of the sensor patterns (WARD and STEINHARDT [692]). Because the individual Doppler filters provide narrowband processing which is equivalent to using only a small sector of the element pattern the spectral effects caused by different sensor patterns is mitigated.

16.1.2.2 Additional delay errors

The effect of different delays in the various array channels (with amplitude and phase errors) can be described as follows

$$s_k(t) = (A + \Delta A_k) \exp\left(\mathrm{j}\omega(t + \Delta t_k) + \frac{2\pi}{\lambda}(x_k)\sin\varphi + \Delta\phi_k\right) \qquad (16.4)$$

where Δt_k is the delay in the k-th channel.

It can be seen from Figure 16.8 that the number of eigenvalues is increased even further (compare with Figure 16.6). Nevertheless, there are areas in the SNIR plot (Figure 16.9) where the SNIR *with* IQ and delay errors is superior to the dashed line (system bandwidth only). In fact, stochastic delay errors can compensate to a certain extent for bandwidth effects and IQ errors.

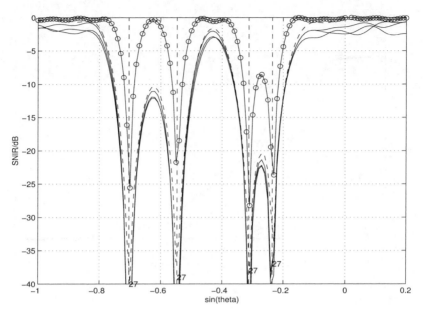

Figure 16.7: Impact of IQ errors on SNIR (optimum jammer suppression): ○ *no errors, zero bandwidth;* dashed line: *2% bandwidth;* solid lines: *IQ errors with* $\sigma = 0.1$ *(By courtesy of U. Nickel)*

The results shown in Figures 16.6–16.9 are based on the optimum processor. Some results on the error sensitivity of the FIR filter processor after (7.12) are given by LIU and PENG [409]. These results suggest that increasing the filter length L may mitigate the error effects. This is in accordance with the results shown in Figure 16.5, where the filter length was increased to compensate for sensor position tolerances. The effect of errors on several other processors are presented by WANG Y. and PENG [681].

16.1.2.3 DC offset error

Offset errors create an additional eigenvalue which gives the impression of an additional interfering source (NICKEL [486]).

16.1.2.4 Tolerances of the filter characteristics

Adaptive jammer nulling is slightly affected by tolerances of the matched filter transmission characteristics in that the SNIR curves run slightly below the optimum. The effect of unequal filter characteristics is equivalent to raising the noise level (NICKEL [486]).

The role of channel errors in space-time processing has briefly be discussed by BARILE *et al.* [35]. They present a few curves which show the decay of the improvement factor versus the error variance. Increasing the number of antenna elements mitigates the effect of channel errors.

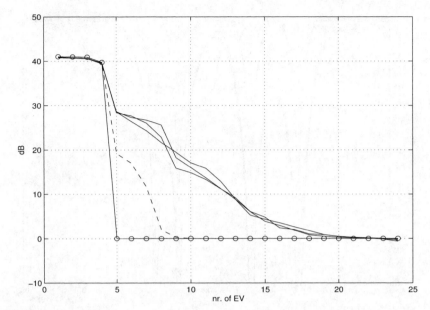

Figure 16.8: Impact of IQ errors on the eigenspectrum (optimum jammer suppression): ○ *no errors, zero bandwidth;* dashed line: *2% bandwidth;* solid lines: *IQ errors with* $\sigma = 0.1$, *delay errors with* $B\tau = 0.1$ *(By courtesy of U. Nickel)*

16.1.2.5 *Mutual coupling*

Mutual coupling between the array elements can degrade the STAP performance significantly if not properly accounted for (FRIEL and PASALA [167]). The achievable SNIR is lowered over the whole Doppler range. The clutter ridge is broadened which results in a broadened clutter notch of the STAP filter. This in turn means degradation of detection of slow targets. If the mutual coupling matrix is known the STAP performance can be restored. However, coupling has to be known to a distance of about 4λ. The effect of mutual coupling is relevant for relatively small arrays. It does not influence the performance of large arrays significantly. From the viewpoint of mutual coupling beamspace STAP algorithms appear to be more robust than techniques working in the element space.

16.1.3 Channel equalisation

16.1.3.1 *Array calibration*

LE CHEVALIER *et al.* [380] have proposed a calibration technique for airborne array radar based on comparison of returns due to a discrete clutter scatterer. One array element is used as reference. Between the reference element and the other elements a well-known phase relation exists which is related to the direction of the clutter scatterer (or calibration source). This technique requires knowledge of the direction of the

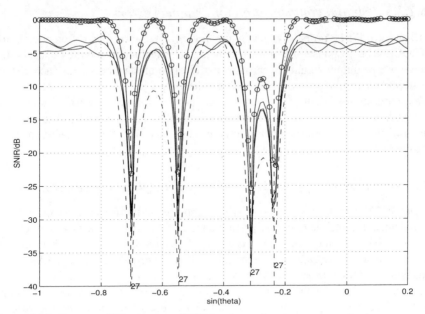

Figure 16.9: Impact of IQ errors on SNIR (optimum jammer suppression): ○ *no errors, zero bandwidth;* dashed line: *2% bandwidth;* solid lines: *IQ errors with* $\sigma = 0.1$, *delay errors with* $B\tau = 0.1$ *(By courtesy of U. Nickel)*

calibration scatterer. It cannot be applied if no dominant scatterer is available, for instance in homogeneous clutter. BICKERT *et al.* [46] applied corrections to the AER-II data (4 channel array) including zero off-set subtraction, compensation for undesired component in the transfer functions of the individual channels, and adjustment of the transfer functions of different array channels.

16.1.3.2 Steering vector optimisation

LEE C. C. and LEE J. H. [383] consider the effect of steering vector errors for the case where the signal is included in the array covariance matrix, that is,

$$\mathbf{R} = \mathbf{Q}_i + \mathbf{Q}_n + \mathbf{Q}_s \tag{16.5}$$

Weighting vectors are optimised in such a way that the effect of steering errors is minimised. Two approaches have been discussed: 1. a weighting vector involving the inverse of $\mathbf{R}$, 2. a weighting vector involving only the noise subspace of $\mathbf{R}$. It is shown numerically that both techniques yield considerable improvement over the case that the steering vector is used with perturbations included. In the case of low SNIR the second technique based on orthogonal projection performs better than the first one.

16.1.3.3 Spatial compensation for channels errors

WU and LI [731] propose a technique for compensation for differences in the FREQUENCY responses of array channels. The technique requires a calibration source whose output signal (linear chirp) is fed into all channels. For each channel a FIR filter is optimised so as to equalise the frequency responses of all channels. The optimisation procedure is based on the discrete Fourier transform. The residual mismatch after equalisation in magnitude and phase is analysed. The effect of equalisation filter length, bandwidth-time product, and the effect of data windowing is discussed.

16.1.3.4 Ender's approach: frequency-FREQUENCY calibration

ENDER [127] has proposed a technique for compensating for array errors in order to maximise the clutter rejection performance in SAR systems with a multichannel antenna. The technique is called *adaptive 2-D calibration*. It has been applied successfully to clutter data received by the AER II airborne multichannel SAR system (RÖSSING and SKUPIN [574]). About 10 dB improvement in SCNR has been achieved experimentally by using this calibration technique.[2] This calibration technique operates in the range and Doppler dimension and makes use of the fact that the received data have been digitised at IF (intermediate frequency). Then matched filtering is accomplished by digital filters which offer the potential of adjusting the transmission characteristics.

Let $z^{(k)}(t,T)$ denote the signal of the k-th antenna channel, where t denotes the 'fast time' associated with the delay of the incoming wave, and T is the 'slow time' given by the sequence of echo pulses. Let $Z^{(k)}(\omega,\Omega)$ be the 2-D Fourier transform of $z^{(k)}(t,T)$. ω denotes the baseband signal frequency while Ω is the Doppler frequency.

Consider a stationary scene (that is, no moving targets present). Then the spectra of channels k and l are related as follows:

$$Z^{(k)}(\omega,\Omega) = h^{(k,l)}(\omega) Z^{(l)}(\omega,\Omega) H^{(k,l)}(\Omega) \qquad (16.6)$$

where $h^{(k,l)}(\omega)$ means the ratio between the transfer functions of the k-th and l-th channel and $H^{(k,l)}(\Omega)$ the ratio between the k-th and l-th Doppler transfer functions which is

$$H^{(k,l)}(\Omega) = \exp\{j\Omega\Delta T(k,l)\} \frac{D^{(k)}(u(\Omega))}{D^{(l)}(u(\Omega))} \qquad (16.7)$$

$D^{(k)}(u(\Omega))$ denotes the two-way directivity pattern of the k-th antenna transmit–receive pair, $u(\Omega)$ is the directional cosine belonging to the clutter Doppler frequency Ω, and $\Delta T(k,l)$ is the time shift given by the azimuthal displacement between the phase centres of the two antenna pairs normalised by the platform velocity.

Let us call the k-th channel a reference channel. Then the l-th channel can be adjusted to the reference channel by minimising the expression

$$\int | Z^{(k)}(\omega,\Omega) - h^{(k,l)}(\omega) Z^{(l)}(\omega,\Omega) H^{(k,l)}(\Omega) |^2 \, d\omega d\Omega \qquad (16.8)$$

---

[2] Private communication by J. Ender.

by variation of $h^{(k,l)}(\omega)$ and $H^{(k,l)}(\Omega)$. The solution is given by a pair of integral equations

$$h^{(k,l)}(\omega) = \frac{\int (Z^{(l)}(\omega,\Omega)H^{(k,l)}(\Omega))^* Z^{(k)}(\omega,\Omega) d\Omega}{\int |Z^{(l)}(\omega,\Omega)H^{(k,l)}(\Omega)|^2 d\Omega} \quad (16.9)$$

$$H^{(k,l)}(\Omega) = \frac{\int (h^{(k,l)}(\omega)Z^{(l)}(\omega,\Omega))^* Z^{(k)}(\omega,\Omega) d\omega}{\int |h^{(k,l)}(\omega)Z^{(l)}(\omega,\Omega)|^2 d\omega}$$

This solution can be approximated by the following iteration

$$\tilde{Z}_{n+1}(\omega,\Omega) := \tilde{Z}_n(\omega,\Omega)\frac{\int \tilde{Z}_n^*(\omega,\Omega)Z^{(k)}(\omega,\Omega) d\Omega}{\int |\tilde{Z}_n(\omega,\Omega)|^2 d\Omega} \quad (16.10)$$

$$\tilde{Z}_{n+2}(\omega,\Omega) := \tilde{Z}_{n+1}(\omega,\Omega)\frac{\int \tilde{Z}_{n+1}^*(\omega,\Omega)Z^{(k)}(\omega,\Omega) d\omega}{\int |\tilde{Z}_{n+1}(\omega,\Omega)|^2 d\omega}$$

where $\tilde{Z}_n$, $n = 1, 2, \ldots$ means the iteratively improved calibrated version of $Z^l(\omega,\Omega)$. Notice that the calibrated signal is updated in an alternating fashion in both of the two frequency domains. This means that errors in both the FREQUENCY responses and the Doppler responses[3] of the individual channels is compensated for.

It should be noted that this technique is based on a two-dimensional frequency domain approach which implies that the data sequences received by the radar are sufficiently long (see also Section 9.5). It has been successfully applied to data received with the AER II multichannel SAR in the sidelooking configuration (ENDER [128, 129]).

A modification of this technique for use with forward looking radar is an open issue.

16.1.3.5 *Estimation of the array manifold*

ENDER [128, 129] used a technique for estimating the array manifold (i.e., a complete set of steering vectors) for use with SAR, that is, with long data sequences. The procedure runs as follows:

1. Take the azimuthal Fourier transform of the recorded clutter data.

2. Calculate spatial covariance matrices for all Doppler frequencies according to

$$\hat{\mathbf{Q}}(f) = \frac{1}{R}\sum_{i=1}^{R} \mathbf{c}_i(f)\mathbf{c}_i^*(f) \quad (16.11)$$

where R denotes a certain range interval and $\mathbf{c}_i(f)$ are vectors including spatial clutter + noise data at frequency f.

---

[3]Due to differences between sensor patterns.

3. Take for each frequency the eigenvector associated with the largest eigenvalue as steering vector.

When this array manifold is used for imaging, target location, etc., all effects due to antenna channel errors, distortion effects caused by the antenna environment, and platform motion effects on the steering vector are taken care of. The technique has been used successfully with a sideways looking array. For all other array configurations there are two angles of arrivals associated with one Doppler frequency which means that the spatial (frequency dependent) covariance matrices exhibit two clutter eigenvalues. It is not obvious how the eigenvectors associated with the two dominating eigenvalues can be used to estimate the angles of arrival.

Once a certain grid of steering vectors has been estimated a finer grid can be generated by interpolation (SCHMIDT [592], WORMS [725]).

16.1.3.6 Combined covariance matrix/beamformer estimation

ROBEY and BARANOSKI [572] developed an iterative procedure for simultaneous estimation of the clutter covariance matrix and the steering vector. The authors describe a technique for estimating the array manifold that is supposed to operate with short data sequences. The algorithm is based on the *expectation-maximisation* principle.

The following steps have to be carried out:[4]

1. **Initial values**: Determine a set of initial beamformers $\mathbf{b}_i^p$ so as to cover the whole angular domain[5] (i denotes a certain clutter patch at a certain azimuth angle). Determine a set of clutter powers for all clutter patches P_i^p. p denotes the iteration step.

2. **Space-time clutter manifold**: Form space-time clutter vectors by use of beamformers $\mathbf{b}_i^p$ and associated Doppler filter vectors $\mathbf{d}_i$: $\mathbf{a}_i^p = \mathbf{d}_i \otimes \mathbf{b}_i^p$.

3. **Covariance matrix**: Estimate the space-time covariance matrix from the measured set of data vectors $\hat{\mathbf{Q}} = \frac{1}{K}\sum_{k=1}^{K} \mathbf{c}_k \mathbf{c}_k^*$.

4. **Approximate response of single clutter patch**: Form the approximate covariance matrix for an individual clutter patch i: $\mathbf{Q}_i^p = P_i^p \mathbf{a}_i^p \mathbf{a}_i^{*p} + P_n \mathbf{I}$. It is assumed that the white noise power P_n is known.

5. **Approximate total clutter response**: Form the approximate covariance matrix for the total of clutter arrivals: $\mathbf{Q}^p = \sum_{i=1}^{I} P_i^p \mathbf{a}_i^p \mathbf{a}_i^{*p} + P_n \mathbf{I}$.

6. **Iteration**: Calculate the conditional expected value of the covariance matrix due to the i-th angular direction according to

$$\mathbf{Q}_i^{p+1} = \mathbf{Q}_i^p - \mathbf{Q}_i^p[(\mathbf{Q}^p)^{-1} - (\mathbf{Q}^p)^{-1}\hat{\mathbf{Q}}(\mathbf{Q}^p)^{-1}]\mathbf{Q}_i^p \qquad (16.12)$$

7. **Transform into the Doppler domain**: $\mathbf{S}_i^{p+1} = \mathbf{F}\mathbf{Q}_i^{p+1}\mathbf{F}^*$ according to (9.1), (9.3).

---
[4]We use here our nomenclature. We describe the algorithm in a bit more detail than given in [572].
[5]Compare with the auxiliary channel processor, Section 5.3.

8. **Beamformer update**: Find the updated beamformer vector $\mathbf{b}_i^{p+1}$ as an eigenvector associated with the maximum eigenvalue of the *spatial* submatrix of $\mathbf{S}_i^{p+1}$ associated with the i-th Doppler frequency.

9. **Clutter power**: $P_i^{p+1} = \mathbf{a}_i^{*(p+1)} \mathbf{Q}_i^{p+1} \mathbf{a}_i^{p+1}$ where $\mathbf{a}_i^{p+1} = \mathbf{d}_i \otimes \mathbf{b}_i^{p+1}$.

10. Do the calculations for all i.

11. Iterate by setting $p := p + 1$.

The computational workload of this algorithm is quite high, however, it converges rapidly. Moreover, calibration of the array might not be required continuously but eventually only once at the beginning of an operational phase. Some 'real world effects' such as mutual coupling between elements, backlobe effects, or non-homogeneity of clutter data may degrade the calibration accuracy.[6]

16.2 Range dependence of clutter Doppler

16.2.1 Impact on adaptation and filtering

The amount of data required for adaptation is of the order of magnitude of $2NM$ for the optimum processor (Chapter 4) and $2LK$ for the space-time FIR filter processor after Chapter 7 (see REED *et al.* [557]).

Basically the adaptation of the clutter filter (e.g., updating the space-time covariance matrix)[7] can be done by use of clutter echo data either gathered from the time or range dimensions or both. In the case of sidelooking array radar the range dimension is the preferable choice because the clutter Doppler is independent of range.

In the case of a forward looking radar the clutter Doppler depends strongly on range which requires that the clutter filter has to be updated while running through the data along the range dimension. This causes additional computational load for the adaptation.

However, we are faced with another, even more stringent dilemma. On the one hand the amount of data associated with a sufficient narrow Doppler band may not be sufficient for adaptation, so that losses in SNIR may occur. On the other hand, if data are taken from a large number of range elements, all of them being associated with different Doppler frequencies, we can expect a broadening of the clutter notch according to the Doppler bandwidth in the data which results again in degradation of slow target detection. As a compromise the incoming data may be subdivided into range segments whose widths increase with range according to Figure 3.6 (LIAO *et al.* [399]).

To evade this dilemma one may think of gathering clutter data from the time dimension (echo sequence) separately for each range increment. This may work as long as the radar dwell time is sufficiently long to produce the required amount of data.

---

[6]Private communication by E. Baranoski.
[7]There are other well-known techniques for direct calculation of the filter from the clutter data, such as updating the inverse, or the gradient technique, see BRENNAN *et al.* [63].

The dwell time of the radar in a certain direction, however, may depend on operational requirements beyond the scope of adaptive MTI processing. For example, efficient use of the radar power by matching to the radar range equation leads to the requirement that the dwell time is short at near range and long at far range. However, at near range we have the problem of range dependence of the clutter Doppler and, hence, shortage of data in the range dimension. Furthermore, the time dimension is not usable for adaptation in the case of a staggered PRF. In this case FIR filters have coefficients that vary during adaptation.

In summary, adaptive techniques and algorithms are required which achieve clutter suppression based on a limited amount of data samples. Orthogonal projection algorithms have proven to achieve some advantage over standard adaptive techniques based on the full covariance matrix. Of course, reducing the order (not only the rank!) of the signal subspace, as has been done in Chapters 5, 6, 7 and 9, is the most efficient way of reducing the required amount of data. Single snapshot techniques (e.g., KOC and TANIK [354], PARK S. and SARKAR [524]) may also play a role in rapidly adapting filters.

16.2.2 Doppler compensation

Some improvement can be obtained by Doppler compensation. Based on a priori knowledge of the radar–target geometry or measurements made by an inertial navigation system the Doppler dependence can approximately be compensated for (see BORSARI [58] and KREYENKAMP [369]). This technique is based on the correction of a single frequency. Therefore, it can be applied only to range unambiguous radar data. In the thesis by LAPIERRE [376] a full Doppler-angle compensation technique for bistatic STAP radar based on the registration of Doppler direction (Lissajou) curves is described. In another publication by LAPIERRE et al. [377] two methods for Doppler compensation in non-sidelooking STAP radar are described. ZULCH [752] has pointed out that in space-based GMTI radar the effect of earth rotation has to be compensated for.

The principle of *adaptive* Doppler compensation has been proposed and verified with experimental radar data by KREYENKAMP and KLEMM [370]. Data were obtained from the AER II experiment (ENDER [127]). The following steps have to be conducted:

- Estimate the clutter Doppler frequency versus range via FFT analysis for a given look direction.

- Transform the received data for Doppler compensation.

- Estimate the space-time clutter covariance by using compensated clutter data from various range bins according to

$$\hat{Q} = \frac{1}{R} \sum_{i=1}^{R} c_i c_i^* \qquad (16.13)$$

- Invert the estimated clutter covariance matrix.
- Suppress clutter components for the whole visible range by multiplying the echo data by the inverse of the clutter covariance matrix.
- Integrate target signal energy using a beamformer cascaded with a Doppler filter bank.

16.2.2.1 Transform of echo data

Suppose a set of space-time echo vectors $\mathbf{x}(r)$ is received by a forward looking array radar (r denotes a certain range bin). The Doppler frequency varies from range bin to range bin as was explained before. Let r_1 denote the cell under test and $r_{21},\ldots,r_{2k},\ldots r_{2K}$ the range of available training data. We look for linear range-dependent transforms $\mathbf{T}(r_{2k})$ so that the sequence

$$\mathbf{x}(r_1) = \mathbf{T}^*(r_{2k})\mathbf{x}(r_{2k}) \qquad (16.14)$$

has a range-independent Doppler frequency associated with range r_1. Each of the $NM \times 1$ data vectors has the form

$$\mathbf{x}(r_{2k}) = \begin{pmatrix} \mathbf{x}_s(r_{2k})\exp[j2\pi f_\mathrm{D} T(r_{2k})] \\ \mathbf{x}_s(r_{2k})\exp[j2\pi 2 f_\mathrm{D} T(r_{2k})] \\ \vdots \\ \mathbf{x}_s(r_{2k})\exp[j2\pi M f_\mathrm{D} T(r_{2k})] \end{pmatrix} \qquad (16.15)$$

where $\mathbf{x}_s$ denotes a spatial subvector containing direction-of-arrival weights, m is the time index, T the pulse repetition interval, N the number of array elements, and M the number of echo pulses.

Then the range-dependent Doppler correction matrix $\mathbf{T}(r_{2k})$ in (16.14) is given by

$$\mathbf{T}(r_{2k}) = \begin{pmatrix} \mathbf{D}_1 & & & & 0 \\ & \ddots & & & \\ & & \mathbf{D}_m & & \\ 0 & & & \ddots & \\ & & & & \mathbf{D}_M \end{pmatrix} \qquad (16.16)$$

where the $N \times N$ submatrices $\mathbf{D}_m$ have diagonal form

$$\mathbf{D}_m = \begin{pmatrix} t_m(r_{2k}) & & 0 \\ & \ddots & \\ 0 & & t_m(r_{2k}) \end{pmatrix} \qquad (16.17)$$

and

$$t_m(r_{2k}) = \mathrm{e}^{-j2\pi(f_\mathrm{D}(r_{2k}) - f_\mathrm{D}(r_1))mT} \quad \text{for } k=1,\ldots,K \qquad (16.18)$$

Notice that the compensation coefficients on the diagonal of $\mathbf{D}_m$ are constant.

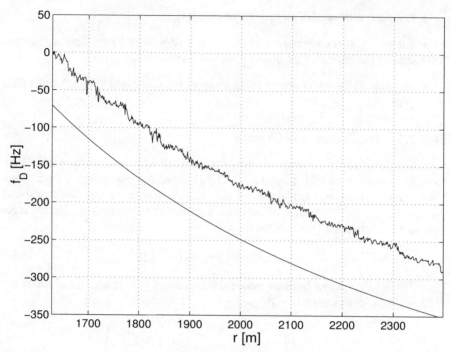

Figure 16.10: Estimated clutter Doppler (averaging over four array channels)

After Doppler compensation the compensated clutter data can be used to estimate the space-time covariance matrix according to (16.13) which, after inversion, becomes the required clutter filter matrix. Notice that the Doppler compensation requires only NM (or KL, depending on the STAP algorithm used) complex multiplications which can easily be carried out in real time.

16.2.2.2 Estimation of the clutter Doppler

If all relevant parameters (aircraft position, direction, velocity, attitude) are known with sufficient precision the range dependence of the clutter Doppler could be calculated from (3.20) and inserted into the transforms given by (16.16). Under ideal conditions this would result in a perfect compensation for the Doppler gradient.

In practice the flight parameters (position, direction, velocity, attitude) are not precisely known. An aircraft has its own motion which depends also on environmental parameters such as wind speed. The accuracy of positioning and velocity estimation of navigation aids is limited so that the aircraft flight parameters are not precisely known. This issue leads us to the concept of adaptive Doppler transforms.

The range-dependent clutter Doppler frequency can be obtained from clutter data by Fourier analysis (using the FFT) and selection of the frequency associated with the largest response. Such Doppler estimates will be corrupted by noise. Since we are

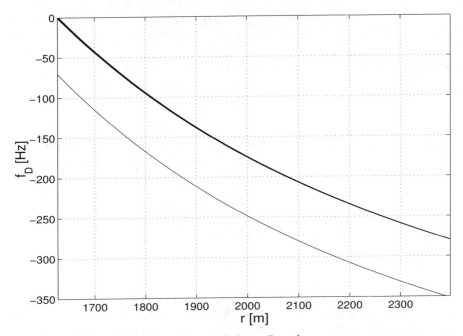

Figure 16.11: Least squares estimation of clutter Doppler

dealing mainly with short range the clutter-to-noise ratio can be expected to be quite high. Additional mitigation of errors due to noise can be achieved by averaging the clutter Doppler spectra over different array channels. Of course, a moving target in one of the range bins would lead to Doppler error. Simulations have shown that the estimation accuracy depends heavily on the number of processed echo pulses. On the one hand, the FFT increases the clutter-to-noise ratio which mitigates noise errors by coherent integration. On the other hand, with increasing number of echoes the spectral resolution increases which leads to more accurate Doppler estimates.

Figure 16.10 shows a numerical example. The Doppler frequency has been plotted versus range. The upper curve is based on measured AER II data. Since the radar looked out of the rear door of the carrier aircraft we obtain a negative Doppler slope. The lower (smooth) curve describes the theoretical Doppler frequency based on GPS information on the aircraft position. Except for a Doppler off-set the curves agree quite well. The off-set is within the measurement accuracy of the GPS receiver.

Least squares approximation is another way of estimating the range dependence of the clutter Doppler. The idea behind this concept is that the range dependency of the Doppler frequency is basically known from the relation

$$f_\mathrm{D} = \frac{2v_\mathrm{p}}{\lambda} \cos\varphi \sqrt{1 - \frac{H^2}{r_s^2}}. \qquad (16.19)$$

Therefore, the Doppler frequency can be approximated by

$$f_\mathrm{D}(r) \approx \frac{2v}{\lambda} - \frac{v}{\lambda}\frac{H^2}{r^2} - \frac{v}{4\lambda}\frac{H^4}{r^4} \qquad (16.20)$$

where $v = v_\mathrm{p} \cos\varphi \cos\vartheta$. Let us define a parameter vector $\mathbf{a} = [a_1, a_2, a_3]^\mathrm{T}$ with $a_1 = \frac{2v}{\lambda}$, $a_2 = -\frac{vH^2}{\lambda}$ and $a_3 = -\frac{vH^4}{4\lambda}$ and a data matrix of the form

$$\mathbf{X} = \left[1, \ldots, 1; r_{21}^{-2}, \ldots, r_{2K}^{-2}; r_{21}^{-4}, \ldots, r_{2K}^{-4}\right]^\mathrm{T}$$

where $[r_{21}, \ldots, r_{2K}]$ denote the range bins associated with the available data. Furthermore $\mathbf{f}_\mathrm{D} = [f_\mathrm{D}(r_{21}), \ldots, f_\mathrm{D}(r_{2K})]^\mathrm{T}$ is the vector of Doppler frequencies calculated by taking the temporal Fourier transform of the echo data at range r_{2k}. The vector

$$\hat{\mathbf{a}} = (\mathbf{X}^T\mathbf{X})^{-1}\mathbf{X}^T\mathbf{f}_\mathrm{D} \qquad (16.21)$$

is then a least squares match for the Doppler frequency

$$\hat{f}_\mathrm{D}(r_k) = \hat{a}_1 + \hat{a}_2 \frac{1}{r_k^2} + \hat{a}_3 \frac{1}{r_k^4} \quad \text{for all } k = 1, \ldots, K \qquad (16.22)$$

An example is shown in Figure 16.11. The fat curve denotes the Doppler frequency as obtained with least squares matching from AER II clutter data. The lower curve shows the theoretical curve based on GPS information. Again we notice an off-set which is caused by the error of the navigation system.

The effect of Doppler compensation is illustrated by simulations shown in Figure 16.12. Comparison of compensated and uncompensated processing reveals the dramatic improvement of the proposed techniques. The uncompensated curve in Figure 16.12 exhibits an extremely broad clutter notch so that detection of slowly moving objects becomes questionable. Under these conditions it is questionable if space-time adaptive processing brings any advantage compared with conventional processing (beamforming + temporal processing). Recall that the broadening of the clutter notch happens because the uncompensated training samples cover a certain Doppler bandwidth. Even outside the clutter notch the compensated processor clearly works better. Of course, the difference between compensated and uncompensated performance is greater at short range than at far range. Outside the clutter notch we find a loss in SNIR of about 2 dB. This loss is caused mainly by the limited number of training samples rather than by imperfect Doppler compensation.

In the following four examples we demonstrate the effectiveness of the adaptive Doppler compensation technique by applying to clutter data measured with the AER II multichannel SAR system.

In this case we have no means of calculating the optimum SNIR because the exact space-time clutter covariance matrix is unknown. Instead, the best possible estimate of the covariance matrix has been obtained by averaging over all available data (quasi-optimum covariance matrix). The data for calculating the quasi-optimum covariance matrix have also been Doppler compensated. The upper curves in Figures 16.13 and 16.14 denote the improvement factor (IF = $\mathrm{SNIR}_\mathrm{output}/\mathrm{SNIR}_\mathrm{input}$) achieved by the quasi-optimum processor.

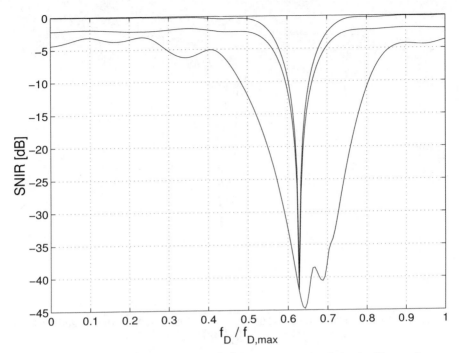

Figure 16.12: Doppler compensation at short range (simulation). Top to bottom: optimum compensated processing, estimated covariance matrix ($K = 100$ samples) with Doppler compensation, without Doppler compensation ($N = M = 15$, $H = 700$ m, $r_1 = 900$ m, $r_{2k} = 901$ to 1000 m)

The lower curves have been calculated for the optimum processor according to (1.3) with the space-time covariance matrix estimated from 800 training samples. In Figure 16.13 no Doppler compensation was applied. We notice some losses in the pass band and a considerably broadened clutter notch. Since our targets of interest (ground objects such as vehicles) show up very close to the clutter notch we recognise a considerable degradation compared with the quasi-optimum compensated processor. After Doppler compensation (Figure 16.14) the quasi-optimum IF curve is well approximated. In particular, we find a very narrow clutter notch.

Figures 16.15 and 16.16 show two examples which are typical for application with SAR. To obtain high geometrical resolution the CPI (coherent processing interval MT) has to be large. Whenever the CPI is large the frequency channels of the Fourier transform tend to become uncorrelated. This results in a block diagonal space-time covariance matrix. The inverse of this matrix is given by the inverses of the various spatial submatrices at all Doppler frequencies (frequency-dependent spatial processing, see Chapter 9). The space-time processing reduces therefore to a frequencywise spatial filtering.

This case is illustrated in Figures 16.15 (no Doppler compensation) and 16.16 (with Doppler compensation). The effect of Doppler correction on the width of the clutter

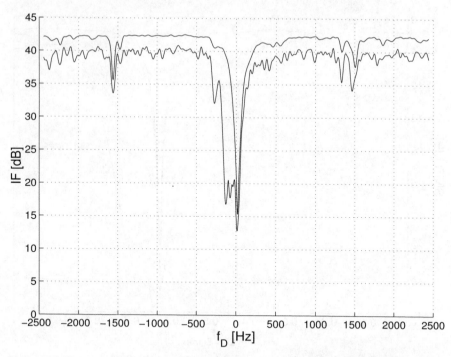

Figure 16.13: Doppler compensation at short range (AER II data). Upper curve: optimum compensated processing, lower curve: estimated covariance matrix ($K = 100$ samples) without Doppler compensation ($N = M = 15$, $H = 700$ m, $r_1 = 900$ m, $r_{2k} = 901$ to 1000 m)

notch and, hence, on the capability of detecting slow moving targets is obvious.

16.2.2.3 Doppler-direction compensation

As has been noted already in section 12.2.4 in bistatic or multistatic radar configurations not only the clutter Doppler but also the receive direction changes with range. In this case Doppler compensation as illustrated before is not sufficient. Instead, Doppler and angle compensation is necessary as has been pointed out by HIMED [266]. A treatment of the Doppler-direction compensation problem based on Doppler-direction curve alignment can be found in the thesis by LAPIERRE [376].

16.3 Aspects of implementation

16.3.1 Comparison of techniques in terms of computational complexity

In Tables 15.1 and 15.3 the numbers of floating point operations (complex multiplications) for the various parts of the adaptive clutter filter and target doppler

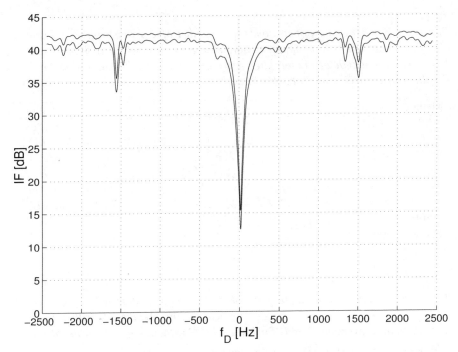

Figure 16.14: Improvement through Doppler compensation (AER II data): Upper curve: quasi-optimum as before ($K = 1024$), lower curve: estimated covariance with Doppler compensation ($M = 100$, $K = 800$)

match are given. Table 15.1 summarises those operations not strictly connected with the range sampling frequency. They are therefore called 'slow' operations.

Table 15.3 gives the numbers of operations required for transform, clutter and Doppler filtering. These operations have to be carried out for each range increment and are therefore called 'fast' operations.

For estimating the clutter covariance matrix (adaptation) we assumed an average over twice the order of the clutter covariance matrix. This value has been shown by REED *et al.* [557] to be sufficient to approximate the optimum SNIR within 3 dB.

For filter calculation we used the number of operations required for a matrix inverse. For time domain techniques the Akaike algorithm for inverting block Toeplitz matrices (AKAIKE [12]) was assumed which requires $8N^3 M$ operations (GOVER and BARNETT [208]). For frequency domain techniques and space-time transform techniques $(NM)^3$ operations have been assumed for matrix inversion.

All figures given are rough indicators for the complexity of the various processing steps. More accurate numbers depend on the actual algorithms chosen. It should be noted that K and L are usually small compared with N and M (typically $N, M = 30$, $K, L = 3, \ldots, 5$). Tables 15.2 and 15.4 give some numerical examples for those parameters used mainly in this book ($K = L = 5$; $N = M = 24$).

The computational load required for adapting the CSM subspace processor

Table 16.1 Number of slow operations (N = number of array elements M = Doppler filter length or number of doppler channels; K = number of antenna channels; L = temporal length of clutter filter; *by use of the Akaike algorithm for block Toeplitz matrix inversion; **only for large M; ***without eigenvector decomposition)

Method	Section	Adaptation	Filter calculation
OAP*	4.2.1	$2(N^3 M^2)^*$	$8N^3 M$
AEP***	5.2	$2(N+M)^3 + 2(N+M)M$	$(N+M)^3 + M(N+M)^2$
SAS*	6.2.1	$2(K^3 M^2)$	$8K^3 M$
ASFF*	7.1.3	$2(K^3 L^2)$	$8K^3 L$
ACP	5.3	$2(N+M)^3 + 2(N+M)M$	$(N+M)^3 + M(N+M)^2$
ASEP	9.3	$2(KL)^3 + 2M(KL)^2$	$(KL)^3 + M(KL)^2$
FDFF	9.4	$2(K^3 L^2)$	$8K^3 L$
FDSP**	9.5	$2K^3 M$	$(K^3)M$

Table 16.2 Numerical example for Table 15.1 (slow operations): N = 24, M = 24, K = 5, L = 5; *by use of the Akaike algorithm for block Toeplitz matrix inversion; **only for large M; ***without eigenvector decomposition

Method	Section	Adaptation	Filter calculation
OAP	4.2.1	$382 \cdot 10^{6*}$	$2.6 \cdot 10^6$
AEP***	5.2	243488	165888
SAS*	6.2.1	144000	24000
ASFF*	7.1.3	6250	5000
ACP	5.3	243488	165888
ASEP	9.3	61250	30625
FDFF	9.4	6250	5000
FDSP**	9.5	6000	3000

Table 16.3 Fast operations (operations per range gate); **only for large M

Method	Section	Data Transform	Filtering	Doppler filtering
OAP	4.2.1	0	$(NM)^2$	0
AEP	5.2	$(N+M)NM$	$(N+M)M$	0
SAS	6.2.1	0 (RF)	KM^2	0
ASFF	7.1.3	0 (RF)	KLM	$\frac{M}{2} \log_2 M$
ACP	5.3	$(N+M)NM$	$(N+M)M$	0
ASEP	9.3	$K \frac{M}{2} \log_2 M$	KLM	0
FDFF	9.4	$2K \frac{2M}{2} \log_2 2M$	KM	$K \frac{M}{2} \log_2 M$
FDSP**	9.5	$K \frac{M}{2} \log_2 M$	KM	0

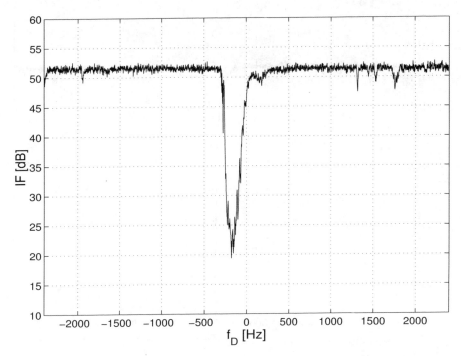

Figure 16.15: Space-frequency processing, with Doppler compensation

(GOLDSTEIN and REED [198]) has been analysed by HALE and WELSH [241]. The authors found that the number of training samples required is about 2.5× the number of degrees of freedom which corresponds roughly to the number predicted by REED et al. [557]. This indicates that a reduction of the system dimension also reduces the amount of training data which is of particular interest in forward looking radar.

16.3.1.1 *Real-time implementations of STAP algorithms*

It is obvious that several of the algorithms compared in Tables 15.1–15.4 are capable of real-time clutter supression. GARNHAM [177] discusses trade-offs between MTI for space-borne radar and the area coverage rate. His analysis is based on DPCA because the author feels that STAP cannot be implemented by reason of computational load. Unless the PRF is perfectly matched to the subarray spacing DPCA will be inferior to STAP. Clutter suppression using one of the previously mentioned techniques can be accomplished in real-time.

Several authors describe STAP systems implemented on general purpose computers or dedicated hardware or give advice for such implementations. The details of these implementations are beyond the scope of this book. For interested readers we give here a list of references rather than going into detail:
MANSUR [434], BROWN and LINDERMAN [70], LINDERMAN and LINDERMAN

516 Special aspects of airborne MTI radar

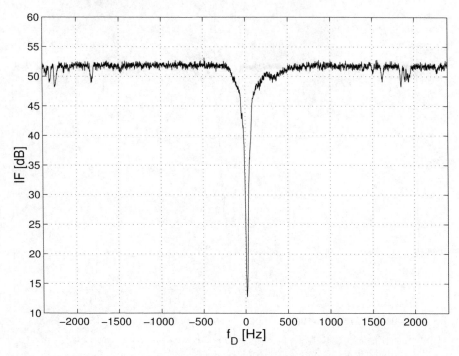

Figure 16.16: Space-frequency processing (effect of Doppler compensation)

[403], LITTLE and BERRY [405], MARTINEZ and MACPHEE [439], NOHARA *et al.* [502], SAMSON *et al.* [581], TEITELBAUM [640], WANG H. and ZHANG [674], PUGH and ZULCH [547], FARINA and TIMMONERI [156], RABIDEAU and KOGON [549], CHOUDHARY *et al.* [92], FRENCH *et al.* [165] (application to knowledge-based STAP), SU and BAO [621] (parallel architecture), DAMINI and BALAJI [110].

16.3.2 Comparison of pre- and post-Doppler architectures

There is a never ending discussion on the right choice among the various processor options. In the following we try to collect a number of aspects which may be help the system designer to find the appropriate solution to his individual problem.

16.3.2.1 *Properties of pre-Doppler STAP processors*

Several pre-Doppler techniques have been described in the literature. We focus here on the space-time FIR filter treated in Chapter 7.

- If the received echo sequence is stationary (no staggering, no platform acceleration) the filter adaptation (estimation of the covariance matrix) has to be carried out only in a segment of the data record, with the segment size given by the temporal filter dimension L.

Table 16.4 Fast operations (operations per range gate); *only for large M; **unless beamforming is done digitally; ***total number of operations. When pipelining the data only 25 operations/range gate are required for each PRI.

Method	Section	Data Transform	Filtering	Doppler filtering
OAP	4.2.1	0	331776	0
AEP	5.2	27648	1152	0
SAS	6.2.1	0 (RF)**	14400	0
ASFF	7.1.3	0 (RF)	600***	$\frac{M}{2}\log_2 M$
ACP	5.3	27648	1152	0
ASEP	9.3	300	600	0
FDFF	9.4	300	1440	300
FDSP*	9.5	300	120	0

- Because of the previously mentioned property all the data in the echo sequence can be used to average the covariance estimate in the time dimension (averaging the clutter covariance matrix over time). Therefore, the number of range bins usually used for averaging can be reduced. This may have advantages if the clutter statistics are strongly range dependent. Range dependence of clutter returns may be caused by heterogeneity of the clutter, by range dependence of the clutter Doppler in all array configurations different from sidelooking, and by the dependence of the clutter power with range (radar equation).

- If the clutter echo sequence is non-stationary (e.g., due to platform acceleration during maneuvres, or due to PRI staggering) the FIR filter can be matched to the non-stationary echo sequence by continuous re-adaptation.

- The temporal dimension of the space-time FIR filter is nearly independent of the length of the echo sequence.

16.3.2.2 Properties of post-Doppler STAP processors

- Post-Doppler processors operating on Doppler segments have to be adapted separately for each Doppler channel. This requires that adaptation is carried out separately for all Doppler bins.

- Because of the aforementioned property post-Doppler processing can be confined to a Doppler interval of interest. This can, for example, save computing power in tracking applications where the target Doppler is approximately known.

- The computational complexity of multi-Doppler bin techniques such as the JDL-GLR (WANG and CAI [668]) and hybrid techniques such as the ASEP (Chapter 9) can be reduced by exploiting the well-known matrix inversion lemma (see 5.5.1.3) for economic inversion of the space-frequency covariance matrices.

- Frequency dependent spatial processing can be a very cost efficient technique, Tables 16.1-16.4. It requires, however, long data sequences as occur in SAR or HPRF radar.

- In regions of strong clutter Doppler variations adaptation for each individual Doppler cell as required by post-Doppler techniques can be advantageous.

16.3.3 Effect of short-time data processing

As has been shown in Chapter 13 suppression of broadband jammers is basically a space-time problem so that a space-time filter can achieve jammer suppression by placing nulls in the antenna directivity pattern. However, it can be shown that in practice even a conventional beamformer, cascaded with an adaptive *temporal filter*, can also provide jammer suppression to a certain extent.

In Chapter 11 we assumed that the jammer produces *white* noise so that temporal samples are uncorrelated. In practice one usually has only a limited amount of data available so that only a short-time correlation can be estimated. Such a short-time correlation function normally shows some correlation values different from zero although the expected value would be equal to zero. The apparent jammer spectrum is not entirely white.

This short-time correlation can enable some jammer suppression when used for adapting a temporal filter provided that this filter is applied only to the set of data where the short-time correlation was estimated from ('primary data'). Calculations have shown that under such conditions the conventional temporal processor may achieve much better clutter suppression than that suggested by the lower curves in Figures 11.11 and 11.12. The adaptive *temporal* filter exploits the apparent short-time correlation for jammer suppression.

However, one has to be very careful about how SNIR is calculated from short samples of data, to avoid drawing misleading conclusions. The apparent increase in SNIR for short data samples can indicate an increase in the false alarm rate due to the fact that residual jamming and noise can correlate with the matched filter.

16.3.4 Inclusion of signal in adaptation

In most parts of this book we assumed that the signal-free clutter covariance matrix is known. In practice the clutter covariance matrix can be corrupted by the desired target signal. Then we have the situation of the minimum variance (MV) spectral estimator (1.103). It has been shown by COX [106] that the optimum processor based on the interference + noise covariance matrix after (1.3), with no signal included, has the resolution properties of the conventional beamformer (or signal match according to (1.102)). The MV estimator which is based on the interference + *signal* covariance matrix has *high-resolution* properties. For comparison recall Figure 1.13. Notice the difference between the asterisks and circles curves. The problem of inclusion of the signal in the covariance matrix has also been discussed by GRIFFITHS [210], HUDSON [275], and MONZINGO and MILLER [466].

For the purpose of target detection by radar the processor based on the signal-free covariance matrix is preferable. In normal radar operation a limited number of beams (Doppler filters) is formed in such a way that the mainlobes of their directivity patterns (transfer characteristics) overlap so that the whole angular (Doppler) domain is covered. A target which is not in the centre of the mainlobe (perfect match between target signal and beamformer) will be detected with slightly reduced detection probability. However, if a mismatched signal is included in the covariance matrix it will be treated by the processor as interference and will be suppressed.

In anti-jamming applications this problem can be circumvented by designing a *generalised sidelobe canceller* (GLSC) with 'blocking matrix' (see Chapters 5, 6 and 9). The blocking matrix places a null in the look direction of the array pattern and thus enables signal-free adaptation of the clutter covariance matrix. The simplest way to find a blocking matrix is to form dipoles out of adjacent array sensors, with the null being steered in the look direction. Then the signal component in the covariance matrix is reduced so that suppression of the signal is mitigated. For example, the sparse LCMV processor by SCOTT and MULGREW [594] or the generalised sidelobe canceller by SU and ZHOU [620] are based on blocking matrices.

However, the concept of a blocking matrix works properly only if the blocking matrix is perfectly orthogonal to the direction of the desired signal. In fact, it can be concluded from the considerations in Section 1.2.3.2 that the GLSC with blocking matrix is identical to the fully adaptive processor (OAP). In the case of mismatch of the blocking matrix to the desired signal, the losses due to signal cancellation will occur. For the use of a blocking matrix including the effect of steering angle errors see , LEE C.-C. and LEE J. H. [384].

Using blocking matrices for suppressing the target signals in auxiliary reference channels is always a compromise between signal and clutter cancellation:

- **Signal cancellation**: On the one hand it is desirable to have an angle/Doppler notch close to the look direction and centre frequency of the actual Doppler filter where the auxiliary channels are blind against the target signal. The width of the notch can be adjusted by weighting the projection matrix (6.24), with $\sin x/x$ coefficients according to (6.25). The notch takes care of the mismatch between the look direction/centre frequency of the Doppler filter, and the target direction/velocity.

- **Clutter cancellation**: A broadened notch in the auxiliary channels in the look direction, i.e., in the mainbeam clutter direction, leads to losses in SCNR close to the mainbeam clutter Doppler. This is equivalent to a degradation in slow target detection.

- **Space-time blocking matrix**: For space-time adaptive sidelobe cancellers the blocking matrix has to be spatial-temporal. Pure spatial blocking matrices fail because a coherent target signal can still be cancelled by the adaptive filter in the temporal dimension, see Chapter 6.

Temporal cancellation does not happen if the target signal is below the noise level. However, if the clutter level is well above the noise but the signal is below, there is no need at all for a blocking matrix.

JENKINS and MORELAND [279] analyse a two-source case which is typical for applications in communications. The first is considered the signal while the second one is interference. It is shown that the performance of the 'optimum' processor[8] is greatly

---

[8] The authors refer to the optimum technique as the *Gram–Schmidt technique*.

degraded by inclusion of the signal in the covariance matrix. In fact, in the case of signal inclusion the processor which is optimum for a clutter + noise covariance matrix is not optimum at all. The authors suggest instead using the eigenvector associated with the signal eigenvalue as the array weighting. Results are presented for the special case of two signals, one being the desired signal, the other one being the jammer. Since eigenvectors are mutually orthogonal this weighting provides perfect cancellation of the interference. In communications, such a technique might be applied because the 'signal eigenvalue' can be identified using the transmitted information in the signal. Applications to radar appear to be questionable.

HAIMOVICH et al. [238], and BERIN and HAIMOVICH [45] discuss the effect of signal inclusion in the presence of pointing errors of the beamformer.[9] The authors show that the eigencanceller is more robust against pointing errors than the optimum processor in the case that the interference covariance matrix was estimated from a finite set of training data.

There are two ways of getting a signal-free covariance matrix. Either the area[10] from which the data are taken for updating the covariance matrix is chosen so large that the target does not play a significant role. This may be in contradiction with the requirements for small sample size due to the Doppler dependency with range in the case of forward looking radar.

The second possibility is to exclude the range cell under test from the adaptation. Several training strategies which avoid inclusion of the target in the adaptation are discussed by BORSARI and STEINHARDT. These techniques are all based on a sidelooking array configuration [57]:

- **Sliding window training**: The covariance matrix is estimated from a certain range window and slides through the whole visible range. The cell under test and eventually the adjacent cells are left out. As the covariance matrix is updated for every range cell this technique is very cost intensive.

- **Range segmentation**: Training in one range segment and filter in another one following the training segment. This technique assumes that the data in the training segment have the same statistics as those in the filter segment. The computational cost is moderate.

- **Sliding hole**: The data dyadics of space-time clutter returns $\mathbf{c}(R)\mathbf{c}^*(R)$ are stored for as many range cells as required for adaptation. Then the covariance matrix can easily be updated by subtracting the dyadic belonging to the cell under test and adding the dyadic belonging to the previous cell. The filter is applied to each gate in the training region. One dyadic/range gate has to be computed.

- **Sliding hole/vector freeze technique**: The range is segmented into a 'slide' and a 'freeze' region. The covariance matrix is estimated only once in the 'slide region' and is kept constant in the 'freeze' region. This economic technique can be used if the clutter background is sufficiently homogeneous.

---

[9]In a practical radar situation there is always a pointing mismatch because the search raster is usually based on the beamwidth.

[10]This may mean range or time (pulse-to-pulse).

16.3.5 Homogeneity of clutter background

The numerical results presented in this book are based on the assumption that the space-time covariance matrix $\mathbf{Q}$ is known. A good estimate of $\mathbf{Q}$ can be obtained as long as the clutter background is homogeneous. In practice the clutter background may consist of a superposition of a homogeneous part and single strong scatterers which may even exhibit a Doppler frequency (vehicles in the antenna sidelobe area). It has been shown by BLUM et al. [50] that DPCA may perform better than space-time adaptive processing in a strongly inhomogeneous environment. Non-homogeneity of clutter plays a major issue in the design of adaptive space-time clutter filters, see WANG H. et al. [673]. FABRIZIO and TURLEY [139] modify the generalised sidelobe canceller by introducing a deemphasis factor to compensate for inhomogeneities of the clutter.

CFAR techniques can be applied to keep the false alarm rate constant (KELLY [286], ROBEY et al. [571], GOLDSTEIN et al. [201], GAU and REED [178], REED and GAU [559], CHEN and REED [90], RICHMOND [567]).

A general discussion of various CFAR techniques has been published by WEBER et al. [700]. An adaptive MTI scheme with robustness against non-homogeneity has been proposed by YONG et al. [735]. CONTE et al. [98] discusses CFAR receivers for detection of overresolved (i.e., extended) targets. The effect of heterogeneous clutter on STAP performance has been discussed in some detail by MELVIN [453].

16.3.5.1 *Orthogonal projection FIR filters*

Another way to cope with strong stationary reflectors in the scene is to use an orthogonal projection type of processing. This technique was suggested in Section 4.2.2. Compared with the optimum processor such techniques are suboptimum, but have advantages in heterogenous clutter environments. Since such filters form an exact null in the Doppler-azimuth characteristics (Figure 4.5) they are capable of cancelling all kinds of stationary clutter regardless of the clutter power. This property is different from the optimum adaptive processor. The optimum processor forms a null whose depth depends on the *average* clutter level. Strong echoes whose power exceeds the average clutter power level will not be sufficiently cancelled.

The space-time FIR filter presented in Chapter 7 can in principle be modified so that a perfect null in the Doppler-azimuth characteristics is generated. The procedure is as follows:

- estimation of the $KL \times KL$ clutter covariance matrix $\mathbf{Q}_T$ in the transformed domain;

- eigendecomposition

$$\mathbf{Q}_T = \mathbf{E}\mathbf{\Lambda}\mathbf{E}^* = \mathbf{E}^{(c)}\mathbf{\Lambda}^{(c)}\mathbf{E}^{(c)*} + \mathbf{E}^{(n)}\mathbf{\Lambda}^{(n)}\mathbf{E}^{(n)*}$$

  according to (1.44);

- omitting the clutter eigenvalues and inverting the noise eigenvalues gives the orthogonal projection matrix

$$\mathbf{Q}_{OP} = \mathbf{E}^{(n)}\mathbf{\Lambda}^{(n)-1}\mathbf{E}^{(n)*}$$

- insert the first K columns of $\mathbf{Q}_{OP}$ into the matrix $\tilde{\mathbf{K}}$ in (7.12). The filter operation is given by (7.13).

This procedure involves the eigendecomposition of the estimated space-time covariance matrix, and it is questionable if under real conditions (array errors) such an exact null is obtainable. MATHEW et al. [442] developed a technique for estimating the eigenvectors and eigenvalues adaptively. The resulting filter is somewhat related to the 'eigenfilters' described by MAKHOUL [432].

It should be noted that the so-obtained orthogonal projection filter is strongly related to the spectral estimation technique proposed by TUFTS and KUMARESAN [655, 656, 657]. The authors found that focusing on the principal eigenvalues of the data matrix of a finite length time series leads to stable estimates of the prediction error filter and, hence, of the frequency components in the data.[11] A single snapshot FIR filter technique based on orthogonal projection has been described by PARK S. and SARKAR [524], see Section 16.5.2.1. sc Francos et al. [164] suggests a technique based on the two-dimensional Wold decomposition which can in principle operate on the data of a single range gate.

ENDER [126] uses *frequency dependent spatial processing* (see Chapter 9) based on orthogonal projection for clutter cancellation in SAR images. While after *optimum* filtering a SAR image of the stationary background is still noticeable (undernulling) the clutter is perfectly cancelled with the orthogonal projection technique.

16.3.5.2 Processing of undernulled clutter

To cope with the beforementioned effect of undernulling KREITHEN and STEINHARDT [366] developed a technique based on a comparison of the null depth (clutter power after adaptive filtering) and the clutter power as received by a beamformer, both of them matched to a certain range-Doppler cell.

16.3.5.3 Knowledge based space-time adaptive filtering

It has been pointed out by STEINHARDT and GUERCI [618] that STAP performance, regardless of the algorithm chosen, may be degraded strongly by errors in the radar equipment as well as by deviations of the statistical behaviour of the actual clutter echoes from assumed models. To improve the STAP performance, a priori knowledge should be introduced in the target detection process as much as possible.

For example, pre-filtering of inhomogeneities controlled by a clutter map or a non-homogeneity detector can improve target detection capability (ANTONIK et al. [20, 21], OGLE et al. [509, 510]), RANGASWAMY et al. [553, 555], LI et al. [593], TEIXEIRA et al. [641]), TINSTON et al. [645], WANG et al. [687]).

In 2001 J. Guerci of DARPA launched the KASSPER (Knowledge aided sensor and signal processing and expert reasoning) project. A summary of the KASSPER project can be found in GUERCI and BARANOSKI [229]. According to the authors the real breakthrough in STAP performance is not that prior knowledge can be useful.

---

[11] Recall that the space-time FIR filter analysed in Chapter 7 was based on the prediction error filter concept.

This is obvious and has been known for long time. The issue to be answered by the KASSPER project is how to use prior knowledge in STAP implemented in a real-time high performance embedded computer (HPEC) because of conflicts between memory access and the HPEC principle.

More detailed information on KASSPER and a challenge data file can be found in the KASSPER workshop proceedings ([230, 231, 232]).

The simulations presented by COOPER and DICKMAN [99] are based on results achieved with the APS-145 AEW radar and parameters taken from the ADS-18S array (18 Yagi elements in the UHF band). An environmental processor identifies the various kinds of clutter (land, sea, noise only). The authors discuss the problem of whether a single STAP architecture or separate STAP solutions should be applied in a heterogeneous clutter + jammer environment. It turns out that a single STAP architecture performs as well as four different architectures. The use of the environmental processor to designate areas of high clutter level results in the highest value of clutter rejection performance. These clutter data are selected for adaptation. The use of some long-range samples improves the ability of the single STAP solution to null jammers without sacrificing clutter rejection performance. WEINER et al. [702] utilise terrain and land feature data for improving the clutter covariance estimate.

BERGIN et al. [42] discuss the impact of high densities of ground targets on the performance of space-time adaptive clutter suppression based on simulations as well as evaluations of MCARM radar data. Inclusion of a priori knowledge about roads in the processing have proven to yield an improvement of about 15 dB. In related paper (BERGIN et al. [43]) the authors set up an interference covariance model consisting of a known (knowledge) and an unknown component. The known component performs so-called coloured loading (in contrast to diagonal loading) so as to compensate for known interference.

The FRACTA algorithm has proven to excise clutter outliers from the training data and thus helps improving STAP performance in scenarios composed of heterogeneous clutter and dense target environment (BLUNT and GERLACH [52], GERLACH and BLUNT [183]).

CAPRARO et al. [85] use terrain data to select secondary training data for the adaptation.

GOODMAN and GURRAM [207] use a priori knowledge of homogeneous clutter areas to average a realtime SAR image. From the averaged SAR image de-emphasis coefficients are derived which are used to weight the individual clutter samples.

LEGTERS and GUERCI [388] exploit the a priori knowledge that target and clutter returns are approximately plane waves. The spatial and Doppler frequencies of radar echoes are analysed by matching with a plane wave model.

The KA-STAP architecture by MELVIN et al. [458] consists of a pre-filter, a matched filter and a lower order STAP stage. Both the first two components as well as the selection of the training data for the STAP stage are steered by a priori knowledge such as clutter map and digital elevation map.

PAGE et al. [520] use multiple CPI data to form a clutter reflectivity map which in turn can be used to support knowledge aided STAP.

HIEMSTRA et al. [261] expand the concept of the multistage Wiener filter based on the conjugate gradient method (CG-MWF) by introducing a priori knowledge about the interference scenario.

16.3.5.4 Non-homogeneity detection

Another way of coping with non-homogeneous clutter data is to use non-homogeneity detection (WICKS et al. [708], MELVIN et al. [448]). This technique is used to include 'homogeneous data' into the estimation of the space-time clutter covariance matrix and to discard 'non-homogeneous' samples. The advantage of this processing has been proven by the authors, using MCARM data (see Section 1.1.6). Another sample discard technique ('GUSSM') has been proposed by WU R. and BAO [29]. MELVIN [450] used eigenvector decomposition to model various types of clutter non-homogeneities. ADVE et al. [6] evaluated MCARM data by combining measured steering vectors and non-homogeneity detection. MELVIN and WICKS [451] demonstrate how to integrate different kinds of non-homogeneity detectors in a practical STAP system. Several adaptive detection algorithms for STAP radar have been proposed for handling problems of non-homogeneous clutter (MELVIN et al. [452]). RABIDEAU and STEINHARDT [550] propose two techniques for data adaptive training to cope with clutter inhomogeneities. The *power selected training* technique (PST) selects those range bins showing the strongest clutter power for estimating the clutter covariance matrix. In the *power selected de-emphasis* technique certain range intervals with strong clutter power are attenuated so as to equalise for the inhomogeneities in clutter power. OGLE et al. [509, 510] propose a multistage nonhomogeneity detector integrated in a multistage Wiener filter clutter canceller.

16.3.6 Non-adaptive space-time filtering

The advantage of adaptive filters is their capability of matching to changing conditions or a priori unknown parameters. There are several unknown parameters which an adaptive filter can cope with, for instance platform velocity, antenna attitude (roll angle), Doppler-range dependency for forward looking radar, and clutter bandwidth.[12]

By using non-adaptive filters the computational load for adaptation is saved. However, some precautions have to be taken to cope with variations of the aforementioned parameters. There are several possible philosophies for tackling this problem:

16.3.6.1 Design of a robust filter

A robust filter can be designed by broadening the clutter notch. For unknown clutter fluctuations this means broadening the clutter notch by a *constant* amount for all Doppler frequencies/azimuth angles. This can easily be done by calculating a model covariance matrix using (3.31), with (2.53) for example. Of course, instead of the simple clutter fluctuation model (2.53) other models may be used.

In order to cope with variations of the platform velocity the clutter notch has to be broadened *in proportion to the Doppler frequency*. For a sidelooking radar for example,

---

[12]Non-adaptive space-time filters do not work in the case of additional jamming.

the clutter notch is unchanged at F = 0 in Figure 4.2 but broadened proportional to | F |. The clutter trench will be broader at the upper and lower ends of Figure 4.2 than in the middle. For the forward looking case the maximum broadening has to be applied in the middle while no broadening is required at the left and right end. Some considerations on tapering the covariance matrix to achieve robustness, for instance against undernulled clutter discretes, have been studied by GUERCI [226].

A covariance matrix with such properties can be designed by superposition of a number of covariance matrices based on different platform velocities. Since the filter has been made robust against platform velocity variations by broadening the clutter notch the penalty for using a non-adaptive filter is a degradation in slow target detection. We do not intend to enter the discussion about the trade-off between robustness of the filter and degradation of slow target detection.

16.3.6.2 Choice of filters

An alternative approach may be to prepare in advance a choice of optimum filters which altogether cover the possible range of parameter variations. The advantage of this solution is that no broadening of the clutter notch is required. Adapting the clutter means making the right selection from the available filter candidates. This selection process replaces the adaptation.

Such techniques have been proposed and analysed by FARINA [153], FARINA *et al.* [154, 155, 157]. Since the selection process is a non-linear operation the authors call such techniques 'non-linear filtering'. In [153] a choice among several linear filters with different widths of the clutter notch is made. In this way the clutter notch can be matched to the clutter bandwidth due to internal clutter fluctuations (see Sections 2.5.1.1, 3.4.3.1). The filter with the minimum clutter output power is selected. This procedure is carried out for all Doppler frequencies of interest. The output at the Doppler frequency with maximum output is compared with a detection threshold. LOMBARDO and COLONE [418] use such filter selection technique to cope with inhomogeneous clutter.

16.3.6.3 Deterministic composition of the clutter covariance matrix

TECHAU *et al.* [637] describe an optimisation technique for the case that the underlying clutter background is known. The covariance matrix is composed of a deterministic part based on a priori knowledge of the background topography so as to circumvent problems associated with clutter estimation in heterogeneous backgrounds, and receiver noise. The optimised processing leads in essence to the optimum processor (1.3).

16.3.7 Further limitations

16.3.7.1 Dependence on CNR

One of the major issues in the design of a radar is: how many channels are required for achieving near optimum processing at minimum cost. The number of antenna channels

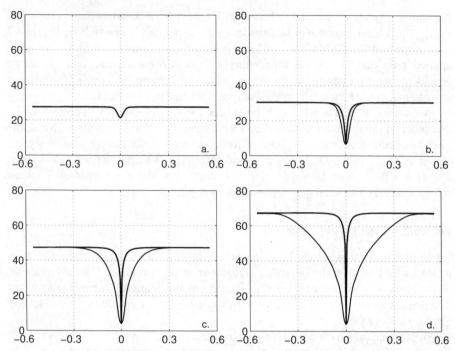

Figure 16.17: Dependence of the IF on CNR ($N = 24$, fat = STAP, thin = BF + temporal filter), CNR/dB = a. -20; b. 0; c. 20; d. 40

is one of the cost driving factors. Before deciding between a single or multichannel antenna the potential benefit achieved by a multichannel antenna should be carefully considered.

The dependence of the improvement factor on the CNR has already been discussed in Figure 4.16. The superiority of space-time adaptive processing over conventional beamforming plus temporal clutter filtering depends significantly on the CNR at the single element and single echo. As a rule of thumb one can state that STAP is worth doing whenever the clutter power at the single element is large compared with the noise power.

This is illustrated in Figures 16.17 and 16.18. The upper curves denote the improvement factor achieved by means of STAP while the lower curves stand for beamforming plus temporal clutter filtering. It can be noticed that the differences between both curves inreases with CNR (particularly in Figure 16.17c and 16.17d). The effect is less stringent if a larger antenna is used (16.17).

Increasing the spatial and temporal resolution to infinity would mean that clutter is received only from a discrete point reflector on the ground with one single clutter Doppler frequency; in this case a conventional beamformer cascaded with a Doppler filter optimum achieves near optimum performance. No STAP is required.

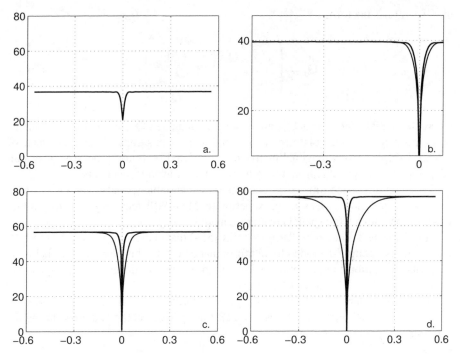

Figure 16.18: Dependence of the IF on CNR ($N = 192$, fat = STAP, thin = BF + temporal filter), CNR/dB = a. -20; b. 0; c. 20; d. 40

16.3.7.2 Near-field obstacles

The effect of near-field obstacles has been addressed by BARILE et al. [35, 36]. For example, the wing of an aircraft may be such an obstacle. This kind of clutter is quite stressing because it tends to spread over the entire azimuth-Doppler plane. This has the effect that the clutter notch is broadened and the improvement factor reduced. The effect of an aircraft wing has been studied by SURESH BABU et al. [627] using MCARM clutter data.

A technique for suboptimum compensation for near-field scatterers has been described by BARILE et al. [37]. The columns of a rectangular planar array are used for spatial near-field scatterer rejection. After combination of the elements in each column the resulting horizontal array is used for space-time adaptive clutter suppression.

16.3.7.3 Manoeuvering radar platform

The effect of radar platform manoeuvres have been addressed by RICHARDSON [564, 343, 565]. Rapid changes of the radar platform attitude may change the radar-jammer-clutter geometry from pulse to pulse. In that case the space-time jammer covariance matrix assumes the form

$$\mathbf{Q_j} = \begin{pmatrix} \mathbf{Q}_1 & & & \\ & \mathbf{Q}_2 & & 0 \\ & 0 & \ddots & \\ & & & \mathbf{Q}_M \end{pmatrix} \quad (16.23)$$

where the spatial submatrices $\mathbf{Q}_m$ are no longer equal, i.e., $\mathbf{Q}_m \neq \mathbf{Q}_n$ due to a change of the geometry between platform, clutter and jammer positions. Since the inverse of (16.23) is also a block diagonal matrix the associated optimum processor means an adaptation of the spatial anti-jamming filter to the platform manoeuvers.

Based on this space-time covariance matrix (or variable spatial covariance matrix) jammer cancellation proves to be superior to MTI/ABF techniques or to jammer cancellation with a fixed spatial covariance matrix.

It has also been shown by RICHARDSON [564] that clutter suppression with STAP techniques is strongly degraded by platform manoeuvers. This effect can probably be mitigated by motion compensation of the radar beam, that is, during the manoeuver the beam position is fixed relative to the earth.

Space-time adaptive FIR filters as have been discussed in Chapter 7 would have to be adapted for each PRI. This leads to a FIR filter whose coefficients vary during the filtering period.

16.3.7.4 Mismatched steering vectors

Any processor involves in one or the other way a space-time steering vector which has to match to the expected target signal in angle and Doppler. As a first approach the steering vector is usually designed based on an idealised array antenna. The array elements are assumed to be point shaped, mutual coupling and other effects are ignored.

The performance of the STAP processor can significantly be improved if the idealised steering vectors are replaced by measured steering vectors. Moreover, the design of rank reduced processors may make use of measured steering vectors. ADVE and WICKS [4] have shown that the performance of the JDL-GLR algorithm (WANG and CAI [668]) can be improved by using measured steering vectors for the Doppler-beamspace transform.

ENDER [125, 126, 127, 129, 131] uses frequency dependent spatial filtering (Section 9.5) for detection of moving targets with the AER-II multichannel SAR. The beamformer is estimated from experimental clutter data. First the Doppler dependent spatial covariance matrices are estimated. The eigenvector associated with the largest eigenvalue reflects the clutter portion of the received echoes and is used as a 'measured' steering vector.

16.3.7.5 Atmospheric effects

Arrays operating near the ground must contend with the effects of radiative solar heating of the earth which results in atmospheric turbulences. These turbulences may lead to phase fluctuations of the phases of the received echo signals. NAGA and WELSH

[472] found from simulations that the phase perturbations, and the degrading effect on the SNIR, increases with frequency, but is normally negligible.

16.4 Adaptive algorithms

In this section a brief overview of the most common algorithms for adaptive interference rejection is given. We refer here to the existing literature rather than discussing the algorithms in detail.

16.4.1 Approximations of the optimum processor

In this section some techniques are quoted which approximate the optimum processor according to (1.3).

16.4.1.1 Sample matrix inversion (SMI)

Recall that most of the clutter (or jammer) suppression techniques described in the previous chapters have been based on the inverse of the clutter covariance matrix. This fact gives the motivation to talk about *adaptive* techniques because the filter is calculated from the received clutter data.

In practice the covariance matrix is not known and has to be estimated from the data. The well-known maximum likelihood estimator for a covariance matrix is

$$\hat{\mathbf{Q}} = \frac{1}{Q} \sum_{q=1}^{Q} \mathbf{c}\mathbf{c}^* \qquad (16.24)$$

where $\mathbf{c}$ is the training sample vector which may include spatial, temporal, space-time, space-TIME or space-time-TIME samples.

In applications where $\mathbf{Q}$ is centro-hermitian (i.e., hermitian with respect to both of the diagonals of $\mathbf{Q}$) a forward-backward estimate of the form

$$\hat{\mathbf{Q}}_{\text{FB}} = \hat{\mathbf{Q}} + \mathbf{J}\hat{\mathbf{Q}}^T\mathbf{J} \qquad (16.25)$$

may be used. The FB covariance estimator offers advantages in the performance of high-resolution estimators such as the MV estimator[13] (JANSSON and STOICA [278]). Because of the equivalence between the MV estimator and the optimum processor some improvement in clutter rejection performance can be expected.

If the training samples have been generated in a transformed domain $\mathbf{c}_T = \mathbf{T}\mathbf{c}$ (see Chapters 5 and 6) then the covariance matrix becomes

$$\hat{\mathbf{Q}}_T = \frac{1}{Q} \sum_{q=1}^{Q} \mathbf{c}_T \mathbf{c}_T^* = \mathbf{T}^* \hat{\mathbf{Q}} \mathbf{T} \qquad (16.26)$$

The number of operations required is QN^2 where N is the order of the covariance matrix.

---
[13] See Section 1.3.2.

The adaptive weighting is then computed from (16.24) or (16.26) by taking the inverse of $\hat{\mathbf{Q}}$ or $\hat{\mathbf{Q}}_T$. REED et al. [557] have shown that the adaptive system achieves a performance roughly 3 dB below optimum SCNR when the number of training vectors Q is twice the order of the covariance matrix: $Q \geq 2N$.

The convergence of any clutter reduction technique based on the sample covariance matrix can be accelerated by forwards–backwards averaging (ZATMAN and MARSHALL [736]). This is possible in the time domain if the PRF is constant (no staggering), and in the spatial domain if the array sensor positions are symmetric. The concept of forwards–backwards averaging was originally introduced by BURG [79, 80] in the context of maximum entropy spectral analysis. In space-time clutter suppression forwards–backwards averaging can be done in both the temporal and spatial dimensions, thus achieving a quadruple use of data samples.

Recall that in the case of constant PRI the echo sequences can be considered to be stationary so that the covariance matrix becomes Toeplitz in the time dimension. Then only a block column of the space-time covariance matrix has to be estimated, and the inverse can be computed by the algorithm of AKAIKE [12] for inverting block Toeplitz matrices. A related algorithm has been addressed by PILLAI et al. [539]. GOVER and BARNETT [208, 209] have extended this concept for use with ill-conditioned Toeplitz matrices which might occur in low noise situations.[14] For the space-time FIR filter after Chapter 7 only the first block column of the inverse of $\mathbf{Q}$ or $\mathbf{Q}_T$ is required.

GERLACH and KRETSCHMER [180] compare the statistical behaviour of the SMI technique for concurrent[15] and sliding window operation. It turns out that both techniques have the same statistical properties and, therefore, the same convergence rate. However, the output data of the concurrent mode are highly correlated which may cause difficulties for subsequent signal detection.

The numerical stability of the SMI covariance matrix estimates can be improved by 'diagonal loading' (loaded SMI = LSMI) which means that a certain amount of 'artificial noise' power is added to the diagonal of the covariance matrix estimate (CARLSON [87], GANZ et al. [176]). GIERULL [189] has shown that the loading factor α should be $\lambda^{(n)} < \alpha < \lambda_{\min}^{(i)}$ where $\lambda^{(n)}$ is the noise eigenvalue and $\lambda_{\min}^{i}$ the minimum interference eigenvalue of the interference + noise covariance matrix.

Another issue is to avoid mismatch problems in adaptive beamforming due to limited training sample support. Tapering the covariance matrix can be a remedy against mismatch errors. GUERCI [226] presents a framework in which both the diagonal loading and the covariance tapering are included.

KIM et al. [287] optimise the loading factor to minimise the sample support required. PILLAI et al. [540, 541] propose spatial and temporal data smoothing techniques for reducing the requirements in training data support.

PILLAI and PILLAI [542, 543] use the principle of 'relaxed convex projection' to enhance a covariance matrix estimated from a small number of training samples. Two corrections are applied: 1. the estimated covariance matrix is forced to become block Toeplitz, 2. the noise eigenvalues are made equal. Considerable improvement in STAP performance was demonstrated. In [543] PILLAI et al. use spatial and temporal

---

[14] So-called conjugate-Toeplitz are covered as well.
[15] The same data are used for adaptation and filtering.

forward-backward smoothing techniques to enhance the covariance matrix estimate.

16.4.1.2 Updated inverse

When the sample matrix inversion technique is applied to a large quantity of data (for instance, a large number of range gates) the estimated covariance matrix may be efficiently computed in the fashion of a moving average

$$\hat{\mathbf{Q}}_q = (1-\alpha)\hat{\mathbf{Q}}_{q-1} + \alpha \mathbf{c}_q \mathbf{c}_q^* \tag{16.27}$$

where α is the 'forget' factor. If the matrix $\hat{\mathbf{Q}}_{q-1}^{-1}$ is known then the updated inverse can be computed directly using the new data vector as follows

$$\hat{\mathbf{Q}}_q^{-1} = \frac{1}{1-\alpha}\hat{\mathbf{Q}}_{q-1}^{-1} - \frac{\alpha}{(1-\alpha)}\frac{\hat{\mathbf{Q}}_{q-1}^{-1}\mathbf{c}_q\mathbf{c}_q^*\hat{\mathbf{Q}}_{q-1}^{-1}}{(1-\alpha+\alpha\mathbf{c}_q^*\hat{\mathbf{Q}}_{q-1}^{-1}\mathbf{c}_q)} \tag{16.28}$$

This computation requires slightly above N^2 complex multiplications to compute $\hat{\mathbf{Q}}_q^{-1}$ given $\hat{\mathbf{Q}}_{q-1}^{-1}$.

It has been shown by BRENNAN et al. [63] that the sample matrix inverse and the updated inverse methods are equivalent in terms of adaptation convergence. XIONG et al. [733] use this technique in conjunction with a data discarding algorithm for suppression of non-homogeneous clutter.

16.4.1.3 Steepest descent techniques

Steepest descent techniques[16] have originally been used to solve systems of linear equations iteratively. WIDROW et al. [709] have modified this technique for use with antenna arrays in the presence of directional interference (jamming). In contrast to the orginal steepest descent algorithm the modified version approximates the optimum weighting given by (1.3), via stochastic approximation directly from the incoming data vectors, i.e., without estimating the sample covariance matrix. The algorithm is as follows

$$\begin{aligned}\mathbf{w}_{q+1} &= \mathbf{w}_q + \mu[d_q\mathbf{x}_q - \mathbf{x}_q\mathbf{x}_q^*\mathbf{w}_q] \\ &= \mathbf{w}_q + \mu[d_q - y_q]\mathbf{x}_q\end{aligned} \tag{16.29}$$

where μ is the loop gain which controls convergence and residual loop noise. d_q is the desired (or pilot) output signal while y_q denotes the actual output signal. This algorithm is equivalent to the control loops proposed by APPLEBAUM [22].

BRENNAN et al. [63] compare the steepest descent algorithm with the sample matrix inverse and the updated inverse techniques. It turns out that the steepest descent algorithm needs only $\propto N$ operations per iteration while the updated inverse requires $\propto N^2$ and the SMI $\propto N^2 + N^3$ operations. However, simulations have shown that the number of iterations required for convergence is very high so that the total number

---

[16] Also referred to as *gradient techniques*.

of calculations is much higher for the steepest descent algorithm (16.29) than for the aforementioned techniques ((16.29) + matrix inversion, or (16.28)).

GRIFFITHS [211] modifies the steepest decent algorithm in order to accelerate the convergence. To achieve this he replaces the pilot signal by the cross-variance between the input data vector and the desired signal

$$\mathbf{w}_{q+1} = \mathbf{w}_q + \mu[\mathbf{q}_{dx} - \mathbf{x}_q \mathbf{x}_q^* \mathbf{w}_q] \qquad (16.30)$$

where $\mathbf{q}_{dx}$ is the cross-variance between d_q and $\mathbf{x}_q$. Notice that in (16.29) only one scalar value $d_q - y_q$ is used for updating of the weight vector $\mathbf{w}_q$ while in (16.30) the difference of two vectors $\mathbf{q}_{dx} - \mathbf{x}$ is calculated. This results in a higher convergence rate. In practice the target signal is assumed to be deterministic so that the cross-variance vector becomes equal to a beamformer

$$\mathbf{q}_{dx} = \mathbf{b}(\varphi) \qquad (16.31)$$

If, accordingly, the desired signal in (16.29) is chosen $d_q = \exp(j\omega t)\mathbf{1}^*\mathbf{b}(\varphi)$ with $\mathbf{1}^*$ being (1 1 1 ... 1) then after reaching a steady state the 'desired signal' d_q provides beamforming in the direction φ.

FROST [168] considers the problem of constraint adaptation. This constraint may for instance be used to prevent signal cancellation in the look direction.

WIDROW and MCCOOL et al. [710] compare two different steepest descent algorithms: the least squares (LMS) algorithm after (16.29) and the 'Differential Steepest Descent' (DSD) algorithm. Furthermore, linear random search is considered. Numerical experiments have demonstrated that the LMS algorithm is the most efficient in terms of convergence rate.

LI and CHEONG [395] present a cost saving algororithm based on the block Toeplitz with Toeplitz block structure of the clutter covariance matrix. Such technique is, due to the nature of Toeplitz matrices, restricted to the case of linear uniformly spaced arrays and equidistant transmit pulses.

16.4.2 QR-decomposition

Several authors propose modifying the fully adaptive space-time processor in such a way that an implementation in parallel arithmetic is possible. This technique has been described by FARINA [149, p. 144] for jammer cancellation and has been extended by FARINA and TIMMONERI [150] for the purpose of space-time clutter rejection. A STAP architecture based on QR-decomposition for real-time operation has been described by MANSUR [434].

The authors propose a QR-decomposition of the received data in order to triangularise the data matrix. The efficiency of such techniques relies on the fact that a system of linear equations involving a triangular matrix can be solved easily by back-substitution. The triangular form of the matrix can be achieved by repeated application of Givens rotations (GOLUB and VAN LOAN[205, p. 201]) or Householder reflections (GOLUB and VAN LOAN [205, p. 194]).

Let

$$\mathbf{X} = (\begin{array}{ccccc} \mathbf{x}_1 & \mathbf{x}_2 & \ldots & \mathbf{x}_q & \ldots & \mathbf{x}_Q \end{array}) \qquad (16.32)$$

be the matrix of Q space-time vectors of received echo data samples. q is the dimension in which the data vectors are generated. In practice q may denote range or time or both. Notice that the maximum likelihood estimator of the clutter covariance matrix is

$$\hat{\mathbf{Q}} = \mathbf{X}\mathbf{X}^* \tag{16.33}$$

An approximation of the optimum processor (1.3) is defined by

$$\mathbf{X}\mathbf{X}^*\mathbf{w} = \mathbf{s} \tag{16.34}$$

Multiplying the data matrix with a product of Givens rotations gives

$$\prod_j \mathbf{G}_j \mathbf{X}^* = \mathbf{G}\mathbf{X}^* = \mathbf{C} \tag{16.35}$$

where $\mathbf{C}$ has the form

$$\mathbf{C} = \begin{pmatrix} \mathbf{R} \\ \mathbf{0} \end{pmatrix} \tag{16.36}$$

$\mathbf{R}$ is an upper triangular matrix and $\mathbf{0}$ a zero matrix. Then (16.34) becomes

$$\mathbf{C}^*\mathbf{C}\mathbf{w} = \mathbf{s} \tag{16.37}$$

since $\mathbf{G}$ is unitary. This equation can be solved by a two-step back-substitution procedure. First the equation

$$\mathbf{C}^*\mathbf{a} = \mathbf{s} \tag{16.38}$$

with $\mathbf{a} = \mathbf{C}\mathbf{w}$ is solved for $\mathbf{a}$. In a second step the equation

$$\mathbf{C}\mathbf{w} = \mathbf{a} \tag{16.39}$$

is solved for $\mathbf{w}$. The residual output power is

$$P_{out} = \mathbf{w}^*\mathbf{C}^*\mathbf{C}\mathbf{w} = \mathbf{a}^*\mathbf{a} \tag{16.40}$$

In this way the (asymptotic) optimum weight vector is calculated without forming and inverting a covariance matrix.

This algorithm has the property that it can be implemented by a systolic array based parallel architecture which has the potential of real-time processing (WARD et al. [689]).

The above technique has been tested using NRL airborne multichannel data (see Section 1.1.6, item 5) by FARINA et al. [151]. The performance of the space-time processing algorithm has been tested against ground and sea clutter, and jamming. The results compare very well with the performance predicted by theory. Systolic arrays appear to be a feasible solution for real-time cancellation of airborne clutter.

The application of a vector lattice algorithm for implementation on a systolic array has been discussed by TIMMONERI et al. [644] for narrowband radar. The paper by FARINA et al. [152] gives an extension of this principle to wideband radar.

BOLLINI et al. [56] compared the QR algorithm with the inverse QR algorithm (IQR) with respect to implementation as a systolic array using CORDIC processors. The IQR-based algorithm promises some saving in computational load.

16.4.3 Orthogonal projection algorithms

Orthogonal projection techniques belong to the class of subspace techniques. A detailed description of the use of subspace techniques is given by NICKEL [493]. This section gives a brief overview of interference suppression techniques that are based on orthogonal projection. The principle of orthogonal projection has been described briefly in Section 1.2.2. The processor is given by

$$\mathbf{w} = \mathbf{P}\mathbf{s} \qquad (16.41)$$

where $\mathbf{P}$ is a projection matrix and $\mathbf{s}$ is the steering vector. The motivation for using orthogonal projection techniques instead of optimum processing based on a sample covariance matrix is as follows: Orthogonal projection techniques are based on the interference subspace of the interference + noise covariance matrix. The interference eigenvectors dominate the covariance matrix and are less effected by noisy data than the eigenvectors associated with the small noise eigenvalues (HAIMOVICH and BARNESS [236]). The rapid convergence has been proved using MOUNTAINTOP data (HAIMOVICH and BERIN [240]). FUTERNIK and HAIMOVICH [170] have shown that orthogonal projection ('eigencanceller') is superior to the SMI and the GLR (KELLY [286], ROBEY et al. [571]) when applied in inhomogeneous clutter, especially close to clutter edges.

16.4.3.1 Gram–Schmidt orthogonalisation

The Gram–Schmidt orthogonalisation procedure has been successfully used for adaptive sidelobe cancellers. LIU H. et al. [406] extend this principle to the use with a fully adaptive array. This requires a modification of the receiver architecture in that a steering vector has to be introduced. Like the Gram–Schmidt sidelobe canceller the fully adaptive processor can be implemented as a systolic array. All Gram–Schmidt based processors are characterised by good numerical stability. PARK Y. C. et al. [527] use the Gram–Schmidt orthogonalisation to design an adaptive MTI filter with rapid convergence.

The reiterative median cascaded canceller (PICCIOLO and GERLACH [538, 184, 537]) is a robust variant of the Gram-Schmidt processor. It is robust against outliers and non-homogeneous clutter.

16.4.3.2 Multistage Wiener filter

The multistage Wiener filter (GOLDSTEIN et al. [203]) performs an orthogonalisation of the covariance matrix by extending the concept of the GSC (generalised sidelobe canceller). The original GSC structure is decomposed into a sequence of one-step Wiener filters. The authors have proven that this technique is very efficient. The rank of the nested system may be chosen much smaller than for the GSC. There is no need for eigendecomposition. The efficiency has been proven using a target signal injected into MOUNTAINTOP (TITI [646], TITI and MARSHALL [647]) and MCARM (see SURESH BABU et al. [626, 627] clutter data. The rank required for the MCARM data was much higher than for the MOUNTAINTOP data, probably, because in the

MOUNTAINTOP experiment the interference is dominated by a few strong jammers, thus requiring fewer degrees of freedom.

16.4.3.3 Hung–Turner method

The technique proposed by HUNG and TURNER [276] assumes that the data contain interference only, that is, neither signal nor noise. The Gram–Schmidt algorithm (or the QR or SVD algorithms) is used to orthonormalise the data matrix. Then instead of the data matrix $\mathbf{X}$ one has the matrix $\mathbf{Y}$ so that the projection matrix becomes

$$\mathbf{P} = \mathbf{I} - \mathbf{X}(\mathbf{X}^*\mathbf{X})^{-1}\mathbf{X}^* = \mathbf{I} - \mathbf{Y}\mathbf{Y}^* \tag{16.42}$$

The weight vector can be computed iteratively via the relation

$$\mathbf{w}_q = \mathbf{w}_{q-1} - (\mathbf{y}_q^*\mathbf{s})\mathbf{y}_q \quad q = 1,\ldots,Q \tag{16.43}$$

starting with $\mathbf{w}_0 = \mathbf{s}$. $\mathbf{y}_q$ is the q-th orthonormalised data vector.

Another possibility is to work directly with the received data $\mathbf{X}$ by using the identity

$$\mathbf{X}(\mathbf{X}^*\mathbf{X})^{-1}\mathbf{X}^* = \mathbf{Y}\mathbf{Y}^* \tag{16.44}$$

In this way the computations required for the Gram–Schmidt procedure are saved. Instead, the term $(\mathbf{X}^*\mathbf{X})^{-1}$ has to be computed.

Finally, the most elegant method is again given by using the QR-decomposition as described in the previous section. Then the inversion of $(\mathbf{X}^*\mathbf{X})$ reduces to an inversion of a triangular matrix which is an explicit procedure.

The Hung–Turner method requires that the number of data vectors Q is smaller than the number of antenna elements, otherwise the orthogonal projection nulls the whole signal vector space. The question is how many data vectors should be included in the data matrix $\mathbf{X}$ which spans the interference subspace. On the one hand a sufficient number of data vectors is required for statistical stability, on the other hand, by taking too many data vectors the signal subspace will be too small which results in signal loss.

Let us assume that $\mathbf{X}$ be composed of Q data vectors. According to HUNG and TURNER [276] the vector $\mathbf{x}_{Q+1}$ falls into the interference subspace if the modulus of the residual vector after orthogonal projection is small, that is, $\|\mathbf{P}\mathbf{x}_{Q+1}\|^2 = \mathbf{x}_{Q+1}^*\mathbf{P}\mathbf{x}_{Q+1}$ is small. Other methods for testing the dimension of the interference subspace are given by NICKEL [481, 493] and BABU et al. [628]. Some numerical evaluation is given in NICKEL [492]. XU et al. [734] describe fast algorithms for determining the interference subspace based on the Lanczos algorithm (GOLUB and VAN LOAN [205, p. 475]). Such technique may particularly be used to track the subspace dimension in applications with time-varying signal characteristics. LEE and LEE [386] describe a technique for constructing the signal subspace by decomposing the array into subarrays. The procedure is computationally very efficient.

For a statistical performance analysis of the Hung–Turner orthogonal projection technique see GIERULL [189]. Other subspace techniques are briefly described by NICKEL [492]. A detailed description of subspace estimation techniques has been given by GIERULL [188]. Fast techniques for subspace estimation have been published by BÜHRING [76, 77] and recently by GIERULL [190].

GERSHMAN et al. [185] modify the Hung–Turner algorithm by adding some constraints of the form $\mathbf{w}^*\mathbf{a}(\theta_k) = 0$. If the angles θ_k are chosen close together the nulls created by orthogonal projection are broadened so that wideband jammers can be supressed. Application of this technique may be appropriate for anti-jamming because only the sidelobe area is significantly modified.[17] It should not be used for space-time clutter rejection since we are strongly interested in a *narrow* clutter notch for the sake of low-speed target detection. The right way of handling wideband clutter is space-time-TIME processing as has been proposed for cancellation of clutter and terrain scattered jammers by FANTE and TORRES [142], and MALAS et al. [433]. CERUTTI-MAORI [89] presents a K-space description of optimum STAP for clutter suppression in wideband radar. It has also been demonstrated in Chapter 13 of this book that space-TIME processing results in narrow jammer nulls in the array directivity pattern. SELIKTAR et al. [600, 601] used a suboptimum 'beam augmented' STAP architecture which showed good performance in clutter + terrain scattered interference suppression. The processor was applied to MOUNTAINTOP data (TITI [646], TITI and MARSHALL [647]).

It has been pointed out by ZATMAN [738] that the Hung–Turner technique is a generalised eigencanceller (or orthogonal projection technique, see for example Section 4.2.2 and HAIMOVICH and BAR-NESS [236]). The rank of the eigencanceler is based on the number of dominant eigenvalues of $\mathbf{Q}$ while the order of the Hung–Turner technique is determined by the number of data vectors. The authors demonstrated that the Hung–Turner method is a cost-effective technique whenever the number of available training samples is low. The Hung–Turner algorithm tends to generate higher sidelobes than the eigencanceler.

16.4.3.4 *The matrix transformation projection (MTP) technique*

GIERULL'S [190, 191] MTP (matrix transformation projection) method is an orthogonal projection technique for interference suppression and azimuth estimation with a sensor array. The algorithm has almost optimum interference cancellation properties. The projection matrix becomes

$$\hat{\mathbf{P}} = \mathbf{I} - \hat{\mathbf{S}}(\hat{\mathbf{S}}^*\hat{\mathbf{S}})^{-1}\hat{\mathbf{S}}^* \qquad (16.45)$$

where

$$\hat{\mathbf{S}} = \mathbf{X}(\mathbf{X}^*\mathbf{E}_1 + \mathbf{X}^*\mathbf{E}_2) \qquad (16.46)$$

$\mathbf{X}$ is an $N \times K$ matrix whose K columns are data snapshot vectors. The matrices $\mathbf{E}_1$ and $\mathbf{E}_2$ are permutation matrices of dimension $N \times K$, i.e., parts of unit matrices with randomly rearranged columns in such a way that the columns of both matrices are mutually orthogonal.

The MTP technique works with small training sample size. Compared with eigenvector based techniques it needs much fewer operations to generate the projection matrix. Filtering is based on the full order, i.e., requires a matrix vector product of order NM for each range increment.

---

[17] Of course, only for sidelobe jammers.

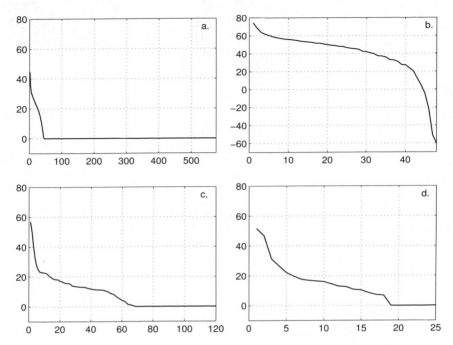

Figure 16.19: Clutter eigenspectra (eigenvalue power [dB] vs number of eigenvalues) for different processing techniques: a. optimum processor, Section 4.2.1 ($N = 24$, $M = 24$); b. auxiliary channel processor, Section 5.3 ($N = M = 24$, $C = 48$); c. symmetric auxiliary sensor processing, Section 6.2.1 ($N = 24$, $K = 5$, $M = 24$); d. auxiliary sensor/echo processor, Section 9.3 ($N = M = 24$, $K = L = 5$)

16.4.3.5 Convergence

HAIMOVICH and BAR-NESS [236] have shown that the projection technique ('eigencanceller') is superior to the optimum (SMI) processor when the number of data snapshots used to estimate the interference covariance matrix is small. The eigencanceller utilises only the principal eigenvalues associated with the interfering sources while the optimum processor is based on the total of eigenvectors. The principal eigenvectors are less affected by noise fluctuations than the smaller eigenvectors associated with the receiver noise. In that sense the projection matrix (1.47) is more stable than the version (1.46).

Two different versions of orthogonal projection have been compared with the optimum (SMI) processor by HAIMOVICH [237]. The two orthogonal processors differ in that one of them (minimum power eigencanceller) takes the eigenvalues into account while the other (minimum weighting vector norm) uses the interference eigenvectors only. The minimum norm eigencanceller outperforms the minimum power eigencanceller and the SMI technique as well in terms of convergence if only a finite set of training data is given.

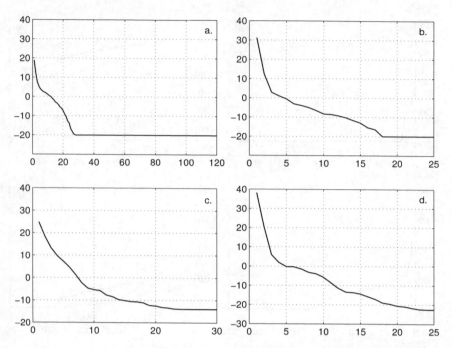

Figure 16.20: *Clutter eigenspectra (eigenvalue power [dB] vs number of eigenvalues) for FIR filter based processing techniques: a. full array processing, Section 7.1.1 ($N = 24$, $L = 5$); b. symmetric auxiliary sensor processing, Section 7.1.3 ($K = 5$, $L = 5$); c. disjoint subarrays, Section 7.1.3 ($N = 24$, $K = 6$, $L = 5$); d. overlapping subarrays, Section 7.1.3 ($K = 5$, $L = 5$)*

16.4.3.6 A general remark on orthogonal projection techniques

The obvious advantage of the orthogonal projection techniques is that the computation of the projection matrix from a limited set of interference data (clutter, jamming) is simplified as compared with SMI techniques. However, the adaptation of the interference filter matrix is only one aspect of interference rejection. It should be noted that in a stationary interference environment updating of the filter matrix is a slow process.

The fast operation in radar interference cancellation is the filtering of all range gates. In this respect the addressed techniques do not offer any advantages because the order of the matrix is the same as the original covariance matrix, that is, it is determined by the number of sensors and echo pulses.

Another aspect follows from Figures 16.19 and 16.20. The eigenspectra for eight different space-time processing schemes which have been treated in Chapters 4–7 are shown. It can be noticed that the eigenspectrum of the optimum processor (Figure 16.19a) shows the largest amount of multiple noise eigenvalues of all processors compared (total number of eigenvalues is $24 \times 24 = 576$). Next is the full array FIR filter (Figure 16.20a) where $L = 5$ echoes have been assumed ($24 \times 5 = 120$

eigenvalues).

The eigenspectra in Figures 16.19b–d and 16.20b–d show different transform based processors as have been described in the sections indicated in the figure captions. It can be noticed that the ratio of the numbers of noise and clutter eigenvalues is significantly reduced. As long as there is a large noise plateau rank reducing subspace techniques can be efficient. This is, however, not verified for any of the transform based techniques.

All of the transforms described in Chapters 5–7 focus on the clutter power while reducing the *order* of the covariance matrix. This property can be viewed as a the reason for 'compression' of the eigenspectrum, that is, the number of noise eigenvalues is reduced.

Rank reduction techniques are useful to reduce the computational cost of the filter adaptation. However, *order* reducing techniques[18] are required for cost and time efficient *filtering* of the data. This important aspect is overlooked in most of the literature.

The *order* reducing techniques can of course be used in conjunction with orthogonal projection techniques. It is, however, a question whether subspace techniques are worth implementing if the size of the covariance matrix is just 4×4 ($K = L = 2$) or 9×9 ($K = L = 3$). Also, exploiting the temporal stationarity of radar echoes by use of the Akaike algorithm [12][19] is not taken into account when comparisons between orthogonal projection and SMI is done.

An obvious advantage of the orthogonal projection technique may lie in the fact that they produce perfect nulls for the interference. This property may be useful in a clutter environment with strongly varying power.

16.4.3.7 *Non-Gaussian clutter*

All linear techniques described above are optimum only if the clutter obeys a gaussian distribution. TSAKALIDES and NIKIAS [651, 652] recognised by using MOUNTAINTOP data (TITI [646], TITI and MARSHALL [647]) that certain types of clutter are non-gaussian but obey an α stable disribution which may have larger tails than the gaussian distribution. It was shown by the authors that a 'least mean p-norm STAP' algorithm is superior to the minimum variance filter according to (1.103). Simulations have shown that in gaussian clutter both techniques perform equally well.

MICHELS *et al.* [462] present an analysis of two processors (PAMF and N-PAMF) operating in inhomogeneous compound gaussian (K-distributed) clutter. It was found that the N-PAMF method offers advantages especially when the sample size of training data is small. MICHELS *et al.* [462, 464] also give some results on the application of STAP to compound-gaussian clutter.

---

[18]Linear space-time transforms, see Chapter 5; spatial transforms: subarrays or auxiliary sensors, see Chapter 6; FIR filter approach, see Chapter 7; frequency and beamspace domain techniques, Chapter 9.
[19]For inverting block-Toeplitz matrices.

16.5 Alternative processor concepts

Most of the space-time processing architectures discussed in Chapters 4, 5, 6, 7, and 9 have been analysed by the author. Remarks on several processors proposed by other authors that are related to the processors discussed have been inserted into the individual chapters where appropriate. However, there are several processor concepts which do not really fit into the above scheme. A brief overview of those techniques is given in this section.

16.5.1 Least squares predictive transform

The concept of least squares prediction is well-known from the Wiener filter theory and also from adaptive coding. Transform coding is a well-known technique for data reduction in communications. Both techniques can be applied to adaptive space-time processing (GUERCI and FERIA [220, 221, 222, 223], GUERCI et al. [224]). A predictive transform is used to reduce the signal vector space down to the required number of degrees of freedom. This results in a reduction of the number of operations required for adaptive clutter cancellation. The performance of this linear prediction STAP processor (LP-STAP) approaches the performance of the optimum processor very closely. When the LP processor is trained with a limited number of samples it may perform better than the SMI technique (finite sample size version of optimum processor) because of the higher convergence rate.

16.5.2 Direct data domain (D^3) approaches

Strong clutter discretes may lead to false alarms in adaptive clutter filter systems. Adaptation is based on averaging over a certain amount of training data so that the response of the filter is optimum for the average clutter power level. In the following we describe briefly some direct data domain approaches which are based solely on the data from the cell under test.

16.5.2.1 Deterministic eigenvalue approach

PARK S. and SARKAR [524, 526] propose a technique for target detection based on solving the generalised eigenvalue problem

$$(\mathbf{X} - \alpha \mathbf{S})\mathbf{w} = \mathbf{0} \tag{16.47}$$

where the matrix

$$\mathbf{X} = \begin{pmatrix} \mathbf{x}_1 & \mathbf{x}_2 & \cdots & \mathbf{x}_L \\ \mathbf{x}_2 & \mathbf{x}_3 & \cdots & \mathbf{x}_{L+1} \\ \vdots & & \cdots & \vdots \\ \mathbf{x}_L & & \cdots & \mathbf{x}_M \end{pmatrix} \tag{16.48}$$

includes N-dimensional data vectors whose elements are the individual array output samples, and

$$\mathbf{S} = \begin{pmatrix} \mathbf{s}_1 & \mathbf{s}_2 & \cdots & \mathbf{s}_L \\ \mathbf{s}_2 & \mathbf{s}_3 & \cdots & \mathbf{s}_{L+1} \\ \vdots & & \cdots & \vdots \\ \mathbf{s}_L & & \cdots & \mathbf{s}_M \end{pmatrix} \qquad (16.49)$$

includes the corresponding vectors of the desired signal. Notice that the data and signal vectors are arranged in such a way that the weighting vector $\mathbf{w}$ operates as a space-time FIR filter along the time dimension. L is the temporal filter length and M the number of coherently processed data.

Since this technique operates on a single snapshot of primary data, problems associated with non-homogeneity of clutter or blinking jammers are avoided. Especially in application where only a few data are available this technique might be useful, for example, for forward looking array radar at short range.

Notice that the eigenvalue problem has to be solved separately for each individual range gate. This may be a problem for application in real-time. Of course, this technique can also be applied after a spatial transform according to Chapter 6. In this case the number of elements N is replaced by the number of antenna output channels K.

The look direction is incorporated in the constraint matrix $\mathbf{S}$. The constraint matrix may also include several constraints so as to reduce the sensitivity to mismatch with the signal (CARLO et al. [86]). In another paper (PARK S. and SARKAR [525]) the authors have extended their technique to multiple look directions. This might be of interest for passive radar or sonar operation. Moreover, such a multiple constraint matrix may be useful for use as Doppler filter bank.

A class of deterministic LS-algorithms for single snapshot signal detection has been described by SARKAR et al. [586]. The performance has been verified by use of MCARM data with synthetic and real target signals injected. It has been shown that single snapshot processing schemes may be superior to stochastic techniques based on the space-time clutter covariance matrix. DONG and BAO [119] combine an adaptive filter against homogeneous clutter with a direct data domain technique against clutter discretes.

16.5.2.2 Hybrid space-time adaptive processing algorithm

The hybrid space-time adaptive processing algorithm by ADVE et al. [5] uses a similar, non-statistical one-snapshot approach for suppression of jammers or strong clutter discretes before cancellation of the homogeneous part of the clutter. In ADVE et al. [10, 9] and HALE et al. [246] a compound processing scheme consisting of a non-statistical direct data domain (D^3) part and a statistical processing element is presented. The D^3 approach minimises the interference within the cell under test in a least squares sense while maximising the gain of the array at the look angle and Doppler. A non-homogeneity detector decides whether the D^3 or the adaptive branch is more appropriate for the actual data.

16.5.2.3 *Gaussian clutter modelling*

The clutter rejection technique described in the previous section has the desirable property that only data from the test cell are used. Therefore, strong clutter discretes showing up in neighbouring cells have no influence on the estimation of the clutter covariance matrix. LIN and BLUM [402] propose a technique in which the clutter power spectrum[20] is modelled by a certain number of gaussian humps arranged along the angle-Doppler trajectory. By using a least squares fit with the data under test the space-time covariance matrix in the gaussian model can be estimated. Simulations and experiments with MCARM data (SURESH BABU *et al.* [626, 627]) have shown a detection performance superior to conventional adaptive techniques based on secondary training data from other range cells. The drawback of this technique is the supposedly high number of operations per range gate.

16.5.3 Frequency hopping

Frequency hopping is a common way of mitigating the effect of narrowband jamming in military radar applications. REED and GAU [560] have described a technique of integrating incoherently the output signals of multiple reduced-rank test statistics operating on different CPIs at different frequencies. This requires that several pulse bursts are dedicated to a certain azimuth-range cell object on the ground. Considerable gain in detection probability can be achieved.

16.6 Summary

This chapter contains some notes and issues concerning the implementation of adaptive space-time or space-frequency filters for clutter suppression in airborne MTI radar. A comparison of the computational load of all techniques discussed in this book is given in Tables 15.1–15.4. The key findings and issues can be summarised as follows

1. **Tolerances of sensor positions** degrade the performance of those processors which are based on equispaced arrays. This concerns mainly those processor architectures using spatial transform techniques with linear or rectangular planar arrays (see Chapter 6). The optimum processor and the auxiliary space-time channel processors are not sensitive to errors of the sensor positions. In conclusion we can state that the dramatic reduction of computational expense offered by the spatial transforms has to be paid for in terms of high accuracy of the antenna channels.

2. **Antenna channel errors** of various kinds in connection with the radar system bandwidth have been analysed by NICKEL [486]. IQ errors cause an increase of interference eigenvalues while delay errors may compensate for them to a certain extent. The increase in the number of eigenvalues has to be countered by additional degrees of freedom of the processor. The impact of frequency

---

[20]See, for example, Figures 3.16, 3.18, 3.19, etc.

disparity on the performance of STAP radar has been pointed out by WANG *et al.* [675].

3. There are **channel equalisation** techniques to compensate for all kinds of differences between channels. ENDER's technique adjusts both the frequency responses in the range and the echo time dimensions in the frequency domain. This technique requires long echo data sequences as is common in SAR systems. Channel error compensation for short data sequences is an open issue.

4. Before adapting the clutter suppression filter it is advantageous to remove the range dependency of the clutter Doppler by adaptive Doppler compensation.

5. **Inclusion of the signal** in the adaptation of the clutter suppression filter may degrade target detection especially when adaptation is done with a small amount of data. In this case the range cell under test has to be excluded from the adaptation which requires some additional memory as well as some arithmetic operations.

6. The benefit achieved by space-time adaptive clutter rejection techniques depends significantly on the CNR at the single antenna element. It may not be worthwhile to design a multichannel antenna when the expected clutter power is of the order of magnitude of the noise power.

Appendix

Sonar applications

In the following we give some examples of the application of space-time adaptive processing to sonar problems. We will address briefly the problem of suppression of reverberation.[1] Secondly we will touch upon matched field processing techniques which can be used for *passive* estimation of source parameters. We omit any introduction to sonar principles. For an introduction to sonar fundamentals the reader is referred to the book of URICK [659]. The content of this chapter follows closely a paper by KLEMM [311]. All figures have been taken from this publication. The application of space-time filtering for application with active sonar is also addressed by JAFFER [277]. Several articles on spatial array processing in a shallow water sound channel have been published by KLEMM [293, 294, 296, 297, 299, 310]. This appendix focuses on the possibilities of space-time processing under the boundary conditions of an acoustic waveguide. Some applications of space-time processing to sonar are given in KLEMM ed. [347]. Adaptive suppression of reverberation in active sonar by means of STAP has been demonstrated on the basis of measured sonar data by MAIWALD *et al.* [430]. PILLAI *et al.* [545] use space-frequency processing for cancellation of ambient ship noise.

A.1 Introduction

Active sonar, although being an echo technique like radar, differs from radar in several aspects:

- The spatial and temporal variability of the ocean normally does not permit coherent processing of coherent pulse trains. The low velocity of sound propagating in water (about 1500 m/s) and the long achievable range (up to several hundreds of kilometres) lead to echo durations of the order of magnitude of *minutes*. This is much larger than the usual correlation time limited by surface waves and other irregularities of the acoustic channel. In contrast to radar, the Doppler frequency due to a moving target can be retrieved from the time history

---

[1] The sonar equivalent of clutter.

Table A.1 Sonar parameters

source frequency	200 Hz
water depth	100 m
number of modes	9
white noise power at sensor level	−30 dB
source range	6000 m
source depth	50 m
sensor spacing	5 m
depth of horizontal array	40 m
sensor depths of vertical array	40–95 m
relative velocity target/array	15 m/s
number of hydrophones	10
look direction of horizontal array	endfire
sound velocity profile (water)	0 m 1525 m/s
	20 m 1500 m/s
	40 m 1502 m/s
subbottom parameters	density 2 g/cm^3
	attenuation 0.5 dB/λ
	sound speed 1600 m/s

of one individual echo (2.14). Following the nomenclature used earlier we are talking in this appendix only about space-TIME processing.

- Receiver noise is normally very weak. The dominating noise component is *ambient* noise which originates from various natural (e.g., surface waves) and man-made sources (ship traffic). Ambient noise is more or less directive and broadband. For simplicity we will assume in the analysis below that the noise is white in the spatial and temporal dimensions. Simple models for ambient noise are given by CRON and SHERMAN [108].

- As stated above we consider the case of a shallow water sound channel. Under such waveguide conditions sound propagation can be described in terms of normal modes. The results presented below are based on the normal mode sound propagation model SNAP[2], see JENSEN and FERLA [280].

A.2 Signal processing in the modal environment

All the signal processing techniques summarised in Chapter 1 can in principle be used in a modal environment. For the purpose of theoretical analysis we have first to formulate covariance matrices and signal replicas (steering vector) on the basis of normal mode sound propagation.

---
[2] SNAP: Saclantcen Normal Mode Acoustic Propagation Model.

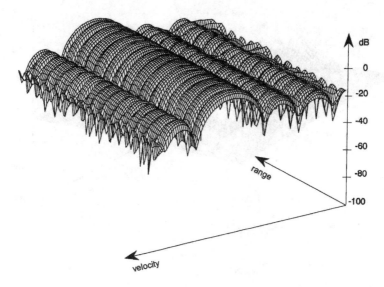

Figure A.1: Range/velocity estimation with horizontal array, signal match (SM). (©1993 IEEE)

The sound pressure due to a narrowband source can be described in terms of normal modes (the vertical eigenfunctions of the acoustic waveguide)

$$P(\zeta, z, r) = \sum_{q=1}^{Q} p_q(\zeta, z, r) \qquad (A.1)$$

$$= \exp[-j(\omega t - \pi/4)] \sum_{q=1}^{Q} A_q \exp(jk_q r)$$

where

$$A_q = p_0 \frac{\omega \rho^2}{H} \sqrt{\frac{1}{8\pi r}} \frac{\mu(\zeta)\mu(z)}{\sqrt{k_q} \exp(-\alpha_q r)} \qquad (A.2)$$

The various quantities are as follows: p_0 source strength; ρ water density; H water depth; r range between source and receiver; z receiver depth; ζ source depth; α_q modal attenuation coefficient; $\mu(.)$ normal mode function; Q number of modes; k_q modal wavenumbers; ω frequency.

A.2.1 Signal models

A.2.1.1 Covariance matrix across sensor array

Each of the sensors of an array receives an acoustic signal as given by (A.1) and (A.2). The various modal components can be included in a $N \times M$ mode-sensor transform

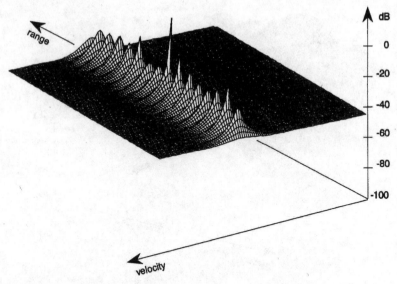

Figure A.2: Range/velocity estimation with horizontal array, MV estimator. (©1993 IEEE)

containing the q-th modal wave components at the n-th sensor:

$$\mathbf{P} = \begin{pmatrix} p_{11}(\zeta, z_1, r_1) & p_{12}(\zeta, z_1, r_1) & \cdots & p_{1Q}(\zeta, z_1, r_1) \\ p_{21}(\zeta, z_2, r_2) & p_{22}(\zeta, z_2, r_2) & \cdots & p_{2Q}(\zeta, z_2, r_2) \\ \vdots & \vdots & \cdots & \vdots \\ p_{N1}(\zeta, z_N, r_N) & p_{N2}(\zeta, z_N, r_N) & \cdots & p_{NQ}(\zeta, z_N, r_N) \end{pmatrix} \quad (A.3)$$

where p_{nq} is the contribution of the q-th mode to the n-th sensor. For a linear vertical array the elements of $\mathbf{P}$ depend on the sensor depths

$$p_{nq} = p_q(\zeta, z_n, r) \quad (A.4)$$

while for a linear horizontal array one gets

$$p_{nq} = p_q(\zeta, z, r_n) \quad (A.5)$$

where

$$r_n = r_0 + d_n \cos \varphi \quad (A.6)$$

with φ being the angle of the source relative to the array axis. d_n denotes the position of the n-th sensor in the array and r_0 is the distance between the source and the first sensor.

The horizontal modal wavenumbers k_q are associated with vertical modal angles of arrival in the following way

$$k_q^{(h)} = k_0 \cos \beta_q \quad (A.7)$$

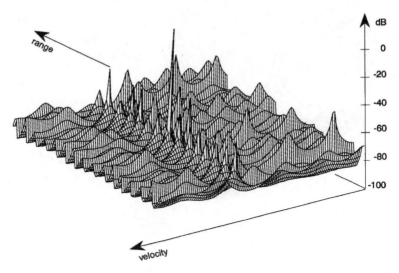

Figure A.3: Range/velocity estimation with horizontal array, ME estimator. (©1993 IEEE)

and the vertical wavenumbers are

$$k_q^{(v)} = k_0 \sin \beta_q \qquad (A.8)$$

where $k_0 = \frac{2\pi}{\lambda}$. This means that the total of modal arrivals with different vertical wavenumbers is interpreted by a linear horizontal array as *azimuthal* spread. The width of the angular spread of an individual arrival is proportional to the direction of arrival $\sin \varphi$. At broadside, incoming waves appear to be planar, even in the vertical, while arrivals coming from any other direction appear to be spread out in angle. This spread increases as the direction approaches endfire.

On the other hand the modal spread of the signal can be exploited for matched field processing even with a horizontal array. As the modal spread is a maximum in the endfire direction the best results in matched field processing can be expected at endfire.

The summation of modal components in (A.1) means that the modes are entirely coherent. In this case the signal covariance matrix across the array becomes a dyadic of the following form

$$\mathbf{R} = \mathbf{PP}^* \qquad (A.9)$$

For simplicity we assume that the ambient noise is white (which is verified only under special conditions). Then we get

$$\mathbf{R} = \mathbf{PP}^* + P_w \mathbf{I} \qquad (A.10)$$

with P_w being the white noise power. As stated before the covariance matrix given by (A.9) is based on the assumption that the modes are entirely coherent which requires

550 Sonar applications

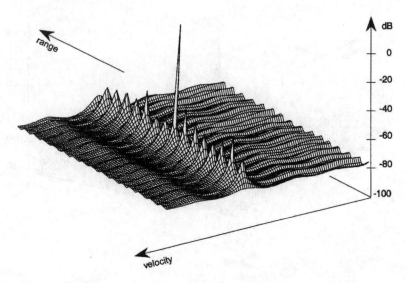

Figure A.4: Range/velocity estimation with horizontal array, TME estimator. (©1993 IEEE)

an ideal rigid acoustic channel. In practice there are some fluctuations, particularly through surface waves. Such fluctuations cause decorrelation between normal modes. To describe such effects an intermode correlation matrix can be introduced so that (A.9) becomes

$$\mathbf{R} = \mathbf{PQP}^* + P_\mathrm{w}\mathbf{I} \qquad (A.11)$$

where $\mathbf{Q}$ is the $M \times M$ inter-mode correlation matrix. According to CLAY [93] the inter-mode correlation coefficients can be described as

$$q_{ik} = \exp[-(k_i - k_k)^2 r^2 \sigma^2 / (2H^2)] \qquad (A.12)$$

where σ is the rms wave height and k_i, k_k are horizontal modal wavenumbers.

A.2.1.2 Steering vector

The steering vector (or signal replica) is simply a coherent sum over the normal mode contributions

$$\mathbf{h}(\Theta) = \left(\sum_{q=1}^{Q} p_{nq}(\zeta, z_n, r_n) \right) \qquad (A.13)$$

where Θ may run over source parameters such as r, z, φ, ω, and parameters of the acoustic channel or the environment.

For conventional matching according to (1.102) we get

$$P_\mathrm{SM}(\Theta) = \frac{\mathbf{h}^*(\Theta)\mathbf{R}\mathbf{h}(\Theta)}{\mathbf{h}^*(\Theta)\mathbf{h}(\Theta)} \qquad (A.14)$$

Signal processing in the modal environment

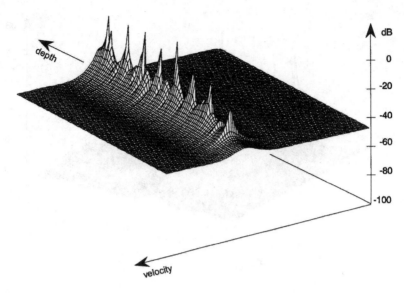

Figure A.5: Depth/velocity estimation with horizontal array, MV estimator. (©1993 IEEE)

The MV spectral estimator according to (1.104) is

$$P_{MV}(\Theta) = [\mathbf{h}^*(\Theta)\mathbf{R}^{-1}\mathbf{h}(\Theta)]^{-1} \qquad (A.15)$$

The ME spectral estimator according to (1.105) is

$$P_{ME}(\Theta) = \frac{1}{\mathbf{h}^*(\Theta)\mathbf{rr}^*\mathbf{h}(\Theta)} \qquad (A.16)$$

where $\mathbf{r}$ is the first column of $\mathbf{R}^{-1}$. Notice that

$$\mathbf{hh}^* = \mathbf{PP}^* \qquad (A.17)$$

so that the above spectral estimators can also be written as

$$\begin{aligned} P_{SM}(\Theta) &= \frac{\operatorname{tr}[\mathbf{P}(\Theta)\mathbf{P}^*(\Theta)\mathbf{R}]}{\operatorname{tr}[\mathbf{P}(\Theta)\mathbf{P}^*(\Theta)]} \\ P_{MV}(\Theta) &= [\operatorname{tr}(\mathbf{P}(\Theta)\mathbf{P}(\Theta)^*\mathbf{R}^{-1})]^{-1} \\ P_{ME}(\Theta) &= [\operatorname{tr}(\mathbf{P}(\Theta)\mathbf{P}(\Theta)^*\mathbf{rr}^*)]^{-1} \end{aligned} \qquad (A.18)$$

These estimators have been analysed by the author [294, 296] for estimation of range and depth of acoustic sources by exploiting the modal interference pattern. Later on such techniques have been referred to as *matched field processing*. Readers interested in a tutorial on matched field processing are referred to the paper by BAGGEROER and KUPERMAN [26].

552 Sonar applications

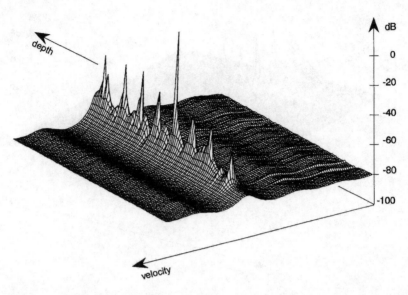

Figure A.6: Depth/velocity estimation with horizontal array, TME estimator. (©1993 IEEE)

A.2.2 Extension to space-time matched field processing

The extension of the signal models given above to space-time quantities is straightforward. The rationale for space-time matched field processing is the potential of simultaneous position and velocity estimation of acoustic sources. Introducing time samples at instants γmT ($m = 1,\ldots,M$, T = sampling interval) in (A.1) gives

$$P(\zeta, z, r_m, v) = \sum_{q=1}^{Q} p_q(\zeta, z, r_m, mT) \qquad (A.19)$$

$$= \exp[-j(\omega t - \pi/4 - \gamma mT)] \sum_{q=1}^{Q} A_q \exp(jk_q r)$$

where the index $m = [1, M]$ indicates the time dependence of an object moving at radial velocity v. γ takes care of one or two-way propagation. $\gamma = 2$ in the case of active and $\gamma = 1$ in the case of a passive system. Then the $NM \times Q$ mode-sensor-time transform becomes

$$\mathbf{P} = \begin{pmatrix} \mathbf{P}_1 \\ \vdots \\ \mathbf{P}_m \\ \vdots \\ \mathbf{P}_M \end{pmatrix} \qquad (A.20)$$

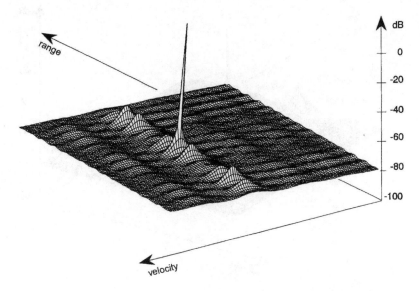

Figure A.7: Range/velocity estimation with vertical array, TME estimator. (©1993 IEEE)

where the time-dependent submatrices are given by

$$\mathbf{P}_m = \begin{pmatrix} p_{11m}(\zeta, z_1, r_1) & p_{12m}(\zeta, z_1, r_1) & \cdots & p_{1Qm}(\zeta, z_1, r_1) \\ p_{21m}(\zeta, z_2, r_2) & p_{22m}(\zeta, z_2, r_2) & \cdots & p_{2Qm}(\zeta, z_2, r_2) \\ \cdot & \cdot & \cdots & \cdot \\ \cdot & \cdot & \cdots & \cdot \\ p_{N1m}(\zeta, z_N, r_N) & p_{N2m}(\zeta, z_N, r_N) & \cdots & p_{NQm}(\zeta, z_N, r_N) \end{pmatrix} \quad (A.21)$$

For a linear vertical array the elements of $\mathbf{P}_m$ are

$$p_{nqm} = p_q(\zeta, z_n, r_{nm}, v) \quad (A.22)$$

with

$$r_{nm} = r + \gamma v m T \quad (A.23)$$

where vmT is the range variation due to the radial target (or sonar platform) velocity v. For a linear horizontal array we get, similar to (A.5),

$$p_{nqm} = p_q(\zeta, z, r_{nm}, v) \quad (A.24)$$

with

$$r_{nk} = r + d_n \cos\varphi + \gamma v m T \quad (A.25)$$

Then the space-time steering vector becomes

$$\mathbf{h}(\Theta) = \left(\sum_{q=1}^{Q} p_{nq}(\zeta, z_n, r_{nm}) \right) \quad (A.26)$$

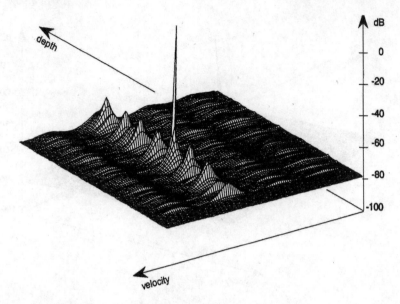

Figure A.8: Depth/velocity estimation with vertical array, TME estimator. (©1993 IEEE)

where the parameter vector Θ now includes the velocity v.

Once we have obtained the space-time covariance matrix given by (A.20), (A.21), and (A.10), and the space-time steering vector (A.26) we can calculate spectra according to (A.14), (A.15) or (A.16).

In Chapter 7 a space-time FIR filter has been described which provided near-optimum clutter rejection at very low cost. As this filter uses the first block column of the space-time covariance matrix it is strongly related to the ME spectral estimator. In fact, in accordance with (7.5), we can formulate a space-time spectral estimator of the following form

$$P_{\text{TME}}(\Theta) = [\mathbf{h}^*(\Theta)\tilde{\mathbf{K}}\tilde{\mathbf{K}}^*\mathbf{h}(\Theta)]^{-1} \qquad (A.27)$$

where $\tilde{\mathbf{K}}$ is the first block column of $\mathbf{R}$ with dimensions $NM \times N$. 'TME' stands for *temporal* maximum entropy. This estimator has the properties of the MV estimator in the spatial dimension. It functions, however, as an ME estimator in the temporal domain.

A.3 Active sonar application: suppression of reverberation

As discussed in Chapters 2 and 3 reflections from a stationary background (clutter) received by a moving radar are Doppler shifted. The Doppler shift due to an individual scatterer is proportional to the angle between the position of the scatterer and the platform velocity vector (2.35). While we assumed for radar applications that the

Active sonar application: suppression of reverberation 555

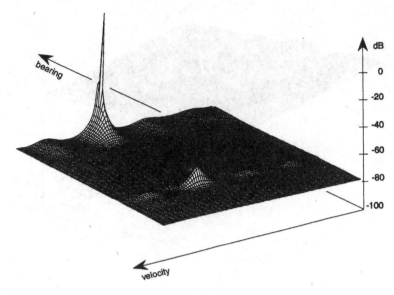

Figure A.9: Bearing/velocity estimation with horizontal array, TME estimator. (©1993 IEEE)

various arrivals coming from all directions are plane waves (free space propagation) we are now faced with propagation of reverberation in a waveguide. In terms of normal modes, the reverberation wave field is a superposition of

- arrivals coming from all possible directions similar to (2.34),

- with each arrival being a superposition of horizontal plane waves propagating at different vertical modal angles (A.19).

The Doppler shift of each individual modal arrival is again proportional to the sine of the angle of arrival regardless if this angle is composed of an azimuth component and a modal vertical angle. Therefore, in the case of a sidelooking array all arrivals will show up on the diagonal of a φ–f_D plot like Figure 3.18, or on a circle like in Figure 3.19, in the case of a forward looking array. The superposition of various modes can result in a modulation of the reverberation spectrum, however, suppression of reverberation via space-*TIME* processing is in principle a similar problem as clutter rejection via space-*time* processing in radar. Of course, the environmental conditions in a real ocean including effects such as internal motion, ambient noise, surface waves, volume scattering, etc., may degrade the reverberation suppression performance. Predictions on the achievable performance are not possible based on the simple models used here. Experimental verification is required.

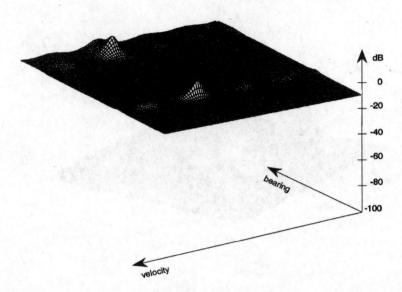

Figure A.10: Bearing/velocity estimation with horizontal array, beamformer based TME estimator. (©1993 IEEE)

A.4 Estimation of target position and velocity

In Figures A.1–A.10, we show some numerical examples which illustrate the principle of space-time matched field processing for estimation of the position and velocity of moving acoustic sources. The modal field due to a moving source has been modelled by (A.10), where the elements of the sensor-mode-time transform **P** are given by (A.19).

The numerical examples shown below are based on the SNAP model (JENSEN and FERLA [280]) with the choice of parameters listed in Table A.1.

In Figure A.1, a range-velocity spectrum of a moving target is shown. The plot is based on conventional signal-matching (SM) according to (A.14). A horizontal array was assumed. It can be noticed that in the velocity dimension we have the expected $\frac{\sin x}{x}$. This happens because in the target model we used only a single Doppler frequency due to the relative motion between target and sonar. As can be seen from Figure A.1 the conventional signal match is not capable of resolving the target in the range dimension.

The MV estimator after (A.15), however, shows a distinct peak at the target range (Figure A.2). Notice that there are much stronger sidelobes in the range than in the Doppler dimension. This reflects the fact that the interpretation of the coherent modal field by a certain power spectral estimator is more difficult than the search for a single Doppler line (or the position of a point target by an array). It was found by the author [294, 296] that for passive localisation by field matching high-resolution techniques have to be applied.

The performance of the ME estimator according to (A.16) on the same problem is shown in Figure A.3. As one can see the peak at the target position/velocity is even

Estimation of target position and velocity 557

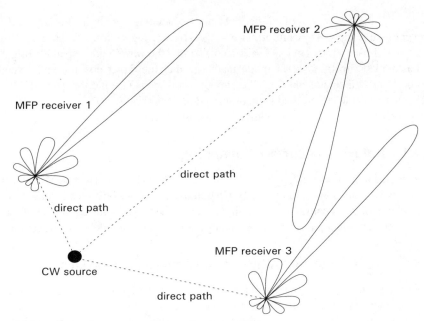

Figure A.11: *Multistatic CW system* (©1993 IEEE)

higher than in the previous example. However, the ME power estimator is well known for creating spurious sidelobes which can be noticed clearly in this example.

Figure A.4 shows for comparison the same situation, now generated by use of the TME estimator according to (A.27). As can be seen the TME estimator achieves the maximum peak-to-sidelobe level.

In Figures A.5 and A.6, a *horizontal* array is used for estimating target depth. The TME estimator again shows a clear distinction between a peak at the true target position/velocity and the sidelobes while the MV response is ambiguous.

In Figures A.7 and A.8, a *vertical* array in conjunction with the TME estimator has been used to estimate range/velocity and depth/velocity. Comparing these spectra with Figures A.6 and A.4 it appears that a vertical array is more suitable for matched field processing than a horizontal array. Of course, as the mode functions are vertical a vertical array is more adapted to mode matching than a horizontal one.

In Figures A.9 and A.10 the TME estimator (A.27) is used to estimate the target bearing. Recall that the modal spread causes some uncertainty about the source direction. Our particular interest is the behaviour of an undersampled array in a modal field. As is well known an array exhibits an ambiguous response if the sensor spacing is greater than half the wavelength. Let us consider the general case where the angle(s) under which the ambiguous responses appear are different from the true source direction φ_t and from the angle symmetric to the array broadside direction $180°-\varphi_t$. All other directions are associated with different modal structures which can be exploited to distinguish between the true position and the ambiguous responses.

In Figure A.9 we applied perfect mode matching while in Figure A.10 a conventional plane wave beamformer was used as steering vector. The spacing of the array was chosen to be 5 m ($\lambda/2 = 3.75$ m) so that the wave field is spatially slightly undersampled. In the case of mode matching the true target position and velocity can be clearly distinguished from the ambiguous response (in the foreground). If no information on the modal field is incorporated in the steering vector (Figure A.10) the array shows two ambiguous responses of equal strength.

A.4.1 Influence of surface fluctuations

As matched field processing relies on highly sensitive techniques such as MV or TME estimators it can be expected that any irregularities in the acoustic channel lead to a degradation of the localisation capability. It has been shown by the author [296] that a rms surface wave height of 1% of the channel depth causes significant degradation.

A.4.2 Application: a multistatic CW surveillance system

The matched field processing approach discussed in this section may be used to create a bi- or multistatic source–sensor configuration, see Figure A.11. The environment is illuminated by a CW source. The direct paths are cancelled through spatial nulling. If the configuration is in motion the nulling procedure has to be carried out adaptively. The reflected paths are evaluated in the receivers by use of space-time matched field processing.

A.5 Summary

Two applications of space-time processing for sonar applications have been presented. The numerical examples are based on a shallow water sound propagation model which calculates the acoustic pressure in terms of normal mode functions.

Suppression of reverberation (the sonar equivalent of clutter) by means of space-TIME processing to achieve a narrow reverberation notch seems to be possible. Reverberation propagates in the shallow water waveguide in normal modes like any other acoustic wave. Each mode is associated with its individual vertical angle of arrival which in turn is associated with an individual Doppler frequency if the sonar is on a moving platform. The resulting power spectrum will look similar to the clutter spectra shown in Figures 3.18 and 3.19.

The second application is simultaneous position and Doppler estimation using high-resolution space-time matched field processing techniques. Results have been presented for range/Doppler and depth/Doppler estimation by horizontal and vertical hydrophone arrays. Best performance is achieved by using the TME estimator which is strongly related to the space-time FIR filter in Chapter 7.

Bibliography

[1] ABRAMOVICH, Y. I., SPENCER, N. K., and ANDERSON, S. J.: 'Stochastic-constraints method in nonstationary hot-clutter cancellation Part I: Fundamentals and supervised training applications', *IEEE Trans. AES*, Vol. 34, No. 4, 1998, pp. 1271-1292

[2] ABRAMOVICH, Y. I., SPENCER, N. K., and ANDERSON, S. J.: 'Stochastic-constraints method in nonstationary hot-clutter cancellation Part II: Unsupervised training applications', *IEEE Trans. AES*, Vol. 36, No. 1, January 2000, pp. 132-150

[3] D'ADDIO, E., DI BICEGLIE, M., and BOTTALICO, S.: 'Detection of moving objects with airborne SAR', *Signal Processing*, Vol. 36, 1994, pp. 149-162

[4] ADVE, R. S., and WICKS, M. C.: 'Joint domain localized processing using measured spatial steering vectors', *IEEE RADARCON'98*, 11-14 May 1998, Dallas, TX, pp. 165-170

[5] ADVE, R. S., HALE, T. B., and WICKS, M. C.: 'A two-stage hybrid space-time adaptive processing algorithm', *IEEE National Radar Conference*, 20-22 April 1999, Boston, MA, pp. 279-284

[6] ADVE, R. S., HALE, T. B., and WICKS, M. C.: 'Transform domain localized processing using measured steering vectors and non-homogeneity detection', *IEEE National Radar Conference*, 20-22 April 1999, Boston, MA, pp. 285-290

[7] ADVE, R. S., and WICKS, M. C.: 'Joint domain localized adaptive processing using nonorthogonal steering vectors', *Digital Signal Processing*, Vol. 9, 1999, pp. 36-44

[8] ADVE, R. S, HALE, T. B., and WICKS, M. C.: 'Practical joint domain localised adaptive processing in homogeneous and nonhomogeneous environments. Part 1: Homogeneous environments', *IEE Proc. Radar, Sonar and Navigation*, Vol. 147, No. 2, April 2000, pp. 57-65

[9] ADVE, R. S, HALE, T. B., and WICKS, M. C.: 'Practical joint domain localised adaptive processing in homogeneous and nonhomogeneous environments. Part 2: Nonhomogeneous environments', *IEE Proc. Radar, Sonar and Navigation*, Vol. 147, No. 2, April 2000, pp. 66-74

[10] ADVE, R., WICKS, M. C., B., and ANTONIK, P.: 'Ground moving target indication using knowledge based space-time adaptive processing', *IEEE RADAR2000*, 8-12 May 2000, Alexandria, VA, pp. 735-740

[11] ADVE, R., SCHNEIBLE, R., and McMILLAN, R.: 'Adaptive space/frequency processing for distributed aperture radars', *2003 IEEE Radar Conference*, May 5-8, 2003, Huntsville, AL, pp. 160-164

[12] AKAIKE, H.: 'Block Toeplitz Matrix Inversion', *SIAM Journal of Applied Mathematics*, Vol. 24, 1973, pp. 234-241

[13] ALBAREL, G., TANNER, J. S., and UHLMANN, M.: 'The trinational AMSAR programme: CAR active antenna architecture', *Radar'97*, 14-16 October 1997, Edinburgh, Scotland, pp. 344-347

[14] ANDERSON, D. B.: 'A microwave technique to reduce platform motion and scanning noise in airborne moving target radar', *IRE WESCON Conv. Record*, Vol. 2, pt. 1, 1958, pp. 202-211

[15] ANDERSON, T. W.: *An Introduction to Multivariate Statistical Analysis*, John Wiley, New York, 1957

[16] ANDERSON, K.: 'Adaptive Doppler filtering applied to modern air traffic control radars', *IEEE 2004 Radar Conference*, 26-29 April 2004, Philadelphia, PA

[17] ANDREWS, G. A.: 'A Detection Philosophy for AMTI Radar', *IEEE Int. Radar Conf.*, Arlington, VA, 1975, pp. 111-116

[18] ANDREWS, G. A.: 'Evaluation of Airborne Doppler Processors', *EASCON 77*, pp. 4-5A–4-5G

[19] ANDREWS, G. A.: 'Radar pattern design for platform motion compensation', *IEEE Trans. on Antennas and Propagation*, Vol. AP-26, No. 4, July 1978, pp. 566-571

[20] ANTONIK, P., SCHUMAN, H., LI, P., MELVIN, W., and WICKS, M.: 'Knowledge-based space-time adaptive processing', *Proc. IEEE National Radar Conference*, 1997, Syracuse, NY, pp. 372-377

[21] ANTONIK, P., SCHUMAN, H. K., MELVIN, W. L., and WICKS, M. C.: 'Implementation of knowledge based control for space-time adaptive processing', *Proc. IEE Radar'97*, 14-16 October 1997, Edinburgh, Scotland, pp. 478-482

[22] APPLEBAUM, S. P.: 'Adaptive arrays', *IEEE Transactions on Antennas and Propagation*, Vol. 24, No. 5, September 1976, pp. 585-598

[23] AUMANN, H. M., WARD, J., and WILLWERTH, F. G.: 'Inverse displaced phase center antenna for aircraft motion emulation', *14. Conference on Antenna Measurement and Techniques*, Columbo, OH, October 1992, pp. 1-6

[24] AYOUB, T. F., PUGH M. L., and HAIMOVICH.: 'Space-time adaptive processing for high PRF radar', *Proc. IEE Radar'97*, 14-16 October 1997, Edinburgh, Scotland, pp. 468-472

[25] AYOUB, T. F., HAIMOVICH, A. M., and PUGH, M. L .: 'Reduced-rank STAP for high PRF radar', *IEEE Trans. AES*, Vol. 35, No. 3, July 1999, pp. 953-962

[26] BAGGEROER, A. B., and KUPERMAN, W.: 'Matched field processing in ocean acoustics', in: J. M. F.MOURA,I. M. G. Lourtie (editors), *Acoustic signal processing for ocean exploration*, NATO ASI Series, Vol. 388, Kluwer Academic Publishers, Dordrecht, 1993, pp. 79-114

[27] BALAJI, B., and GIERULL, C.H.: 'Theoretical analysis of small sample size behaviour of eigenvector projection technique applied to STAP ', *IEEE 2002 Radar Conference*, Long Beach, CA, 22-25 April 2002, pp. 373-377

[28] BAO, Z., LIAO, G., Wu, R., ZHANG, Y., and WANG, Y.: 'Adaptive spatial-temporal processing for airborne radars', *Chinese Electronics Journal*, Vol. 2, No. 1, 1993, pp. 2-7

[29] BAO, Z., LIAO, G., and ZHANG, Y.: 'Partially adaptive space-time processing for clutter suppression in Airborne Radars', *Proc. IEEE International Radar Conference*, Alexandria, VA, 1995, pp. 120-124

[30] BAO Z., WU S., LIAO G., and XU Z.: 'Review of reduced rank space-time adaptive processing for airborne radars', *Proc. International Conference on Radar (ICR'96)*, 8-10 October 1996, Beijing, China, pp. 766-769

[31] BARANOSKI, E. J.: 'Improved pre-Doppler STAP algorithm for adaptive nulling in airborne radars', *IEEE Proc. 29th ASILOMAR Conference on Signals, Systems and Computers*, Pacific Grove, CA, 30 Oct.–2 Nov. 1995, pp. 1173-1177

[32] BARANOSKI, E. J.: 'Constraint optimization for partially adaptive subspace STAP algorithms', *ASILOMAR Conf. on Signal, Systems and Computers*, Pacific Grove, CA, 1-4 Nov. 1998, pp. 1527-1531

[33] BARBAROSSA, S., and PICARDI, G.: 'Predictive adaptive moving target indicator', *Signal Processing*, Vol. 10, 1986, pp. 83-97

[34] BARBAROSSA, S., and FARINA, A.: 'Space-time-frequency processing of synthetic aperture radar signals', *IEEE Trans. AES*, Vol. 30, No. 2, April 1994, pp. 341-357

[35] BARILE, E. C., FANTE, R. L., and TORRES, J. A.: 'Some limitations on the effectiveness of airborne adaptive radar', *IEEE Trans. AES*, Vol. 28, No. 4, October 1992, pp. 1015-1032

[36] BARILE, E. C., FANTE, R. L., GUELLA, T. P., and TORRES, J. A.: 'Performance of space-time adaptive airborne radar', *Proc. IEEE National Radar Conf.*, Cambridge, MA, 1993, pp. 173-175

[37] Barile, E. C., GUELLA, T. P., and LAMENSDORF, D.: 'Adaptive antenna space-time processing techniques to suppress platform scattered clutter for airborne radar', *IEEE Trans. AES*, Vol. 31, No. 1, January 1995, pp. 382-389

[38] BAYLISS, E. T.: 'Design of monopulse antenna difference patterns with low sidelobes', *Bell Systems Technical Journal*, Vol. 47, 1968, pp. 623-640

[39] BELL, K. L., VAN TREES, H. L., and GRIFFITHS, L. J.: 'Adaptive beampattern control using quadratic constraints for circular arrays', *ASAP 2000 Workshop*, 13-14 March 2000, MIT Lincoln Laboratory, pp. 43-48

[40] BERGER, S. D., and WELSH, B. M.: 'Selecting a reduced-rank transformation for STAP - a direct form perspective', *IEEE RADARCON'98*, 11-14 May 1998, Dallas, TX, pp. 177-182

[41] BERGER, S. D., and WELSH, B. M.: 'Selecting a reduced rank transformation for STAP – a direct form perspective', *IEEE Trans. AES*, Vol. 35, No. 2, April 1999, pp. 722-729

[42] BERGIN, J. S., TECHAU, P. M., MELVIN, W. L., and GUERCI, J. R.: 'GMTI STAP in target-rich environments: site specific analysis ', *IEEE 2002 Radar Conference*, Long Beach, CA, 22-25 April 2002, pp. 391-395

[43] BERGIN, J.S., TEIXERA, C.M., TECHAU, P.M., and GUERCI, J.R.: 'STAP with knowledge-aided data pre-whitening ', *IEEE 2004 Radar Conference*, 26-29 April 2004, Philadelphia, PA

[44] BERGIN, J.S., TEIXEIRA, C.M., and TECHAU, P.M.: 'Multi-resolution signal processing techniques for airborne radar', *IEEE 2004 Radar Conference*, 26-29 April 2004, Philadelphia, PA

[45] BERIN, M. O., and HAIMOVICH, A. M.: 'Signal cancellation effects in adaptive radar mountaintop data-set', *ICASSP'96*, Atlanta, GA, 1996

[46] BICKERT, B., ENDER, J., and KLEMM, R.: 'Verification of adaptive signal enhancement algorithms based on STAP techniques using four channel AER-II radar data in a forward looking air-to-ground GMTI mode', *Proc. TIWRS 2003*, Elba, Italy, 15-18 September 2003

[47] BICKERT, B.: 'Multi channel STAP/GMTI with forward looking radar, comparison of space-time adaptive (STAP) processors', *FGAN FHR Report Nr. 77*, FGAN, Wachtberg, 2004

[48] BILLINGSLEY, B., FARINA, A., GINI, F., GRECO, M. V., and LOMBARDO, P.: 'Impact of experimentally measured Doppler spectrum of ground clutter on MTI and STAP', *Proc. IEE Radar'97*, 14-16 October 1997, Edinburgh, Scotland, pp. 290-294

[49] BIRD, J. S., and BRIDGEWATER, A. W.: 'Performance of space-based radar in the presence of earth clutter', *IEE Proc., Pt. F*, Vol 131, No. 5, August 1984, pp. 491-500

[50] BLUM, R. S., MELVIN, W. L., and WICKS, M.: 'An analysis of adaptive DPCA', *IEEE National Radar Conference*, Ann Arbor, MI, 13-16 May 1996, pp. 303-308

[51] BLUM, R., S., and MCDONALD, K. F.: 'Analysis of STAP Algorithms for cases with mismatched steering and clutter statistics ', *IEEE Trans. SP*, Vol. 48, No. 2, February 2000, pp. 301-310

[52] BLUNT, S.D., and GERLACH, K.: 'Efficient robust AMF using the enhanced FRACTA algorithm: results from KASSPER I & II ', *IEEE 2004 Radar Conference*, 26-29 April 2004, Philadelphia, PA

[53] BOCHE, H., and SCHUBERT, M.: 'On the narrowband assumption for array signal processing', *International Journal on Electronics and Communications (AEÜ)*, Vol. 53, No. 2, 1999, pp. 117-120

[54] BODEWIG, E., *Matrix Calculus*, North-Holland Publishing Company, Amsterdam, 1959

[55] BOJANCZYK, A.W., and MELVIN, W.L.: 'Simplifying the computational aspects of STAP', *IEEE RADARCON'98*, 11-14 May 1998, Dallas, TX, pp. 123–128

[56] BOLLINI, P., CHISCI, L., FARINA, A., GIANELLI, M., TIMMONERI, L., and ZAPPA, G.: 'QR versus IQR algorithms for adaptive signal processing: performance evaluation for radar applications', *IEE Proc. Radar, Sonar and Navigation*, Vol. 143, No. 5, October 1996, pp. 328-340

[57] BORSARI, G., K., and STEINHARDT, A. O.: 'Cost-efficient training strategies for space-time adaptive processing algorithms', *29th Asilomar Conference on Signals, Systems and Computers*, Pacific Grove, 30 October – 2 November 1995, pp. 650-654

[58] BORSARI, G. K.: 'Mitigating effects on STAP processing caused by an inclined array', *IEEE RADARCON'98*, 11-14 May 1998, Dallas, TX, pp. 135-140

[59] BOS, van den, A.: 'Complex gradient and Hessian', *IEE Proc. on Vision, Image and Signal Processing*, Vol. 141, No. 6, December 1994, pp. 380-382

[60] BRANDWOOD, D. H.: 'A complex gradient operator and its application in adaptive array theory', *IEE Proc.*, Vol 130, Pts. F and H, No. 1, February 1983, pp. 11-14

[61] BRAUNREITER, D. C., SCHMITT, H. A., and CHEN, H.-W.: 'Theoretical and experimental results on applications of wavelet transform to RF STAP and real time optical compensation', *Proc. NATO-IRIS*, Vol. 43, No. 2, July 1999, pp. 89-97

[62] BRENNAN, L. E., and REED, I. S.: 'Theory of adaptive radar', *IEEE Trans. AES*, Vol. 9, No 2, March 1973, pp. 237-252

[63] BRENNAN, L. E., MALLETT, J. D., and REED, I. S.: 'Adaptive arrays in airborne MTI', *IEEE Trans. on Antennas and Propagation*, Vol. AP-24, No. 5, 1976, pp. 607-615

[64] BRENNAN, L. E., and MALLET, J. D.: 'Efficient simulation of external noise incident on arrays', *IEEE Trans. on Antennas and Propagation*, Vol. AP-24, September 1976, pp. 740-741

[65] BRENNAN, L. E., and STAUDAHER, F. M.: 'Subclutter visibility demonstration', *Technical Report RL-TR-92-21, Adaptive Sensors Incorporated*, March 1992

[66] BRENNAN, L. E., STAUDAHER, F., and PIWINSKY, D. J.: 'Comparison of space-time adaptive processing approaches using experimental airborne radar data', *IEEE National Radar Conference*, April 1993, Boston, MA, pp. 176-185

[67] BRENNER, A., and ENDER, J.G.H.: 'First experimental results achieved with the new very wideband SAR system PAMIR ', *EUSAR 2002*, 4-6 June 2002, Cologne, Germany

[68] BRENNER, A., and ENDER, J.G.H.: 'Airborne SAR Imaging with Subdecimeter Resolution ', *Proc. EUSAR 2004*, 25-27 May 2004, Ulm, pp. 267-270

[69] BROWN, R. D., and WICKS, M. C.: 'A space-time adaptive processing approach for improved performance and affordability', *IEEE National Radar Conference*, Ann Arbor, MI, 13-16 May 1996, pp. 321-326

[70] BROWN, R., and LINDERMAN, R.: 'Algorithm development for an airborne real-time STAP demonstration', *Proc. IEEE National Radar Conference*, 1997, Syracuse, NY, pp. 331-336

[71] BROWN, R. D., LITTLE, M. O., SCHNEIBLE, R. A., and WICKS, M. C.: 'Application of space time adaptive processing (STAP) in airborne bistatic scenarios', *RADAR'99*, 17-21 May 1999, Brest, France

[72] BROWN, R. D., SCHNEIBLE, R. A., WICKS, M. C., WANG, H., and ZHANG, Y.: 'STAP for clutter suppression with sum and difference beams', *IEEE Transactions AES*, Vol. 36, No. 2, April 2000, pp. 634-646

[73] BRYN, F.: 'Optimum signal processing of three-dimensional arrays operating on Gaussian signal and noise', *JASA*, Vol 34, No. 3, 1962, pp. 289-297

[74] BUCKLEY, K. M.: 'Spatial/Spectral filtering with linearly constrained minimum variance beamformers', *IEEE Trans. ASSP*, Vol. 35, No. 3, March 1987, pp. 249-266

[75] BÜHRING, W., and KLEMM, R.: 'Ein adaptives Filter zur Unterdrückung von Radarstörungen mit unbekanntem Spektrum' (An adaptive filter for suppression of clutter with unknown spectrum), *FREQUENZ*, Vol. 30, No. 9, September 1976 (in German), pp. 238-243

[76] BÜHRING, W.: 'Adaptive orthogonal projection for rapidly converging interference suppression', *Electronics Letters*, Vol. 14, 1978, pp. 515-516

[77] BÜHRING, W.: 'Improving convergence of adaptive antenna beamforming by orthogonal pre-processing', *Proc. EUSIPCO-80*, 16-19 September, 1980, Lausanne, Switzerland, pp. 185-186

[78] BÜRGER, W., BICKERT, B., and KLEMM, R.: 'Nulling properties of the mean square space-time adaptive FIR filter', *Pro. EUSAR 2004*, 25-27 May 2004, Ulm, pp. 537-540

[79] BURG, J. P.: 'A new analysis technique for time series data', *Proc. NATO Advanced Study Institute*, Enschede, 1968

[80] BURG, J. P.: 'Maximum entropy spectral analysis', *Proc. NATO Advanced Study Institute*, Enschede, 1968

[81] CALDWELL, J.T., and HALE, T.: 'Space-time adaptive processing for forward looking arrays', *IEEE 2004 Radar Conference*, 26-29 April 2004, Philadelphia, PA

[82] CALVARY, Ph., and JANER, D.: 'Spatio-temporal coding for radar array processing', *IEEE ICASSP*, 12-15 May 1998, Seattle, WA, pp. 2509-2512

[83] CAPON, J.: 'High resolution frequency wavenumber spectrum analysis, *Proc. IEEE*, Vol. 57, No. 8, 1969, pp. 1408-1418

[84] CAPON, J., GREENFIELD, R. J., and KOLKER, R. J.: 'Multidimensional maximum-likelihood processing of a large aperture seismic array', *Proc. IEEE*, Vol. 55, No. 5, 1967, pp. 192-211

[85] CAPRARO, C.T., WEINER, D.D., and WICKS, M.C.: 'Improved STAP performance using knowledge-aided secondary data selection', *IEEE 2004 Radar Conference*, 26-29 April 2004, Philadelphia, PA

[86] CARLO, J.T., SARKAR, T.K., and WICKS, M.C.: 'A least squares multiple constraint direct data domain approach for STAP', *2003 IEEE Radar Conference*, May 5-8, 2003, Huntsville, AL, pp. 431-438

[87] CARLSON, B. D.: 'Covariance matrix estimation errors and diagonal loading in adaptive arrays', *IEEE Trans. AES*, Vol. 24, No. 4, July 1988, pp. 397-401

[88] CERUTTI-MAORI, D.: 'Performance analysis of multistatic configurations for spaceborne SAR/MTI based on the auxiliary beam approach', *Proc. EUSAR 2004*, , 25-27 May 2004, Ulm, pp. 631-634

[89] CERUTTI-MAORI, D.: 'Optimum STAP in K-Space for wideband SAR/MTI-radar', *Frequenz*, Vol. 56, 2002, pp. 239-243

[90] CHEN, W.-S., and REED, I .: 'A new CFAR detection test for radar', *Digital Signal Processing*, Vol. 1, 1991, pp. 198-214

[91] CHEN, P., and BEARD, J. K.: 'Bistatic GMTI experiment for airborne platforms', *IEEE RADAR2000*, 8-12 May 2000, Alexandria, VA, pp. 42

[92] CHOUDHARY, A., LIAO, W.-K., WEINER, D., VARSHNEY, P., LINDERMAN, R., LINDERMAN, M., and BROWN, R.: 'Design, implementation and evaluation of parallel pipelined STAP on parallel computers', *IEEE Transactions AES*, Vol. 36, No. 2, April 2000, pp. 528-548

[93] CLAY, C. S.: 'Effect of slightly irregular boundary on the coherence of waveguide propagation', *JASA*, Vol. 36, pp. 833-837

[94] COE, D. J., and WHITE, R. G.: 'Experimental moving target detection results from a three-beam airborne SAR', *AEU*, Vol. 50, No. 2, March 1996

[95] COE, D. J., and WHITE, R. G.: 'Experimental moving target detection results from a three-beam airborne SAR', *Proc. EUSAR'96*, 26-28 March 1996, Königswinter, Germany, pp. 419-422 (VDE Publishers)

[96] COMPTON, R. T. jr.: 'The bandwidth performance of a two-element adaptive array with tapped delay-line processing', *IEEE Transactions on Antenna and Propagation*, Vol. AP-36, No. 1, January 1988, pp. 5-14

[97] COMPTON, R. T. jr.: 'The relationship between tapped delay-line and FFT Processing in Adaptive Arrays', *IEEE Transactions on Antenna and Propagation*, Vol. AP-36, No. 1, January 1988, pp. 15-26

[98] CONTE, E., DE MAIO, A., and RICCI, G.: 'Space-time adaptive radar detection of distributed targets', *IEEE RADAR2000*, 8-12 May 2000, Alexandria, VA, pp. 614-618

[99] COOPER, R. M., and DICKMAN, S. N.: 'A comparison of low PRF STAP architectures', *Proc. of the Conf. on Adaptive Antennas*, 7-8 November 1994, Melville, NY 11747, pp. 143-148

[100] COOPER, R. M.: 'Space time adaptive processing for a carrier based airborne early warning radar system', *Proc. IEEE National Radar Conference*, 1997, Syracuse, NY, pp. 66-71

[101] CORBELL, P.M., and HALE, T.B.: '3-dimensional STAP performance analysis using the cross-spectral matrixSteering vector mismatch: analysis and reduction ', *IEEE 2004 Radar Conference*, 26-29 April 2004, Philadelphia, PA

[102] COUTTS, S. D.: 'Passive localization of moving emitters using out-plane-multipath', *IEEE Transactions AES*, Vol. 36, No. 2, April 2000, pp. 584-595

[103] COVAULT, C.: 'Joint-Stars patrols Bosnia', *Aviation Week & Space Technology*, February 1996, pp. 44-49

[104] COVAULT, C.: 'Space-based radars drive advanced sensor technologies', *Aviation Week & Space Technology*, 5 April 1999, pp. 49-50

[105] COX, H.: 'Optimum arrays and the Schwartz inequality', *JASA*, Vol. 45, No. 1, 1969, pp. 228-232

[106] COX, H. 'Sensitivity considerations in adaptive beamforming', *NATO Advanced Study Institute on Signal Processing*, Loughborough, UK, 1973, pp. 17-32

[107] COX, H.: 'Resolving power and sensitivity to mismatch of optimum array processors', *JASA*, Vol. 54, No. 3, 1973, pp. 771-785

[108] CRON, B. F., and SHERMAN, C. H.: 'Spatial correlation function for various noise models', *JASA*, Vol. 34, No. 11, November 1962, pp. 1732-1736

[109] CURLANDER, J. C., and MCDONOUGH, R. N., *Synthetic Aperture Radar*, John Wiley & Sons, Inc., 1991

[110] DAMINI, A., and BALAJI, B.: 'Real-time STAP', *EUSAR 2002*, 4-6 June 2002, Cologne, pp. 641-645

[111] DAVIS, M. E.: 'Space based moving target detection challenges', *IEE radar 2002*, 15-17 October 2002, Edinburgh, Scotland, pp. 143-147

[112] DAY, J. K.: 'Space-time adaptive processing from an airborne early warning radar perspective', *IEEE Proc. 29th ASILOMAR Conference on Signals, Systems and Computers*, Pacific Grove, CA, 30 Oct.–2 Nov. 1995, pp. 1187-1192

[113] De GREVE, S., LAPIERRE, F. D., and VERLY, J. G.: 'Canonical framework for describing suboptimum radar space-time adaptive processing (STAP) Techniques.', *IEEE 2004 Radar Conference*, 26-29 April 2004, Philadelphia, PA

[114] DESMÉZIÈRES, A., BERTAUX, N., and LARZABAL, P.: 'Space-time polarization processing for propagation channel identification', *SEE PSIP'99*, Paris, France, 18-19 January 1999, pp. 101-106

[115] DILLARD, G. M.: 'Recursive computation of the discrete Fourier transform, with applications to a pulse-Doppler radar system', *Comput. & Elect. Engng.*, Vol 1, 1973, pp. 143-152

[116] DILLARD, G. M.: 'Signal-to-noise ratio loss in an MTI cascaded with coherent integration filters', *IEEE Int. Radar Conference*, Arlington, VA, 1975, pp. 117-122

[117] DOGANDZIC, A., and NEHORAI, A.: 'Cramer-Rao bounds for estimating range, velocity and direction with a sensor array', *IEEE SAM2000 Workshop*, 16-17 March 2000, Cambridge, MA

[118] DOHERTY, J. F.: ' Suppression of terrain scattered jamming in pulse compression radar', *IEEE Transactions on Signal Processing*, Vol. 2, No. 1, January 1995, pp. 4-6

[119] DONG, R., and BAO, Z.: 'Direct data domain STAP algorithm for airborne radar applications ', *CIE International Conference on Radar*, 15-18 October 2001, pp. 770-772

[120] DONG, Y., TAO, R., ZHOU, S., and WANG, Y.: 'Joint STAP-WVD based SAR slowly moving target detection and imaging', *IEEE RADAR2000*, 8-12 May 2000, Alexandria, VA, pp.499-503

[121] EATON, C. J., and COE, D. J.: 'Motion of ground moving targets and implications for aided recognition', *IEE radar 2002*, 15-17 October 2002, Edinburgh, Scotland, pp. 370-374

[122] ENDER, J.: 'Detectability of slowly moving targets using a multi-channel SAR with an along-track antenna array', *SEE/IEE SAR Conference*, Paris, May 1993

[123] ENDER, J.: 'AER – ein experimentelles Mehrkanal-SAR', *DGON Radar Symposium 1993*, Neubiberg, Germany, pp. 120-125 (in German)

[124] ENDER, J.: 'Signal Processing for multichannel SAR applied to the experimental SAR system AER', *Intern. Conference on Radar*, Paris, May 1994, pp. 220-225

[125] ENDER, J.: 'Azimutpositionierung bewegter Ziele mit Mehrkanal-SAR (Azimuth positioning of moving targets with multi-channel SAR)', *URSI-Conference*, Kleinheubach, Germany, 1994, pp. 341-358 (in German)

[126] ENDER, J.: 'Detection and estimation of moving target signals by multi-channel SAR', *Proc. EUSAR'96*, 26-28 March 1996, Königswinter, Germany, pp. 411-417. Also: *AEU*, Vol. 50, March 1996, pp. 150-156

[127] ENDER, J.: 'The airborne experimental multi-channel SAR system ΛER-II', *Proc. EUSAR'96*, 26-28 March 1996, Königswinter, Germany, pp. 49-52

[128] ENDER, J.: 'Mehrkanalverfahren: Mit dem flugzeuggetragenen System 'AER-II' erzielte experimentelle Ergebnisse' (Multichannel SAR: results obtained with the airborne AER-II system), *9th DGON Radarsymposium*, 8-10 April 1997, Stuttgart, Germany, pp. 309-320 (in German)

[129] ENDER, J.: 'Experimental results achieved with the airborne multi-channel SAR system AER-II', *Proc. EUSAR'98*, 25-27 May 1998, Friedrichshafen, Germany, pp. 315-318

[130] ENDER, J.: 'Anti-jamming adaptive filtering for SAR imaging', *DGON IRS'98*, 15-17 September 1998, Munich, Germany

[131] ENDER, J.: 'Space-time processing for multichannel synthetic aperture radar', *IEE ECEJ*, special issue on STAP, Vol. 11, No. 1, February 1999, pp. 29-38

[132] ENDER, J., and BRENNER, A.: 'PAMIR - a wideband phased array SAR/MTI system', *EUSAR 2002*, 4-6 June 2002, Cologne, Germany

[133] ENDER, J., and KLEMM, R.: 'Airborne MTI via digital filtering', *Proc. IEE*, Pt. F, Vol. 136, No. 1, 1989, pp. 22-28

[134] ENDER, J., and KLEMM, R.: 'Festzielunterdrückungsfilter für bewegte Sensorgruppen – Auswirkung suboptimaler Abtastung '(Clutter suppression filter for moving sensor arrays – effect of suboptimum sampling), *Proc. AAST* (Aachener Symposium für Signaltheorie), September 1990, pp. 70-76 (in German)

[135] ENDER, J., and SAALMANN, O.: 'Eine aktive phasengesteuerte Gruppenantenne für Mehrkanal-SAR' (An active phased array antenna for multichannel SAR), *ITG Fachtagung Antennen*, Dresden, Germany, April 1994, pp. 109-114 (in German)

[136] ENDER, J. H. G, and Brenner, A. R.: 'PAMIR - a wideband phased array SAR/MTI system ', *IEE Proc. RSN*, June 2003

[137] ENTZMINGER, J. N., FOWLER, C. A., and KENNEALLY, W. J.: 'JointSTARS and GMTI: past, presence and future', *IEEE Transactions AES*, Vol. 35, No. 2, April 1999, pp. 748-761

[138] EVANS N., and LEE P.: 'The RADARSAT-2&3 topographic mission ', *Proc. EUSAR 2002*, Cologne, 4-6 June, 2002, pp. 37-39

[139] FABRIZIO, G.A., TURLEY, M.D.: 'An advanced implementation for surveillance radar systems ', *Workshop on Statistical Signal Signal Processing*, 6-8 August 2001, Singapore, pp. 134-137

[140] FANTE, R. L.: 'Cancellation of specular and diffuse jammer multipath using a hybrid adaptive array', *IEEE Trans. AES*, Vol. 27, No. 5, September 1991, pp. 823-837

[141] FANTE, R. L., BARILE, R. C., GUELLA, T. P., and TORRES, J. A.: 'Performance of space-time adaptive airborne radar', *IEEE National Radar Conference*, April 1993, Boston, MA, pp. 173-175

[142] FANTE, R. L., and TORRES, J. A.: 'Cancellation of diffuse jammer multipath by an airborne adaptive radar', *IEEE Trans. AES*, Vol. 31, No. 2, April 1995, pp. 805-820

[143] FANTE, R. L., DAVIS, R. M., and GUELLA, T. P.: 'Wideband cancellation of multiple mainbeam jammers', *IEEE Transactions on Antennas and Propagation*, Vol. 44, No. 10, October 1996, pp. 1402-1413

[144] FANTE, R. L.: 'Adaptive space-time radar', *Journal of the Franklin Institute*, Vol. 335B, No. 1, 1998, pp. 1-11

[145] FANTE, R. L.: 'Ground and airborne target detection with bistatic adaptive space-based radar', *IEEE National Radar Conference*, 20-22 April 1999, Boston, MA, pp. 7-11

[146] FANTE, R. L .: 'Ground and airborne target detection with bistatic adaptive space-based radar', *IEEE AES Magazine*, October 1999, pp. 39-44

[147] FANTE, R. L.: 'Synthesis of adaptive monopulse patterns', *IEEE Trans. AP*, Vol. 47, No. 5, May 1999, pp. 773-774

[148] FANTE, R. L., and VACCARO, J. J.: 'Wideband cancellation of interference in a GPS receive array', *IEEE Transactions AES*, Vol. 36, No. 2, April 2000, pp. 549-564

[149] FARINA, A., *Antenna based signal processing techniques for Radar Systems*, Artech-House, Norwood, MA, 1992

[150] FARINA, A., and TIMMONERI, L.: 'Space-time processing for AEW radar', *Proc. RADAR 92*, Brighton, UK, 1992, pp. 312-315

[151] FARINA, A., GRAZIANO, R., and TIMMONERI, L.: 'Adaptive space-time processing with systolic algorithm: experimental results using recorded live data', *Proc. IEEE International Radar Conference*, Alexandria, VA, 1995, pp. 595-602

[152] FARINA, A., SAVERIONE, A., and TIMMONERI, L.: 'MVDR vectorial lattice applied to space-time processing for AEW radar with large instantaneous bandwidth', *IEE Proc. Radar, Sonar Naviagtion*, Vol. 143, No. 1, February 1996, pp. 1-6

[153] FARINA, A.: 'Linear and non-linear filters for clutter cancellation in radar systems', *Signal Processing*, Vol. 59, 1997, pp. 101-112

[154] FARINA, A., LOMBARDO, P., and CARAMANICA, F.: 'Non-linear non-adaptive clutter cancellation for airborne early warning radar', *Proc. IEE RADAR'97*, 14-16 Oct. 1997, Publication No. 449, pp. 420-424

[155] FARINA, A., LOMBARDO, P., and PIRRI, M.: 'Nonlinear nonadaptive space-time processing for airborne early warning radar', *IEE Proc. Radar, Sonar and Navigation*, Vol. 145, No. 1, February 1998, pp. 9-18

[156] FARINA, A., and TIMMONERI, L.: 'Real-time STAP techniques', *IEE ECEJ*, special issue on STAP, Vol. 11, No. 1, February 1999, pp. 13-22

[157] FARINA, A., LOMBARDO, P., and PIRRI, M.: 'Nonlinear STAP processing', *IEE ECEJ*, special issue on STAP, Vol. 11, No. 1, February 1999, pp. 41-48

[158] FARINA, A., and LOMBARDO, P.: 'Space-time techniques for SAR', in: KLEMM, R. ed., *Applications of space-time adaptive processing*, IEE publishers, London, UK, 2004

[159] FARINA, A., and TIMMONERI, L.: 'Cancellation of clutter and e.m. interference with STAP algorithms. Application to live data aquired with a ground based phased array radar ', *IEEE 2004 Radar Conference*, 26-29 April 2004, Philadelphia

[160] FARINA, A., and LOMBARDO, P. .: 'Adaptive array of antennas and STAP', *IEEE International Symposium on phased array systems and technology*, 14-17 October 2003, Boston, MA

[161] FAUBERT, D., VINEBERG, K. A., and LIGHTSTONE, L.: 'Clutter cancellation with a two-phase center airborne radar', *RADARCON 90*, Adelaide, Australia, April 1990, pp. 45-50

[162] FENNER, D. K., and HOOVER jr, W. F.: 'Test results of a space-time adaptive processing system for airborne early warning radar', *IEEE National Radar Conference*, Ann Arbor, MI, 13-16 May 1996, pp. 88-91

[163] FERRIER, J. M., CARRARA, B., and GRANGER, P.: 'Antenna subarray architectures and antijamming constraints', *International Conference on Radar*, Paris, France, 3-6 May 1994, pp. 466-469

[164] FRANCOS, J. M., FU, W., and NEHORAI, A.: 'Interference estimation and mitigation for STAP using the two-dimensional Wold decomposition parametric model', *ASAP 2001*, MIT Lincoln Laboratory, 13-14 March 2001

[165] FRENCH, M.C., SUH, J.,DAMOULAKIS, J., and CRAGO, S.P.: 'Novel signal processing architectures for knowledge-based STAP algorithms ', *IEEE 2004 Radar Conference*, 26-29 April 2004, Philadelphia, PA

[166] FRIEDLANDER, B.: 'A subspace method for space time adaptive processing', *IEEE Trans. SP*, Vol 53, No. 1, January 2005, pp. 74-82

[167] FRIEL, E. M., and PASALA, K. M.: 'Effects of mutual coupling on the performance of STAP antenna arrays', *IEEE Transactions AES*, Vol. 36, No. 2, April 2000, pp. 518-527

[168] FROST, O. L.: 'An algorithm for linearly constrained adaptive array processing', *Proc. IEEE*, Vol. 60, No. 8, August 1972, pp. 926-935

[169] FUHRMANN, D. R., and RIEKEN, D. W.: 'Array calibration for circular-array STAP using clutter scattering and projection matrix fitting', *ASAP 2000 Workshop*, 13-14 March 2000, MIT Lincoln Laboratory, pp. 79-84

[170] FUTERNIK, A., and HAIMOVICH, A. M.: 'Performance of adaptive radar in range-heterogeneous clutter', *Journal of the Franklin Institute*, Vol. 335B, No. 1, 1998, pp. 71-89

[171] GABEL, R. A., KOGON, S. M., and RABIDEAU, D. J.: 'Algorithms for mitigating terrain-scattered interference', *IEE ECEJ*, special issue on STAP, Vol. 11, No. 1, February 1999, pp. 49-56

[172] GABRIEL, W. F.: 'Spectral analysis and adaptive array superresolution techniques', *Proc. IEEE*, Vol. 68, No. 6, June 1980, pp. 654-666

[173] GABRIEL, W. F.: 'Using Spectral Estimation Techniques in Adaptive Processing Antenna Systems', *IEEE Trans. AP*, Vol AP-34, No. 3, March 1986, pp. 291-300

[174] GAFFNEY, J.B., GUTTRICH, G., SURESH BABU, B.N., and TORRES, J.A.: 'Performance comparison of fast-scan GMTI STAP architectures ', *IEEE Radar Conference*, 2001, pp. 252-257

[175] GALATI, G. (Editor), *Advanced radar techniques and systems*, IEE, London, 1993

[176] GANZ, M. W., MOSES, R. L., and WILSON, S. L.: 'Convergence of the SMI and the diagonally loaded SMI algorithms with weak interference', *IEEE Transactions on Antennas and Propagation*, Vol. 38, No. 3, March 1990, pp. 394-399

[177] GARNHAM, J. W.: 'Application of digital beamforming for look-down, clutter limited, MTI radar – with specific application to space based radar', *IEEE Aerospace Conference*, 1-8 February 1998, Snowmass at Aspen, USA (Piscataway, NJ), pp. 349-357

[178] GAU, Y. L., and REED, I. S.: 'An improved reduced-rank CFAR space-time adaptive radar detection algorithm', *IEEE Trans. SP*, Vol. 46, No. 8, August 1998, pp. 2139-2146

[179] GENELLO, G. J. WICKS, M., and SOUMEKH, M.: 'Some interesting aspects of adaptive airborne phased array radar: Achieving MTD using SAR', *RADAR'99*, 17-21 May 1999, Brest, France

[180] GERLACH, K., and KRETSCHMER, F. E. jr.: 'Convergence properties of Gram-SCHMIDT and SMI adaptive algorithms: Part II', *IEEE Trans. AES-27*, No. 1, January 1990, pp. 83-91.

[181] GERLACH, K., and STEINER, M. J.: 'An adaptive matched filter that compensates for I, Q mismatch errors', *IEEE Trans. SP*, Vol. 45, No. 12, December 1997, pp. 3104-3107

[182] GERLACH, K., and PICCIOLO, M. L.: 'Airborne/spacebased radar STAP using a structured covariance matrix ', *IEEE Trans. AES*, Vol. 39, No. 1, January 2003, pp. 269-281

[183] GERLACH, K., and BLUNT, S.: 'Efficient reiteration censoring of robust STAP using the FRACTA algorithm ', *Intern. Conf. on Radar*, 3-5 September 2003, Adelaide, Australia, pp. 57-61

[184] GERLACH, K., and PICCIOLO, M.L.: 'Robust STAP using reiterative censoring', *2003 IEEE Radar Conference*, May 5-8, 2003, Huntsville, AL, pp. 244-251

[185] GERSHMAN, A. B., SEREBRYAKV, G. V., and BOEHME, J. F.: 'Constrained Hung-Turner adaptive beam-forming algorithm with additional robustness to wideband and moving jammers', *IEEE Transactions on Antennas and Propagation*, Vol. 44, No. 3, March 1996, pp. 361-367

[186] GHOUZ, H. H. M., ELGHANI, F. I. A., and QUTB, M. M.: 'Adaptive space-time processing for interference suppression in phased array radar systems (Part-1: Search radar)', *URSI 17th National Radio Science Conference*, 22-24 Feb. 2000, Minufiya University, Egypt, pp. B8.1-B8.8

[187] GHOUZ, H. H. M., ELGHANI, F. I. A., and QUTB, M. M.: 'Adaptive space-time processing for interference suppression in phased array radar systems (Part-2: Tracking radar)', *URSI 17th National Radio Science Conference*, 22-24 Feb. 2000, Minufiya University, Egypt, pp. B9.1-B9.7

[188] GIERULL, C. H.: 'Schnelle Signalraumschätzung in Radaranwendungen' ('Fast signal space estimation in radar applications'), Dr.-Ing. Thesis, Shaker Verlag, Aachen, Germany, 1995 (in German)

[189] GIERULL, C. H.: 'Performance analysis of fast projections of the Hung-Turner type for adaptive beamforming', *Signal Processing*, 50, 1996, pp. 17-28

[190] GIERULL, C. H.: 'A fast subspace estimation method for adaptive beamforming', *AEU*, Vol. 51, No. 4, 1997, pp. 196-205

[191] GIERULL, C. H.: 'Angle estimation for small sample size with fast eigenvector-free subspace method', *IEE Proc. Radar, Sonar and Navigation*, Vol. 146, No. 3, June 1999, pp. 126-132

[192] GIERULL, C., and LIVINGSTONE, C.: 'SAR-GMTI Concept for RADARSAT-2', in: KLEMM, R. ed., *Applications of space-time adaptive processing*, IEE publishers, London, UK, 2004

[193] GIERULL, C.H., and BALAJI, B.: 'Application of fast projection techniques without eigenanalysis to STAP ', *IEEE RADARCON 2002*, 22-25 April 2002, Long Beach, CA

[194] GODARA, L. C.: 'Application of the Fourier transform to broadband beamforming', *JASA*, Vol. 98, No. 1, July 1995, pp. 230-240

[195] GOGGINS, W. B., and SLETTEN, C. J.: 'New concepts in AMTI radar', *Microwave Journal*, January 1974, pp. 29-35

[196] GOJ, W. W.: 'Synthetic aperture radar and electronic warfare', ARTECH-HOUSE, Boston and London, 1993

[197] GOLDSTEIN, J. S., ZULCH, P. A., and REED, I. S.: 'Reduced rank space-time adaptive radar processing', *ICASSP'96*, Atlanta, GA, 1996, pp. 1173-1176

[198] GOLDSTEIN, J. S., and REED, I. S.: 'Subspace selection for partially adaptive sensor array processing', *IEEE Trans. AES*, Vol. 33, No. 2, April 1997, pp. 539-543

[199] GOLDSTEIN, J. S., and REED, I. S.: 'Theory of partially adaptive radar', *IEEE Trans. AES*, Vol. 33, No. 4, October 1997, pp. 1309-1325

[200] GOLDSTEIN, J. S., KOGON, S. M., REED, I. S., WILLIAMS, D. B., and HOLDER, E. J. E.: 'Partially adaptive radar signal processing: the cross-spectral approach', *IEEE ASILOMAR-29*, Pacific Grove, CA, 30 Oct.-2 Nov. 1996, pp. 1383-1387

[201] GOLDSTEIN, J.S., REED, I.S., ZULCH, P.A., and MELVIN, W.L.: 'A multistage STAP CFAR detection technique', *IEEE RADARCON'98*, 11-14 May 1998, Dallas, TX, pp. 111-116

[202] GOLDSTEIN, J.S., REED, I.S., and SCHARF, L.: 'A multistage representation of the Wiener filter based on orthogonal projections', *IEEE Trans. IT*, Vol. 44, No. 7, November 1998, pp. 2943-2959

[203] GOLDSTEIN, J.S., and REED, I. S.: 'Multistage partially adaptive STAP CFAR detection algorithm', *IEEE Trans. AES*, Vol. 35, No. 2, April 1999, pp. 645-661

[204] GOLDSTEIN, J.S., GUERCI, J.R., and REED, I. S.: 'Advanced concepts in STAP', *IEEE RADAR2000*, 8-12 May 2000, Alexandria, VA, pp. 699-704

[205] GOLUB, G.H., and VAN LOAN, C.F., *Matrix computations*, Johns Hopkins University Press, Baltimore and London, 2nd edition, 1989

[206] Goodman, N.A., Stiles, J.M.: 'A general signal processing algorithm for MTI with multiple receive apertures ', *IEEE Radar Conference*, 2001, pp. 315-320

[207] GOODMAN, N.A., and GURRAM, P.R.: 'STAP training through knowledge aided predictive modeling ', *IEEE 2004 Radar Conference*, 26-29 April 2004, Philadelphia, PA

[208] GOVER, M.J., and BARNETT, S.: 'Inversion of certain extensions of Toeplitz matrices', *Journal of Mathematical Analysis and Applications*, Vol. 100, 1984, pp. 339-353

[209] GOVER, M.C.J., and BARNETT, S.: 'Inversion of Toeplitz matrices which are not strongly non-singular', *IMA Journal of Numerical Analysis*, No. 5, 1985, pp. 101-110

[210] GRIFFITHS, L. J.: 'A comparison of multidimensional Wiener and maximum likelihood filters for antenna arrays', *Proc. IEEE*, November 1967, pp. 2045-2047

[211] GRIFFITHS, L. J.: 'A Simple adaptive algorithm for real-time processing in antenna arrays', *Proc. IEEE*, Vol. 57, No. 10, Oct. 1969, pp. 1696-1704

[212] GRIFFITHS, L. J.: 'Linear constraints in hot clutter cancellation', *ICASSP'96*, Atlanta, GA, 1996, pp. 1181-1184

[213] GRIFFITHS, L. J.: 'Multiple-pulse STAP adaptation prior to radar Doppler processing', *30th Asilomar Conference*, 3-6 Nov. 1996, Pacific Grove, CA, pp. 394-398

[214] GRIFFITHS, L. J., TECHAU, P. M., BERGIN, J. S., and BELL, K. L.: 'Space-time adaptive processing in airborne radar systems', *IEEE RADAR2000*, 8-12 May 2000, Alexandria, VA, pp. 711-716

[215] GRÖGER, I.: 'Antenna pattern shaping', *International Conference on Radar*, Paris, 1989, pp. 283-287

[216] GRÖGER, I., SANDER, W., and WIRTH, W. D.: 'Experimental phased array radar ELRA with extended flexibility', *Radar 90*, Arlington, VA, 1990, pp. 286-290

[217] GROSS, L.A., and HOLT, H.D.: 'AN/APY-6 realtime surveillance and targeting radar development', *Proc. NATO/IRIS Conference*, 19-23 October 1998, paper G-3

[218] GRUENER, W., TOERNIG, J. P., and FIELDING, P. J.: 'Active-electronically-scanned-array based radar system features', *Radar'97*, 14-16 October 1997, Edinburgh, Scotland, pp. 339-343

[219] GU, Z., BLUM, R. S., MELVIN, W. L., and WICKS, M. C.: 'Comparison of STAP algorithms for airborne radar', *Proc. IEEE National Radar Conference*, 1997, Syracuse, NY, pp. 60-65

[220] GUERCI, J. R., and FERIA, E. H.: 'Least squares predictive transform space-time array processing for adaptive airborne MTI radar', *Proc. IEEE 6th Workshop on Statistical Signal and Array Processing*, University of Victoria, British Columbia, October 1992, pp. 65-69

[221] GUERCI, J. R., and FERIA, E. H.: 'Optimal hybrid space-time processing', *Proc. of the Conf. on Adaptive Antennas*, 7-8 November 1994, Melville, NY 11747, pp. 107-111

[222] GUERCI, J. R., and FERIA, E. H.: 'Predictive-transform space-time processing for airborne MTI radar', *Proc. IEEE NAECON'1994*, Dayton, OH, 23-27 May 1994, pp. 47-56

[223] GUERCI, J. R., and FERIA, E. H.: 'Application of a least squares predictive-transform modeling methodology to space-time adaptive array processing', *IEEE Transactions on Signal Processing*, Vol. 44, No. 7, July 1996, pp. 1825-1833

[224] GUERCI, J. R., PILLAI, S. U., and KIM, Y. L.: 'Efficient space-time adaptive processing for airborne MTI-mode radar', *ICASSP'96*, Atlanta, GA, 1996

[225] GUERCI, J. R., GOLDSTEIN, J. S., ZULCH, P. A., and REED, I. S.: 'Optimal reduced-rank 3D-STAP for joint hot and cold clutter mitigation', *IEEE National Radar Conference*, 20-22 April 1999, Boston, MA, pp. 119-124

[226] GUERCI, J. R.: 'Theory and application of covariance matrix tapers for robust adaptive beamforming', *IEEE Trans. SP*, Vol. 47, No. 4, April 1999, pp. 977-985

[227] GUERCI, J. R., GOLDSTEIN, J. S., and REED, I. S.: 'Optimal and adaptive reduced-rank STAP', *IEEE Transactions AES*, Vol. 36, No. 2, April 2000, pp. 647-663

[228] GUERCI, J.R.: *Space-time adaptive processing for radar*, Artech House, Boston, 2003

[229] GUERCI, J.R., and BARANOSKI, E.J.: 'An Overview of Knowledge-Aided Adaptive Radar at DARPA', *IEEE Signal Processing Magazine*, Special issue on 'Knowledge Based Systems for Adaptive Radar Detection, Tracking and Classification', January 2006

[230] GUERCI, J.: 'Knowledge-aided sensor signal processing and expert reasoning (KASSPER)', *KASSPER 02 Workshop Proceedings* KASSPER 02 Workshop, April 3, 2002, Washington DC 23-27

[231] GUERCI, J.: 'KASSPER: A Framework for Next Generation Adaptive Sensors', *KASSPER 03 Conference Proceedings* KASSPER 03 Workshop, April 3, 2002, Las Vegas, NV 23-27

[232] GUERCI, J.: 'Knowledge-aided sensor signal processing and expert reasoning', *KASSPER 04 Workshop Proceedings* KASSPER 04 Workshop, 5-7 April , 2004, Clearwater, FL 23-27

[233] GUYVARCH, J. P.: 'Radar spatio-temporel à codes de phase optimisés', *RADAR'99*, 17-21 May 1999, Brest, France

[234] GUPTA, I. J., and KSIENSKI, A. A.: 'Effect of mutual coupling on the performance of adaptive arrays', *IEEE Trans. AP*, Vol. 31, No. 5, September 1983, pp. 785-791

[235] HÄNDEL, G. F.: 'On the History of MUSIC', *IEEE Signal Processing Magazine*, March 1999, p. 13

[236] HAIMOVICH, A. L., and BAR-NESS, Y.: 'An Eigenanalysis interference canceler', *IEEE Trans. Signal Processing*, Vol. 39, No. 1, January 1991, pp. 76-84

[237] HAIMOVICH, A.: 'The eigencanceller: adaptive radar by eigenanalysis methods', *IEEE Trans. AES*, Vol. 32, No. 2, April 1996, pp. 532-542

[238] HAIMOVICH, A. M., PUGH, M. L., and BERIN, M. O.: 'Training and signal cancellation in adaptive radar', *IEEE 1996 National Radar Conference*, Ann Arbor, 13-16 May 1996, pp. 124-129

[239] HAIMOVICH, A. M., PECKHAM, C., AYOUB, T., GOLDSTEIN, J. S., and REED, I. S., 'Performance analysis of reduced rank STAP', *Proc. IEEE National Radar Conference*, 1997, pp. 42-47

[240] HAIMOVICH, A. M., and BERIN, M.: 'Eigenanalysis-based space-time adaptive radar: performance analysis', *IEEE Trans. AES*, Vol. 33, No. 4, October 1997, pp. 1170-1179

[241] HALE, T., and WELSH, B.: 'Secondary data support in space-time adaptive processing', *IEEE RADARCON'98*, 11-14 May 1998, Dallas, TX, pp. 183-188

[242] HALE, T.: 'Operating the cross spectral metric algorithm with limited secondary data support', *RADAR'99*, 17-21 May 1999, Brest, France

[243] HALE, T.B., TEMPLE, M.A., WICKS, M.C., RAQUET, J.F., and OXLEY, M.E.: ' Performance charcterisation of hybrid STAP architecture incorporating elevation interferometry ', *IEE Proc. Radar, Sonar and Navigation*, Vol. 149, No. 2, April 2002, pp. 77-82

[244] HALE, T. B., TEMPLE, M. A., RAQUET, J. F.,OXLEY, M. E., and WICKS, M. C.: 'Localized three-dimensional adaptive spatial-temporal processing for airborne radar', *IEE radar 2002*, 15-17 October 2002, Edinburgh, Scotland, pp. 191-195

[245] HALE, T. B., TEMPLE, M. A., RAQUET, J. F.,OXLEY, M. E., and WICKS, M. C.: 'Elevation interferometric STAP using a thinned planar array ', , *IEEE 2002 Radar Conference*, Long Beach, CA, 22-25 April 2002, pp. 408-414

[246] HALE, T., TEMPLE, M., and WICKS, M.: 'Target detection in herogeneous airborne radar interference using 3D STAP', *2003 IEEE Radar Conference*, May 5-8, 2003, Huntsville, AL, pp. 252-257

[247] HANLE, E.: 'Pulse chasing with bistatic radar-combined space-time filtering', *Signal Processing II: Theories and Applications*, Elsevier Science Publishers B. V. (North Holland), 1983, pp. 665-668

[248] HANLE, E.: 'Survey of bistatic and multistatic radar', *IEE Proc.*, Vol. 133, Pt. F, No. 7, December 1986, pp. 587-595

[249] HANLE, E.: 'Polarimetric suppression of clutter at off-broadside directions', *International Conference on Radar*, Paris, 1994, pp. 586-591

[250] HANLE, E.: 'Adaptive chaff suppression by polarimetry with planar phased arrays at off-broadside', *IEEE Radar International Conference*, Alexandria, VA, 1995, pp. 108-112

[251] HANLE, E.: 'Polarimetric Antenna Influences with Synthetic Aperture Radar at Off-broadside', *EUSAR'96*, Königswinter, Germany, 1996, pp. 121-124 (VDE Publishers)

[252] HARTNETT, M.P., and DAVIS, M.E.: 'Multi-channel signal subspace processing methods for SAR-MTI', *2003 IEEE Radar Conference*, May 5-8, 2003, Huntsville, AL, pp. 133-138

[253] HAYKIN, S. (Editor), *Array Signal Processing*, Prentice-Hall Inc., Englewood Cliffs, NJ, 1985

[254] HAYKIN, S., *Adaptive Filter Theory*, Prentice-Hall, Englewood Cliffs, NJ, 1986

[255] HAYSTEAD, J.: 'JSTARS – real-time warning and control for surface warfare', *Defense Electronics*, July 1990, pp. 31-39

[256] HAYWARD, S.: 'Effects of motion on adaptive arrays', *IEE Proc. on Radar, Sonar and Navigation*, Vol. 144, No. 1, 1997, pp. 15-20

[257] HERBERT, G. M.: 'Space-time adaptive processing (STAP) for wideband airborne radar', *IEEE RADAR2000*, 8-12 May 2000, Alexandria, VA, pp. 620-625

[258] HERBERT, G. M., and RICHARDSON, P. G.: 'On the benefits of space-time adaptive processing (STAP) in bistatic airborne radar', *IEE radar 2002*, 15-17 October 2002, Edinburgh, Scotland, pp. 365-369

[259] HERSEY, F., MELVIN, W.L., McCLELLAN, J.H.: 'Registration-based range-dependent compensation in airborne bistatic radar STAP ', *EURASIP Signal Processing*, Vol 84 (9), September 2004, pp. 1481-1500

[260] HIEMSTRA, J.D.: 'Colored diagonal loading ', *IEEE RADARCON 2002*, 22-25 April 2002, Long Beach, CA

[261] HIEMSTRA, J.D., ZOLTOWSKI, M.D., and GOLDSTEIN, J.S.: 'Recursive and knowledge-aided implementations of the multi-stage Wiener filter', *2003 IEEE Radar Conference*, May 5-8, 2003, Huntsville, AL, pp. 46-50

[262] HIMED, B., and MELVIN, W. L.: 'Analysing space-time adaptive processors using measured data', *31st ASILOMAR Conference*, 2-5 Nov. 1997, Los Alamitos, CA, pp. 930-935

[263] HIMED, B., MICHELS, J. H., and ZHANG, Y.: 'STAP performance for bistatic phased array radar applications', *IEEE Southeastern Symposium on System Theory*, 5-7 March 2000, Tallahassee, FL

[264] HIMED, B., and MICHELS, J. H.: 'Performance analysis of the multi-stage Wiener filter', *IEEE RADAR2000*, 8-12 May 2000, Alexandria, VA, pp. 729-734

[265] HIMED, B., MICHELS, J. H., and ZHANG, Y.: 'Bistatic STAP performance analysis in radar applications ', *IEEE Radar Conference*, 2001, pp. 198-203

[266] HIMED, B.: 'Effects of bistatic clutter dispersion on STAP systems', *IEE radar 2002*, 15-17 October 2002, Edinburgh, Scotland, pp. 360-364

[267] HIMED, B., WICKS, M. C., and GENELLO, G. J.: 'Accounting for array effects in joint-domain localized STAP processing', *IEE radar 2002*, 15-17 October 2002, Edinburgh, Scotland, pp. 186-190

[268] HIMED, B., ZHANG, Y., and HAJJARI, A.: 'STAP with Angle-Doppler compensation for bistatic airborne radars', *IEEE 2002 Radar Conference*, Long Beach, CA, 22-25 April 2002, pp. 311-317

[269] HIMED, B., WICKS, M.C., and ZULCH, P.: 'A new constrained joint-domain localized approach for airborne radar', *IEEE RADARCON 2002*, 22-25 April 2002, Long Beach, CA

[270] HIPPLER, J., and FRITSCH, B.: 'Calibration of the Dornier SAR with trihedral corner reflectors', *Proc. EUSAR'96*, 26-28 March 1996, Königswinter, Germany, pp. 499-503 (VDE Publishers)

[271] HOCHWALD, B. M., MARZETTA, T. L., RICHARDSON, T. J., SWELDENS, W., and URBANKE, R.: 'Systematic design of unitary space-time constellation', *IEEE Trans. Information Theory*, Vol. 46, No. 6, September 2000, pp. 1962-1973

[272] HOFFMAN, A., and KOGON, S. M.: 'Subband STAP in wideband radar systems', *IEEE SAM Workshop*, 15-16 March 2000, Cambridge, MA, pp. 256-260

[273] HOOGEBOOM, P., VAN HALSEMA, D., HERPFER, E., MARTIN, F., FOURNET, P., PERTHUIS, D. CANAFOGLIA, G., and GIUNTI, M.: 'SOSTAR-X, a high performance radar demonstrator for airborne ground surveillance', *Proc. EUSAR2000*, 23-25 May 2000, Munich, Germany, pp. 825-827 (VDE Publishers)

[274] HOVANESSIAN, S. A., *Introduction to Synthetic Array and Imaging Radars*, Artech-House, 1980

[275] HUDSON, J. E., *Adaptive array principles*, Peter Peregrinus Ltd, Stevenage and New York, 1981

[276] HUNG, E. K., and TURNER, R. M.: 'A fast beamforming algorithm for large arrays', *IEEE Trans. AES*, Vol. 19, No. 4, July 1983, pp. 598-607

[277] JAFFER, A. G.: 'Constrained partially adaptive space-time processing for clutter suppression', *ASILOMAR'94*, 28th Asilomar Conference on Signal, Systems and Computers, 3 Oct.– 2 Nov. 1994, Pacific Grove, CA, pp. 671-676

[278] JANSSON, M., and STOICA, P.: 'Forward only and forward-backward sample covariances – a comparative study', *Signal Processing*, Vol. 77, 1999, pp. 235-245

[279] JENKINS, R. W., and MORELAND, K. W.: 'A comparison of the eigenvector weighting and Gram-SCHMIDT adaptive antenna techniques', *IEEE Trans. AES*, Vol. 29, No. 2, April 1993, pp. 568-575

[280] JENSEN, F., and FERLA, M.: 'SNAP: The Saclantcen normal mode acoustic propagation model', *SACLANTCEN Memorandum*, SM-121, 1979

[281] JIANG, N, Wu, R., LIU, G., and LI, J.: 'Clutter suppression and moving target parameter estimation for airborne phased array radar', *Electronics Letters*, 2 March 2000, Vol. 36, No. 5, pp. 456-457

[282] JOHANNISSON, B., and STEEN, L.: 'Partial digital beamforming using antenna subarray division', *International Conference on Radar*, Paris, France, 3-6 May 1994, pp. 63-67

[283] JOUNY, I. I., and CULPEPPER, E.: 'Modeling and mitigation of terrain scattered interference', *IEEE Antennas and Propagation Symposium*, 18-23 June 1995, Newport Beach, pp. 455-458

[284] KADAMBE, S., and OWECHKO, Y.: 'Computation reduction in space-time adaptive processing (STAP) of radar signals using orthogonal wavelet decompositions', *ASILOMAR* Conference on Signals, Systems and Computers, 29 Oct. - 1. Nov. 2000, pp. 641-645 23-27

[285] KEALEY, P.G., and FINLEY, I.P.: 'Comparison of the radar clutter cancellation performance of post- and pre-Doppler STAP for ground moving target identification from an experimental airborne surveillance radar ', *IEEE 2004 Radar Conference*, 26-29 April 2004, Philadelphia, PA

[286] KELLY, E. J.: 'An adaptive detection algorithm', *IEEE Trans. AES*, Vol. 22, No. 1, March 1986, pp. 115-127

[287] KIM, Y. L., PILLAI, S. U., and GUERCI, J. R.: 'Optimal loading factor for minimal sample support space-time adaptive radar', *Proc. ICASSP*, 12-15 May 1998, Seattle, WA, pp. 2505-2508

[288] KIM, S.-J., ILTIS, R.A.: 'STAP for GPS receiver synchronisation', *IEEE Trans. AES*, Vol. 40, No. 1, January 2004, pp. 132-144

[289] KIRSCHT, M.: 'Detection, velocity estimation and imaging of moving targets with single-channel SAR', *EUSAR'98*, 25-27 May 1998, Friedrichshafen, Germany, pp. 587-590 (VDE Publishers)

[290] KLEMM, R.: 'Suppression of jammers by multiple beam signal processing', *IEEE International Radar Conference*, Arlington, VA, 1975, pp. 176-180

[291] KLEMM, R.: 'Suboptimal simulation and suppression of clutter signals', *IEEE Trans. AES*, Vol. 12, No. 2, March 1976, pp. 210-212

[292] KLEMM, R.: 'Adaptive clutter suppression in step scan radars', *IEEE Trans. AES*, Vol. 14, No. 4, July 1978, pp. 685-688

[293] KLEMM, R.: 'Horizontal array gain in shallow water', *Signal Processing*, Vol 2, 1980, pp. 347-360

[294] KLEMM, R.: 'Use of generalized resolution methods to locate sources in random dispersive media', *IEE Proc. Pt. F*, Vol. 127, No. 1, February 1980, pp. 34-40

[295] KLEMM, R.: 'High resolution analysis of non-stationary data ensembles', *Proc. EUSIPCO-80*, 1980, North Holland, Amsterdam, pp. 711-714

[296] KLEMM, R.: 'Range and depth estimation by line arrays in shallow water', *Signal Processing*, Vol. 3, 1981, pp. 333-344

[297] KLEMM, R.: 'Low-error bearing estimation in shallow water', *IEEE Trans. AES*, Vol. 18, No. 4, July 1982, pp. 352-357

[298] KLEMM, R.: 'Suboptimum clutter suppression for airborne phased array radars', *IEE RADAR 82*, London, 1982, pp. 473-476

[299] KLEMM, R.: 'Problems of spatial signal processing in shallow water', *Issues in Acoustic Signal/ImageProcessing*, NATO ASI Series, Vol. F1, Springer Verlag, 1983, pp. 113-137

[300] KLEMM, R.: 'Adaptive clutter suppression for airborne phased array radar',*Proc. IEE*, Vol. 130, No. 1, February 1983, pp. 125-132

[301] KLEMM, R.: 'Some Properties of Space-Time Covariance Matrices', *Proc. International Conference on Radar*, Paris, 1984, pp. 357-361

[302] KLEMM, R.: 'Adaptive airborne MTI: an auxiliary channel approach', *Proc. IEE, Pt. F*, June 1987, pp. 269-276

[303] KLEMM, R.: 'New airborne MTI techniques', *Proc. IEE RADAR 87*, London 1987, pp. 380-384

[304] KLEMM, R.: 'Airborne MTI via subgroup processing', *Proc. International Colloquium on Radar*, Paris, 1989, pp. 43-47

[305] KLEMM, R.: 'Adaptive airborne MTI with two-dimensional motion compensation', *IEE Proc., Pt. F*, December 1991, pp. 551-558

[306] KLEMM, R.: 'Antenna design for airborne MTI', *Proc. Radar 92*, October 1992, Brighton, UK, pp. 296-299

[307] KLEMM, R.: 'Effect of time-space clutter filtering on SAR resolution', *Proc. ISSSE 92*, Paris, September 1992, pp. 462-465

[308] KLEMM, R.: 'Adaptive air- and spaceborne MTI under jamming conditions', *IEEE National Radar Conference*, Boston, MA, April 1993, pp. 167-172

[309] KLEMM, R.: 'Bewegtzielentdeckung mit nicht äquidistanten luft- und raumgestützten Antennengruppen', *DGON Radarsymposium 1993*, Neubiberg, Germany, pp. 444-449 (in German)

[310] KLEMM, R.: 'Detection of slow targets by a moving active sonar', *in: Acoustic Signal Processing for Ocean Exploration*, NATO ASI Series, Vol. 388, Kluwer Academic Publishers, Dordrecht, 1993, pp. 165-170

[311] KLEMM, R.: 'Interrelations Between Matched-Field Processing and Airborne MTI Radar', *IEEE Journal of Oceanic Engineering*, Vol. 18, No. 3, July 1993, pp. 168-180

[312] KLEMM, R.: 'Fehlereinflüsse in Gruppenantennen für luft- und raumgestütztes MTI-Radar' (Effect of array antenna errors in air- and spaceborne MTI radar), *ITG Fachtagung Antennen*, Dresden, April 1994, pp. 103-108 (in German)

[313] KLEMM, R.: 'Entfernungsmehrdeutigkeiten in luft- und raumgestütztem MTI-Radar' (Range ambiguities in air- and spaceborne MTI radar), *URSI Tagung 1994, Kleinheubach*, Germany (in German)

[314] KLEMM, R.: 'Effect of multiple-time around clutter on an adaptive space-borne MTI radar', *L'Onde Electrique*, Vol. 74, No. 3, May-June 1994, pp. 18-23, also: *International Conference on Radar*, Paris 1994, pp. 121-126

[315] KLEMM, R.: 'Adaptive airborne MTI: comparison of sideways and forward looking radar', *IEEE International Radar Conference*, Alexandria, VA, May 1995, pp. 614-618

[316] KLEMM, R.: 'Multidimensional signal processing for airborne MTI radar', *Proceedings Entretiens Science et Défence*, La Villette, France, 24-25 January 1996, pp. 335-345

[317] KLEMM, R.: 'Forward looking radar/SAR: clutter and jammer rejection with STAP', *Proc. EUSAR'96*, 26-28 March 1996, Königswinter, Germany, pp. 485-488 (VDE Publishers)

[318] KLEMM, R.: 'Real-time adaptive MTI, part I: space-time processing', *Proc. CIE International Conference on Radar*, 8-10 October 1996 Beijing, China, pp. 755-760

[319] KLEMM, R.: 'Real-time adaptive MTI, part II: space-frequency processing', *Proc. CIE International Conference on Radar*, 8-10 October 1996, Beijing, China, pp. 430-433

[320] KLEMM, R.: 'STAP for circular forward looking array antennas', *Proc. Radar'97*, 14-16 October 1997, Edinburgh, Scotland

[321] KLEMM, R.: 'Adaptive airborne MTI with tapered antenna arrays', *IEE Proc. Radar, Sonar and Navigation*, special issue on *Antenna array processing techniques*, Vol. 145, No. 1, February 1998, pp. 3-8

[322] KLEMM, R.: 'Interrelated problems in space-time processing for SAR and ISAR', *IEE Proc. Radar, Sonar and Navigation*, special issue on *Spectral applications in radar*, Vol. 145, No. 5, October 1998

[323] KLEMM, R., *Space-time adaptive processing - principles and applications*, 1st edition, IEE publishers, Stevenage, Herts., UK, 1998

[324] KLEMM, R., *Principles of Space-time adaptive processing*, 2nd edition, IEE publishers, Stevenage, Herts., UK, 2002

[325] KLEMM, R. (Editor), *IEE ECEJ*, Special issue on STAP, Vol. 11, No. 1, February 1999

[326] KLEMM, R.: 'Introduction to space-time adaptive processing', *IEE ECEJ*, special issue on STAP, Vol. 11, No. 1, February 1999, pp. 5-12

[327] KLEMM, R.: 'STAP with staggered PRF', *RADAR'99*, 17-21 May 1999, Brest, France

[328] KLEMM, R.: 'The role of models in STAP research', *SEE PSIP'99*, 18-19 January 1999, Paris, France, pp. 128-133

[329] KLEMM, R.: 'Effect of temporal decorrelation on the performance of space-time adaptive processors', *Int. Journal of Electronics and Communications (AEÜ)*, Vol. 53/6, December 1999, pp. 371-378

[330] KLEMM, R.: 'Comparison between monostatic and bistatic antenna configurations for STAP', *IEEE Trans. AES*, Vol. 36, No. 2, April 2000, pp. 596-607

[331] KLEMM, R.: 'Effect of bistatic radar configurations on STAP', *Proc. EUSAR2000*, 23-25 May 2000, Munich, Germany, pp. 817-820 (VDE Publishers)

[332] KLEMM, R.: 'Cramer-Rao analysis of reduced order STAP processors', *IEEE RADAR2000*, 8-12 May 2000, Alexandria, VA, pp.584-589

[333] KLEMM, R.: 'Ambiguities in bistatic STAP radar', *Proc IGARSS 2000*, Honolulu, HI, 24-28 July 2000

[334] KLEMM, R.: 'Space-time adaptive FIR filtering with staggered PRI', *ASAP 2001*, MIT Lincoln Laboratory, Lexington, MA, 13-15 March 2001

[335] KLEMM, R.: 'Effect of aircraft crabbing on sidelooking STAP radar', *EUSAR 2002*, Cologne, Germany, 4-6 June 2002

[336] KLEMM, R.: 'STAP for non-stationary clutter echo sequences', *EUSAR 2002*, Cologne, Germany, 4-6 June 2002

[337] KLEMM, R .: 'Effect of ambiguities on GMTI radar', *IEE Proc. Radar 2002*, Edinburgh, Scotland, 2002, pp.148-152

[338] KLEMM, R.: 'A planar antenna for omnidirectional GMTI radar ', *Radar 2002*, 15-17 October 2002, Edinburgh, Scotland, pp.177-180

[339] KLEMM, R., and, ENDER, J.: 'Two-dimensional filters for radar and sonar applications', in: *Signal Processing V* (EUSIPCO Barcelona), Elsevier Science Publisher B. V., 1990, pp. 2023-2026

[340] KLEMM, R., and ENDER, J.: 'New aspects of airborne MTI', *Proc. IEEE Radar 90*, Arlington, VA, 1990, pp. 335-340

[341] KLEMM, R., and ENDER, J.: 'Multidimensional filters for moving sensor arrays', *Proc. IASTED on Signal Processing and Digital Filtering*, June 1990, Lugano, Switzerland, pp. 9-12

[342] KLEMM, R., and ENDER, J.: 'Two-dimensional signal processing for airborne MTI', *AGARD Conference Proc.* Vol. 501, 6-9 May 1991, Ottawa, Canada, pp. 18-1 – 18-7

[343] KLEMM, R. (Editor), *Digest of the IEE Colloquium on STAP*, 6 April 1998, IEE, London, UK

[344] KLEMM, R.: 'STAP for range ambiguous spacebased radar: bistatic configurations', ', *Proc. EUSAR 2004*, 25-27 May 2004, Ulm, Germany

[345] KLEMM, R.: 'STAP for range ambiguous spacebased radar: monostatic configurations', ', *Proc. EUSAR 2004*, 25-27 May 2004, Ulm, Germany

[346] KLEMM, R.: 'Issues in GMTI design for forward looking airborne radar ', *Radar 2004*, 19-21 October 2004, Toulouse, France

[347] KLEMM, R.:'STAP with omnidirectional antenna arrays', in KLEMM, R., ed.: *Applications of space-time adaptive processing*, IEE Publishers, London, 2004

[348] KLEMM, R.: 'Large adaptive antennas for ground surveillance radar ', *RTO SET/STI symposium on "Smart and Adaptive Antennas"*, 7-9 April 2003, Chester

[349] KLEMM, R.: 'Radar aspects of ground target tracking ', *RTO SET symposium on "Target tracking and sensor data fusion for military observation systems"*, 7-9 April 2003, Budapest

[350] KLEMM, R., ed.: *Applications of space-time adaptive processing*, IEE Publishers, London, 2004

[351] KLEMM, R.: 'Tilted omnidirectional array antennas in range ambiguous STAP radar ', *Signal Processing*, 84, 2004, pp. 1581-1592

[352] KLEMM, R.: 'Ground target tracking with STAP radar: The sensor ', in KLEMM, R. (ed.): *Applications of space-time adaptive processing*, IEE Publishers, 2004, Chapter 24, pp. 467-500

[353] KNOTT, E. W., SHAEFFER, J. F., and TULEY, M. T., *Radar Cross Section*, Artech House, Dedham, MA, 1985

[354] KOC, A. T., and TANIK, Y.: 'Direction finding with a uniform circular array via single snapshot processing', *Signal Processing*, Vol. 56, 1997, pp. 17-31

[355] KOCH, W.: 'Retrodiction for Bayesian multiple hypothesis/multiple target tracking in densely cluttered environment', *SPIE Conference on Signal and Data Processing of Small Targets*, Vol. 2759, 1996, pp. 429-440

[356] KOCH, W., and VAN KEUK, G.: 'Multiple hypothesis track maintenance with possible unresolved measurement', *IEEE Trans. AES*, Vol. 33, No. 3, July 1997, pp. 883-892

[357] KOCH, W., and KLEMM, R .: 'Ground target tracking with STAP radar', *IEE Proc. Radar, Sonar and Navigation*, Vol. 148/3, June 2001, pp.173-185

[358] KOCH, W .: 'Effect of Doppler ambiguities on GMTI tracking', *IEE Proc. Radar 2002*, Edinburgh, Scotland, 2002, pp.153-157

[359] KOGON, S. M., WILLIAMS, D. B., and HOLDER, E. J.: 'Beamspace techniques for hot clutter cancellation', *Proc. IEEE Int. Conf. on Acoustics, Speech and Signal Processing*, Vol. 2, May 1996, pp. 1177-1180

[360] KOGON, S. M., WILLIAMS, D. B., and HOLDER, E. J.: 'Exploiting coherent multipath for mainbeam jammer suppression', *Proc. IEE Radar Sonar and Navigation*, 1998

[361] KOGON, S. M., HOLDER, E. J., and WILLIAMS, D. B.: 'Mainbeam jammer suppression using multipath returns', *31st ASILOMAR Conference*, 2-5 Nov. 1997, Los Alamitos, CA, pp. 279-283

[362] KOGON, S. M., and ZATMAN, M. A..: 'Bistatic STAP for Airborne Radar Systems', *ASAP 2000*, Workshop, MIT Lincoln Laboratory, 13-14 March 2000, pp. 1-6

[363] KOGON, S. M., and ZATMAN, M.: 'Techniques for Range-ambiguous clutter mitigation in space-based radar systems', in: KLEMM, R. ed., *Applications of space-time adaptive processing*, IEE publishers, London, UK, 2003/4, to be published

[364] KOGON, S.M.: 'Adaptive array processing techniques for terrain scattered interference mitigation', *Ph.D. thesis*, Georgia Institute of Technology, December 1996

[365] KRAUT, S., HARMANCI, K., and KROLIK, J.: 'Space-time adaptive processing for over-the-horizon spread-Doppler clutter mitigation', *IEEE SAM2000 Workshop*, 15-16 March 2000, Cambridge, MA, pp. 245-249

[366] KREITHEN, D., and STEINHARDT, A. O.: 'Target detection in post-STAP undernulled clutter', *IEEE Proc. 29th ASILOMAR Conference on Signals, Systems and Computers*, Pacific Grove, CA, 30 Oct.-2 Nov. 1995, pp. 1203-1207

[367] KRETSCHMER, F. F., and LIN, F.-L. C.: 'Effects of main tap position in adaptive clutter processing', *IEEE International Radar Conference*, Arlington, VA, 1985

[368] KREYENKAMP, O.: 'Clutter covariance modelling for STAP in forward looking radar', *DGON International Radar Symposium 98*, 15-17 September 1998, Munich, Germany

[369] KREYENKAMP, O.: 'Bewegtzielentdeckung fuer flugzeuggetragenes Radar mit linearer Gruppenantenne durch Raum-Zeit-Filterung (Moving target detection for airborne radar with linear array antenna using space-time filtering)', Dr.-Ing. thesis, Ruhr-Universitaet Bochum, Bochum, Germany, 2000 (in German)

[370] KREYENKAMP, O., and KLEMM, R.: 'Doppler compensation in forward looking STAP radar', *IEE Proc. Radar, Sonar and Navigation*, 2001, pp. 253-258

[371] KRIKORIAN. D.K.V., and ROSEN, R.A.: 'Acceleration compensation by matched filtering ', *IEEE RADARCON 2002*, 22-25 April 2002, Long Beach, CA

[372] KUZMINSKIY, A.M., and HATZINAKOS, D.: 'Semiblind estimation of spatio-temporal filter coefficients based on a training-like approach', *IEEE Signal Processing Letters*, Vol. 5, No. 9, September 1998, pp. 231-233

[373] LACOMME, Ph., MARCHAIS, J.-C., HARDANGE, J.-Ph., and NORMANT, E., *Air- and Spaceborne Radar Systems: An Introduction*, SkyTech Publishing, Mendham, NJ, 2001

[374] LACOMME, Ph.: 'New trends in airborne phased array radars ', *IEEE Symposium on Phased Array Systems and Technology 2003*, 14-17 October 2003, Boston, pp. 17-22

[375] LANDAU, H. J., and POLLAK, H. O.: 'Prolate spheroidal wave functions, Fourier analysis and uncertainty-III: The dimension of the space of essentially time- and band-limited signals', *Bell System Technical Journal*, July 1962, pp. 1295-1336

[376] LAPIERRE, F.: 'Registration-based range-dependent compensation in airborne bistatic radar STAP ', *Ph.D. thesis, Université de Liège*, 2004

[377] LAPIERRE, F., VAN DROOGENBROEK, M., and VERLY, J.G.: 'New methods for handling the range dependence of the clutter spectrum in non-sidelooking monostatic STAP radars', *IEEE ICASSP*, 6-10 April 2003, Hong Kong, pp. v-73 - v-76

[378] LAPIERRE, F.D., VERLY, J.G., and VAN DROOGENBROEK, M. : 'New solutions to the problem of range dependece in bistatic STAP radars', *2003 IEEE Radar Conference*, May 5-8, 2003, Huntsville, AL, pp. 452-459

[379] LEATHERWOOD, D.A., and MELVIN, W.L.: 'Configuring a sparse aperture antenna for spaceborne MTI radar', *2003 IEEE Radar Conference*, May 5-8, 2003, Huntsville, AL, pp. 139-148

[380] LE CHEVALIER, F., SAVY, L., and DURNIEZ, F.: 'Clutter calibration for space-time airborne MTI radars', *Proc. International Conference on Radar (ICR'96)*, 8-10 October 1996, Beijing, China, pp. 82-85

[381] LE CHEVALIER, F.: 'Future concepts for electromagnetic detection: from space-time-frequency resources management to wideband radars', *RADAR'99*, 17-21 May 1999, Brest, France

[382] LEE, F. W., and STAUDAHER, F.: 'NRL adaptive array flight test data base', *Proc. of the IEEE Adaptive Antenna Systems Symposium*, Melville, New York, November 1992

[383] LEE, C.-C., and LEE, J.-H.: 'Robust adaptive array beamforming under steering vector errors', *IEEE Transactions on Antennas and Propagation*, Vol. 45, No. 1, January 1997, pp. 168-175

[384] LEE, C.-C., and LEE, J.-H.: 'Eigenanalysis interference cancellers with robust capabilities', *Signal Processing*, Vol. 58, 1997, pp. 193-202

[385] LEE, C.-C., and LEE, J.-H.: 'Eigenspace-based adaptive array beamforming with robust capabilities', *IEEE Trans.* AP-45, No. 12, December 1997, pp. 1711-1716

[386] LEE, C.-C., and LEE, J.-H.: 'Efficient computation of the eigensubspace for array signal processing', *IEE Proc. Radar, Sonar and Navigation*, Vol. 146, No. 5, October 1999, pp. 235-242

[387] LEGG, J. A., and GRAY, D. A.: 'SAR moving target parameter estimation accuracy', *EUSAR'98*, 25-27 May 1998, Friedrichshafen, Germany, pp. 287-290, (VDE Publishers)

[388] LEGTERS, G.R., and GUERCI, J.R.: 'Physics-based airborne GMTI radar signal processing ', *IEEE 2004 Radar Conference*, 26-29 April 2004, Philadelphia, PA

[389] LESTURGIE, M.: 'Use of STAP techniques to enhance the detection of slow targets in shipborne HFSWR', *Intern. Conf. on Radar*, 3-5 September 2003, Adelaide, Australia, pp. 504-509

[390] LEVINSON, N.: 'The Wiener RMS (root mean square) error criterion in filter design and prediction', *J. Math. Phys.*, Vol. XXV, 1947, pp. 261-278

[391] LEWIS, B. L., and EVINS, J. B. 'A new technique for reducing radar response to signals entering antenna sidelobes', *Trans. IEEE Trans. Antenna and Propagation*, Vol. AP-31, No. 6, November 1983, pp. 993-996

[392] LI, J., LIU, Z.-S., and STOICA, P.: '3-D target feature extraction via interferometric SAR', *IEE Proc. Radar, Sonar and Navigation*, Vol. 144, No. 2, April 1997, pp. 71-80

[393] LI, J., LIU, G., NANZHI, J., and STOICA, P.: 'Moving target feature extraction for airborne high-range resolution phased-array radar', *IEEE Trans. SP*, Vol. 49, No. 2, February 2001, pp. 277-289

[394] LI, T., SIDIROPOULOS, N. D., and GIANNAKIS, G. B.: 'PARAFAC-STAP for the UESA radar', *ASAP 2000 Workshop*, 13-14 March 2000, MIT Lincoln Laboratory, pp. 49-54

[395] LI, Y., and CHEONG, C. H.: 'Block Toeplitz with Toeplitz block covariance matrix for space-time adaptive processing ', *IEEE 2002 Radar Conference*, Long Beach, CA, 22-25 April 2002, pp. 311-317

[396] LIAO, G., BAO, Z., and ZHANG, Y. 'Adaptive space-time processing for airborne radars based on Doppler pre-processing', *Journal of Xidian University*, Vol. 20, December 1993, pp. 1-6

[397] LIAO, G., BAO, Z., and ZHANG, Y.: 'On clutter DoF for phased array airborne early warning radars', *Chinese Electronics Journal*, Vol. 10, No. 4, October 1993, pp. 307-314

[398] LIAO, G., BAO, Z., and ZHANG, Y.: 'Doppler pre-filtering based adaptive array processing in airborne radars', *Int. Conf. on Radar*, Paris, May 1994, pp. 48-53

[399] LIAO G., BAO Z., and XU, Z.: 'Evaluation of adaptive space-time processing for inclined sideways looking array airborne radars', *Proc. International Conference on Radar (ICR'96)*, 8-10 October 1996, Beijing, China, pp. 78-81

[400] LIGHTSTONE, L., FAUBERT, D., and REMPEL, G.: 'Multiple phase centre DPCA for airborne radar', *IEEE National Radar Conference*, March 1991, Los Angeles, CA, pp. 36-40

[401] LIN, W.-T., and YU, K.-B.: 'Adaptive beamforming for wideband jamming cancellation', *Proc. IEEE National Radar Conference*, 1997, Syracuse, NY, pp. 82-85

[402] LIN, X., and BLUM, R. S.: 'Robust STAP algorithms using prior knowledge for airborne radar application', *Signal Processing*, Vol. 79, 1999, pp. 273-287

[403] LINDERMAN, M. H., and LINDERMAN, R. W.: 'Real-time STAP demonstration on an embedded high performance computer', *Proc. IEEE National Radar Conference*, 1997, Syracuse, NY, pp. 54-59

[404] LIPPS, R., CHEN, V., and BOTTOMS, M.: 'Advanced SAR GMTI techniques ', *IEEE 2004 Radar Conference*, 26-29 April 2004, Philadelphia, PA

[405] LITTLE, M., O., and BERRY, W. P.: 'Real-time multichannel airborne radar measurements', *Proc. IEEE National Radar Conference*, 1997, Syracuse, NY, pp. 138-142

[406] LIU, H., GHAFOR, A., and STOCKMANN, P. H.: 'Application of Gram-SCHMIDT algorithm to fully adaptive arrays', *IEEE Trans. AES*, Vol. 28, No. 2, April 1992, pp. 324-334

[407] LIU, Q.-G., PENG, Y.-N., and MA, Z.-E.: 'A simple solution for clutter suppression in airborne phased radar', *IEEE International Symposium on Antennas and Propagation*, Ann Arbor, MI, 1993, pp. 1892-1895

[408] LIU, Q.-G., PENG, Y.-N., Sun, X., and MA, ZHANG-ER: 'A space-time transformation approach for adaptive clutter suppression in airborne radar', *IEEE International Symposium on Circuits and Systems*, Chicago, IL., 3-6 May 1993, pp. 247-250

[409] LIU, Q.- G., and PENG, Y.-N.: 'Analysis of array errors and a short-time processor in airborne phased array radars', *IEEE Transactions on Aerospace and Electronic Systems*, Vol. 32, No. 2, April 1996, pp. 587-597

[410] LIU, G., and LI, J.: 'Moving target detection via airborne HRR phased array radar', *IEEE Trans. AES*, Vol. 37, No. 3, July 2001

[411] LO, Y. T., LEE, S. W., and LEE, Q. H.: 'Optimization of directivity and signal-to-noise ratio of an arbitrary antenna array', *Proc. IEEE*, Vol. 54, No. 8, August 1966, pp. 1033-1045

[412] LOMBARDO, P.: 'Optimum multichannel SAR detection of slowly moving targets in the presence of internal clutter motion', *Proc. International Conference on Radar (ICR'96)*, 8-10 October 1996, Beijing, China, pp. 321-325

[413] LOMBARDO, P.: 'DPCA processing for SAR moving targets detection in the presence of internal clutter motion and velocity mismatch', *Proc. SPIE*, Vol. 2958, 23-26 September 1996, pp. 50-61

[414] LOMBARDO, P.: 'Data selection strategies for radar space-time adaptive processing', *IEEE RADARCON'98*, 11-14 May 1998, Dallas, TX, pp. 201-206

[415] LOMBARDO, P., GRECO, M., GINI, F., FARINA, A., BILLINGSLEY, J.B.: 'Impact of clutter spectra on radar performance prediction ', *IEEE Trans. AES*, Vol. 37, No. 3, July 2001, pp. 1022-1038

[416] LOMBARDO, P., COLONE, F.: 'A dual adaptive channel STAP Scheme for target detection and DOA estimation ', *Intern. Conf. on Radar*, 3-5 September 2003, Adelaide, Australia, pp. 115-118

[417] LOMBARDO, P., and COLONE, F.: 'An alternating transmit approach for STAP with short antenna arrays', *IEEE 2004 Radar Conference*, 26-29 April 2004, Philadelphia, PA

[418] LOMBARDO, P., COLONE, F.: 'Non-linear STAP filters based on adaptive 2D-FIR filters', *2003 IEEE Radar Conference*, May 5-8, 2003, Huntsville, AL, pp. 51-58

[419] LONG, M.W.: 'Radar reflectivity of Land and sea', 3rd edition, Artech House, Boston, MA, 2001

[420] LUKIN, A. N., and UDALOV, V. P.: 'Estimating the velocity of a two-point radiation source in the Fresnel zone with space-time signal processing', *Radio and Communication Technology*, Vol. 3, No. 6, 1998, pp. 87-92

[421] LUMINATI, J.E., and HALE, T.B.: 'Steering vector mismatch: analysis and reduction ', *IEEE 2004 Radar Conference*, 26-29 April 2004, Philadelphia, PA

[422] MA, E. M., and ZAROWSKI, C. J.: 'On lower bounds for the smallest eigenvalue of a Hermitian Positive-definite matrix', *IEEE Transactions on Information Theory*, Vol. 41, No. 2, March 1995, pp. 539-540

[423] MADURASINGHE, P., and CAPON, S.: 'A signal subspace technique for computing weights of an airborne phased array ', *Intern. Conf. on Radar*, 3-5 September 2003, Adelaide, Australia, pp. 67-70

[424] MAHAFZA, B. R., KNIGHT, D. L., and AUDEH, N. F.: 'Forward-looking SAR imaging using a linear array with transverse motion', *Proc. IEEE SOUTHEASTCON*, 4-7 April 1993

[425] MAHAFZA, B. R., and HEIFNER, L. H.: 'Multitarget detection using synthetic sampled aperture radars', *IEEE Trans. AES*, Vol. 31, No. 3, July 1995, pp. 1127-1132

[426] MAHER, J., ZHANG, Y., and WANG, H.: 'A performance evaluation of $\Sigma\Delta$-STAP approach to airborne surveillance radars in the presence of both clutter and jammers', *Proc. IEE Radar'97*, 14-16 October 1997, Edinburgh, Scotland, pp. 305-309

[427] MAHER, J., and LYNCH, D.: 'Effects of clutter modeling in evaluating STAP processing for space-based radars', *IEEE RADAR2000*, 8-12 May 2000, Alexandria, VA, pp. 565-570

[428] MAILLOUX, R. J., *Phased array antenna handbook*, Artech House, Boston and London, 1994

[429] MAILLOUX, R. J.: 'Array failure correction with a digitally beamformed array', *IEEE Transactions on Antennas and Propagation*, Vol. 44, No. 12, December 1996, pp. 1543-1550

[430] MAIWALD, D., BENEN, S., and SCHMIDT-SCHIERHORN, H.: 'Space-time signal processing for surface ship towed active sonar', in: KLEMM, R. ed., *Applications of space-time adaptive processing*, IEE publishers, London, UK, 2004

[431] MAKHOUL, J.: 'Linear prediction: a tutorial review', *Proc. IEEE*, Vol. 63, No. 4, April 1975, pp. 561-580

[432] MAKHOUL, J.: 'On the Eigenvalues of Symmetric Toeplitz Matrices', *IEEE Trans. ASSP*, Vol. ASSP-29, No. 4, August 1981, pp. 868-872

[433] Malas, J.A., Pasala, K.M., and Westerkamp, J.: 'Automatic target classification of slow moving ground targets in clutter', *IEEE Trans. AES*, Vol. 40, No. 1, January 2004, pp. 190-205

[434] MANSUR, H. H.: 'Space/time adaptive processing architecture implementations using a high performance scalable computer', *Proc. IEEE National Radar Conference*, 1997, Syracuse, NY, pp. 325-330

[435] MAO, Y.H., LIN, D., and LI, W.: 'An adaptive MTI based on maximum entropy spectrum estimation principle', *International Radar Colloquium*, Paris, France, 1984, pp. 103-107

[436] MARDIA, K. V., KENT, J. T., and BIBBY, J. M.: 'Multivariate analysis', Academic Press, 1997

[437] MARSHALL, D. F.: 'A two step adaptive interference nulling algorithm for use with airborne sensor arrays', *Proc. of the Seventh Workshop on Statistical Signal and Array Processing*, Quebec City, Canada, 26-29 June 1994, pp. 301-304

[438] MARTINSEK, D., GOLDSTEIN, R.: 'Bistatic radar experiment', *EUSAR'98*, 25-27 May 1998, Friedrichshafen, Germany, pp. 31-34 (VDE Publishers)

[439] MARTINEZ, D. R., and MACPHEE, J. V.: 'Real-time testbed for space-time adaptive techniques', *Proc. of the Conf. on Adaptive Antennas*, 7-8 November 1994, Melville, NY 11747, pp. 135-142

[440] MATHER, J. L.: 'Resolution of correlated multipath using the incremental multi-parameter (IMP) algorithm', *Proc. IASTED on Signal Processing and Digital Filtering*, June 1990, Lugano, Switzerland, pp. 229-232

[441] MATHER, J. L., REES, H. D., and SKIDMORE, I. D .: 'Adaptive clutter and jammer cancellation for element-digitised airborne radar', *33rd Asilomar Conference*, Pacific Grove, CA, 24-27 October 1999, pp. 92-97

[442] MATHEW, G., and REDDY, V. U.: 'Adaptive Estimation of Eigensubspace', *IEEE Transactions on Signal Processing*, Vol. 43, No. 2, February 1995, pp. 401-411

[443] MCDONALD, K. F., and BLUM, R. S .: 'Performance characterisation of STAP algorithms with mismatched steering and clutter statistic', *ASILOMAR Conference on Signals, Systems and Computers*, 29 Oct. - 1. Nov. 2000, pp. 646-650 23-27

[444] McDONALD, K.F., RAGHAVAN, R., and FANTE, R.: 'Lessons learned through the implementation of space-time adaptive algorithms for GPS reception in jammed environments', *IEEE PLANS 2004*, 26-29 April 2004, Monterey, CA, pp. 418-428

[445] MEDLIN, G.: 'Digital filter designs for the pre-filter MTI technique with SAR', *Proc. IEEE/AIAA/NASA 9th Digital Avionics Systems Conference*, October 1990, pp. 595-601

[446] MEDLIN, G. W., and ADAMS, J. W.: 'Improved filters for moving target indication with synthetic aperture', *Int. Conference on Radar*, Paris, 1989, pp. 392-397

[447] MEISL, P., THOMPSON, A. A., and LUSCOMBE, A. P.: 'RADARSAT-2 mission: overview and development status', *EUSAR 2000*, Munich, Germany, 23-25 May 2000, pp. 373-376

[448] MELVIN, W. L., WICKS, M. C., and BROWN, R. D.: 'Assessment of Multichannel Airborne radar measurements for analysis and design of space-time processing architectures and algorithms', *IEEE 1996 National Radar Conference*, Ann Arbor, MI, 13-16 May 1996, pp. 130-135

[449] MELVIN, W. L., and HIMED, B.: 'Comparative analysis of space-time adaptive algorithms with measured airborne data', *Proc. 7th International Signal Processing Applications and Technology Conference*, Boston, 7-10 October 1996, pp. 1479-1483 (Miller Freeman, San Francisco)

[450] MELVIN, W. L.: 'Eigenbased modeling of nonhomogeneous airborne radar environments', *IEEE RADARCON'98*, 11-14 May 1998, Dallas, TX, pp. 171-176

[451] MELVIN, W. L., and WICKS, M. C.: 'Improving practical space-time adaptive radar', *IEEE National Radar Conference*, 1997, pp. 48-53

[452] MELVIN, W. L., GUERCI, J. R., CALLAHAN, M. J., and WICKS, M. C.: 'Design of adaptive detection algorithms for surveillance radar', *IEEE RADAR2000*, 8-12 May 2000, Alexandria, VA, pp. 608-613

[453] MELVIN, W. L.: 'Space-time adaptive radar performance in heterogeneous clutter', *IEEE Transactions AES*, Vol. 36, No. 2, April 2000, pp. 621-633

[454] MELVIN, W. L.: 'Space-time adaptive processing and adaptive arrays: special collection of papers', *IEEE Transactions AES*, Vol. 36, No. 2, April 2000, pp. 508-509

[455] MELVIN, W.L., CALLAHAN, M.J., and WICKS, M.C.: 'Adaptive clutter cancellation in bistatic radar ', *ASILOMAR*, 34, 29 Oct.-01 Nov. 2000, Pacific Grove, pp. 1125-1130

[456] MELVIN, W. L., CALLAHAN, M. J., and WICKS, M. C.: 'Bistatic STAP: Application to airborne radar ', *IEEE 2002 Radar Conference*, Long Beach, CA, 22-25 April 2002, pp. 1-7

[457] MELVIN, W.: ' A STAP overview', *IEEE A&E System Magazin*, Vol. 19, No. 1, January 2004, pp. 19-35

[458] Melvin, W.L., SHOWMAN, G.A., and GUERCI, J.R.: 'A knowledge-aided GMTI detection architecture ', *IEEE 2004 Radar Conference*, 26-29 April 2004, Philadelphia, PA

[459] MERIGEAULT, S., CHAMOUARD, E., and SAILLARD, J.: 'Space-time processing for detection of slow moving targets', *RADAR'99*, 17-21 May 1999, Brest, France

[460] MERMOZ, H.: 'Filtrage adaptif et utilisation optimal d'une antenne', *NATO Advanced Study Institute on Signal Processing, with Emphasis on Underwater Acoustics*, Grenoble, 1964, pp. 160-299 (in French)

[461] MEYER-HILBERG, J., and BICKERT, B.: 'Real-time STAP as a key technology for subclutter moving target detection', *Proc. EUSAR2000*, 23-25 May 2000, Munich, Germany, pp. 821-824 (VDE Publishers)

[462] MICHELS, J. H., HIMED, B., and RANGASWAMY, M.: 'Evaluation of the normalized parameter adaptive matched filter STAP test in airborne radar clutter', *IEEE RADAR2000*, 8-12 May 2000, Alexandria, VA, pp. 769-774

[463] MICHELS, J. H., RANGASWAMY, M., and HIMED, B.: 'Performance of STAP tests in compound-gaussian clutter', *IEEE SAM Workshop*, 15-16 March 2000, Cambridge, MA, pp. 250-255

[464] MICHELS, J. H., HIMED, B., and RANGASWAMY, M.: 'Performance of STAP tests in Gaussian and compound-Gaussian clutter', *Digital Signal Processing*, Vol. 10, 2000, pp. 309-324

[465] MICHELS, J.H., ROMAN, J.R., and HIMED, B.: 'Beam control using the parametric adaptive matched filter STAP approach', *2003 IEEE Radar Conference*, May 5-8, 2003, Huntsville, AL, pp. 405-412

[466] MONZINGO, R. A., and MILLER, T. W., *Introduction to adaptive arrays*, John Wiley & Sons, 1980

[467] MOO, P.: 'GMTI performance of $\Sigma\Delta$-STAP for a forward-looking radar', *IEEE Radar Conference*, 2001, pp. 258-263

[468] MOORE, T.D., and GUPTA, I.J.: 'The effect of interference power and bandwidth on space-time adaptive processing', *IEEE Antennas and Propagation Int. Symposium*, 22-27 June 2003, Columbus, OH, pp. 53-56

[469] MORGAN, C., JAROSZWSKI, S., and MOUNTCASTLE, P.: 'Terrain height estimation using GMTI radar', *IEEE 2004 Radar Conference*, 26-29 April 2004, Philadelphia, PA

[470] MOUNTCASTLE, P.D.: 'New implementation of the Billingsley clutter model for GMTI data cube generation', *IEEE 2004 Radar Conference*, 26-29 April 2004, Philadelphia, PA

[471] MURRAY, D. J., COE, D. J., and WHITE, R. G.: 'Experimental MTI with spaceborne geometries', *IEE Colloquium on Radar Interferometry*, 11 April 1997, London, UK, pp. 3/1–3/6

[472] NAGA, V. D., and WELSH, B. M.: 'Atmospheric induced errors in space-time adaptive processing', *PIERS Workshop*, 20-22 July 1998, Baveno, Italy, pp. 211-213

[473] NAGEL, D.: 'Aufzeichnung und Auswertung von HPRF-Clutterdaten mit einem Multimode Bordradar der Nächsten Generation (Recording and evaluation of HPRF clutter data with a multimode radar of the next generation)', *DGON Radarsymposium*, 8-10 April 1997, Stuttgart, Germany, pp. 221-231 (in German)

[474] NAKHMANSON, G. S.: 'Space-time processing of broadband signals reflected from moving targets in coherent multiposition measuring systems', *Telecommunication and Radio Engineering*, Vol 48, No. 2, 1993, pp. 118-127

[475] NATHANSON, F. E., *'Radar design principles'*, McGraw-Hill Book Company, New York etc., 1969

[476] NAYEBI, M. M., AREFF, M. R., and BASTANI, M. H.: 'Detection of coherent radar signal with unknown Doppler shift', *IEE Proc. Radar, Sonar and Navigation*, Vol. 143, No. 2, April 1996, pp. 79-86

[477] NELANDER, L.: 'Deconvolution approach to terrain scattered interference mitigation', *IEEE RADARCON 2002*, 22-25 April 2002, Long Beach, CA

[478] NGUYEN, H.N., HIEMSTRA, J.D., and GOLDSTEIN, J.S.: 'The reduced rank multistage Wiener filter for circular array STAP', *2003 IEEE Radar Conference*, May 5-8, 2003, Huntsville, AL, pp. 66-70

[479] NICKEL, U.: 'Superresolution by spectral line fitting', *Signal processing II: theory and applications*, North Holland, 1983, pp. 645-648 (Conf. Record EUSIPCO-83)

[480] NICKEL, U.: 'Angular superresolution with phased array radar: a review of algorithms and operational constraints', *IEE Proc. Pt. F*, Vol. 134, No. 1, Feb. 1987, pp. 7-10

[481] NICKEL, U.: 'Some properties of fast projection methods of the Hung-Turner Type', *Signal Processing III (Proc. EUSIPCO)*, 1986, pp. 1165-1168

[482] NICKEL, U.: 'Angle estimation with adaptive arrays and its relation to superresolution', *IEE Proceedings*, Vol. 134, Pt. H, No. 1, February 1987, pp. 77-82

[483] NICKEL, U.: 'Algebraic formulation of KUMARESAN-TUFTS superresolution method, showing relation to ME and MUSIC methods', *IEE Proc.*, Vol. 135, Pt. F, No. 1, February 1988, pp. 7-10,

[484] NICKEL, U.: 'Subgroup configurations for interference suppression with phased array radar', *International Conference on Radar*, Paris, 1989, pp. 82-86

[485] NICKEL, U.: 'A corrected monopulse estimation method for adaptive arrays', *IEE Radar 92*, Brighton, UK, 1992, pp. 324-327

[486] NICKEL, U.: 'On the influence of channel errors on array signal processing methods', *Archiv der Elektrischen Übertragung*, Vol. 47, No. 4, 1993, pp. 209-219

[487] NICKEL, U. 'Radar target parameter estimation with antenna arrays', in HAYKIN, LITVA and SHEPHERDS (eds), *Radar Array Processing* (Springer Series in Information Sciences, Vol. 25), Springer Verlag, New York, 1993

[488] NICKEL, U.: 'Monopulse estimation with adaptive arrays', *IEE Proc.*, Pt. F, Vol. 140, No. 5, October 1993, pp. 303-308

[489] NICKEL, U.: 'Subarray Configurations for Digital Beamforming with Low Sidelobes and Adaptive Interferences Suppression', *IEEE International Radar Conference*, Alexandria, VA, 8-11 May 1995, pp. 714-719

[490] NICKEL, U.: 'Relationships between fully adaptive arrays and sidelobe canceler configurations', TPSP Working Paper, FGAN-FFM, 20.5.96, private communication

[491] NICKEL, U.: 'Monopulse estimation with subarray-adaptive arrays and arbitrary sum and difference beams', *IEE Proc. Radar, Sonar, Navigation*, Vol. 143, No. 4, August 1996, pp. 232-238

[492] NICKEL, U.: 'Fast subspace methods for radar applications', *Proc. SPIE'97 Conference*, San Diego, CA, 27 July–1 August 1997

[493] NICKEL, U.: 'On the application of subspace methods for small sample size', *AEU*, Vol. 51, No. 6, 1997, pp. 279-289

[494] NICKEL, U.: 'Performance of corrected adaptive monopulse estimation', *IEE Proc. Radar, Sonar and Navigation*, Vol. 146, No. 1, February 1999, pp. 17-24

[495] NICKEL, U.: 'Superresolution and jammer suppression with broadband arrays for multi-function radar', in: KLEMM, R. ed., *Applications of space-time adaptive processing*, IEE publishers, London, UK, 2004

[496] NICKEL, U.: 'Corrected space-time adaptive monopulse ', *International Radar Symposium*, 30 September - 2 October 2003, Dresden, pp. 609-614

[497] NICKEL, U.: 'Performance analysis of space-time adaptive monopulse', *SIGNAL PROCESSING*, Vol. 84, 2004, pp. 1561-1579

[498] NICKEL, U.: 'An overview of generalised monopulse estimation ', *IEEE AES Magazine*, 2005

[499] NICOLAU, E., and ZAHARIA, D., *Adaptive arrays*, Elsevier Science Publishing Company, New York, 1989

[500] NOHARA, T. J., SCARLETT, P., and EATCOCK, B. C.: 'A radar processor for space-based radar', *IEEE National Radar Conference*, April 1993, Boston, MA, pp. 12-16

[501] NOHARA, T. J.: 'Comparison of DPCA and STAP for space-based radar', *Proc. IEEE International Radar Conference*, Alexandria, VA, 1995, pp. 113-119

[502] NOHARA, T. J., PREMJI, A., and WEBER, P.: 'Space-based radar signal processing demonstrator: preliminary experimental results', *IEEE International Symposium on Phased Array Systems and Technology*, 15-18 Oct. 1996, Boston, MA, pp. 307-312

[503] NOHARA, T.: 'Design of a space-based radar signal processor', *IEEE Trans. AES*, Vol. 34, No. 2, April 1998, pp. 366-377

[504] NOHARA, T.J., WEBER, P., PREMJI, A., and BHATTACHARYA, T.: 'Airborne ground moving target indication using non-sidelooking antennas', *IEEE RADARCON'98*, 11-14 May 1998, Dallas, TX, pp. 269-274

[505] NORDEBO, S., CLAESSON, I., and NORDHOLM, S.: 'Broadband adaptive beamforming: a design using 2-D spatial filters', *IEEE International Symposium on Antennas and Propagation*, Ann Arbor, MI, 1993, pp. 702-705

[506] NORDWALL, B.: 'Radar on S-3B: overland demo', *Aviation Week & Space Technology*, 22 April 1996, pp. 62-63

[507] NORDWALL, B.: 'Astor builds on U-2's advanced SAR system', *Aviation Week & Space Technology*, 30 August 1999, pp. 54-55

[508] ODENDAAL, J. W., BARBARD, E., and PISTORIUS, C. W.: 'Two-dimensional superresolution radar imaging using the MUSIC algorithm', *IEEE Trans. AP*, Vol. 42, No. 10, October 1994, pp. 1386-1391

[509] OGLE, W.C., NGUYEN, H.N., GOLDSTEIN, J. S., ZULCH, P.A., and Wicks, M.C.: 'Nonhomogeneity detection and the multistage Wiener filter ', *IEEE RADARCON 2002*, 22-25 April 2002, Long Beach, CA

[510] OGLE, W.C., NGUYEN, H.N., TINSTON, M.A., GOLDSTEIN, J.S., ZULCH, P.A., and WICKS, M.C.: 'A multistage nonhomogeneity detector', *2003 IEEE Radar Conference*, May 5-8, 2003, Huntsville, AL, pp. 121-125

[511] ONG, K.P.: 'Signal processing for airborne bistatic radar ', *PhD Thesis*, June 2003, University of Edinburgh

[512] OPPENHEIM, A. V., and SCHAFER, R. W., *Digital signal processing*, Prentice-Hall Inc., Englewood Cliffs, NJ, 1975

[513] PAINE, A. S.: 'Minimum variance monopulse technique for an adaptive phased array radar', *IEE Proc. Radar, Sonar and Navigation*, Vol. 145, No. 6, Dec. 1998, pp. 374-379

[514] PAINE, A. S.: 'Application of the minimum variance monopulse technique to space-time adaptive processing', *IEEE RADAR2000*, 8-12 May 2000, Alexandria, VA, pp. 596-601

[515] PAINE, A.S.: 'Space-time adaptive processing for domain factorised element-digitised array radar ', *CIE International Conference on Radar*, 15-18 October 2001, pp. 758-764

[516] PAINE, A. S.: 'Optimal adaptive processing for domain factorised element-digitised array radar', *IEE Proc. Radar, Sonar and Navigation*, Vol. 148, No. 2, April 2001, pp. 81-88 @@

[517] PAINE, A. S.: 'A comparison of partially adaptive STAP techniques for airborne element digitised phased array radar', *IEE radar 2002*, 15-17 October 2002, Edinburgh, Scotland, pp. 181-185

[518] PAPOULIS, A., *The Fourier integral and its applications*, McGraw-Hill, New York, 1962

[519] PADOS, D. A., TSAO, T., MICHELS, J. H., and WICKS, M. C.: 'Joint domain space-time adaptive processing with small training data sets', *IEEE RADARCON'98*, 11-14 May 1998, Dallas, TX, pp. 99-104

[520] PAGE, D., SCARBOROUGH, S., OWIRKA, G., and CROOKS, S.: 'Improving knowledge-aided STAP performance using past CPI data', *IEEE 2004 Radar Conference*, 26-29 April 2004, Philadelphia, PA

[521] PARK, H.-R., and WANG, H.: 'An adaptive polarization-space-time processor for radar system', *IEEE International Symposium on Antennas and Propagation*, Ann Arbor, MI, 1993, pp. 698-701

[522] PARK, H.-R., LI, J., and WANG, H.: 'Polarization-space-time domain generalized likelihood ratio detection of radar targets', *Signal Processing*, Vol. 41, 1995, pp. 153-164

[523] PARK, H.-R., LEE, C. J., and WANG, H.: 'Adaptive polarimetric processing for surveillance radar', *Digest IEEE Symposium on Antennas and Propagation*, Seattle, 20-24 June 1994, pp. 148–151

[524] PARK, S., and SARKAR, T.K.: 'A deterministic eigenvalue approach to space-time adaptive processing', *IEEE International Symposium on Phased Array Systems and Technology*, 15-18 Oct. 1996, Boston, MA, pp. 372–375

[525] PARK, S., and SARKAR, T. K.: 'Prevention of signal cancellation in an adaptive nulling problem', *Proc. IEEE National Radar Conference*, 1997, Syracuse, NY, pp. 191-195

[526] PARK, S., and SARKAR, T. K.: 'A modified generalized eigenvalue approach to space-time adaptive processing', *IEEE Antennas and Propagation Society International Symposium*, 21-26 June 1998, Piscataway, NJ, pp. 1292–1295

[527] PARK, Y. C., SUNG, H. J., and YOUN, D. H.: 'Adaptive eigenfilter implementation optimum MTI processor', *Electronics Letters*, Vol. 32, No. 18, 29 August 1996, pp. 1654-1655

[528] PARK, C. Y., SUNG, H. J., and YOUN, D. H.: 'An optimum space-time MTI processor for airborne radar', *Journal of Electrical Engineering and Information Science*, Vol. 3, No. 5, 1998, pp. 677-685

[529] PARKER, P.: 'Space-time autoregressive filtering for matched subspace STAP ', *IEEE Trans. AES*, Vol. 39, No. 2, April 2003, pp. 510-520

[530] PARKER, P.: 'Short CPI STAP for airborne radar ', *IEEE RADARCON 2002*, 22-25 April 2002, Long Beach, CA

[531] PARKER, P.: 'Performance of the space-time AR filter in non-homogeneous clutter ', *Intern. Conf. on Radar*, 3-5 September 2003, Adelaide, Australia, pp. 67-70

[532] PAULRAJ, A. J., and LINDSKOG, E.: 'Taxonomy of space-time processing for wireless networks', *IEE Proc. Radar, Sonar and Navigation*, Vol. 145, No. 1, Feb. 1998, pp. 25-31

[533] PECKHAM, C. D., HAIMOVICH, A. M., AYOUB, T. F., GOLDSTEIN, J. S., and REED, I. S.: 'Reduced rank STAP performance analysis', *IEEE Transactions AES*, Vol. 36, No. 2, April 2000, pp. 664-676

[534] PERRY, R. P., DIPIETRO, R. C., and FANTE, R. L.: 'SAR detection and imaging of moving targets', *EUSAR'98*, 25-27 May 1998, Friedrichshafen, Germany, pp. 579-582 (VDE Publishers)

[535] PETTERSON, M.: 'Focusing of moving targets in an ultra-wide band SAR GMTI system', *Proc. EUSAR2000*, 23-25 May 2000, Munich, Germany, pp. 837-840 (VDE Publishers)

[536] PETTERSON, M.: 'Detection of moving targets in wideband SAR', *IEEE Trans. AES*, Vol 40, No. 3, July 2004, pp. 780-796

[537] PICCIOLO, M.L., and GERLACH, K.: 'Median cascaded canceller using reiterative processing', *2003 IEEE Radar Conference*, May 5-8, 2003, Huntsville, AL, pp. 71-78

[538] PICCIOLO, M.L.: 'A robust loaded reiterative median cascaded canceller ', *IEEE 2004 Radar Conference*, 26-29 April 2004, Philadelphia, PA

[539] PILLAI, S. U., LEE, W. C., and GUERCI, J.: 'Multichannel space-time adaptive processing', *IEEE Proc. 29th ASILOMAR Conference on Signals, Systems and Computers*, Pacific Grove, CA, 30 Oct.–2 Nov. 1995, pp. 1183-1186

[540] PILLAI, S. U., GUERCI, J. R., and KIM, Y. L.: 'Generalized forward/backward subaperture smoothing techniques for sample starved STAP', *IEEE ICASSP'98*, 12-15 May 1998, Seattle, WA, pp. 2501-2504

[541] PILLAI, S. U., KIM, Y. L., and GUERCI, J. R .: 'Generalized Forward/backward subaperture smoothing techniques for sample starved STAP', *IEEE Trans. SP*, Vol. 48, No. 12, December 2000, pp. 3569-3574

[542] PILLAI, S. U., and PILLAI, S. R.: 'Projection Approach for STAP', *IEEE 2002 Radar Conference*, Long Beach, CA, 22-25 April 2002, pp. 397-401

[543] PILLAI, S.U., GUERCI, J.R., and PILLAI, S.R.: 'Performance analysis of data sample reduction techniques for STAP ', *IEEE Symposium on Phased Array Systems and Technology 2003*, 14-17 October 2003, Boston, pp. 565-570

[544] PILLAI, S.U., GUERCI, J.R., and Pillai, S.R.: 'Efficient tapering methods for STAP ', *IEEE 2004 Radar Conference*, 26-29 April 2004, Philadelphia, PA

[545] PILLAI, S.U., GUERCI, J.R., and PILLAI, S.R.: 'Wideband STAP for passive sonar', *IEEE OCEANS 2003*, 22-26 September 2003, San Diego, CA, pp. P2814-P2818

[546] PISARENKO, V. F.: 'The retrieval of harmonics from covariance functions', *Geophysics Journal of Royal Astronomical Society*, Vol. 33, 1973, pp. 347-366

[547] PUGH, M. L., and ZULCH, P. A.: 'RLSTAP algorithm development tool for analysis of advanced signal processing techniques', *IEEE ASILOMAR-29*, Pacific Grove, CA, 30 Oct.–2 Nov. 1996, pp. 1178-1183

[548] RABIDEAU, D.: 'Multistage cancellation of terrain scattered jamming and conventional clutter', *IEEE ICASSP'98*, 12-15 May 1998, Seattle, WA, pp. 2001-2004

[549] RABIDEAU, D. J., and KOGON, S. M.: 'A signal processing architecture for space-based GMTI radar', *IEEE National Radar Conference*, 20-22 April 1999, Boston, MA, pp. 96-101

[550] RABIDEAU, D. J., and STEINHARDT, A. O.: 'Improved adaptive clutter cancellation through data-adaptive training', *IEEE Trans. AES*, Vol. 35, No. 3, July 1999, pp. 879-891

[551] RABIDEAU, D. J.: 'Clutter and jammer multipath cancellation in airborne adaptive radar', *IEEE Transactions AES*, Vol. 36, No. 2, April 2000, pp. 566-583

[552] RANGASWAMY, M., and MICHELS, J. H.: 'Performance analysis of space-time adaptive processing in non-gaussian radar clutter backgrounds', *RADAR'99*, 17-21 May 1999, Brest, France

[553] RANGASWAMY, M., HIMED, B., MICHELS, J. H.: 'Performance analysis of the homogeneity detector for STAP applications ', *IEEE Radar Conference*, 2001, pp. 193-197

[554] RANGASWAMY, M.: 'An Overview of space-time adaptive processing for radar ', *Intern. Conf. on Radar*, 3-5 September 2003, Adelaide, Australia, pp. 45-50

[555] RANGASWAMY, M., MICHELS, J. H., and HIMED, B..: 'Statistical analysis of the non-homogeneity detector for STAP applications ', *Digital Signal Processing* 14, 2004, pp. 253-267

[556] REED, C.W., VAN WECHEL, R., JOHNSTON, I., BAEDER, B., and HOGAN, E.: '$FaSTAP^{TM}$: A scalable anti-jam architecture for GPS', *IEEE PLANS 2004*, 26-29 April 2004, Monterey, CA, pp. 496-502

[557] REED, I. S., MALLETT, J. D., and BRENNAN, L. E.: 'Rapid convergence rate in adaptive arrays', *IEEE Trans. AES*, Vol. AES-10, No. 6, November 1974, pp. 853-863

[558] REED, I.S., GAU, Y.L., and Truong, T. K.: 'CFAR detection and estimation for STAP radar', *IEEE Trans. AES*, Vol. 34, No. 3, July 1998, pp. 722-735

[559] REED, I., and GAU, Y. L.: 'A fast CFAR detection space-time adaptive processing algorithm', *IEEE Trans. SP*, Vol. 47, No. 4, April 1999, pp. 1151-1154

[560] REED, I. S., and GAU, Y. L.: 'Noncoherent summation of multiple reduced-rank test statistics for frequency-hopped STAP', *IEEE Trans. SP*, Vol. 47, No. 6, June 1999, pp. 1708-1711

[561] REES, H. D., and SKIDMORE, I.D.: 'Adaptive attenuation of clutter and jamming for array radar', *IEE Proc. Radar, Sonar and Navigation*, Vol. 145, No. 4, August, 1998, pp. 193-198

[562] RICHARDSON, P. G.: 'Analysis of the adaptive space time processing technique for airborne radar', *IEE Proc. Radar, Sonar, Navig.*, Vol 141, No. 4, August 1994, pp. 187-195

[563] RICHARDSON, P. G., and HAYWARD, S. D.: 'Adaptive space-time processing for forward looking radar', *Proc. IEEE International Radar Conference*, Alexandria, VA, 1995, pp. 629-634

[564] RICHARDSON, P. G.: 'Effects of manoeuvre on space-time adaptive processing performance', *Proc. IEE Radar'97*, 14-16 October 1997, Edinburgh, Scotland, pp. 285-289

[565] RICHARDSON, P. G.: 'Space-time adaptive processing for manoeuvring airborne radar', *IEE ECEJ*, special issue on STAP, Vol. 11, No. 1, February 1999, pp. 57-63

[566] RICHARDSON, P. G.: 'STAP covariance matrix structure and its impact on clutter plus jamming suppression solutions', *IEE Electronics Letters*, Vol. 37, No. 2, 18 January 2001, pp. 118-119

[567] RICHMOND, C. D.: 'The theoretical performance of a class of space-time detection and training strategies for airborne radar', *ASILOMAR Conf. on Signal, Systems and Computers*, Pacific Grove, CA, 1-4 Nov. 1998, pp. 1327-1331

[568] RIJACZEK, A. W.: *Principles of high-resolution radar*, Artech-House, Boston and London, 1996

[569] RINGEL, M. B.: 'An advanced computer calculation of ground clutter in an airborne pulse doppler radar', *NAECON '77 Record*, pp. 921-928

[570] RIVKIN, P., ZHANG, Y., and WANG H.: 'Spatial adaptive pre-suppression of wideband jammers in conjunction with STAP: a sideband approach', *Proc. International Conference on Radar (ICR'96)*, 8-10 October 1996, Beijing, China, pp. 439-443

[571] ROBEY, F. C., FUHRMANN, D. R., KELLY, E. J., and NITZBERG, R.: 'A CFAR adaptive matched filter detector', *IEEE Trans. AES*, Vol. 28, No. 1, January 1992, pp. 208-216

[572] ROBEY, F. C., and BARANOSKI, E. J.: 'Full space-time clutter covariance estimation', *Proc. IEEE ICASSP*, Detroit, 1995, pp. 1896-1899

[573] ROBINSON, J. W. C.: 'Multiple invariance spatio-temporal spectral analysis', *Proc. ICASSP99*, 15-19 May 1999, Phoenix, AZ, pp. 1532-1535

[574] RÖSSING, L., and SKUPIN, U.: 'Multi-channel SAR/MTI processing - simulation and real SAR data results', *Proc. EUSAR'96*, 26-29 March 1996, Königswinter, Germany, pp. 289-292 (VDE Publishers)

[575] RÖSSING, L., and ENDER, J. H. G.: 'The multi-channel SAR system 'AER-II - state of the program and new results', *German Radar Symposium 2000*, BerLIN, Germany, 11-12 October 2000

[576] ROMAN, J. R., DAVIS, D. W., and MICHELS, J. H.: 'Multichannel models for airborne phased array clutter', *Proc. IEEE National Radar Conference*, 1997, Syracuse, NY, pp. 72-75

[577] ROMAN, J. R., RANGASWAMY, M., DAVIS, D. W., ZHANG, Q., HIMED, B., and MICHELS, J. H.: 'Reduced rank STAP performance analysis', *IEEE Transactions AES*, Vol. 36, No. 2, April 2000, pp. 677-692

[578] ROUCOS, S. E., and CHILDERS, D. G.: 'A two-dimensional maximum entropy spectral estimator', *IEEE Trans. on Information Theory*, Vol. IT-26, No. 5, September 1980, pp. 554-560

[579] ROY, R., PAULRAJ, A., and KAILATH, T.: 'ESPRIT - a subspace rotation approach to estimation of parameters of cisoids in noise', *IEEE Trans. ASSP*, Vol. 34, No. 5, October 1986, pp. 1340-1342

[580] RUSS, J.A., CASBEER, D.W., and SWINDLEHURST, A. L.: 'STAP detection using space-time autoregressive filtering', *IEEE 2004 Radar Conference*, 26-29 April 2004, Philadelphia, PA

[581] SAMSON, J. R., GRIMM, D., MORILL, K., and ANDRESEN, T.: 'STAP performance on a PARAGON/TOUCHSTONE system', *IEEE National Radar Conference*, Ann Arbor, MI, 13-16 May 1996, pp. 315-320

[582] SANDER, W.: 'Beamforming with phased array antennas', *IEEE International Conference on Radar*, London, 1982, pp. 403-407

[583] SANDER, W.: 'Monitoring and calibration of active phased arrays', *Proc. IEEE International Radar Conference*, Arlington, VA, 1985, pp. 45-51

[584] SANYAL, P. K., BROWN, R. D., LITTLE, M. O., SCHNEIBLE, R. A., and WICKS, M. C.: 'Space-time adaptive processing bistatic airborne radar', *IEEE National Radar Conference*, 20-22 April 1999, Boston, MA, pp. 114-118

[585] SARKAR, T. K., NAGRAJA, S., and WICKS, M. C.: 'A deterministic direct data domain approach to signal estimation utilizing nonuniform and uniform 2-D arrays', *Digital Signal Processing*, Vol. 8, 1998, pp. 114-125

[586] SARKAR, T. K., WANG, H., PARK, S., ADVE, R., KOH, J., KIM, K., ZHANG, Y., WICKS, M., and BROWN, R. D.: 'A deterministic least-squares approach to space-time adaptive processing (STAP)', *IEEE Trans. AP*, Vol. 49, No. 1, January 2001, pp. 91-102

[587] SCHARF, L.: 'Detection, estimation and time series analysis', Addison-Wesley Publishing Company, 1991

[588] SCHINDLER, J. K., STEYSKAL, H., and FRANCHI, P.: 'Pattern synthesis for moving target detection with TECHSAT21 - a distributed space-based radar system', *IEE radar 2002*, 15-17 October 2002, Edinburgh, Scotland, pp. 375-379

[589] SCHLEHER, C., *MTI and pulsed doppler radar*, Artech House, Boston and London,1991

[590] SCHMIDT, R. O.: 'Multiple emitter location and signal parameter estimation', *IEEE Trans. on Antennas and Propagation*, Vol. AP-34, No. 3, March 1986, pp. 276-280

[591] SCHMIDT, R. O., and FRANKS, R. E.: 'Multiple source DF signal processing: an experimental system', *IEEE Trans. AP*, Vol AP-34, No. 3, March 1986, pp. 281-290

[592] SCHMIDT, R. O.: 'Multilinear array manifold interpolation', *IEEE Transactions on Signal Processing*, Vol. 40, No. 4, April 1992, pp. 857-866

[593] LI, P., SCHUMAN, H., MICHELS, J.H., and HIMED, B.: 'Space-time adaptive processing (STAP) with limited sample support ', *IEEE 2004 Radar Conference*, 26-29 April 2004, Philadelphia, PA

[594] SCOTT, I., and MULGREW, B.: 'The sparse LCMV beamformer design for suppression of Ground Clutter in Airborne Radar', *IEEE Transactions on Signal processing*, Vol. 43, No. 12, December 1995, pp. 2843-2851

[595] SEDWICK, R. J., HACKER, T. L., and MARAIS, K.: 'Performance analysis for an interferometric space-based GMTI system', *IEEE RADAR2000*, 8-12 May 2000, Alexandria, VA, pp. 689-694

[596] SEDYSHEV, Y. N., GORDIENKO, V. N., MAHER, J., and LYNCH, D.: 'The coherent bistatic radar with multi-stage space-time adaptive processing of signals and jamming', *IEEE RADAR2000*, 8-12 May 2000, Alexandria, VA, pp. 329–334

[597] SEE, C.- M. S., COWAN, C. F. N., and NEHORAI, A.: 'Spatio-temporal channel identification and equalisation in the presence of strong co-channel interference', *Signal Processing*, Vol. 78, 1999, pp. 127-138

[598] SEED, R. G., FLETCHER, A. S., and ROBEY, F. C.: ' STRAAP: Space-time-radiating array adaptive processing', *IEEE Conference on Phased Array Antennas*, Boston, MA, 2003, p. 136-141

[599] SELIKTAR, Y., WILLIAMS, D. B., and MCCLELLAN, J. H.: 'Evaluation of Partially Adaptive STAP Algorithms on the Mountain Top Data Set', *ICASSP'96*, Atlanta, GA, 1996, pp. 1169-1172

[600] SELIKTAR, Y., WILLIAMS, D. B., and HOLDER, E. J.: 'Beam-augmented space-time adaptive processing', *Proc. ICASSP99*, 15-19 May 1999, Phoenix, AZ, pp. 1581-1584

[601] SELIKTAR, Y., WILLIAMS, D. B., and HOLDER, E. J .: 'Beam-augmented STAP for joint clutter and jammer multipath mitigation', *IEE Proc. Radar, Sonar and Navigation*, Vol. 147, No. 5, October 2000, pp. 225-232

[602] SHAW, G. A., and McAULAY, R. J.: 'The application of multichannel signal processing to clutter suppression for a moving platform radar', *IEEE ASSP Spectrum Estimation Workshop 2*, Tampa, FL, November 1983, pp. 308-311

[603] SHERMAN, C. H., and CRON, B. F.: 'Spatial correlation functions for various noise models', *JASA*, Vol. 34, No. 11, November 1962, pp. 1732-1736

[604] SHNITKIN, H.: 'A unique JOINT STARS phased array antenna', *Microwave Journal*, January 1991, pp. 131-141

[605] SHNITKIN, H.: 'Joint Stars phased array antenna', *IEEE NAECON'1994*, Dayton, OH, 23-27 May 1994, pp. 142-150

[606] SHOR, S. W. W.: 'Adaptive technique to discriminate against coherent noise in a narrow-band system', *JASA*, Vol. 39, January 1966, pp. 74-78

[607] SHOWMAN, G.A., MELVIN, W.L., and ZYWICKI, D.J.: 'Application of the Cramér-Rao lower bound for bearing estimation to STAP performance studies ', *IEEE 2004 Radar Conference*, 26-29 April 2004, Philadelphia, PA

[608] SHOWMAN, G.A., MELVIN, W.L., and Belenkii, M.: 'Performance evaluation of two polarimetric STAP architectures', *2003 IEEE Radar Conference*, May 5-8, 2003, Huntsville, AL, pp. 59-65

[609] SKOLNIK, M., *Radar handbook*, 1st edn., McGraw-Hill, New York, 1970

[610] SKOLNIK, M., *Radar handbook*, 2nd edn., McGraw-Hill, New York, 1990

[611] SKOLNIK, M., *Introduction to Radar Systems*, 3rd edn., McGraw-Hill, New York, 2001

[612] SMITH, B.E., and HALE, T.B.: 'An analysis of the effect of windowing on selected STAP algorithms ', *IEEE 2004 Radar Conference*, 26-29 April 2004, Philadelphia, PA

[613] SORELIUS, J., MOSES, R. L., SÖDERSTRÖM, T., and SWINDLEHURST, A. L.: 'Effects of nonzero bandwidth on direction of arrival estimators in array signal processing', *IEE Proc. Radar, Sonar and Navigation*, Vol. 145, No. 6, Dec. 1998, pp. 317-324

[614] SOUMEKH, M., and HIMED, B.: 'SAR-MTI processing of multi-channel airborne radar measurements (MCARM) data ', *IEEE 2002 Radar Conference*, Long Beach, CA, 22-25 April 2002, pp. 24-28

[615] SOUMEKH, M., and HIMED, B.: 'Multi-channel signal subspace processing methods for SAR-MTI', *2003 IEEE Radar Conference*, May 5-8, 2003, Huntsville, AL, pp. 126-132

[616] SPAFFORD, L. J.: 'Optimum radar signal processing in clutter', *IEEE Trans. on Information Theory*, Vol. IT-14, No. 5, September 1968, pp. 734-743

[617] STEINHARDT, A. O., and PULSONE, N. B.: 'Subband STAP processing, the fifth generation', *IEEE SAM Workshop*, 15-16 March 2000, Cambridge, MA, pp. 1-6

[618] STEINHARDT, A., and GUERCI, J.: 'STAP for radar: what works, what doesn't, and what's in store', *IEEE 2004 Radar Conference*, 26-29 April 2004, Philadelphia, PA

[619] STOICA, P., HÄNDEL, P., and SÖDERSTRÖM, T.: 'Study of Capon method for array signal processing', *Circuits, Systems, Signal Processing*, Vol. 14, No. 6, 1995, pp. 749-770

[620] SU, J., and ZHOU, Y. Q.: 'Adaptive clutter suppression for airborne array radars using clutter subspace approximation', *RADAR 92*, Brighton, UK, 1992, pp. 155-158

[621] SU, T., and BAO, Z.: 'Parallel processor in space-time adaptive signal processing ', *CIE International Conference on Radar*, 15-18 October 2001, pp. 783-787

[622] SUN, X., PENG, Y.-N., and LIU, Q. G.: 'Suboptimum approach for adaptive clutter suppression in airborne phased array radar', *IEE Electronic Letters*, No. 17, August 1993, pp. 1551-1553

[623] SUN, X., PENG, Y.-N., LIU, Q. G., and LU D.-J., 'Several ways for clutter suppression in airborne radar', *Proc. IEEE Symposium on Antennas and Propagation*, Seattle, 20-24 June 1994, pp. 152-155

[624] SUNG, H. J., PARK, Y. C., and YOUN, D. H.: 'An optimum space-time MTI processor for airborne radar', *IEEE ICASSP*, 12-15 May 1998, Seattle, WA, pp. 2033-2036

[625] SURESH BABU, B. N., TORRES, J. A., and LAMENSDORF, D.: 'Space-time adaptive processing for airborne phased array radar', *Proc. of the Conf. on Adaptive Antennas*, 7-8 November 1994, Melville, New York 11747, pp. 71-75

[626] SURESH BABU, B. N., TORRES, J. A., and MELVIN, W.L.: 'Processing and evaluation of multichannel airborne radar measurements (MCARM) measured data', *IEEE International Symposium on Phased Array Systems and Technology*, 15-18 Oct. 1996, Boston, MA, pp. 395-399

[627] SURESH BABU, B. N., TORRES, J. A., and GUELLA, T. P. 'Impact of near-field scattering on multichannel airborne radar measurements (MCARM)', *Proc. IEEE National Radar Conference*, 1997, Syracuse, NY, pp. 227-231

[628] SURESH BABU, K. V., YOGANANDAM, Y., and REDDY, V. U.: 'Adaptive estimation of eigensubspace and tracking the directions of arrival', *Signal Processing*, Vol. 68, 1998, pp. 317-339

[629] SWINDLEHURST, A. L., and PARKER, P.: 'Parametric clutter rejection for space-time adaptive processing', *ASAP 2000 Workshop*, , MIT Lincoln Laboratory, Lexington, MA, 13-14 March 2000, pp. 7-12

[630] SWINGLER, D. N.: 'A low complexity MVDR beamformer for use with short observation times', *IEEE Trans. SP*, Vol. 47, No. 4, April 1999, pp. 1154-1160

[631] SYCHEV, M. I.: 'Space-time radio signal processing based on the parametrical spectral analysis', *PIERS Workshop on Advances in Radar Methods*, 20-22 July 1998, Baveno, Italy, pp. 234-236

[632] SYCHEV, M. I.: 'Space-time radio processing based on the parametrical spectral analysis', *RADAR'99*, 17-21 May 1999, Brest, France

[633] TAM, W., and FAUBERT, D.: 'Displaced phase centre antenna clutter suppression in space-based radar applications', *Proc. IEE Radar'87*, IEE conference publication No. 281, pp. 385-389

[634] TANG, Z., WANG, Y., and ZHU, M.: 'STAP in conformal phased array airborne radar ', *CIE International Conference on Radar*, 15-18 October 2001, pp. 952-955

[635] TAYLOR, T. T.: 'Design of line source antennas for narrow beamwidth and low sidelobes', *IEEE Trans. AP*, Vol. 3, January 1955, pp. 16-28.

[636] TAYLOR, T. T.: 'Design of circular apertures for narrow beamwidth and Low Sidelobe', *IRE Trans. AP*, Vol. 8, 1960, pp. 17-22

[637] TECHAU, P. M., GUERCI, J. R., SLOCUMB, T. H., and GRIFFITHS, L. J.: 'Performance bounds for interference mitigation in radar systems', *IEEE National Radar Conference*, 20-22 April 1999, Boston, MA, pp. 12-17

[638] TECHAU, P. M.: 'Effects of receiver filtering on hot clutter mitigation', *IEEE National Radar Conference*, 20-22 April 1999, Boston, MA, pp. 84-89

[639] TECHAU, P. M., BERGIN, J. S., and GUERCI, J. R.: 'Effects of clutter motion on STAP in a hergeneous environment', *IEEE Radar Conference*, 2001, pp. 204-209

[640] TEITELBAUM, K.: 'Mapping Space-Time Adaptive Processing Onto Massively Parallel Processors', *Proc. of the Conf. on Adaptive Antennas*, 7-8 November 1994, Melville, New York 11747, pp. 113-119

[641] TEIXEIRA, C.M., BERGIN, J.S., and TECHAU, P.: 'Adaptive thresholding of non-hmogeneity detection for STAP applications ', *IEEE 2004 Radar Conference*, 26-29 April 2004, Philadelphia, PA

[642] THOMPSON, E. A., and PASALA, K. M.: 'Space-time adaptive processing of curved antenna arrays', *IEEE Antennas and Propagation Society International Symposium*, 13-18 July 1997, Montreal, pp. 2430-2433

[643] THOMPSON, A. A., and LIVINGSTONE, C. E.: 'Moving target performance for RADARSAT-2', *Proc. IGARSS2000 (CD-ROM)*, 24-28 July 2000, Honolulu, HI, USA

[644] TIMMONERI, L., PROUDLER, I. K., FARINA, A., and McWHIRTER, J. G.: 'QRD-based MVDR algorithm for adaptive multipulse antenna array signal processing', *IEE Proc.-Radar, Sonar, Navigation*, Vol. 141, No. 2, April 1994, pp. 93-102

[645] TINSTON, M.A., OGLE, W.C., PICCIOLO, M.L., and GOLDSTEIN, J.S.: 'Classification of training data with reduced-rank generalized inner product ', *IEEE 2004 Radar Conference*, 26-29 April 2004, Philadelphia, PA

[646] TITI, G. W.: 'An Overview of the ARPA/NAVY mountaintop program', *IEEE Adaptive Antenna Symposium*, Melville, NY, 7-8 November 1994

[647] TITI, G. W., MARSHALL, D. F.: 'The ARPA/NAVY mountaintop program: adaptive signal processing for airborne early warning radar', *ICASSP'96*, Atlanta, GA, 1996, pp. 1165-1168

[648] TOBIN, M.: 'Real-time simultaneous SAR/GMTI in a tactical airborne environment', *Proc. EUSAR'96*, 26-28 March 1996, Königswinter, Germany, pp. 63-66 (VDE Publishers)

[649] TOMLINSON, Ph. G.: 'Modeling and analysis of monostatic/bistatic space-time adaptive processing for airborne and space-based radar', *IEEE National Radar Conference*, 20-22 April 1999, Boston, MA, pp. 102-107

[650] TSAKALIDES, P., and NIKIAS, C. L.: 'Space-time adaptive processing in stable impulse interference', *30th Asilomar Conference*, 3-6 Nov. 1996, Pacific Grove, CA, pp. 389-393

[651] TSAKALIDES, P., and NIKIAS, C. L.: 'Robust space-time adaptive processing (STAP) in non-gaussian clutter environments', *IEE Proc. Radar, Sonar and Navigation*, Vol. 146, No. 2, April 1999

[652] TSAKALIDES, P., RASPANTI, R., and NIKIAS, C.: 'Angle/Doppler estimation in heavy-tailed clutter backgrounds', *IEEE Trans. AES*, Vol. 35, No. 2, April 1999, pp. 419-436

[653] TSANDOULAS, G. N.: 'Unidimensionally scanned phased arrays', *IEEE Trans. on Antennas and Propagation*, Vol. AP-28, No. 1, November 1973, pp. 1383-1390

[654] TSAO, T., HIMED, B., and MICHELS, J. H.: 'Effects of interference rank estimation on the detection performance of rank reduced STAP algorithms', *IEEE RADARCON'98*, 11-14 May 1998, Dallas, TX, pp. 147-152

[655] TUFTS, D. W., and KUMARESAN, R.: 'Singular value decomposition and improved frequency estimation using linear prediction', *IEEE Trans. ASSP*, Vol. 30, N0. 4, August 1982, pp. 671-675

[656] TUFTS, D. W., KUMARESAN, R., and KIRSTEINS, I.: 'Data adaptive signal estimation by singular value decomposition of a data matrix', *Proc. IEEE*, Vol. 70, No. 6, June 1982, pp. 684-685

[657] TUFTS, D. W., and KUMARESAN, R.: 'Estimation of frequencies of multiple sinuoids: making linear prediction perform like maximum likelihood', *Proc. IEEE*, Vol. 70, No. 9, September 1982, pp. 975-989

[658] ULABY, F.T., and DOBSON, M.C.: 'Handbook of radar scattering', Artech House, Boston, MA, 1989

[659] URICK, R. J., *Principles of underwater sound*, McGraw-Hill, 2nd edition, 1975

[660] VARADARAJAN, V., and KROLIK, J.L.: 'Space-time interpolation for adaptive arrays with limited training data', *IEEE ICASSP*, 6-10 April 2003, Hong Kong, pp. v-353 - v-356

[661] VAN KEUK, G., and BLACKMAN, S. S.: 'On phased-array radar tracking and parameter control', *IEEE Trans. AES*, Vol. 29, No. 1, January 1993, pp. 186-194

[662] VAN TREES, H. L., *Detection, estimation and modulation theory, Part III*, John Wiley & Sons, Inc., New York, 1971

[663] VOLES, R.: 'New approach to clutter locking', *Proc. IEE*, Vol. 120, No. 11, 1973, pp. 1383-1392

[664] WALKE, C.: 'The impact of inter-cell interference cancellation on the performance of SD/TD/CDMA systems ', in: KLEMM, R. ed., *Applications of space-time adaptive processing*, IEE publishers, London, UK, 2004

[665] WALKE, C., REMBOLD, B.: 'Joint detection and joint pre-distortion techniques for SD/TD/CDMA systems ', *FREQUENZ*, 55, 2001, 7-8, pp. 1-10

[666] WANG, H. S. C.: 'Mainlobe clutter cancellation by DPCA for space-based radars', *IEEE Aerosp. Applications Conference Digest*, Crested Butte, CO, February 1991, pp. 1-28

[667] WANG, H., and CAI, L.: 'On the implementation of the optimum spatial-temporal processor for airborne surveillance systems', *Proc. CIE International Conference on Radar*, 22-24 Oct. 1991, Beijing, China, 1991, pp. 365-368

[668] WANG, H., and CAI, L.: 'On adaptive spatial-temporal processing for airborne surveillance radar systems', *IEEE Trans. AES*, Vol. 30, No. 3, July 1994, pp. 660-670

[669] WANG, H., ZHANG, Y., and WICKS, M.: 'Performance evaluation of space-time processing for adaptive array radar in coherent repeater jamming environments', *Proc. of the Conf. on Adaptive Antennas*, 7-8 November 1994, Melville, New York 11747, pp. 65-69

[670] WANG, H., ZHANG, Y., and ZHANG, Q.: 'A view of current status of space-time processing algorithm research', *Proc. IEEE International Radar Conference*, Alexandria, VA, 1995, pp. 635-640

[671] WANG, H., ZHANG, Y., and ZHANG, Q.: 'An improved and affordable space-time adaptive processing approach', *Proc. International Conference on Radar (ICR'96)*, Beijing, China, 8-10 October 1996, pp. 72-77

[672] WANG, H.: 'An overview of space-time adaptive processing for airborne radars', *Proc. International Conference on Radar (ICR'96)*, 8-10 October 1996, Beijing, China, pp. 789-794

[673] WANG, H., ZHANG, Y., and ZHANG, Q.: 'Lessons learned from recent STAP experiments', *Proc. International Conference on Radar (ICR'96)*, 8-10 October 1996, Beijing, China, pp. 761-765

[674] WANG, H., and ZHANG, Y.: 'A STAP configuration for airborne surveillance radars', *Proc. IEEE National Radar Conference*, 1997, Syracuse, NY, pp. 217-221

[675] WANG, L., WANG, Y., DAI, G., and ZHEN, S.: 'Frequency bandpass disparity of space-time adaptive array system ', *CIE International Conference on Radar*, 15-18 October 2001, pp. 765-769

[676] WANG, X. G.: 'Suboptimal method for time-space signal processing in airborne radar', *Electronic Letters*, Vol. 29, No. 16, August 1993

[677] WANG, Y.-L., PENG, Y.-N., and BAO, Z.: 'The practical approaches to space-time signal processing for airborne array Radar', *Proc. IEEE International Radar Conference*, Alexandria, VA, 1995, pp. 125-130

[678] WANG, Y., and PENG, Y.-N.: 'Space-time joint processing method for simultaneous clutter and jammer rejection in airborne radar', *Electronics Letters*, Vol. 32, No. 3, February 1996, pp. 258-259

[679] WANG Y., BAO Z., and LIAO G.: 'Three united configurations on adaptive spatial-temporal processing for airborne surveillance radar systems', *Proc. ICSP'93*, Beijing, October 1993, pp. 381-386

[680] WANG, Y., and PENG, Y.-N.: 'An effective method for clutter and jamming rejection in airborne phased array radar', *IEEE International Symposium on Phased Array Systems and Technology*, 15-18 Oct. 1996, Boston, MA, pp. 349-352

[681] WANG, Y., and PENG, Y.-N.: 'Configuration and performance analysis of space-time adaptive processor for airborne radar', *Proc. IEEE National Radar Conference*, 1997, Syracuse, NY, pp. 343-348

[682] WANG, Y.-L., PENG, Y.-N., and BAO, Z.: 'Space-time adaptive processing for airborne radar with various array orientations', *IEE Proc. Radar, Sonar and Navigation*, Vol. 144, No. 6, December 1997, pp. 330-340

[683] WANG, Y.-L., BAO, Z., and PENG, Y.-N.: 'STAP with medium PRF mode for non-sidelooking airborne radar', *IEEE Transactions AES*, Vol. 36, No. 2, April 2000, pp. 609-620

[684] WANG, Y., and CHEN, J.: 'Robust STAP approach in nonhomogeneous clutter environments ', *CIE International Conference on Radar*, 15-18 October 2001, pp. 753-757

[685] WANG Z., and BAO, Z.: 'Airborne clutter suppression without desired Doppler frequency', *Proc. SITA'87*, 19-21 November 1987, Tokyo, Japan, pp. EE2-5-1

[686] WANG Z., and BAO Z.: 'A novel algorithm for optimum and adaptive airborne phased arrays', *Proc. SITA'87*, 19-21 November 1987, Tokyo, Japan, pp. EE2-4-1

[687] WANG, Y.-L., CHEN, J.-W., BAO, Z., and PENG, Y.-N.: 'Robust space-time adaptive processing for airborne radar in no-homogeneous clutter environments ', *IEEE Trans. AES*, Vol. 39, No. 1, January 2003, pp. 70-81

[688] WANG, W., LI, S., and MAO, S.: 'Modified joint domain localized reduced-rank STAP ', *CIE International Conference on Radar*, 15-18 October 2001, pp. 783-787

[689] WARD, C. R., ROBSON, A. J., HARGRAVE, P. J., and McWHIRTER, J. G.: 'Application of a systolic array to adaptive beamforming', *IEEE Proceedings*, Vol. 131, Pt. F, No. 6, October, 1984, pp. 638-645

[690] WARD, J.: 'Space-time adaptive processing for airborne radar', *Technical Report No. 1015*, MIT Lincoln Laboratory, December 1994

[691] WARD, J.: 'Space-Time Adaptive Processing for Airborne Radar', *ICASSP95*, 8-12 May 1995, Detroit, MI, pp. 2809-2812

[692] WARD, J., and STEINHARDT, A. O.: 'Multiwindow post-Doppler space-time adaptive processing', *Proc. of the Seventh Workshop on Statistical Signal and Array Processing*, Quebec City, Canada, 26-29 June 1994, pp. 461-464

[693] WARD, J.: 'Cramer-Rao bounds for target angle and Doppler sstimation with space-time adaptive processing radar', *Proc. 29th ASILOMAR Conference on Signals, Systems and Computers*, 30 October–2 November 1995, pp. 1198-1203

[694] WARD, J.: 'Maximum likelihood angle and velocity estimation with space-time adaptive processing radar', *Proc. ASILOMAR 96*, 4-6 November 1996

[695] WARD, J., BARANOSKI, E. J., and GABEL, R. A.: 'Adaptive processing for airborne surveillance radar', *30th Asilomar Conference*, 3-6 Nov. 1996, Pacific Grove, CA, pp. 566-571

[696] WARD, J., and HATKE, G. F.: 'An efficient rooting algorithm for simultaneous angle and Doppler estimation with space-time adaptive processing radar', *ASILOMAR97*, 2-5 November 1997, Pacific Grove, CA

[697] WARD, J.: 'Space-time adaptive processing with sparse antenna arrays', *ASILOMAR98*, 2-4 November 1998, Monterey, CA, USA

[698] WAX, M., and ANU, Y.: 'Performance analysis of the minimum variance beamformer', *IEEE Transactions on Signal Processing*, Vol. 44, No. 4, April 1996, pp. 928-937

[699] WAX, M., and ANU, Y.: 'Performance analysis of the minimum variance beamformer in the presence of steering vector errors', *IEEE Transactions on Signal Processing*, Vol. 44, No. 4, April 1996, pp. 938-947

[700] WEBER, P., HAYKIN, S., and GRAY, R.: 'Airborne pulse-Doppler radar: false alarm control', *IEE Proc.*, Vol 134, Pt. F, No. 2, April 1987

[701] WEHNER, D.: *High Resolution Radar*, Artech-House, Norwood, MA, 2nd edn, 1995

[702] WEINER, D. D., CAPRARO, G. T., and WICKS, M. C.: 'An approach for utilizing known terrain and land feature data in estimation of the clutter covariance matrix', *IEEE RADARCON'98*, 11-14 May 1998, Dallas, TX, pp. 381-386

[703] WEIPPERT, M.E., Hiemstra, J.D., GOLDSTEIN, J.S., and GARREN, D.A.: 'Efficient implementation of the multistage Wiener filter for multiple beam applications', *IEEE Symposium on Phased Array Systems and Technology 2003*, 14-17 October 2003, Boston, pp. 152-157

[704] WEIPPERT, M.E., HIEMSTRA, J.D., GOLDSTEIN, J.S., ZOLTOWSKI, M.D., and REED, I.: 'Signal dependent reduced-rank multibeam processing ', *Intern. Conf. on Radar*, 3-5 September 2003, Adelaide, Australia, pp. 51-56

[705] WHITE, W. D.: 'Wideband interference cancellation in adaptive sidelobe cancellers', *IEEE Trans. AES*, Vol. 19, No. 6, November 1983, pp. 915-925

[706] WICKS, M., WANG, H., and CAI, L.: 'Adaptive array processing for airborne radar', *IEE RADAR 92*, Brighton, UK, 1992, pp. 159-162

[707] WICKS, M., PIWINSKI, D., and LI, P.: 'Space-time adaptive processing in modern electronic warfare environments', *Proc. IEEE International Radar Conference*, Alexandria, VA, 1995, pp. 609-613

[708] WICKS, M. C., MELVIN, W. L., and CHEN, L.: 'An efficient architecture for nonhomogeneity detection in space-time adaptive processing airborne early warning radar', *Proc. IEE Radar'97*, 14-16 October 1997, Edinburgh, Scotland, pp. 295-299

[709] WIDROW, B., MANTEY, P. E., GRIFFITHS, L. J., and GOODE, B. B.: 'Adaptive antenna systems', *Proc. IEEE*, Vol. 55, No. 12, December 1967, pp. 2143-2159

[710] WIDROW, B., and McCOOL, J. M.: 'A comparison of adaptive algorithms based on the methods of steepest descent and random search', *IEEE Transactions on Antennas and Propagation*, Vol. 24, No. 5, September 1976, pp. 615-637

[711] WILDEN, H., and ENDER, J.: 'The crow's nest antenna – experimental results', *IEEE International Radar Conference*, Arlington, VA, 1990, pp. 280-285

[712] WIRTH, W. D.: 'Feinauflösung im Azimut bei geradlinig bewegten Radarzielen (High azimuth resolution of targets with straight flight path)', *NTZ*, No. 12, 1973, pp. 539-541

[713] WIRTH, W. D.: 'Fast and efficient target search with phased array radars', *Proc. IEEE International Radar Conference*, Arlington, VA, 1975, pp. 198-203

[714] WIRTH, W. D.: 'Suboptimal suppression of directional noise by a sensor array before beamforming', *IEEE Transactions on Antennas and Propagation*, September 1976, pp. 741-744

[715] WIRTH, W. D.: 'High resolution in azimuth for radar targets moving on a straight line', *IEEE Transactions on Aerospace and Electronic Systems*, Vol. 16, No. 1, January 1980, pp. 101-104

[716] WIRTH, W. D.: 'Signal Processing for target detection in experimental phased array radar ELRA', *IEE Proc. Pt. F*, No. 5, 1982, pp. 311-316

[717] WIRTH, W. D.: 'Phased array radar with solid state transmitter', *International Conference on Radar*, Paris, 1984, pp. 141-145

[718] WIRTH, W. D.: 'Omnidirectional low probability of intercept radar', *International Conference on Radar*, Paris, 1989, pp. 25-30

[719] WIRTH, W. D.: 'Energy saving by coherent sequential detection of radar signals with unknown Doppler shift', *IEE Proc. Radar, Sonar and Navigation*, Vol. 142, No. 3, June 1995, pp. 145-152

[720] WIRTH, W. D.: 'Long term coherent integration for a floodlight radar', *IEEE International Radar Conference*, Alexandria, VA, 1995, pp. 698-703

[721] WIRTH, W. D.: 'Sequential detection for energy saving and management', *IEEE International Symposium on Phased Array Systems and Technology (ISPAST)*, Boston, MA, 1996

[722] WIRTH, W. D., *Radar techniques using array antennas*, IEE Publishers, Stevenage, Herts., UK, 2001

[723] WIRTH, W.D.: 'Direction of arrival estimation with multipath scattering by space-time processing ', *SIGNAL Processing*, 84 (2004), pp. 1677-1688

[724] WIRTH, W. D., and NICKEL, U.: 'Beamforming and array processing with active arrays', *IEEE International Symposium on Antennas and Propagation*, Ann Arbor, MI, 1993, pp. 1540-1543

[725] WORMS, J.: 'Superresolution methods and model errors', *Int. Conf. on Radar*, Paris, May 1994, pp. 454-459

[726] WORMS, J.G.: 'Spatial superresolution with conformal array antennas', *IEEE Radar 2000*, 8-12 May, 2000, Alexandria, VA, pp. 723-728

[727] WRIGHT, P. J., and WELLS, M.: 'STAP for airborne target detection using a space-based phased-arrays radar', *IEE ECEJ*, special issue on STAP, Vol. 11, No. 1, February 1999, pp. 23-27

[728] WU, R., and BAO, Z.: 'Training method in space-time adaptive processing with strong generalizing ability', *Proc. IEEE International Radar Conference*, Alexandria, VA, 1995, pp. 603-608

[729] WU, R., BAO Z., and ZHANG Y.: 'Subarray level adaptive spatial-temporal processing for airborne radars', *Proc. ICSP'93*, October 1993, Beijing, China, pp. 391-395

[730] WU, R., BAO, Z., and MA, Y.: 'Control of peak sidelobe level in adaptive arrays', *IEEE Trans. AP*, Vol. 44, No. 10, October 1996, pp. 1341-1347

[731] WU, S., and LI, Y.: 'Adaptive channel equalization for space-time adaptive processing', *Proc. IEEE International Radar Conference*, Alexandria, VA, 1995, pp. 624-628

[732] WU, R., and BAO, Z.: 'Array pattern distortion and remedies in space-time adaptive processing for airborne radar', *IEEE Trans. AP*, Vol. 46, No. 7, July 1998, pp. 963-970

[733] XIONG, J., LIAO G., and WU S.: 'Recursive algorithm of adaptive weight extraction of space-time signal processing for airborne radars', *Proc. International Conference on Radar (ICR'96)*, 8-10 October 1996, Beijing, China, pp. 86-90

[734] XU, G., ZHA, H., GOLUB, G., and KAILATH, Th.: 'Fast algorithms for updating signal spaces', *IEEE Trans. Circuits and Systems*, II. Analog and Digital Processing, Vol. 41, No. 8, August 1994, pp. 537-549

[735] YONG, H., PENG, Y., and WANG, X.: 'Airborne adaptive MTI scheme with preventing the whitening of the target', *IEEE Systems Magazine*, July 1999, pp. 19-21

[736] ZATMAN, M., and MARSHALL, D.: 'Forwards-backwards averaging for adaptive beamforming and STAP', *ICASSP'96*, Atlanta, GA, USA, 1996

[737] ZATMAN, M.: 'The Effect of Bandwidth on STAP', *IEEE Int. Symp. on Antennas and Propagation*, Baltimore, MD, 21-26 July 1996, pp. 1188-1191

[738] ZATMAN, M.: 'Properties of the Hung-Turner projections and their relationship to the eigencanceller', *30th ASILOMAR Conference on Signals, Systems and Computers*, 3-6 Nov. 1996, Pacific Grove, CA, pp. 1176-1180

[739] ZATMAN, M., and BARANOSKI, E.: 'Time delay steering architectures for space-time adaptive processing', *IEEE Antennas and Propagation Society International Symposium*, 13-18 July 1997, Montreal, pp. 2426-2429

[740] ZATMAN, M.: 'Circular array STAP', *IEEE National Radar Conference*, 20-22 April 1999, Boston, MA, pp. 108-113

[741] ZATMAN, M.: 'Circular array STAP', *IEEE Transactions AES*, Vol. 36, No. 2, April 2000, pp. 518-527

[742] ZATMAN, M.: 'Radar resource management for UESA ', *IEEE RADARCON 2002*, 22-25 April 2002, Long Beach, CA

[743] ZEGER, A. E., and BURGESS, L. R.: 'An adaptive AMTI radar antenna array', *IEEE NAECON '74 Record*, pp. 126-132

[744] Zekavat, S.A.R., and Abdi, A.: 'Statistical clutter modeling for airborne radar', *IEEE International Conference on Information Communications and Sognal Processing*, 9-12 September, 1997, Singapore, pp. 466-470

[745] ZHANG, Q., and MIKHAEL, W. B.: 'Estimation of the clutter rank in the case of subarraying for space-time adaptive processing', *Electronics Letters*, Vol. 33, No. 5, 27 February 1997, pp. 419-420

[746] ZHANG, Y., and WANG, H.: 'Further results of $\Sigma\Delta$-STAP approach to airborne surveillance radars', *Proc. IEEE National Radar Conference*, 1997, Syracuse, NY, pp. 337-342

[747] ZHANG, Y., and WANG, H.: 'Further study on space-time adaptive processing with sum and difference beams', *IEEE Antennas and Propagation Society International Symposium*, 13-18 July 1997, Montreal, pp. 2422-2425

[748] ZHANG, Y., MAHER, J., CHANG, H., and WANG, H.: 'Adaptive pre-suppression of jammers for STAP-based airborne surveillance radars', *IEEE RADARCON'98*, 11-14 May 1998, Dallas, TX, pp. 32-37

[749] ZHANG, Y., Hajjari, A., Adzima, L., and Himed, B.: 'Adaptive beam-domain processing for spacebased radars', *IEEE 2004 Radar Conference*, 26-29 April 2004, Philadelphia, PA

[750] ZHOU, L. J., and FENG, S.: 'Detection and imaging of moving targets with SAR based on space-time two-dimensional signal processing', *Chinese Journal of Electronics*, Vol. 5, No. 1, July 1996, pp. 82-87

[751] ZIEGENBEIN, J.: 'Spectral analysis using the Karhunen-Loève transform', *Proc. ICASSP'79*, 1979, Washington DC, pp. 182-185

[752] ZULCH, P., DAVIS, M., and ADZIMA, L.: 'The earth rotation effect on a LEO L-band GMTI SBR and mitigation strategies ', *IEEE 2004 Radar Conference*, 26-29 April 2004, Philadelphia, PA

Glossary

a	noise-to-clutter ratio
$\mathbf{a}$	auxiliary channel vector
ABF	adaptive beamforming
ACP	auxiliary channel processor
AEP	auxiliary eigenvector processor
AEW	airborne early warning
AMSAR	Airborne Multi-role Solid-state Active array Radar
ASEP	auxiliary sensor/echo processor
ASFF	auxiliary sensor FIR filter processor
AWACS	Airborne Warning and Control System
A_r	received signal amplitude
B	bandwidth
B	number of beamformer elements
B_c	clutter bandwidth
B_D	Doppler bandwidth
$\mathbf{b}$	beamformer vector
b_k	beamformer weights
β	look angle relative to array axis
B_s	system bandwidth
c	velocity of light
$\mathbf{c}$	vector of clutter echoes
$\mathbf{c}_F$	vector of clutter spectral components
$\mathbf{c}_T$	transformed vector of clutter echoes
C	number of space-time channels
CALC	constrained averaged likelihood ratio
CFAR	constant false alarm rate
CRB	Cramér–Rao bound
CNR	clutter-to-noise ratio
Coho	coherent oscillator
CPI	coherent processing interval
d	sensor spacing
d_s	subarray displacement
DARPA	Defence advanced research and projects agency
$D(\varphi)$	horizontal sensor directivity pattern
δ_R	direction of receiver motion

δ_T	direction of transmitter motion
Δ_R	width of range bin
$D(\theta)$	vertical sensor directivity pattern
DFB	Doppler filter bank
DFT	discrete Fourier transform
DPCA	displaced phase centre antenna
d_x	sensor spacing in x-direction
d_y	sensor spacing in y-direction
d_z	sensor spacing in z-direction
$\mathbf{e}_i$	unit vector (i-th column of unit matrix)
$E\{\}$	expectation
$E(t)$	envelope of transmitted waveform
φ	azimuth
F	normalised target Doppler frequency
$\mathbf{F}$	DFT matrix, Doppler filter bank
$\mathbf{f}$	vector of DFT or DFB output signals
FAP	(optimum) fully adaptive processor
f_c	carrier frequency
f_D	Doppler frequency
φ_c	clutter angle of arrival
φ_L	look angle
$\Phi_m(v_p, \varphi)$	temporal phase term
$\Psi_i(\varphi)$	spatial phase term
FDFF	frequency domain FIR filter
FDSP	frequency dependent spatial processing
FFT	fast Fourier transform
FIR	finite impulse response
FL	forward looking
f_{Ny}	Nyquist frequency
f_{PR}	pulse repetition frequency, PRF
$FREQUENCY$	frequency associated with TIME (fast radar time)
f_r	relative clutter Doppler frequency, normalised by maximum clutter Doppler frequency
f_s	spatial frequency
$G(.)$	transmit directivity pattern
GLSC	generalised sidelobe canceller
GMTI	ground moving target indication
H	platform altitude above ground
$\mathbf{I}$	identity matrix
IF	improvement factor
IF	intermediate frequency
IIR	infinite impulse response
int$\{\}$	next integer number
ISAR	inverse synthetic aperture radar

rmj	$\sqrt{-1}$
j	jammer index
j_i	i-th jammer signal component
j_{ik}	ik-th element of the Fisher information matrix
$\mathbf{j}$	jammer vector
J	number of jammers
$\mathbf{J}$	Fisher information matrix
K	reduced number of antenna channels
KASSPER	Knowledge aided sensor signal processing and reasoning
L	number of FIR filter taps, temporal dimension of space-time FIR filter
$L(.)$	reflectivity function
LMS	least mean square
λ	wavelength
M	number of echo pulses or temporal samples
MEM	maximum entropy method
MFR	matched filter response
ML	maximum likelihood
MTI	moving target indicator
MVE	minimum variance estimator
μ	factor determining the number of Doppler channels
N	number of antenna elements
$\mathbf{n}$	noise vector
$\mathbf{n}_F$	noise spectrum vector
$\mathbf{n}_T$	transformed noise vector
$\mathbf{N}$	noise covariance matrix
N_e	number of eigenvalues
N_t	number of transmit elements
NIR	noise-to-interference ratio
$\mathbf{O}$	zero matrix
OAP	optimum adaptive processor
OPP	orthogonal projection processor
OUS	overlapping uniform subarray configuration
ω_c	carrier frequency
ω_D	angular Doppler frequency
P_c	clutter power
P_D	detection probability
P_F	false alarm probability
PDR	pulse Doppler radar
P_j	jammer power
P_n	noise power
P_r	received signal power
PRI	pulse repetition interval
PRF	pulse repetition frequency

P_s	signal power
PSF	point spreading function
ψ	crab angle
P_t	transmitted power
$\mathbf{q}$	interference vector
Q	number of data vectors, number of normal modes
$\mathbf{Q}$	interference + noise covariance matrix
$\mathbf{Q}_\mathrm{c}$, $\mathbf{Q}^{(c)}$	clutter covariance matrix
$\mathbf{Q}_\mathrm{i}$	interference covariance matrix
$\mathbf{Q}_\mathrm{j}$	jammer covariance matrix
$\mathbf{Q}_\mathrm{n}$	noise covariance matrix
$\mathbf{Q}_\mathrm{s}$	signal covariance matrix
$\mathbf{Q}_\mathrm{F}$	interference + noise power spectral matrix
$\mathbf{Q}_\mathrm{T}$	transformed interference + noise covariance matrix
R	range
R	number of range increments
$\mathbf{R}$	covariance matrix of signal + interference + noise
r_{90}	quarter wave sampling interval
RF	radio frequency
R_g	ground range
$R_2 w$	two-way slant range
rms	root mean square
ρ	autocorrelation
R_s	slant range
$R(t)$	range
$\mathbf{S}$	signal + clutter covariance matrix
SAEP	subarray based auxiliary echo processor
SAR	synthetic aperture radar
SAS	symmetric auxiliary sensor configuration
SCNR	signal-to-clutter + noise ratio
Σ	DPCA shift operator
$\mathbf{s}$	signal vector
$\mathbf{s}_\mathrm{F}$	signal spectrum vector
$\mathbf{s}_\mathrm{T}$	transformed signal vector
$\mathbf{s}(.)$	steering vector
$\mathrm{sinc}(x)$	$\sin(x)/x$
SL	sidelooking
SM	signal match
SMI	sample matrix inverse
SNR	signal-to-noise ratio
SINR	signal-to-interference + noise ratio
SNIR	signal-to-noise + interference ratio
SOSTAR	Stand Off Surveillance and Target Aquisition Radar
$s_\mathrm{r}(t)$	received signal
ST	space-time
STAP	space-time adaptive processing

Σ	DPCA shift operator
t	time
T	pulse repetition interval
$\mathbf{T}$	space-time transform matrix
$\mathbf{T}_s$	spatial transform matrix
τ	echo delay
τ_t	round trip delay
θ	depression angle
TIME	echo delay time, range time (fast time)
tr	trace of a square matrix
time	pulse-to-pulse time (slow time)
UESA	UHF Electronically Scanned Array
ULA	uniformly spaced linear array
v_c	radial clutter velocity
v_p	platform velocity (x-direction)
v_R	receiver velocity
v_{rad}	radial target velocity
v_T	transmitter velocity
v_t	target velocity
$v_{\tan}$	tangential target velocity
$\mathbf{x}$	vector of received echoes
$\mathbf{x}_F$	spectral vector of received echoes
$\mathbf{x}_T$	transformed vector of received echoes
x_i	x-coordinate of i-th sensor
y	output signal
y_c	correction pattern
y_i	y-coordinate of i-th sensor
z_i	z-coordinate of i-th sensor
*	conjugate complex or conjugate complex transpose
*	convolution
$\otimes$	Kronecker product
$\mathbf{0}$	zero vector

Index

D^3, 540

Abdi, 57
Abramovich, 2
ACP, 170, 349
ACR, 171
across-track
  horizontal, 386
  vertical, 387
adaptation, 303
  complexity, 121
  range dependent, 505
  with signal included, 206
adaptive algorithm, 529
adaptive arrays, 7
adaptivity
  vertical, 321
ADPCA, 238
Adve, 312, 335, 524, 528, 541
AEP, 163, 348
AER II, 13
aircraft
  hull, 337
  wing, 527
Akaike, 236, 366, 513, 530, 539
  algorithm, 229
Albarel, 13, 242
algorithm
  adaptive, 529
ambiguity, 317
  clutter filter response, 411
  coincidence, 101
  Doppler, 317, 329, 410
  left–right, 99
  range, 317, 318
  spatial, 101
  temporal, 101

AMSAR, 13, 242
Anderson, 16, 40, 133
Andrews, 7, 131, 133, 134, 138
angle
  depression, 319
  dive, 73
  of arrival
    modal, 548
  squint, 131
angle-Doppler cell, 311
angular spread, 549
antenna, 241
  array
    full, 224
    crow's nest, 259
    directivity pattern, 413
    elliptical, 203
    monopulse, 354
    multichannel, 413, 526
    phased array, 11
    receive, 327
    transmit, 327
antenna aperture, 230
antenna array
  errors, 491
antenna beam, 79
antenna configuration
    asymmetric, 200
antenna sidelobes, 241
antennas
  planar
    optimum, 202
anti-jamming, 249, 404
Antonik, 522
Anu, 47, 491
aperture
  array, 439

spherical, 259
aperture size
  effective, 251
Applebaum, 531
APY-6, 13
array
  360° azimuthal coverage, 264
  active, 11
  antenna
    size, 230
  broadband, 404
  circular, 237, 352
    ring, 265
  circular planar, 241, 242, 326
  complex, 117
  configuration
    non-linear, 242
  configurations, 241
  conformal, 78, 241, 262, 337
    forward looking, 263
  cylindrical, 263, 337, 360
  displaced ring, 268
  elliptical, 237
  equispaced, 89, 103
  errors, 393
  forward looking, 75, 193, 195, 318, 319, 324, 350, 492
  fully digitised, 237
  fully filled, 282, 396
  geometry, 53
  horizontal, 548
    planar, 267
  horizontal planar
    rectangular, 267
  linear, 75, 175, 237, 241, 273, 278, 318, 320, 343
    equidistant, 360
    equispaced, 214
    forward looking, 180, 231
    sidelooking, 175, 231
  motion
    artificial, 2
  non-uniform polarisation, 291
  octagonal, 270
  omnidirectional coverage, 264
  orientation, 106
  planar, 78, 175, 258, 320, 321, 360
    equidistant, 360
    horizontal, 267
    sidelooking, 175
  planar circular, 343
  planar horizontal, 241
  randomly distributed, 254
  randomly spaced, 241
  randomly thinned, 269
  rectangular, 78
  rectangular planar, 242
  rotation, 282
  sensor, 414
  sidelooking, 75, 193, 195, 214, 270, 318, 322, 350, 418, 492
  size, 141
  systolic, 534
  tapered, 252
  thinned, 258
  undersampled, 419
  uniformly spaced, 269
  vertical, 548, 557
  volume, 78, 259
array axis
  orientation, 203
array structure
  irregular, 253
arrays
  randomly spaced, 257
arrival
  multiple, 325
artificial noise, 177
ASEP, 447
ASFF, 314
asymmetric auxiliary sensors, 200
atmospheric effects, 528
autocorrelation, 420
auxiliary channel, 161
auxiliary channel processor, 170
auxiliary eigenvector processor, 176
auxiliary sensor, 191, 302
auxiliary sensors
  symmetric, 447
AWACS, 13
axis

common time, 237
Ayoub, 321, 323, 324
azimuth
    compression, 410

Böhme, 2
Bühring, 213, 215, 535
Bühring and Klemm, 5, 38, 40
Bürger, 236
Babu, 535
Baggeroer, 551
Balaji, 128, 129, 516
band limitation, 415
bandwidth, 413
    clutter, 61, 71, 109, 118, 146, 168, 172, 199, 234, 279
    Doppler, 61, 505
    effects, 146, 172, 199, 223, 314, 419
    limitations, 116
    relative, 414, 415
    spatial, 116
    system, 111, 114, 147, 168, 172, 175, 199, 233, 440
    target
        Doppler, 410
Bao, 207, 214, 307, 308, 491, 524, 541
Bar-Ness, 26, 534, 536, 537
Baranoski, 149, 159, 221, 239, 504, 522
Barbarossa, 1, 3, 121, 213
Barile, 146, 527
Barnett, 366, 530
Bayliss, 247, 252
    coefficients
        transformed, 252
Bayliss weighting, 255
beam
    auxiliary, 171
beam pattern, 78
    transmit, 327
beam traces, 71, 319
beam-Doppler
    covariance matrix) covariance matrix between the channels, 312

    subspace, 312
beamformer, 164, 220, 302, 418
    corrected, 135
    estimation, 504
    multiple, 171
    primary, 173, 191
    secondary, 173, 183, 191, 243, 258, 365
    vector, 209
beamformer + adaptive temporal filter, 279
beamformer + clutter filter, 353
beamformer vector, 122
beamforming, 303
    multiple, 11, 171
    narrowband, 1
beampattern, 249
    identical, 260
beams
    auxiliary, 209
beamstearing
    phase coefficients, 245
beamsteering
    inertialess, 11
    multiple, 31
Beard, 377
Bell, 265
Berger, 169, 178
Bergin, 307, 523
Berin, 26, 520, 534
Berry, 12, 516
bias, 456
    azimuth, 464
    expected, 464
    parameter estimates, 465
Bickert, 297, 501
Billingsley, 61
Bird, 321
bistatic
    Doppler, 378
bistatic configuration, 378
    horizontal across-track, 396
    tandem, 395
    vertical across-track, 396
bistatic radar, 377
    geometry, 378

Blackman, 11
blind velocity, 279, 334
block Toeplitz matrix, 229, 539
blocking matrix, 174, 519
    auxiliary, 204
    space-time, 174, 519
    spatial, 204, 205, 310, 519
    transform, 204, 205
Blum, 238, 492, 521, 542
Blunt, 523
Boche, 150
Bodewig, 133
Bojanczyk, 159
Bollini, 533
Borsari, 118, 383, 506, 520
Brandwood, 38
Braunreiter, 159
Brennan, 1, 9, 12, 16, 21, 100, 308, 413, 505, 531
Brennan and Reed, 7
Brenner, 13
Bridgewater, 321
Brown, 12, 207, 516
Bryn, 7
Buckley, 413
Buehring, 126
Burg, 5, 37, 38, 47, 49, 213, 215, 530

Cai, 208, 517, 528
Caldwell, 159
calibration
    frequency-FREQUENCY, 502
    spatial, 500
Calvary, 335
cancellation
    jammer + clutter
        simultaneous, 344
    jammer and clutter
        separate, 356
    mainbeam jammer, 413
canceller
    multipath, 371
Capon, 44, 243
Capraro, 523
CARABAS, 87
Carlo, 541

Carlson, 530
Cerutti-Maori, 170, 536
chaff, 59
channel
    acoustic, 545
    auxiliary, 131, 161, 163, 172, 205, 303, 446
        generation of, 177
        space-time, 162, 171
    central, 365
    clutter, 349
    clutter matched, 349
    difference, 131
    Doppler, 165
    main, 131
    number of, 177, 183
    output, 303
    reference, 502, 519
    search, 161, 162, 171
    space-time
        receive, 171
    spatial, 352
    sum, 131
channel equalisation, 500
channels
    auxiliary
        generation of, 176
    number of
        minimum, 365
Chen, 312, 377, 521
Cheong, 532
Childers, 47
chirp
    linear FM, 416
Choudhary, 516
circular array, 481
cluter suppression
    degradation by jamming, 346
clutter, 3, 478
    airborne, 71
    ambiguous
        suppression, 320
    background
        inhomogeneous, 214
    bandwidth, 61, 71, 109, 168, 172, 199, 234, 330

Index 625

beam, 177
broadband, 353
cancellation, 117, 130
   polarimetric, 284
   space-time, 317
   spatial only, 328
chaff, 59
component, 89
covariance matrix, 319
   deterministic composition, 525
   estimation, 300
   signal-free, 518, 520
discrete, 175
Doppler
   bistatic, 378
Doppler filter, 177
Doppler frequency, 71, 72, 103, 167, 171
   range dependence, 151, 377
   relative, 418
   sidelobe, 250
Doppler spectrum, 360
Doppler-azimuth trajectories, 73
eigenvalue, 382
eigenvector, 127, 163
filter, 243
   temporal, 40
filtering
   space-time, 403
fluctuation model, 172
fluctuations, 57, 61, 84, 115, 438
   unknown, 524
ground, 4, 57
heterogeneous, 249
homogeneous, 93
internal motion, 61
model, 104, 331
modelling
   gaussian, 542
motion
   internal, 234
moving, 59, 153
multiple arrival, 319
multiple-time-around, 68, 317, 318, 324, 326
narrowband, 277, 353

notch, 140, 167, 172, 199, 201, 232, 273, 317, 327, 338, 346, 463
   ambiguous, 334
   broadening, 505
   depth, 236
   primary, 319
   secondary, 319
   width, 279
plus jamming, 344
power spectral matrix, 311
primary, 319
rejection, xviii, 86, 122
   range dependent, 318
   space-time, 122
rejection architecture, 250
rejection performance, 115, 248
response
   unambiguous, 335
return
   multiple, 318
returns
   ambiguous, 317
   multiple, 319
sea, 4, 59
spatial frequency, 116
spectrum, 61
   minimum variance, 333
   undersampled, 279
stationary, 214
subspace, 145, 163, 363
suppression, 40, 344, 403, 410
   optimum, 343
   real-time, 299
trajectories, 143
trajectory, 378, 381
   range-Doppler, 324
undernulled, 522
velocity, 59, 153
weather, 4, 59, 268
clutter cancellation, 361, 519
clutter canceller
   space-time adaptive, 311
clutter channel, 349
clutter covariance matrix
   clutter

space-frequency, 308
  estimation, 303
clutter Doppler
  range dependence, 319
clutter echo
  stationary, 215
clutter filter
  auxiliary channel, 357
  space-time, 357
clutter filtering
  temporal, 353
clutter fluctuations, 199
clutter notch, 281
  broadening, 524
clutter primary, 324
clutter rejection, 492
  near-optimum, 163
  space-frequency, 403
  under jamming conditions, 343
clutter resolution, 279
clutter spectrum
  Fourier, 77
  minimum variance, 77
  sidelobe, 241
clutter suppression, 9
  space-time, 213
clutter-to-noise ratio, 22, 145
CNR, 22, 126, 145, 437, 477, 525
Coe, 12, 404
coherent processing interval, 281, 440
coincidence
  phase, 112
coincident phase centre techniques, 209
Colone, 288, 443, 525
colored diagonal loading, 243
compensation
  antenna errors, 305
  for antenna tolerances, 305
complexity
  computational, 228, 512
Compton, 2, 404, 413
computational expense, 366
computing time, 122
configuration

symmetric auxiliary sensors, 191, 201, 360
Conte, 521
convergence, 537
convolution
  fast, 304
  temporal, 304
Cooper, 308, 313, 523
correction
  coefficients, 135
  patterns, 131, 134
correction matrix, 452
correction patterns, 129
correction values, 452
correlation
  full, 112
  function, 414
    space-TIME, 415
  ridge, 92, 93
  short-time, 518
  spatial
    exponential, 214
  temporal
    exponential, 214
correlation function, 413
correlation matrix
  intermode, 550
correlation time
  surface waves, 545
cost, 122
Coutts, 377
covariance matrix
  block column, 218
  block diagonal, 311
  block Toeplitz, 217
  centre row, 303
  clutter, 298, 382
  clutter + noise, 105, 123, 216
  clutter, temporal, 5
  diagonalisation, 42
  estimation, 367, 504
  frequency-dependent space-frequency, 306
  interference + noise, 26, 98
  jammer, 85
  jammer + noise, 33

inverse, 358
noise, 18, 20, 21, 26, 86
normalised, 22
of received signal, 44
partial, 134, 136
signal + noise, 18
singular, 349
space-TIME, 415
space-time, 79
   clutter + jammer + noise, 359
   clutter + noise, 217
covarince matrix, 546
Covault, 12, 13
Cox, 44, 46
CPCT, 97, 117, 209
CPI, 440
crab angle, 71, 74, 76
Cramér–Rao bound, 425, 428, 429
   properties, 430
   relations with monopulse, 430
   validity, 434
cross-term
   spectral, 306
crow's nest antenna, 259, 492
CSM technique, 169
Culpepper, 372
Curlander, 147

Damini, 516
data
   pipelining, 237
   reordering, 238
   secondary, 118
   selection strategy, 93
   signal-free, 174
data segment, 216, 217, 224
   shifted, 218
Davis, 281, 377
Day, 13, 315
De Grève, 159
decorrelation
   sensor-to-sensor, 494
   spatial, 173, 199, 268
   temporal, 63, 110, 173
decorrelation effects, 60, 233
degradation

clutter rejection, 224
slow target detection, 505, 525
degree of freedom, 167, 172, 224, 305, 346, 354
   spatial, 371
   temporal, 371
   vertical, 242, 326
delay, 111
   quarter wavelength, 414
density tapering, 258, 259
depression angle, 321
Desmézieres, 285
detection
   slow target, 301
   slow targets, 182, 199, 346
   threshold, 483
detection probability, 479
DFT, 41
   multichannel, 304
diagonal loading, 177, 349
Dickman, 523
diemansion
   temporal, 234
difference
   pattern, 241, 243, 252, 253
   weighting, 252
difference pattern, 456
   adapted, 451
   slope, 456
   spatial, 450
   temporal, 450
Dillard, 215
dimension
   filter
      spatial, 230
      temporal, 232
   spatial, 213, 233
      removing, 220
   temporal, 213, 233
dipole diagram, 206
direct data domain, 540
direction
   of arrival, 414
directivity pattern, 242
   identical, 276
   normalised, 188

discrete Fourier transform, 41, 297
displacement
    transmit/receive, 378
distance
    between channels, 312
distortion
    PSF, 410
distribution
    alpha stable, 539
    gaussian, 539
diving angle, 73
DoF, 354
Dogandzic, 426
Doherty, 370
Dong, 1, 541
Doppler
    bandwidth, 152
    bistatic, 378, 394
    colouring, 115
    compensation, 506
    discrimination, 3
    effect, 52
    filter, 44, 164
        bank, 122, 172, 183
    frequency
        clutter, 71, 103
        main beam, 131
        relative, 72
        target, 125
    frequency range
        unambiguous, 109
    gradient, 274
    range dependent, 381
    spread
        system, 152
    unambiguous, 132
Doppler channel
    actual, 452
Doppler filter, 303
    length, 215, 232
    weight, 196
Doppler filter bank, 330
Doppler filtering, 303
Doppler frequency
    normalised, 103
Doppler response, 503

ambiguous, 333
Doppler spread, 151, 152
    within range gate, 64
Doppler warping, 118
DOSAR, 13
doublet, 261
DPCA, xviii, 7, 100, 121, 136, 209, 213, 214, 313
    bistatic configuration, 394
    case, 89
    clutter canceller, 71
    condition, 88, 89, 98, 141, 393
    effect, 117, 147, 267
    mode, 112
    principle, 86
    role, 117
    shift operator, 99
    two-pulse canceller, 96
DPCA condition, 282
DRA, 12
Dunsmore, 377
dwell time, 505
DWT, 159
dyadics
    space-time
        clutter, 249

earth rotation, 321
Eaton, 404
echo sequence
    length, 230
eigencanceler, 537
eigencanceller, 520, 536
    minimum norm, 537
    minimum power, 537
eigenfunction
    vertical, 547
eigenspectra
    ASEP, 449
eigenspectrum, 384, 497
eigenvalue
    clutter, 199
        number of, 199
    deterministic approach, 540
    dominant, 497
    jamming, 348

number of, 223, 346
eigenvalues, 22, 98, 444
    additional, 352
    clutter, 98, 99, 145, 234
    noise, 98, 145
    small, 166
eigenvector, 98, 127
    clutter, 163
    decomposition, 145
    matrix, 127
eigenvector canceller, 520
eigenvector transform, 163
element groups, 269
ellipses, 75
ELRA, 187, 258
    antenna, 258
Ender, 1, 2, 9, 11, 13, 16, 78, 147, 221, 237, 238, 267, 307, 405, 422, 480, 502, 503, 506, 522, 528
Ender and Klemm, 9
endfire, 274, 549
energy management, 11
Entzminger, 12
environment
    modal, 546
environmental effects, 437
equivalence
    staggering/acceleration, 335
error
    amplification, 491
    amplitude, 495
    antenna, 491
    array channel, 495
    azimuth
        motion induced, 11
    DC offset, 499
    delay, 491, 495, 498
        stochastic, 498
    IQ, 491, 495, 497
    mitigation, 499
    phase, 495
    random, 491
errors
    channel phase, 496
estimate

    azimuth, 451
    velocity, 451
estimation, 425
    maximum likelihood, 425
    position, 552
    velocity, 440, 552
estimation accuracy
    angular, 439
estimator
    ME, 551, 554, 556
    MV, 551, 556
    SM, 550
    TME, 554, 557
Evans, 14
Evins, 2, 9

Fabrizio, 521
Fante, 2, 371, 377, 413, 450, 536
far range, 319
Farina, 1, 3, 7, 9, 16, 121, 525, 532, 533
fast convolution, 305
fast operations, 513
fast time, 502
Faubert, 97, 129
Feng, 237
Fenner, 12, 307
Feria, 540
Ferla, 546
Ferrier, 180
filter
    anti-clutter, 357
    anti-jamming, 357
    choice of, 525
    coefficient, 215
    digital, 213
        temporal, 213
    dimension
        temporal, 234
    FIR
        computation of, 236
        length, 228
        space-time, 248, 458
        spatial and temporal, 236
        suboptimum, 231
        TIME domain, 419

high pass, 238
IIR
  space-time, 238
jammer + clutter
  cascaded, 361
least squares
  FIR, 215
  space-time, 215
length, 223
  temporal, 410
low pass, 238
matrix, 219
operator, 220
prediction error, 215
robust, 524
space-frequency, 307
space-time, 114, 226
  adaptive, 213
  clutter + jammer, 359
  FIR, 213, 221, 364
spatial, 114, 518
temporal, 114, 277
temporal adaptive, 275
vector, 222
Wiener
  multistage, 534
filter coefficients
  estimation, 306
filter length
  temporal, 275, 278, 412, 421
filter matrix, 39
filtering
  complexity, 122
  range-dependent, 505
  real-time, 223
  space-time
    non-adaptive, 524
FIM, 428
Finley, 297
FIR filter, 462
  coefficients
    time dependent, 337
  fully adaptive, 320
  length, 495
  orthogonal projection, 521
  processor, 336

space-TIME, 419
space-time, 279, 301, 320, 336, 410, 448, 458, 554
  least squares, 337
space-time least squares, 304
spatial, 205
symmetric auxiliary sensor, 494
time varying, 331
variable coefficients, 337
Fisher information matrix, 428
FL, 99
flash lobe, 286
flight direction, 71
flight path, 73
  aligned, 385
  parallel, 386
fluctuations
  clutter, 199
FM chirp, 407
Four linear arrays concept, 265
Fourier transform
  space-time, 297
  two-dimensional, 103, 113, 502
Francos, 522
freeze technique, 520
French, 516
FREQUENCY, 404
frequency
  carrier, 53, 413
  Doppler, 53
  local oscillator, 131
  Nyquist, 223
  rectangular response, 172
  spatial, 237
FREQUENCY response, 503
frequency response
  rectangular, 64, 112, 147
Friedlander, 265
Friel, 500
Fritsch, 13, 16
Frobenius–Schur, 133
Frost, 532
Fuhrmann, 265

Gabel, 2, 9
Gabriel, 45, 47

Gaffney, 180
Galati, 37, 377
Garnham, 515
Gau, 521, 542
Genello, 2
geometry
    bistatic, 385
    radar, 331
Gerlach, 523, 530, 534
Gershman, 536
Ghouz, 59
Gierull, 14, 26, 128, 129, 535, 536
Godara, 404
Goggins, 12
Goldstein, 16, 129, 159, 169, 174, 178, 204, 213, 307, 377, 515, 521, 534
Golub, 532, 535
Goodman, 260, 523
Gover, 366, 530
Gröger, 34, 269
gradient technique, 505, 531
Gram–Schmidt, 519, 534
grating lobe, 184, 258
grating null, 184, 258
Gray, 426
grazing angle, 5
Griffiths, 182, 214, 372, 518, 532
Groeger, 185, 258
Gross, 13, 16
ground clutter, 4
Gruener, 13, 242
GSC, 534
Gu, 238
Guerci, 3, 17, 159, 164, 169, 213, 522, 523, 525, 530, 540
guidance, 11
Gupta, 2, 14
Gurram, 523
Guyvarch, 2

Haendel, 48
Haimovich, 26, 159, 520, 534, 536, 537
Hale, 122, 169, 242, 328, 356, 515, 541

Hamming window
    spatial, 104
    temporal, 104
Hanle, 285, 377
Hartnett, 377
Hatke, 426
Hatzinakos, 2
Haykin, 7, 39
Haystead, 12
Hayward, 10, 16, 117, 123, 383, 407, 413
Herbert, 150, 378
Hersey, 262
Hiemstra, 213, 243, 523
high PRF mode, 272
high-resolution, 492
high-resolution techniques, 44
Himed, 2, 12, 159, 213, 312, 378, 383, 388, 491, 512
Hippler, 13, 16
historical overview, 15
Hochwald, 2
Hoffman, 405
Holt, 13, 16
homogeneity
    clutter background, 521
Hoogeboom, 13
Hovanessian, 406
HPRF, 317, 318
Hudson, 7, 18, 32, 518
Hung, 10, 535
hyperbola, 72

IF, 125
    maximum, 140
    optimum, 225, 497
IF loss, 172
Iltis, 214
implementation
    aspects of, 175, 512
improvement factor, 20, 125, 165, 180, 222
    optimum, 492
impulse response
    rectangular, 63, 112
index

    spatial, 415
    temporal, 415
information
    polarimetric, 284
INS, 335
interference, 23, 25
    simulation, 21
interference reference, 444
interference rejection, 79
interference suppression, 2
internal motion, 4
interpolation, 216
inverse update, 505
ISAR, 2, 11, 56, 403
    combination with SAR, 405
    multichannel, 405
    resolution, 410, 411
isodops, 71, 274, 319, 377, 381

Jaffer, 221, 545
jammer
    airborne, 353
    broadband, 343
    cancellation, 114, 495
    CW, 343
    direction, 415
    distribution
        horizontal, 354
        two-dimensional, 354
    frequency independent, 422
    geometry, 353
    ground, 353
    nulling, 179
    power, 415
    scenario, 346
    spectrum, 518
    suppression, 242
jammer + clutter rejection
    simultaneous, 343
jammer and clutter rejection
    cascaded, 343
jammer cancellation
    optimum, 357
    sideband, 367
jammer nulling, 204
    adaptive, 499

jammer plus clutter, 355
jammer rejection, 492
jammer suppression, 358, 361
    fully adaptive, 357
jammer-to-noise ratio, 354
jamming, 59, 85, 343, 415, 438
    terrain scattered, 2
Janer, 335
Jansson, 529
JDL-GLR, 174, 306, 311, 312, 459, 462, 517, 528
Jenkins, 519
Jensen, 546
Jiang, 213
JNR, 354
Johannisson, 180
JointSTARS, 13, 16
Jouny, 372

Kadambe, 159
KASSPER, 17, 522
Kealey, 297
Kelly, 521
Kim, 214
Kirscht, 3
Klemm, 2, 3, 9, 10, 16, 32, 34, 44, 47, 48, 98, 121, 126, 133, 147, 159, 163, 170, 171, 173, 176, 210, 213, 215, 221, 236–238, 242, 260, 264, 270, 281, 311, 317, 323, 329, 337, 338, 343, 353, 356, 377, 378, 383, 394, 404, 443, 444, 492, 506, 545
Koc, 506
Koch, 2, 3, 17
Kogon, 2, 16, 149, 317, 372, 383, 405, 516
Kraut, 2
Kreithen, 522
Kretschmer, 38, 530
Kreyenkamp, 61, 118, 383, 506
Krikorian, 336
Krolik, 491
Kumaresan, 522
Kuperman, 44, 551

Kuzminskiy, 2

Lacomme, 7, 317
Landau, 182
Lapierre, 388, 506, 512
Le Chevalier, 377, 500
Leatherwood, 335
Lee, 12, 14, 34, 535
Lee and Lee, 26
Lee C. C., 501, 519
Lee J. H., 501, 519
Legg, 426
Legters, 523
Lesturgie, 1
Levinson, 37, 215
Lewis, 2, 9
Li, 45, 48, 213, 214, 502, 522, 532
Liao, 9, 505
Lightstone, 97
limitations, 525
Lin, 38, 404, 542
Linderman, 12, 516
Lindskog, 2
linear predictor, 225
Lipps, 3
Lissajou graph, 78
Lissajou pattern, 381
Little, 12, 516
Liu, 214, 235, 499
Liu H., 534
Liu Q.-G., 128, 177
Livingstone, 14, 17
Lo Y. T., 18
log-likelihood function, 427
Lombardo, 1, 3, 9, 40, 93, 214, 288, 443, 525
look angle, 73, 418, 452
look direction, 71, 73, 273
low probability of intercept, 66
LPRF, 317
Lukin, 3
Luminati, 122
Lynch, 1

Ma, 20
MacPhee, 516

Madurasinghe, 243
Mahafza, 9, 10
Maher, 1, 208
Mailloux, 252, 491
main lobe response, 353
mainlobe
    broadening, 251
Maiwald, 545
Makhoul, 37, 522
Malas, 536
Mallett, 21
Mann, 2
manoeuvering radar platform, 527
Mansur, 516, 532
Mao, 213
Mardia, 428, 434, 435
Marshall, 12, 356, 530, 534, 536, 539
Martinez, 516
Martinsek, 377
matched field processing, 2, 545
matched filter, 44, 52
    parametric adaptive, 215
    response, 404, 421
    space-time, 124
Mather, 44, 49, 328
Mathew, 129, 522
matrix
    block diagonal, 85
    block Toeplitz, 360
    covariance, 242
        clutter, 194
        reduced order, 197
    cross variance, 308
    data, 219
    inverse
        frequency dependent, 175
    inversion, 311, 367
    partitioning, 304
    persymmetric, 360
    power spectral, 298
    range-Doppler, 383
    space-time, 299
    Toeplitz, 360
    transform, 171, 200
        typical, 176
    transformed

634 Index

singular, 351
matrix inversion lemma, 175, 517
matrix transformation projection, 536
maximum entropy, 47
    temporal, 554
maximum likelihood estimation, 425
MCARM, 12, 523
McAuley, 121, 315
McCool, 532
McDonald, 2, 492
McDonough, 147
MDV, 281
Medlin, 403
Meisl, 14, 17
Melvin, 3, 159, 312, 521, 523, 524
MEM, 47
Merigeault, 2
Mermoz, 7
Meyer-Hilberg, 311
MFR, 421
Michels, 14, 213, 215, 539
Mikhael, 100, 182
milestones, 16, 17
Miller, 518
minimum detectable velocity, 281
mismatch
    azimuth, 456
    direction, 519
    Doppler, 519
    random, 462
    signal, 174
    speed, 336
    velocity, 456
modal arrival, 555
modal attenuation, 547
mode
    active, 356
    passive, 356
model
    ground clutter, 57
    jammer, 59
    moving clutter, 59
modes
    coherent, 549
monopulse, 130, 252, 354
    adaptive, 450

antenna, 7
    corrected, 450
monopulse processing
    space-time, 450
monopulse quotient, 450
monopulse radar, 459
Monzingo, 518
Monzingo and Miller, 7, 18
Moo, 256
Moore, 2
Moreland, 519
Morgan, 2
motion
    compensation, 115, 117, 131
    lateral, 268
    platform, 115, 207
    radar platform, 194
    radial, 406
    tangential, 406
        of clutter, 407
        of target, 406
motion compensation, 93, 129
MOUNTAINTOP, 377
Mountcastle, 40
MPRF, 317
MTI, 3, 79
    adaptive, 79
    airborne, 131
        motion compensated, 131
    conventional, 330
    performance, 115, 200
    radar
        airborne, 1
        shipborne, 1
        space-based, 1
    systems design, 79
    temporal, 443
    under jamming conditions, 343
MTP, 536
Mulgrew, 174, 204, 519
Multi-jammer scenario, 353
multipath
    out-of-plane, 377
Murray, 16, 87
mutual coupling, 500
mutual independence, 82

MV spectrum, 417, 419
Myrick, 3

Naga, 528
Nakhmanson, 53
Nathanson, 4, 5
Nayebi, 124
near range, 319
near-field obstacles, 527
Nehorai, 426
Nelander, 2
network
    space-time
        signal matching, 196
    subarray
        beamformer, 183
Nguyen, 266
Nickel, 2, 9, 11, 32, 34, 44, 45, 49, 163, 171, 180, 243, 253, 275, 424, 450, 495, 499, 534, 535
Nicolau and Zaharia, 7
Nikias, 539
NIR, 22
Nohara, 9, 87, 207, 307, 516
noise, 415
    ambient, 1, 546
    artificial, 177, 349
    correlated, 415
    covariance matrix, 415
    normalisation, 243
    power, 415
    reveiver, 546
noise-to-clutter ratio, 22
noise-to-interference ratio, 22
non-adaptive filtering, 524
Nordebo, 372
Nordwall, 12, 13
normal mode function, 547
normal modes, 546
normalisation factor, 244
nose radar, 286
notch
    clutter, 319, 492
    jammer, 492
        broadening, 497
NRL data, 533

nuisance parameter, 427
number of channels, 166
numerical stability, 122
Nyquist, 223
    condition
        spatial, 276
    frequency, 279, 323, 324
    rate, 101, 248
    sampling, 98, 116, 279
        spatial, 91
    spacing, 108
Nyquist sampling, 88, 282

OAP, 165, 172, 455, 462
octant, 261
Odendaal, 48
Ogle, 522, 524
OLPI, 66
Ong, 378
operations
    number of, 367
    saving of, 367
OPP, 127
Oppenheim and Schafer, 42
optimum processing, 193
optimum processor, 9, 18, 24, 182
    subarray level, 481
order reduction, 539
orthogonal
    projection, 25
orthogonal projection, 534
    processor, 171
OTH, 2
OUS, 182
over-the-horizon radar, 2
Owechko, 159

Pados, 312
Page, 523
Paine, 328, 450
PAMF, 215
PAMIR, 13
Papoulis, 61
parameter estimation, 425
parameters, 68
    systems, 439

Park, 3
Park H.-R., 284, 285
Park S., 506, 522, 540, 541
Park Y. C., 20, 225, 534
Parker, 214, 303
Pasala, 262, 500
pattern
    difference, 131
Paulray, 2
Peckham, 164, 170
Peng, 9, 159, 177, 235, 307, 499
Perry, 426
Petterson, 1, 87
phase
    centre
        different, 209
        displacement, 181
    jitter, 491
    quadratic, 407
    spatial, 414
Picardi, 213
Picciolo, 534
Pillai, 236, 530, 545
Pisarenko, 48
planar array, 78
platform
    acceleration, 335
    motion, 112
        compensation, 86
        perturbation, 335
    velocity, 238
platform motion
    radar, 62
platform velocity, 281
point reflector, 237
point spreading function, 411
    azimuthal, 411
pointing error, 520
polarisation, 284
    non-uniform, 286
Pollak, 182
post-Doppler STAP, 517
power
    spectrum, 382
power selected
    de-emphasis, 524

    training, 524
power spectral density, 413
power spectral matrix, 298
pre- vs post-Doppler, 516
pre-Doppler STAP, 516
pre-transform, 219, 303
pre-whitening, 297
prediction error, 38
prediction error filter, 38
PRF, 5, 103, 318, 323
    effect of, 279
    mode, 214
    staggered, 314
PRI, 7, 55, 89, 318
    constant, 338
    staggered, 318, 338
        pulse-to-pulse, 330
probability
    of detection, 479
    of false alarm, 479, 480
processer
    optimum, 334
processing
    adaptive subgroup, 311
    auxiliary channel
        space-time, 351
        spatial, 351
    auxiliary sensor/echo, 447
    broadband, 413
    cascaded, 311
    conventional, 443
    conventional MTI, 272
    frequency dependent spatial, 485
    frequency-dependent spatial, 306
    matched field, 2, 545
    non-adaptive, 121
    one-dimensional, 486
    optimum, 182
    partial adaptive, 486
    polarisation-space-time, 284
    post-Doppler, 486
    real-time, 533
    scheme
        space-time, 228
    space-FREQUENCY, 371, 404
    space-frequency, 297

space-TIME, 11, 371, 404, 536
space-time, 1, 11, 272, 311
  sonar, 545
space-time adaptive, 9
space-time-FREQUENCY, 149
space-time-subband, 153
space-time-TIME, 149, 153, 175,
  404, 536
spatial, 1
spatial dimension, 11
suboptimum, 485
processor
  Σ-Δ, 206, 313
  2-D
    FIR filter, 236
  adaptive, 303
  array channel, 180
  auxiliary channel, 170, 171, 177,
    349, 493
    space-time, 175, 347
    spatial, 349
  auxiliary eigenvector, 163, 176,
    348
  auxiliary sensor
    two-stage, 364
  auxiliary space-time, 314
  auxiliary space-time channel, 300
  beamformer + temporal MTI, 242
  cascaded, 360, 362, 365
  conventional MTI, 242
  eigenvectors, 449
  FIR filter, 321, 499
  frequency domain, 304
  fully adaptive, 121, 172, 231
  hybrid, 541
  jammer and clutter, 365
  JDL-GLR, 313
  LCVM, 519
  LR, 121
  optimum, xviii, 122, 133, 193,
    209, 273, 311, 345, 436,
    442, 492, 499
  optimum adaptive, 165, 242, 455
  optimum spatial, 497
  orthogonal projection, 127, 171
  overlapping subarrays, 183

post-Doppler, 181, 303
post-Doppler STAP, 315
reduced order, 425, 442
scheme
  simplification, 226
simplified, 303
space-frequency, 300
space-time, 121, 171, 242
space-time-TIME, 372
STAP, 378
subarray, 352
  FIR filter, 242
suboptimum, xviii
subspace, 162
symmetric auxiliary sensor/echo,
  301
symmetric auxiliary sensors, 191,
  196, 199, 361
workload, 512
program
  snir2pdf, 480
projection
  matrix, 26, 205
  technique, 237
PSF, 411
PST-GLR, 284
Pugh, 516
pulse
  compression, 403
  Doppler radar, 1, 3, 55
  repetition frequency, 5
  repetition interval, 7, 55, 80, 85
  train, 55
pulse train
  coherent, 545
Pulsone, 405

QR decomposition, 532

Rössing, 502
Rabideau, 2, 149, 516, 524
radar
  airborne, 322
  aircraft nose, 241, 242
  bistatic, 377
  equation, 478

geometry
    airborne, 322
    spaceborne, 321
GMTI
    space-based, 280
    imaging, 413
    manoeuvers, 527
    operation time, 252
    passive mode, 343
    phased array, 11, 413
    sidelooking, 117
    spaceborne, 322
    synthetic aperture, 413
    tracking, 252
    wideband, 268
radar operation
    realistic, 477
radar parameters, 436
radar signal processing, 17
radial velocity, 71
radome, 286
    effects, 286
Rangaswamy, 3, 14, 522
range
    ambiguity, 318
    cell
        under test, 520
    dependence, 76
        clutter Doppler, 318
    estimation, 44
    increment, 73, 79, 164, 223, 229, 318, 320
    independence, 385
    interval
        ambiguous, 320
        unambiguous, 324
    migration, 66
    ring, 62
    segmentation, 520
    target, 52
    unambiguous, 132
    visible, 164
    walk, 61, 110, 146, 150, 235, 441
range bin, 62
range compression, 404
range dependence

    of clutter Doppler, 67
range gate, 151
range-Doppler
    IF matrix, 323
range-Doppler matrix, 153, 323, 369, 383
rank reduction, 539
raster
    half wavelength, 258
Rayleigh distribution, 479
receive
    channel
        identical, 165
    process, 51
    signal, 51
receiver noise power, 246
Reddy, 129
reduction
    number of channels, 166, 172
Reed, 2, 16, 122, 159, 169, 174, 178, 204, 236, 307, 367, 413, 513, 515, 521, 530, 542
Rees, 328
reflection
    diffuse, 371
    specular, 371
reflectivity function, 58
Rembold, 2
replica, 105
resolution
    array, 412
    ISAR, 403
response
    azimuth, 492
    Doppler, 492
reverberation, 545
    suppression, 554
RF domain
    clutter cancellation, 130
Ricean distribution, 479
Richardson, 9, 10, 16, 87, 114, 117, 128, 356, 378, 527, 528
Richmond, 521
Rieken, 265
Rihaczek, 52
ring dipole, 259, 267

Ringel, 57
ripple, 335
    pass band, 235
Rivkin, 367
Robey, 504, 521
Robinson, 3
Roman, 51, 215
Roucos, 47
Roy, 48
Russ, 214

SAEP, 457, 462, 485
sample
    matrix inversion, 529
    size, 141
        spatial, 230
        temporal, 230
    space-TIME, 413
    subgroup
        space-time, 173
sampling
    effects, 142
    irregular, 214
    Nyquist, 143
    spatial, 107, 124, 143
    temporal, 107, 143
sampling effects, 442
Samson, 516
Sander, 241, 258, 491
Sandhu, 2
SAR, 2, 11, 403
    combination with ISAR, 405
    imaging, 238
    inverse, 405
    multichannel, 413
Sarkar, 506, 522, 540, 541
scatterer
    uncorrelated, 62
scenario
    jammer, 350
    jammer + clutter, 362
    jammer and clutter, 345
Scharf, 7
Schindler, 1
Schleher, 5, 18, 20, 317
Schmidt, 48, 504

Schubert, 150
SCNR, 126
Scott, 174, 204, 519
sea clutter, 4
search, 11
    channel, 161, 162, 164
search channel, 444
Sedwick, 1
Sedyshev, 377
See, 2
Seed, 3
Seliktar, 159, 536
semiplane, 223
sensitivity
    mismatch, 336
sensor
    asymmetric auxiliary, 200
    auxiliary, 191
    depth, 548
    directivity patterns, 54
    doublets, 254
    index, 89
    number of, 241
    omnidirectional, 99
    pattern, 103
    spacing, 89
        equidistant, 242
    symmetric
        auxiliary, 192, 226
    symmetric auxiliary, 191
    triplets, 254
sensor pattern
    vertical, 324
sensor patterns, 439
sequential test
    coherent, 64
Shaw, 121, 315
shift operator, 90
Shnitkin, 12
Shor, 7
short-time data processing, 518
shortening
    filter length, 215
Showman, 3, 285, 425
sidelobe
    control, 343

low, 241
spurious, 104
wall, 104
sidelobe canceller, 32, 131, 174, 179, 204, 457
generalised, 519, 534
signal
acoustic, 547
bandwidth
relative, 416
broadband, 420
cancellation, 174, 206, 519
direction, 416
distortion, 422
inclusion, 174, 206
inclusion in adaptation, 518
match, 103, 556
processing
spatial, 11
reference, 162
replica, 546
target, 332
signal-to-clutter + noise ratio, 225
simulation
interference, 21
single snapshot, 506
SINR metric, 169
Skidmore, 328
Skolnik, 4, 5, 7, 9, 10, 16, 55, 64, 86, 98, 117, 129–131, 144, 207, 209, 331, 377, 379, 457
Skupin, 502
SL, 99
slant range
receive, 379
transmit, 379
sliding hole, 520
sliding window, 520
slow operations, 513
slow time, 502
SM, 556
SMI, 529
Smith, 242
smoothing, 216
SNAP, 546, 556
SNIR, 353, 479, 497

loss in, 505
SNR, 437, 477
sonar, 1, 545
active, 554
Sorelius, 441
SOSTAR, 13
Soumekh, 2, 12
space-Doppler characteristics, 71
space-frequency technique, 210
space-TIME
processing, 114, 117
space-time
adaptive processing, 9
clutter rejection, 114
filtering
knowledge based, 522
FIR filter, 305
processing, 114
concept, 115
subspace techniques, 159
vector quantities, 122
spaceborne, 393
spaceborne radar, 280
spacing
interelement, 414
Spafford, 5
sparse array, 280, 393
spectral shaping, 278
spectrum
azimuth-Doppler, 71
clutter, 103, 115
clutter and jammer, 114
continuous, 105
eigen-, 98
Fourier, 77, 103
high-resolution, 105
jammer+noise, 418
jammer-clutter, 357
maximum entropy, 106
minimum variance, 105, 126, 419
MUSIC, 106
power, 382
range-velocity, 556
spot jammer
coherent, 330
spotlight SAR, 11

spread
    modal, 549
SPST-GLR, 284
staggering
    equidistant, 336
    pseudorandom, 332
    quadratic, 334
standard deviation, 455, 463
standard parameters, 68
STAP, 9, 272
    Doppler factored, 307
    radar, 7
    real-time implementation, 515
stationarity
    temporal, 234, 314
Staudaher, 12, 100
Steen, 180
Steepest descent, 531
steepest descent techniques, 531
steering vector, 105, 125, 546
    idealised, 528
    mismatch, 528
    optimisation, 501
    space-TIME, 419
Steinhardt, 181, 302, 306, 405, 498, 520, 522, 524
Stiles, 260
stochastic approximation, 531
Stoica, 529
stop band, 274
Su, 9, 174, 204, 315, 519
subarray, 243, 269, 278, 457
    beam, 180, 261
    beamforming, 258, 261
    checkerboard, 242, 252, 326
    configuration, 241, 248
    dartboard, 242, 252
    directivity pattern, 182, 187
    disjoint, 183, 184, 227, 248
    displacement, 183
    equal, 185
    formation, 252
    forming, 245, 248
    irregular, 242
    level, 241
    non-uniform, 187

normalisation, 247, 251, 253
output, 243, 246, 252
overlapping, 181, 351
planar
    two-dimensional, 267
pre-sum, 191
shape
    different, 261
structure, 275
uniform
    overlapping, 192, 226
subarray beams, 269
subarrays, 180
    irregular, 279
    overlapping, 183, 302
    identical, 360
subgroup
    angle-Doppler, 311
subgrouping
    horizontal, 242
subinterval, 237
submatrix
    spatial, 180, 351, 415
suboptimum techniques, 9
subspace
    clutter, 163, 179
    low order, 360
    processor, 162
    space-time, 159
    space-time transform, 160
subvector
    spatial, 244
Sun, 315, 498
Sung, 20
superresolution, 11
suppression
    broadband jammers, 518
Suresh Babu, 12, 16, 121, 207, 307, 527, 534, 542
surface
    cylindrical, 203
Swindlehurst, 214
Swingler, 180
Sychev, 2
symmetric auxiliary data scheme, 301
symmetric auxiliary sensors, 191

symmetry, 360
synchronisation
    PRF, sensor displacement, velocity, 237
synthetic aperture radar, 210
system bandwidth, 63, 84, 168, 172, 175, 199, 440, 495
system Doppler spread, 67
system transfer function, 419
systolic array, 533

TACCAR, 4, 131
Tam, 129
tandem, 385
Tang, 262
Tanik, 506
tapering, 241, 251, 272, 273, 277
    aperture, 241, 242, 250
    compensation, 243
    effect, 251
    loss, 277
    weight, 244
    weights, 245
target
    Doppler frequency, 165
        normalised, 225
    imaging, 405
    model, 331
    motion, 53, 405
    Rayleigh fluctuating, 453
    signal, 478
    slow, 273
    velocity, 125
target mismatch, 339
target signal
    replica, 308
Taylor weighting, 243, 248, 251
Techau, 2, 61, 149, 525
technology
    digital, 233
Teitelbaum, 516
Teixeira, 522
terrain following, 11
Thompson, 14, 17, 262
TIME, 404
Timmoneri, 1, 532, 533

Tinston, 522
Titi, 12, 16, 129, 169, 307, 356, 372, 377, 534, 536, 539
TME, 554
Tobin, 12
Toeplitz matrix, 47, 215
tolerance, 492
    filter characteristics, 499
    sensor position, 491, 492
Tomlinson, 57
Torres, 371, 536
track, 11
tracking
    ground target, 459
    multiple target, 11
    scenario, 472
        modelling, 460
training samples, 79
trajectora
    clutter, 384
transform
    auxiliary sensor, 362
    data dependent, 164
    data independent, 170
    discrete wavelet, 159
    eigenvector, 163, 445
    mode-sensor-time, 552
    order reducing, 442, 443
    secondary, 362
    sensor-mode-time, 556
    single elements and samples, 173
    space-time, 179, 180, 200, 244, 444
    spatial, 179, 209, 226, 302, 351, 446
    spatial subspace, 359
    subarray, 189
transform techniques, 436
transformed domain, 453
transmission
    alternating, 288
    omnidirectional, 99
transmit
    beam pattern, 103
    beamwidth, 139
    directivity, 99

directivity pattern, 58, 188
process, 51
signal, 51
transmitter above receiver, 387, 393
travel delay effect, 150
Tsakalides, 539
Tsandoulas, 7, 16, 97
Tsao, 128
Tufts, 10, 522
Turley, 521
Turner, 10, 535
two-pulse canceller, 24, 71, 136
two-pulse clutter canceller, 23
two-pulse or three-pulse clutter canceller, 213

Udalov, 3
UESA, 269
ULA, 269
uncertainty
  jammer direction, 419
  spatial, 418
undersampling, 393
  spatial, 88, 143, 144
  temporal, 88, 142
updated inverse, 531
Urick, 545
UWB SAR, 87

Vaccaro, 2
van den Bos, 38
Van Keuk, 3, 11
Van Loan, 532, 535
Van Trees, 51, 53
VAR filter, 213
Varadarajan, 491
vector
  beamformer, 310
  space-time, 299
vector space
  clutter, 362
  reduction, 213
velocity
  blind, 318, 330
    ambiguous, 279
  clutter, 3

estimation, 44
  unambiguous, 330
light, 52
ocean waves, 4
radial, 3, 125
resolution, 439
target, 3, 52
underwater sound, 545
vertical nulling, 321
visible range, 79
Voles, 4

Walke, 2
Wang, 3, 208, 284, 312, 517, 522, 528
Wang H., 9, 206, 208, 256, 347, 516, 521
Wang H. and Cai, 174, 306, 311, 313
Wang H. S. C., 129
Wang X., 307
Wang Y., 9, 159, 177, 307, 499
Wang Y.L., 16
Wang Z., 214
Wang, H., 313
Wang, Y., 308, 313
Wang, Y.-L., 321
Ward, 9, 16, 99–101, 121, 159, 181, 239, 302, 306, 315, 337, 425, 426, 429, 436, 498, 533
wavefront
  parabolic, 209
waveguide
  acoustic, 547
wavenumber
  horizontal, 548
  modal, 547
  vertical, 549
Wax, 47, 491
weather clutter, 4
  suppression, 40
Weber, 2, 521
Wehner, 403
weight
  Doppler filter, 196
weighting
  auxiliary, 209
  Hamming, 77

Weiner, 523
Weippert, 213
Wells, 297
Welsh, 169, 178, 515, 529
White, 12, 404
Wicks, 9, 207, 312, 313, 372, 524, 528
wideband STAP, 536
Widrow, 7, 531, 532
Wilden, 11, 267
Wirth, 11, 31, 44, 64, 66, 171, 179, 269, 403, 404, 495
Worms, 504
Wright, 297
Wu, 502
Wu R., 249, 307, 491, 524

Xiong, 307, 531
Xu, 535

Yong, 521
Yu, 404

Zarowski, 20
Zatman, 13, 17, 78, 149, 168, 265, 317, 383, 530, 536
Zeger, 9
Zekavat, 57
Zhang, 1, 182, 208, 313, 367, 516
Zhang Q., 100
Zhang Y., 206
Zhou, 174, 204, 237, 315, 519
Ziegenbein, 48
Zoltowski, 3
Zulch, 506, 516